Frontiers in Physicochemical Biology

Contributors

Philip H. Bolton

Antony W. Burgess

Lilia Calabrese

Mildred Cohn

A. R. Crofts

E. Domingo

Pierre Douzou

Thomas Early

J. P. Ebel

R. A. Flavell

Irwin Fridovich

Henrik Garoff

Quentin H. Gibson

M. Grunberg-Manago

Osamu Hayaishi

Ari Helenius

Benno Hess

Pierre Joliot

David R. Kearns

Patrick Maurel

A. M. Michelson

B. D. Nageswara Rao

Yu. A. Ovchinnikov

Mark Ptashne

Bernard Pullman

Adelio Rigo

Giuseppe Rotilio

D. Sabo

S. Saphon

Charles A. Sawicki

Harold A. Scheraga

Kai Simons

Mary K. Swenson

T. Taniguchi

Paul O. P. Ts'o

C. Weissmann

Kunio Yagi

Andrew Ziemiecki

Frontiers in
Physicochemical Biology

Proceedings of an International Symposium held in celebration of the Fiftieth Anniversary
of the Institut de Biologie Physico-Chimique (Fondation Edmond de Rothschild)

Paris, May 23–27, 1977

EDITED BY

Bernard Pullman

Institut de Biologie Physico-Chimique
Fondation Edmond de Rothschild
Paris, France

ACADEMIC PRESS   New York San Francisco London 1978

A Subsidiary of Harcourt Brace Jovanovich, Publishers

ACADEMIC PRESS, INC.
111 Fifth Avenue, New York, New York 10003

United Kingdom Edition published by
ACADEMIC PRESS, INC. (LONDON) LTD.
24/28 Oval Road, London NW1   7DX

Library of Congress Cataloging in Publication Data

Main entry under title:

Frontiers in physicochemical biology.

Includes bibliographies.
1. Molecular biology——Congresses. 2. Biological
physics——Congresses. 3. Paris. Institut de biologie
physico—chimique. I. Pullman, Bernard, Date
II. Paris. Institut de biologie physico—chimique.
QH506.F76   574.8'8   78—18258
ISBN 0—12—566960—7

PRINTED IN THE UNITED STATES OF AMERICA

Contents

PART II Physicochemical Aspects of the Mechanisms of Genetic Expression

PART IV Study of Organized Systems

List of Contributors

Numbers in parentheses indicate the pages on which the authors' contributions begin.

Philip H. Bolton (65), Department of Chemistry, University of California, San Diego, La Jolla, California 92093

*Antony W. Burgess** (115), Department of Biophysics, Weizmann Institute of Science, Rehovoth, Israel, and Department of Chemistry, Cornell University, Ithaca, New York 14853

Lilia Calabrese (357), Institute of Biological Chemistry, University of Rome, and Consiglio Nazionale delle Ricerche, Center for Molecular Biology, Rome, Italy

Mildred Cohn (191), Department of Biochemistry and Biophysics, University of Pennsylvania Medical School, Philadelphia, Pennsylvania 19104

A. R. Crofts† (459), Department of Biochemistry, Medical School, University of Bristol, Bristol BS8 1TD, United Kingdom

E. Domingo (167), Institute für Molekularbiologie I, Universität Zürich, 8093 Zürich, Switzerland

Pierre Douzou (421), INSERM, Montepellier, France, and Institut de Biologie Physico-Chimique, 75005 Paris, France

Thomas Early (65), Department of Chemistry, University of California, San Diego, La Jolla, California 92093

J. P. Ebel (213), Laboratoire de Biochimie, Institut de Biologie Moléculaire et Cellulaire du CNRS, 67000 Strasbourg, France

R. A. Flavell (167), Institute für Molekularbiologie I, Universität Zürich, 8093 Zürich, Switzerland

* Present address: Walter and Eliza Hall Institute of Medical Research, Parkville, Australia.

† Present address: Department of Physiology and Biophysics, University of Illinois at Urbana–Champaign, 524 Burrill Hall, Urbana, Illinois 61801.

Irwin Fridovich (269), Department of Biochemistry, Duke University Medical Center, Durham, North Carolina 27710

Henrik Garoff (387), European Molecular Biology Laboratory, D-6900 Heidelberg, Federal Republic of Germany

Quentin H. Gibson (369), Department of Biochemistry, Molecular and Cell Biology, Wing Hall, Cornell University, Ithaca, New York 11853

M. Grunberg-Manago (237), Institut de Biologie Physico-Chimique, 75005 Paris, France

Osamu Hayaishi (283), Department of Medical Chemistry, Kyoto University Faculty of Medicine, Kyoto 606, Japan

Ari Helenius (387), European Molecular Biology Laboratory, D-6900 Heidelberg, Federal Republic of Germany

Benno Hess (409), Max-Planck-Institut für Ernährungsphysiologie, Dortmund, Federal Republic of Germany

Pierre Joliot (485), Institut de Biologie Physico-Chimique, Fondation Edmond de Rothschild, 75005 Paris, France

David R. Kearns (65), Department of Chemistry, University of California, San Diego, La Jolla, California 92093

Patrick Maurel (421), INSERM, Montpellier, France, and Institut de Biologie Physico-Chimique, 75005 Paris, France

A. M. Michelson (309), Institut de Biologie Physico-Chimique, 75005 Paris, France

B. D. Nageswara Rao (191), Department of Biochemistry and Biophysics, University of Pennsylvania Medical School, Philadelphia, Pennsylvania 19104

Yu. A. Ovchinnikov (81), Shemyakin Institute of Bioorganic Chemistry, U.S.S.R. Academy of Sciences, Ul. Vavilova, 32 Moscow 117312, U.S.S.R.

Mark Ptashne (183), The Biological Laboratories, Harvard University, Cambridge, Massachusetts 02138

Bernard Pullman (143), Institut de Biologie Physico-Chimique, Fondation Edmond de Rothschild, 75005 Paris, France

Adelio Rigo (357), Institute of Physical Chemistry, University of Venice, Venice, Italy

Giuseppe Rotilio* (357), Institute of Biological Chemistry, University of Camerino, Camerino, Italy

D. Sabo (167), Institute für Molekularbiologie I, Universität Zürich, 8093 Zürich, Switzerland

---

* Present address: II Cattedra di Chimica Biologica della Facoltà di Scienze Matematiche, Fisiche e Naturali-Istituto di Chimica Biologica-Università di Roma, Rome, Italy.

S. *Saphon* (459), Department of Biochemistry, Medical School, University of Bristol, Bristol BS8 1TD, United Kingdom

Charles A. Sawicki (369), Department of Biochemistry, Molecular and Cell Biology, Wing Hall, Cornell University, Ithaca, New York 14853

Harold A. Scheraga (115), Department of Chemistry, Cornell University, Ithaca, New York 14853

Kai Simons (387), European Molecular Biology Laboratory, D-6900 Heidelberg, Federal Republic of Germany

*Mary K. Swenson** (115), Department of Biophysics, Weizmann Institute of Science, Rehovoth, Israel

T. Taniguchi (167), Institute für Molekularbiologie I, Universität Zürich, 8093 Zürich, Switzerland

Paul O. P. Ts'o ,(3), Division of Biophysics, The Johns Hopkins University, School of Hygiene and Public Health, Baltimore, Maryland 21205

C. Weissmann (167), Institute für Molekularbiologie I, Universität Zürich, 8093 Zürich, Switzerland

Kunio Yagi (299), Institute of Biochemistry, Faculty of Medicine, University of Nagoya, Nagoya 466, Japan

Andrew Ziemiecki (387), European Molecular Biology Laboratory, D-6900 Heidelberg, Federal Republic of Germany

---

* Present address: Israel Oceanographic Institute, Haifa, Israel.

Preface

This volume contains the proceedings of an international symposium held at the Institut de Biologie Physico-Chimique, Fondation Edmond de Rothschild of Paris, in May 1977, to celebrate the fiftieth anniversary of its foundation. In May 1967, we celebrated the fortieth anniversary of our Institute by holding an International Symposium on Molecular Associations in Biology, the proceedings of which were published, under that title, by Academic Press. We hope to have established a tradition of such celebrations every decade.

Fifty years is traditionally an important anniversary. For an institution, it is a test of performance. The Institut de Biologie Physico-Chimique, founded in 1927 by the Baron Edmond de Rothschild, the celebrated philanthropist and Maecenas, inspired by Jean Perrin, was probably the first research institute in the world devoted exclusively to the study of the physicochemical aspects of life. Its creation coincided with the birth of modern biophysics and molecular biology. Its activities during this half a century have been manyfold and diverse, covering at one time or another practically all of the essential directions of exploration in the modern science of life.

This symposium was organized to cover the main topics of present day molecular biology and biophysics, which are also the major disciplines researched at our Institute. I believe that thanks to the excellence of the lectures presented by our invited speakers it has achieved its goal.

It is my pleasant duty to thank all those whose efforts made this memorable meeting possible and successful. Our deepest thanks are due, as they were ten years ago, to the present Baron Edmond de Rothschild, the grandson of the founder, for both his generosity, which made this meeting possible, and his continuous efforts for and devotion to the welfare of our Institute. My thanks are also due the Comité de Direction and the Conseil d'Administration of our Fondation for their enthusiastic support of the idea of this celebration and, in particular, to Professors Hubert Curien and René Wurmser and Drs. Pierre Douzou, Pierre Joliot, Marianne Grunberg-Manago, Michael Michelson, and Sabine Wurmser for their invaluable help in planning the scientific program for the meeting. I would also like to acknowledge the efficient handling of secretarial and related problems by Mrs. de Hauss, Mrs.

xv

Landez, and Mrs. Pencranne. Madame Pullman's constant help in all aspects of the preparation and holding of the symposium went far beyond a wife's obligation.

Above all, I would like to express our gratitude to all our distinguished guests, speakers, session chairmen, and discussants, especially those who came from distant places to share their knowledge with us. Their contributions made the meeting successful. We look forward to seeing them all again at our future symposia and, if possible, at our Centennial.

Bernard Pullman

Part I

THE PHYSICAL CHEMISTRY OF FUNDAMENTAL BIOMOLECULES

CHAIRMAN: P. O. LÖWDIN
Chapters 1 and 2

CHAIRMAN: J. TIGYI
Chapters 3, 4, and 5

1

Current Progress and Future Prospects of the Research on Nucleic Acids

PAUL O. P. TS'O

Division of Biophysics, The Johns Hopkins University, School of Hygiene and Public Health, Baltimore, Maryland

I. INTRODUCTION

I was fortunate to have the opportunity to attend the 40th anniversary of the founding of the Institut de Biologie Physico-Chimique in 1967 and the 50th anniversary of the Institut in 1977. Within those 10 years, enormous strides have taken place in the scientific development in nucleic acid research: for example, the chemical and enzymatic synthesis of a complete tRNA gene, sequence analysis of bacteriophages and animal viruses, the determination of the three-dimensional structure of tRNA in the crystalline state, perfection in forming recombinant DNA that are functionally active, and elucidation of the structural unit (nucleosome or μ body) in interphase chromatin.

In this chapter, I will describe the current progress and future prospects in two areas of nucleic acid research that are being actively pursued in my laboratory. At the end, I shall describe the connection and the interrelationship between these two areas, which may open a new frontier in nucleic acid research.

II. NUCLEAR MAGNETIC RESONANCE STUDIES

There are four programs in this area of research. They will be discussed in the following sections.

3

FRONTIERS IN PHYSICOCHEMICAL BIOLOGY

A. Studies of Short RNA and DNA Helices in Solution

In 1971, our laboratory embarked upon a program for the synthesis of self-complementary ribosyl and deoxyribosyl oligonucleotides that form palindromic helices of one-half of one or one helical turn in solution under the appropriate conditions of ionic strength and temperature. The synthesis has two objectives: (1) The amount must be sufficient for nuclear magnetic resonance (nmr) studies; (2) the synthesis procedure and the products should aid the spectral assignment of the resonances observed in the nmr studies. Between 1972 and 1977, rapid progress was made that drastically reduced the amount of material required for nmr study, and which has greatly improved the organic synthesis procedure for the preparation of these oligonucleotides.

In 1975, we reported a comprehensive study of the proton magnetic resonance on the nonexchangeable protons and on the hydrogen-bonded NH—N protons of the ribosyl ApApGpCpUpU helix in solution (Borer *et al.*, 1975; Kan *et al.*, 1975a). All the resonances of the base proton, ribose–H_1, and the hydrogen-bonded NH—N protons have been assigned with a great deal of certainty. The effects of helix-coil transition on these 20 resonances have been carefully recorded and analyzed in terms of changes in chemical shifts, coupling constants, and linewidth. The effect of salt concentration and oligonucleotide concentration on the helical-coil transition process has also been investigated.

Confronting the challenging task of making spectral assignments for the 17 C—H resonances and the 3 hydrogen-bonded NH—N resonances, we have adopted three major procedures, in addition to the conventional approach of relying on chemical shifts, coupling constants, and the spin lattice relaxation time, T_1.

The first procedure compares the spectra of a sequence-related series of oligonucleotides in which each member of the series is incremented one nucleotide unit from its predecessor. Hence, this aspect is termed "incremental assignment." The shortest member of this series must be previously assigned by other standard methods and the longest member is obviously the molecule of interest. In the present case, ApA, A_2G, A_2GC, A_2GCU, and A_2GCU_2 comprise the series. The spectra of the series of oligomers are usually recorded in low salt, at high temperature, and, if possible, at low strand concentration. In this environment the inter- and intrastrand interactions are greatly reduced, as indicated by the similarity of the oligomer and monomer spectra; in addition, the resonances are narrow and usually better resolved. Each of the oligomers in the series exists as a substantially unstacked single strand; therefore, any major change in the resonance pattern from an oligomer to the one incrementally longer is due to the resonances of the new nucleotide and its shielding effects on the previously present protons. The magnetic field effect of the newly added nucleotide acts principally on its immediate 5' neighbor. The procedure is illustrated in Fig. 1.

The second assignment procedure compares the effect of temperature variations, as demonstrated in Fig. 2. The entire spectrum of the hexamer and those of many oligomers in the series have been recorded over a 0°–90°C range. The interval between temperature points was 2°–4°C in regions where δ changes rapidly with temperature or several protons resonate very close to each other. The interval was ~ 10°C in temperature regimes of little overlap or change in δ. From such data the spectral assignment established at one temperature can be transferred to that at another temperature.

The third procedure involves the assignment of NH—N resonances, which depends on the linewidth measurements. The linewidths of the NH—N resonances of the helical duplex are related to the rates of the following exchange reactions (Crothers et al., 1974):

$$NH_{helix} \underset{k_{ch}}{\overset{k_{hc}}{\rightleftarrows}} NH_{coil} \tag{1}$$

$$NH_{coil} \underset{k_{wc}}{\overset{k_{cw}}{\rightleftarrows}} H_2O \tag{2}$$

The Bloch equations relating the properties of the nmr signals to the various exchange rates for the helix-coil transition in water have been presented by Crothers et al. (1974). Based on the work of Porschke et al. (1973) on A_2GCU_2 and the work of Ravetch et al. (1974) on $A_{3-4}GCU_{3-4}$, as well as the equation of exchange (Kan et al., 1975a), we concluded that $\tau_{cw}^{-1}\tau \simeq 12$, where the average lifetime, τ, is $\tau = \tau_{hc}\tau_{ch}/(\tau_{hc} + \tau_{ch})$ and is dominated by τ_{ch} and where τ_{cw}^{-1} is the exchange rate between NH in the coiled state and the H in water. However, the computations of both τ_{cw}^{-1} and τ^{-1} involve a fair amount of assumption and extrapolation. Therefore, the conclusion that $\tau\tau_{cw}^{-1} \gg 1$ must be regarded as tentative. Nevertheless, we shall proceed with this conclusion that the linewidths of the NH resonances in the current experiment are primarily determined by the lifetime of the helix.

The above conclusions readily lead to the assignment of the NH resonances based on the linewidth data presented in Fig. 3. The NH resonance that has the largest linewidth and the highest sensitivity to thermal effects on the line-broadening and line-shifting is assigned to the NH—N of the two terminal A(1) · U(6) pairs; the NH resonance with the smallest linewidth and the lowest sensitivity to thermal effects is assigned to the NH—N of the two middle G(3) · C(4) pairs; and the NH resonance that has an intermediate linewidth and sensitivity is therefore

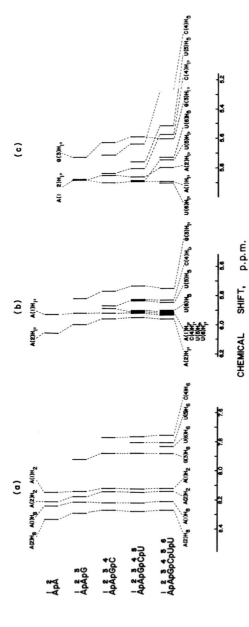

CHEMICAL SHIFT, p.p.m.

Fig. 1 Incremental assignment scheme for all the base and H$_{1'}$ resonances of A$_2$GCU$_2$. The thickness of the lines represents the number of signals contained in a resonance envelope. (a) The assignment of H$_2$, H$_6$, and H$_8$ resonances at ~65°C. (b) The assignment of H$_{1'}$ and H$_5$ resonances at ~65°C. Four resonances are clustered at this temperature. (c) The assignment of H$_{1'}$ and H$_5$ resonances at ~35°C. The four resonances clustered at ~65°C are resolved at this temperature and can be assigned. [Na$^+$] = 0.02 M and c_s (concentration of strand) = 1 mM for each of the spectra except for the hexamer spectrum shown in the region of 5–6 ppm, where c_s = 10 mM and [Na$^+$] = 0.07 M. Chemical shifts are expressed in reference to DSS. Reprinted with permission from Borer *et al.* (1975), *Biochemistry* **14**, 4847. Copyright by the American Chemical Society.

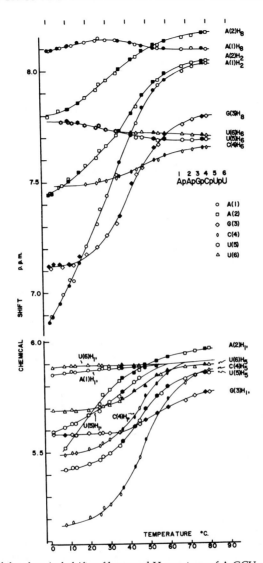

Fig. 2 Plot of the chemical shifts of base and $H_{1'}$ protons of A_2GCU_2 in D_2O (10 mM in strand concentration, 0.01 M sodium phosphate buffer, pD 7.0, 0.07 M Na$^+$) versus temperature. All chemical shifts are expressed in reference to DSS. The filled symbols represent the data from the 100-MHz spectrometer and the open symbols represent the data from the 220-MHz spectrometer. Reprinted with permission from Borer *et al.* (1975), *Biochemistry* **14**, 4847. Copyright by the American Chemical Society.

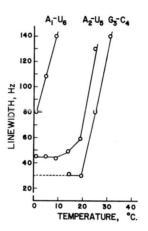

Fig. 3 Plot of the linewidth at half-height of the NH—N resonances of the A_2GCU_2 duplex versus temperature. Reprinted with permission from Kan *et al.* (1975a), *Biochemistry* **14**, 4864. Copyright by the American Chemical Society.

assigned to the two interior $A(2) \cdot U(5)$ pairs. The last assignment is supported by the chemical shift computations shown in Table II, as discussed later.

In analyzing the chemical shift data, a helical duplex of Kendrew's molecular model of A_2GCU_2 was constructed from the coordinates of A'-RNA given by Arnott and co-workers (Arnott and Hukins, 1972; Arnott *et al.*, 1973). Similarly, such a duplex was also constructed from the coordinates of B-DNA recommended by Arnott and Hukins (1973). Based on these models, projections of neighboring protons onto the plane of each base were made from the Kendrew model. Such a projection for the A_2GCU_2 helix in A'-RNA geometry is shown in Fig. 4. These projections now enable us to ascertain the distance of the protons of interest from the base ring.

Additional and valuable information can also be obtained from the coupling-constant data on the $H_{1'}$ resonances during the helix-coil transition and the temperature variation profile, as shown in Fig. 5.

The accuracy of the coupling-constant measurements in these ^1H nmr spectra is estimated to be ± 1 Hz. This causes the measurements of the individual $J_{1'-2'}$ values to be scattered considerably. Nevertheless, much valuable information can be gained from the general trend of the effect of temperature on the $J_{1'-2'}$ values. Figure 5 illustrates the properties of the averages (circles) and ranges (vertical bars) of the $J_{1'-2'}$ values of the five $H_{1'}$ doublets that decrease most rapidly upon decreasing the temperature. The $J_{1'-2'}$ values for the tentatively assigned $U(6)H_{1'}$ resonance (triangles) are plotted separately. At high temperatures the $J_{1'-2'}$ values are close to those observed for the appropriate nucleoside

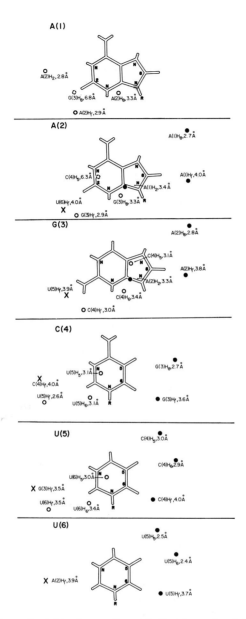

Fig. 4 Projections of neighboring protons onto the plane of each base of the A_2GCU_2 helix (A'-RNA geometry). Projections and distances were taken directly from the Kendrew model. Open circles indicate the protons positioned below the base plane; filled circles indicate the protons positioned above the plane; broken circles indicate protons from a distant neighbor about 6.8 Å away; and crosses indicate $H_{1'}$ atoms in the other strand. The views are normal to the base planes with the free 5'-OH terminus above and the 3' terminus below the base planes. Vertical distances of the protons to the base planes are indicated in parentheses. Reprinted with permission from Borer *et al.* (1975), *Biochemistry* **14**, 4847. Copyright by the American Chemical Society.

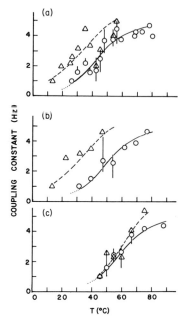

Fig. 5   The $H_{1'}$–$H_{2'}$ coupling constants, $J_{1'-2'}$, for the six ribose $H_{1'}$ resonances of A_2GCU_2 as a function of temperature and counterion concentration: (a) 0.07 M Na⁺; (b) 0.17 M Na⁺; (c) 1.07 M Na⁺. The $J_{1'-2'}$ values for the tentatively identified $U(6)H_{1'}$ signal (---) were plotted separately from the average of the other five resonances (——). The vertical bars indicate the range of values observed for these five signals and the lowest temperature point indicates the temperature at which all the resonances become singlets with $J_{1'-2'} \leq 1.5$ Hz. Reprinted with permission from Borer *et al.* (1975), *Biochemistry* **14**, 4847. Copyright by the American Chemical Society.

monophosphates (e.g., see Ts'o, 1974). Upon decreasing the temperature these doublets become narrow with smaller J values; they merge into a singlet within the resolution of the spectrometer at a sufficiently low temperature. Therefore, the $J_{1'-2'}$ values are less than 1.5 Hz for the furanose ring of each residue in the helical duplex. The low $J_{1'-2'}$ values indicate that the furanose of the helical duplex is most likely in the C-3′-*endo* conformation. It is known from the X-ray diffraction studies that the C-3′-*endo* conformation belongs to the A- and A′-RNA's, whereas the C-2′-*endo* conformation belongs to the B form of nucleic acids, such as B-DNA. Therefore, the data based on $J_{1'-2'}$ clearly favor the notion that the A_2GCU_3 duplex in solution assumes an A or A′ conformation.

    A comparison between the predicted chemical shifts of these 20 proton resonances and the observed resonances was made. The predicted chemical shifts of formation for the A_2GCU_2 sequence in both the A′-RNA and B-DNA geometries are reported in Table I. The computa-

Reprinted with permission from Borer et al. (1975), Biochemistry 14, 4847. Copyright by the American Chemical Society.

TABLE I

Chemical Shifts of Formation of 17 Nonexchangeable Protons in (1) the A_2GCU_2 Duplex as Compared with Calculated Ring-Current Shifts from Two X-Ray Crystallographic Models and (2) the A_2GCU_2 Coils at 90°C[a]

| | | | Duplex form (δ in ppm from DSS) | | | | Coil form (δ in ppm from DSS) | |
| | | | $\Delta\delta_{calc}$[d] | | $\Delta\delta_{obs} - \Delta\delta_{calc}$ | | | |
	δ_{obs}[b]	$\Delta\delta_{obs}$[c]	A'	B	A'	B	δ[e]	$\Delta\delta$[f]
G(3)H_8	7.20	0.93	0.85	0.33	0.08	0.60	7.91	0.13
C(4)H_6	7.56	0.48	0.42	0.21	0.06	0.27	7.77	0.15
C(4)H_5	5.14	0.97	1.07	0.66	−0.10	0.31	5.89	0.19
A(2)H_2	7.50	0.76	0.64	0.44	0.12	0.32	8.16	0.13
A(2)H_8	7.84	0.67	0.79	0.25	−0.12	0.45	8.28	0.27
U(5)H_6	7.95	0.07	0.15	0.18	−0.08	−0.11	7.79	0.11
U(5)H_5	5.43	0.51	0.46	0.31	0.05	0.20	5.87	0.06
A(1)H_2	6.89	1.36	1.05[g]	1.10[g]	0.31[g]	0.26[g]	8.16	0.12
A(1)H_8	8.16	0.22	0.00[g]	0.02[g]	0.22[g]	0.20[g]	8.20	0.16
U(6)H_6	7.95	0.07	0.03	0.03	0.04	0.04	7.81	0.09
U(6)H_5	5.69	0.25	0.20	0.09	0.05	0.16	5.90	0.03
G(3)$H_{1'}$	5.59	0.35	0.08	0.17	0.27	0.18	5.77	0.13
C(4)$H_{1'}$	5.48	0.50	0.04	0.07	0.46	0.43	5.92	0.04
A(2)$H_{1'}$	5.45	0.69	0.10	0.27	0.59	0.42	5.98	0.15
U(5)$H_{1'}$	5.60	0.38	0.02	0.05	0.36	0.33	5.92	0.08
A(1)$H_{1'}$	5.86	0.27	0.00	0.22	0.27	0.05	5.92	0.19
U(6)$H_{1'}$	5.86	0.12	0.00	0.00	0.12	0.12	5.92	0.08

[a] Reprinted with permission from Borer *et al.* (1975), *Biochemistry* **14,** 4847. Copyright by the American Chemical Society.

[b] These are the low-temperature plateau values of δ at 10 mM strand concentration, pD 7.0, 1.07 M Na$^+$. If the 0.17 or 0.07 M Na$^+$ δ versus T (°C) profiles leveled off at low temperature, their plateau values were averaged with the 1.07 M Na$^+$ number to generate the δ_{duplex} value.

[c] $\Delta\delta_{duplex} = \delta_{duplex} - \delta_{mononucleotide}^{0°-5°C}$ is the chemical shift of formation for a proton in the duplex. Appropriate mononucleotide values were selected from the following list (δ in ppm from DSS): pG—H_8, $H_{1'}$, 8.126, 5.936, respectively; pC—H_6, H_5, $H_{1'}$, 8.038, 6.110, 5.979; pA—H_2, H_8, $H_{1'}$, 8.257, 8.512, 6.141; pU—H_6, H_5, $H_{1'}$, 8.015, 5.935, 5.982; and Ap—H_2, H_8, $H_{1'}$, 8.252, 8.381, 6.132, measured in 0.01 M sodium cacodylate buffer, pD 5.9 ± 0.1 in D$_2$O, 1 mM in the appropriate 5' mononucleotide (a 2.0 mM sample of 3'-AMP was also measured). Little or no association of the mononucleotides is expected at these low concentrations. Temperatures were 0°–5°C.

[d] Based on the A'-RNA and B-DNA geometries (Arnott *et al.*, 1973; Arnott and Hukins, 1972) and ring-current isoshielding contours provided by B. Pullman (personal communication; see the discussion in the text).

[e] These are averages of the 90°C chemical shifts of protons in 0.07, 0.17, and 1.07 M Na$^+$ at 10 mM strand concentration, pD 7.0.

[f] $\Delta\delta_{coil} = \delta_{coil} - \delta_{mononucleotide}^{90°C}$ is the chemical shift of formation for a proton in the high-temperature coil form. The following mononucleotide values were used (see footnote b for buffer and nucleotide concentrations): pG—H_8, $H_{1'}$, 8.041, 5.890; pC—H_6, H_5, $H_{1'}$, 7.921, 6.084, 5.960; pA—H_2, H_8, $H_{1'}$, 8.287, 8.547, 6.130; pU—H_6, H_5, $H_{1'}$, 7.905, 5.930, 6.001; and Ap—H_2, H_8, $H_{1'}$, 8.282, 8.359, 6.109.

[g] These calculated values are subject to a correction of ~0.15 ppm due to shielding in end-to-end aggregates.

tion used ring-current isoshielding contours that were mapped in
planes parallel to the bases at the distances shown in Fig. 4. These
contours were generously provided by Professor B. Pullman and Dr. C.
Giessner-Prettre (personal communication) and were calculated by the
procedure given by Giessner-Prettre and Pullman (1970). Such calculations accurately account for the distance effects introduced by the twist-
ing and tilting of bases in the RNA helix as well as next nearest-
neighbor effects.

In comparing the predicted and observed chemical shifts of formation
it is important to set reasonable error limits for the calculations. A major
source of uncertainty in the predictions is that the isoshielding contours
are generated by an approximate molecular orbital calculation. Thus,
agreement between the calculations and predictions within 0.1 ppm is
probably as good as can be expected. An examination of Tables I and II
shows that the calculated $\Delta\delta_{duplex}$ values for 9 of the 11 base protons
agree to within 0.12 ppm of the observed values. The other two calculations are still qualitatively correct; a very large shift is predicted for
$A(1)H_2$ and a small shift for $A(1)H_8$. On the contrary, only 2 of the 11
predictions from the B-DNA model are correct within 0.12 ppm. Thus, a
base stacking geometry similar to that in the A'-RNA model is clearly
favored over the B-DNA conformation according to this analysis.

In a similar manner, the measured chemical shifts of the three NH—N
resonances are compared with the computed values based on the
A'-RNA and B-DNA models, as shown in Table II. There is evidence

TABLE II

Chemical Shifts and the Linewidth at Half-Height of the Hydrogen-Bonded
NH—N Resonances of Guanine and Uracil in $(A_2GCU_2)_2$ at 1°C[a]

	Observed (ppm)	δ_{calc} (ppm)		$\delta_{obs} - \delta_{calc}$ (ppm)		Linewidth at $\omega_{1/2}$ (Hz)
		A'-RNA[b]	B-DNA[c]	A'-RNA	B-DNA	
G(3)NH	13.5[d]	13.3	12.8	0.2	0.7	30
U(5)NH	14.2	14.1	14.1	0.1	0.0	44
U(6)NH	13.2	13.5	14.3	−0.3	−1.1	80

[a] The intrinsic values of the chemical shifts of the NH resonance in the A·U pair and
in the G·C pair are taken to be 14.7 and 13.6 ppm, respectively (Kearns and Shulman,
1974; Kearns, 1976), with the A·U pair in the B-DNA geometry given the same value,
14.6 ppm, as derived for the A·T pair (Patel and Tonelli, 1974). Reprinted from Kan et al.
(1975a), Biochemistry 14, 4864. Copyright by the American Chemical Society.
[b] Calculated values based on the geometry of the A'-RNA helix (Borer et al., 1975).
[c] Calculated values based on the geometry of the B-DNA helix (Borer et al., 1975).
[d] All negative signs are omitted.

from hypochromicity, resonance linewidths, and sample turbidity that suggests the existence of some form of multiple aggregation of the duplexed strands. The association of mononucleosides and mono-nucleotides in water has been studied extensively by nmr and thermodynamic methods (for a review, see Ts'o, 1974). Association persists even at the level of 10 mM AMP, as followed by the concentration dependence of the base and sugar proton resonances. The concentration of end A(1) bases in duplexed A_2GCU_2 was 10 mM in the studies presented here. Therefore, it is not unreasonable that several duplexes might form an end-to-end stacking arrangement similar to that seen for the mononucleotides. This notion is supported by studies with intercalative agents that disrupt this type of association (Kan *et al.*, 1975b). These experiments suggest that this effect may account for 0.1–0.15 ppm for the observed $\Delta\delta_{duplex}$ for $A(1)H_2$ and $A(1)H_8$. This additional shielding increment brings the calculated $\Delta\delta_{duplex}$ values for these A(1) protons within about 0.15 ppm of those observed. Thus, the agreement between the calculated and observed chemical shifts of formation for the base protons is remarkable if this correction based on aggregation is applicable. However, for the data shown in Table I, there is considerable disagreement (0.2–0.5 ppm) between the predicted chemical shifts and the observed chemical shifts of the $H_{1'}$ resonances. Various possibilities were suggested, and this matter is currently under investigation.

In summary, conformational details of the A_2GCU_2 helix in solution can now be ascertained on the basis of the following self-consistent information. (1) The $J_{1'-2'}$ values of all the residues in the helix indicate that each furanose is in a 3'-*endo* conformation. Model building based on the X-ray diffraction data of nucleic acid fibers reveals that the furanose conformation in the A'-RNA (or A-RNA) is 3'-*endo*, whereas the furanose in B-DNA is 3'-*exo* or 2'-*endo*. (2) The chemical shifts of 17 C—H resonances of the A_2GCU_2 helix in solution agree with the computed values based on the geometry of the A'-RNA much better than they agree with those based on the geometry of B-DNA. (3) The chemical shifts of three sets of NH—N resonances representing six base pairs of the A_2GCU_2 helix also agree with the computed values based on the geometry of A'-RNA significantly better than they agree with those based on the geometry of B-DNA. Thus, these data strongly support the conclusion that the A_2GCU_2 helix in solution must assume a conformation closer to that of the A'-RNA than to that of B-DNA. It would be of great interest to synthesize a short DNA helix containing the A_2GCU_2 sequence and to study the conformation of this short DNA helix in solution by proton magnetic resonance (pmr) following the approach outlined above.

Our laboratory has completed the synthesis of a deoxydecanucleotide, d(CCAAGCTTGG), and a series of shorter fragments (E. Leutzinger and P. O. P. Ts'o, unpublished results). This decanucleotide should form a palindromic, self-complementary helix (one complete helical turn for B-DNA) in solution under appropriate conditions. The center of this deoxy helix, (AAGCTT)$_2$, has the same base sequence (except that U has been substituted by T) as the ribosyl helix of (AAGCUU)$_2$. The comparison between the ribosyl and deoxyribosyl helices having similar base sequences would be of great value.

The nmr data on the helix-coil transition of this short helix (A$_2$GCU$_2$)$_2$ contain both thermodynamic and structural information. These helix-coil transition profiles are reports of 20 atoms at separate locations within the helix (including the three NH—N profiles) concerning their magnetic properties during the thermal transition; therefore, in principle, they can provide useful data to assign statistical weights to the various partially bonded states. However, each transition curve is a reflection of the changes in the local magnetic environment of each proton rather than a direct report on the helix-coil populations. In the partially formed duplex the δ value of a proton is determined by its δ value in each microstate, weighted according to the population of the state. Moreover, the value in a particular microstate is determined by shielding influences, which are anisotropically distributed through space.

Thus, it is a very challenging task to disentangle the various factors to obtain meaningful information without more accurate knowledge of the application of nmr theory to nucleic acid research. However, two general conclusions can be reached at present. First, the nmr data follow the expected patterns derived from the optical studies carried out at lower concentrations concerning the effect of concentration and ionic strength on the helix-coil transition of this short helix (Borer et al., 1975). The "average T_m" increases with increasing concentration, showing a linear relationship in a $1/T_m$ versus $-$log concentration (strand) plot, and increases with an increase in ionic strength. Second, the melting of this short helix clearly does not reflect an all-or-nothing pattern. Both C—H resonances and NH—N resonances reveal that, with respect to temperature change, the C$_3$ · C$_4$ pair in the center is more stable than the A$_2$ · U$_5$ pair, which in turn is more stable than the A$_1$ · U$_6$ pair at the end of the helix. These results are equivalent to the fraying of the ends of this short helix. It should be noted that valuable studies on short helices have also been made by others (Patel and Tonelli, 1974, 1975; Arter et al., 1974; Patel, 1976; Early et al., 1977).

B. Theoretical Computation of nmr Data

It is self-evident that a quantitative application of the nmr theory is urgently needed. We, and other nucleic acid nmr chemists, are indebted to Professor B. Pullman and Dr. C. Giessner-Prettre for their contributions to this effort, and we are currently enjoying a fruitful collaboration with them.

We have taken two approaches in our research. In the first approach, the spatial dependence of the ring-current magnetic anisotrophy of the nucleic acid bases has been calculated over a cylindrical domain of 10-Å radius, which extends 8 Å above and below each ring of these bases (Giessner-Prettre *et al.*, 1976). As shown in Fig. 6, separating the bicyclic purine base into two individual cyclic rings allows this spatial dependence to be presented in a series of graphs in cylindrical domains. An example of this graphic approach for the adenine ring is shown in Fig. 7. By this approach, the through-space ring-current effects on the chemical shifts of any resonance for atoms of nucleic acid with known and fixed conformation can be calculated by this first approximation and by the independently determined atomic coordinates. While this graphic approach does not replace the more precise computer-computation approach, it does provide a simple method for manipulating geometric parameters to predict the probable mode of base–base or even base–drug interactions.

The second approach (L. S. Kan, J. R. Kast, D. Y. Ts'o, and P. O. P. Ts'o, unpublished data) utilizes a Fortran-10 computer program (a DEC-10 computer) that prints out the precise Cartesian coordinates and

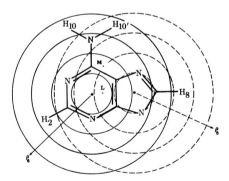

Fig. 6 Concentric contours for each ring of the adenine base spaced at 1.0-Å intervals. The large dot in the center of each ring is the origin of the cylindrical coordinate system for that ring; ρ_5(- - - →) represents the radius for the five-membered ring and ρ_6(——→) represents the radius from the six-membered ring. From Giessner-Prettre *et al.* (1976).

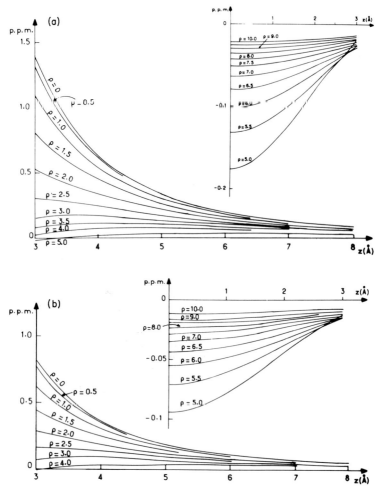

Fig. 7 Graphic approach for adenine ring. (a) $\Delta\delta_6$ versus $|z|$, the vertical distance from the base plane, and ρ_6, the radius from the center of the six-membered ring. (b) $\Delta\delta_5$ versus $|z|$ and ρ_5, the radius from the center of the five-membered ring. From Giessner-Prettre *et al.* (1976).

cylindrical coordinates of all atoms of nucleic acid duplexes having three helical structures (A, A', and B forms) of any specified sequences up to 30 base pairs in length. This program can also calculate the distance between atoms. Programs for other helical structures can be readily added. Thus, the precise coordinates of all atoms of any nucleic acid duplex up to 30 base pairs (such as that shown in Fig. 4) can be ascertained in three or more conformations for use in nmr calculations. We are now able to use the computation program established by Professor

B. Pullman and Dr. C. Giessner-Prettre in conjunction with this computer program to obtain coordinates for the calculation of chemical shifts of all the atoms of known coordinates. It is hoped that various throughspace, electromagnetic effects exerted on the atoms, such as the ring-current anisotrophic effect, the atomic anisotrophic effect, the hydrogen-bonding effect, and the electrostatic effect of the charged group (such as the phosphate), can now be considered individually and collectively in various combinations (Giessner-Prettre et al., 1977) in a quantitative fashion.

Thus, the chemical shifts of all atoms (such as ^{1}H, ^{13}C, and ^{31}P) of a nucleic acid duplex having sequences of up to 30 or more base pairs in these three helical conformations can be calculated for various theoretical analyses. A computer program is now being constructed to produce a nmr spectrum (or a set of nmr spectra) for any sequence of nucleic acid duplexes based on the computations done according to the quantum mechanical considerations on a set of atomic coordinates. Thus, the combination of three computer programs—(1) the program for generating coordinates of all atoms of a nucleic acid duplex of sequences in three or more helical conformations; (2) the program for computing chemical shifts as influenced by through-space, electromagnetic field effects; and (3) the program for producing nmr spectra of a given frequency of measurement (for ^{1}H, 270 or 360 MHz) based on the intrinsic chemical-shift values of the isolated molecules and the extrinsic field effects—can provide a very valuable tool for both theoreticians and the experimentalists in the field of nmr research on nucleic acids. This represents a systematic approach to this problem.

C. Quantitative nmr Studies on the Yeast Transfer RNA^{Phe}

We have attempted to apply the above systematic approach at least in a trial fashion to the study of yeast tRNA^{Phe} in solution (Kan et al., 1977). The two-dimensional structure of yeast tRNA^{Phe} in the proposed cloverleaf model is shown in Fig. 8.

The structure of yeast phenylalanine transfer ribonucleic acid (tRNA^{Phe}) in the crystalline state has been clearly elucidated by X-ray diffraction studies (Kim et al., 1974; Ladner et al., 1975a). Furthermore, the three-dimensional coordinates of all atoms (except hydrogen) in this tRNA molecule have been reported by several laboratories (Quigley et al., 1975; Ladner et al., 1975b; Sussman and Kim, 1976; Stout et al., 1976; see the review by Rich and RajBhandary, 1976). The tRNA^{Phe} structures determined from these two crystal forms are very similar to each other. Therefore, a very crucial question arises: Does the common conformation of tRNA^{Phe} determined in the crystalline state also exist in aqueous

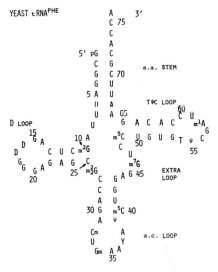

Fig. 8 Cloverleaf structure of bakers' yeast tRNA^{Phe}. a.a., amino acid; a.c., anticodon. Reprinted with permission from Kan *et al.* (1977), *Biochemistry* **16**, 3143. Copyright by the American Chemical Society.

solution? The information obtained from nmr studies can provide a direct and defined answer to this question.

Two spectral regions in the ^{1}H nmr spectrum of tRNA^{Phe} can be investigated for quantitative conformational information. The first region, which will be discussed later (see Tables IV and V and Figs. 16 and 17), contains the very low-field NH—N hydrogen-bonded proton resonances in H_2O. The second region is described in this section, in which the high-field (1–4 ppm) resonances of the methyl/methylene protons from the minor nucleosides in yeast tRNA^{Phe} in D_2O are investigated. The emphasis of this study is not on the base pair region, which can be studied by the NH—N hydrogen-bonded proton resonances, but rather on the loop regions and tertiary structure of the tRNA molecule (Kan *et al.*, 1974; Kastrup and Schmidt, 1975; Reid and Robillard, 1975; Daniel and Cohn, 1975, 1976), especially the D, anticodon, and TΨC loops on tRNA^{Phe}.

In addition to the intact tRNA^{Phe} molecule, four important fragments have also been investigated. These fragments are not only very useful in the assignment of the methyl/methylene resonances of the intact tRNA^{Phe}, but also provide information on the conformation of the whole molecule of tRNA^{Phe}.

Finally, by taking advantage of developments in the distance dependence of the ring-current magnetic effect of the bases described above

(Giessner-Prettre *et al.*, 1976) and in the refined atomic coordinates of yeast tRNA^{Phe} in the crystalline state (A. Rich, personal communication), a quantitative comparison between the calculated shielding effects ($\Delta\delta$) and the observed shielding effects was made. This study suggests that the conformation of yeast tRNA^{Phe} in aqueous solution is grossly, but not totally, identical to that determined in the crystalline state, especially in the TΨC and D regions.

The first step for assignment of these high-field resonances from tRNA^{Phe} and its fragments is to compare their spectra to those from monomers at high temperature. All modified mononucleosides (and mononucleotides) found in tRNA^{Phe} molecules were investigated separately at high temperatures, and their chemical shift data of methyl or methylene proton resonances are marked on the top row in Fig. 9. This knowledge from the mononucleosides (mononucleotides) and the sequence data provide the basis for the unambiguous assignment of the methyl/methylene resonances in tRNA^{Phe} fragments, which were then used for the assignment of the spectra for the whole tRNA molecule.

A few selected 220-MHz ^{1}H nmr spectra of the intact tRNA^{Phe} with 0.01 M MgCl$_2$ from 40° to 98.5°C are shown in Fig. 10. The identity of every signal at 98.5°C is marked on the top of the peak and is extended to the signals measured at lower temperatures by the dotted lines. At 98.5°C (Fig. 10a), the spectral patterns and chemical shifts of tRNA^{Phe} with Mg^{2+} are very similar to those of tRNA^{Phe} without Mg^{2+} at 79°C (Kan *et al.*, 1974). This temperature differential of 20°C in reaching the same state of denaturation is due presumably to the presence of Mg^{2+}, which stabilizes the native form of tRNA^{Phe}. At 80°C (Fig. 10b), while the tRNA^{Phe} in Mg^{2+} is still in the denatured state, the signals from m$_2^2$G and m^{2}G begin to shift upfield, and all signals begin to broaden, except those from the Y base. At 73°C, tRNA^{Phe} is in the middle or near the low end of the native–denatured transition, and a considerable amount of broadening and shifting to high field takes place for all the resonances, though less for those from the Y base. At 40°C, only the resonances from the Y base, D and D', m^{2}G$_{10}$, the T base, and perhaps the m^{5}C are visible in the native tRNA^{Phe} spectrum measured with 0.01 M Mg^{2+}. The temperature transition profiles of the chemical shifts of these resonances are shown in Figs. 11, 12, and 13.

The intact tRNA^{Phe} in the presence of Mg^{2+} was also investigated at a frequency of 360 MHz (Fig. 14). In this sample, the internal standard, *tert*-butanol, was not added so that the spectral region near 1.1–1.4 ppm would not be masked by the signal from *tert*-butanol. [The spectral positions of the resonances were calibrated from the Y signal determined at 200 MHz; the chemical shifts of the Y signal in the intact tRNA

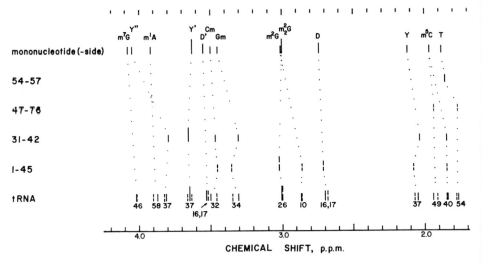

Fig. 9 Schematic plot of the assignment of methyl/methylene resonance signals in intact bakers' yeast tRNA^{Phe} with or without $0.01\,M$ Mg^{2+} at high temperature (98.5°/79°C). The solid vertical lines represent ^{1}H nmr signals that were taken from a sample with $0.01\,M$ phosphate buffer (pH 7), and broken vertical lines represent the ^{1}H nmr signals from the same buffer solution with additional $0.01\,M$ MgCl$_2$. The relative heights of these lines reflect the number of protons of each peak; namely, the tallest line represents six protons, medium height represents four, and the shortest line represents three protons. The methyl/methylene ^{1}H nmr resonances of mononucleotides (mononucleosides) are shown in the first (top) row. Most of them contribute a singlet from the methyl group and the names of the modified bases are given above the resonances. But the Y base shows four (sometimes three) methyl singlets, which are symbolized by Y = C$_{11}$—Me, Y′ = NCOOMe and CCOOMe, and Y″ = N$_3$—Me. Only the D base contains two methylene triplets, namely, D = C$_5$-methylene and D′ = C$_6$-methylene. The spectra of these monomers have been taken within a temperature range of 70°–80°C. The last row shows the ^{1}H nmr signals of these modified bases in intact tRNA^{Phe} (with or without Mg^{2+}). These assigned signals (see the text) are identified by the nucleotide sequential numbers directly below each peak. There are four tRNA^{Phe} fragments (identified by nucleotide sequential numbers for the first and last nucleotides) listed on the first and last rows. The temperatures at which the spectra were taken are the following: 70°C for fragment 54–57, 90°C for fragment 47–76, 63°C for fragment 31–42, and 93°C for fragment 1–45. Measurements were made at sufficiently high temperature to abolish all the secondary structures of these fragments. Usually, the larger the fragments, the higher the temperature needed to reach this condition. Reprinted with permission from Kan *et al.* (1977), *Biochemistry* **16**, 3143. Copyright by the American Chemical Society.

have been shown previously to be insensitive to temperature variation (Kan *et al.*, 1974).] Due to a higher resolution power and sensitivity of the 360-MHz spectrometer, not only was the signal-to-noise ratio of the spectra improved, but the low-field resonances of m^{7}G, m^{1}A, and Y″ became much more evident (Fig. 14). The spectral study at 360 MHz provided the important results listed below.

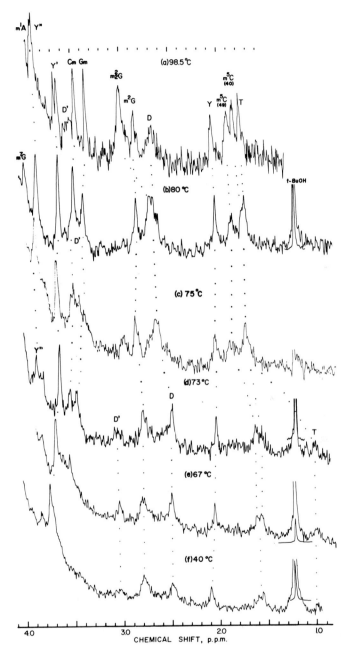

Fig. 10 The 220-MHz nmr spectra of high-field proton resonances in the region of bakers' yeast tRNA[Phe] in a temperature range of 40° to 98.5°C. The experimental conditions are: acquisition time, 0.8 second; 8 blocks (100 transients in one block) for spectrum a, 5 blocks (200 transients/block) for spectra b and c, 6 blocks (200 transients/block) for spectrum d, and 10 blocks (200 transients/block) for spectra e and f. Reprinted with permission from Kan *et al.* (1977), *Biochemistry* **16,** 3143. Copyright by the American Chemical Society.

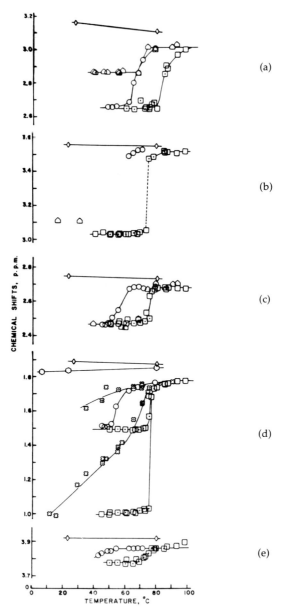

Fig. 11 Chemical shift data of the methyl proton resonances from (a) $m_2^2G_{26}$, (b) $D'_{16,17}$, (c) $D_{16,17}$, (d) T_{54}, and (e) m^1A_{58} in the monomer, the intact tRNA^{Phe}, and its fragments versus temperature. ◇, monomer; ⬡, fragment 54–57; △, fragment 1–45 in 0.01 M MgCl$_2$; ✿, fragment 47–76 in 0.01 M MgCl$_2$; ◯, yeast tRNA^{Phe}; ☐, yeast tRNA^{Phe} in 0.01 M MgCl$_2$. ✿, ⊙, and ⊡ mean that the data were obtained using a 360-MHz spectrometer. Reprinted with permission from Kan *et al.* (1977), *Biochemistry* **16,** 3143. Copyright by the American Chemical Society.

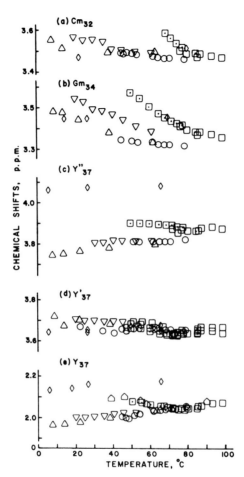

Fig. 12 Chemical shift data of the methyl proton resonances from (a) Cm_{32}, (b) Gm_{34}, (c) Y''_{37}, (d) Y'_{37}, and (e) Y_{37} of the monomer, intact tRNA^{Phe}, and fragments 31–42 and 1–45 versus temperature, where △ represents fragment 31–42 and ▽ represents fragment 31–41 in $0.1\,M$ NaCl and $0.1\,M$ MgCl$_2$. The remaining symbols have the same meaning as in Fig. 11. Reprinted with permission from Kan *et al.* (1977), *Biochemistry* **16**, 3143. Copyright by the American Chemical Society.

1. Comparing the nine spectra collected from 85° to 50°C (five are shown in Fig. 14), it is clear that linewidths of several resonances, such as m²G, D, m⁵C, and T, are narrow at 90°C (the denatured state), broad at intermediate temperature (69–75°C), and then narrow again at 50°C [the native state (see Figs. 11 and 13)]. The broadness of these resonances at the transition temperature (75°C) suggests the presence of multiple conformational states for each resonance at this temperature. In the approximation of a two-state transition, the product of the lifetime

24   PAUL O. P. TS'O

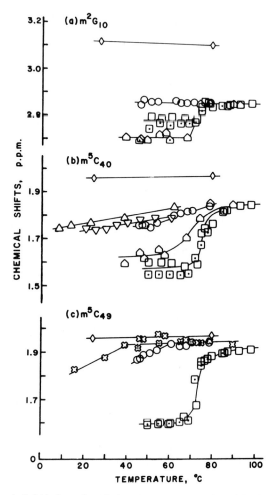

Fig. 13 Chemical shift data of methyl proton resonances from (a) m²G₁₀, (b) m⁵C₄₀, and (c) m⁵C₄₉ of the monomer, intact tRNA^Phe, and its fragments versus temperature. All the symbols have the same meaning as in Fig. 11. Reprinted with permission from Kan *et al.* (1977), *Biochemistry* **16,** 3143. Copyright by the American Chemical Society.

(τ) and the chemical shift differences ($\Delta\nu$) between the two states for each resonance is approximately $\frac{1}{2}\Pi$ (2 to 2$^\frac{1}{2}$) (Pople *et al.*, 1959). In other words, the transition has an intermediate exchange rate measured at 360 MHz at 75°C. However, there is no obvious broadening effect on the m²G, D, and m²₂G methyl signals in the spectrum observed at 220 MHz at 75°C (Fig. 10). This observation indicates that, when measured at 220 MHz, the exchange rate between the two states was sufficiently fast to avoid the broadening or splitting.

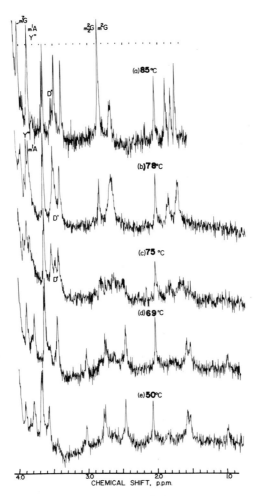

Fig. 14 The 360-MHz nmr spectra of high-field proton resonances in the region of bakers' yeast tRNA^Phe in a temperature range of 50 to 85°C. These spectra were taken under the following conditions: acquisition time, 1 second; 512 transients for spectrum a and 1024 transients for spectra b, c, d, and e. Reprinted with permission from Kan *et al.* (1977), *Biochemistry* **16**, 3143. Copyright by the American Chemical Society.

2. The dramatic shifting and splitting of the methyl resonance of T upon renaturation at low temperature now became much more obvious. After the initial broadening at 78°C (Fig. 14c), a small portion of the T methyl signal appears at 1.0 ppm at 75°C. This peak clearly becomes larger at temperatures below 69°C, whereas the other portion of T could resonate at the same position of the two m^5C's (~1.55 ppm). Thus, the areas of the signal from C—CH$_3$ of the Y base (~2.08 ppm), from the two

m^{5}C's (~1.55 ppm), and from T (~1.0 ppm) were integrated and compared at temperatures below 60°C. If the area of the well-resolved Y signal is taken as unity, then the areas in the peaks at 1.55 and 1.00 ppm are 2.4 ± 0.2 and 0.6 ± 0.2, respectively. This analysis supports the notion that, in the native state, the methyl group of T resonates both at 1.00 ppm (~60%) and at 1.55 ppm (~40%).

3. The signal at 3.03 ppm (50°C or in the native state) is also evident in the 360-MHz spectrum (Fig. 14e). Information from the studies at both 220 and 360 MHz suggests that this signal belongs to D', which clearly appears at 98°C in Fig. 10a or at 85°C in Fig. 14a and shifts progressively upfield when the temperature is lowered to ~80°C (Figs. 10b and 14b). During the temperature transition from 78°C (Fig. 14b) to 69°C (Fig. 14d), the D' signal located at 3.5 ppm (75°C) disappeared with the simultaneous emergence of a distinct peak at 3.03 ppm (69°C).

4. The spectral data at 360 MHz provide additional information about the m$_2^2$G signal. At 75°C, the m$_2^2$G resonance is broad in both 220- and 360-MHz spectra. At 69° and 50°C, the spectra at 360 MHz (Figs. 14d and 14e) provide fairly concrete information that the m$_2^2$G resonance is a broad signal centered around 2.6–2.7 ppm.

5. Much more useful information about m^{1}A which is clearly visible with a narrow linewidth from 85° to 50°C and with an upfield shift of less than 0.2 ppm from high to low temperature. As for the m^{7}G resonance, it is a sharp line at 85°C, becomes broadened with a small upfield shift at 75°C, and then is masked by the edge of the HDO peak, which shifts slightly upfield at lower temperature. Although the m^{7}G resonance cannot be traced precisely at temperatures below 75°C, it is unlikely that this signal is shifted dramatically upfield since all the upfield resonances at 50°C can be adequately accounted for (Fig. 14).

6. Finally, when the intact tRNA^{Phe} is in its native state (Fig. 14e, 50°C with 0.01 M Mg^{2+}), the following resonances are clearly evident in the 360-MHz spectrum: T (at two spectral positions), m^{5}C's, D (perhaps D'), m^{2}G, Cm, Gm, m^{1}A, Y, Y', and Y"—resonances from a total of nine bases. The relatively narrow linewidth of these resonances indicates that these methyl or methylene groups have a fair degree of motional freedom for their relaxation. The chemical shift data of these groups from the nine bases should provide certain useful information about the conformation of the native intact tRNA^{Phe} in solution. These chemical shift data are given in Table III.

In discussing these results, two points should be emphasized. The first point concerns the linewidths at half-height ($\Delta\nu_{\frac{1}{2}}$) of the methyl/methylene resonances in the helix-coil transition between 50° and 85°C measured at 360 MHz. As shown in Fig. 15, the $\Delta v_{\frac{1}{2}}$'s can be divided into three categories according to the temperature effects.

The first category includes the methyl signals from Cm, Gm, and Y

TABLE III

Chemical Shifts of the Methyl and Methylene Resonances from Intact Yeast tRNA[Phe a]

Minor bases	Chemical shifts (δ, ppm)[b]			
	tRNA			
	Denatured (85°–98.5°C)	Native (~50°C)	Monomer[c]	$\Delta\delta$[d]
m^2G_{10}	2.86	2.74	3.11	0.37
$D_{16,17}$ (C_5)	2.68	2.47	2.74	0.27
(C_6)	3.52	3.03[e]	3.56	0.53
$m^2_2G_{26}$	3.00	2.64[f]	3.10	0.46
Cm_{32}	3.46	3.58[f]	3.50	−0.08
Gm_{34}	3.35	3.57	3.44	−0.13
Y_{37} (CCH_3)	2.07	2.08	2.16	0.08
(NCH_3)	3.87	3.90	4.07	0.17
($COOCH_3$)	3.63	3.66	3.64	−0.02
($COOCH_3$)	3.66	3.68	3.67	−0.01
m^5C_{40}	1.85	1.56	1.95	0.39
m^7G_{46}	4.00[g]	—	4.09	—
m^5C_{49}	1.91	1.58	1.95	0.37
T_{54}	1.78	1.49, 1.01	1.88	0.87, 0.39
m^1A	3.86	3.78	3.93	0.15

[a] In 0.35–0.5 mM, 0.01 M MgCl$_2$, 0.01 M potassium phosphate buffer, pH 6.5–7.0. Reprinted with permission from Kan et al. (1977), Biochemistry **16,** 3143. Copyright by the American Chemical Society.

[b] Negative sign omitted.

[c] These monomer data were taken between 63° and 80°C.

[d] $\Delta\delta = \delta_{tRNA, native state} - \delta_{monomer}$.

[e] Tentative assignment.

[f] These data were taken at 68°C; all others were taken at 50°C.

[g] These data were obtained at 85°C in a 360-MHz nmr spectrometer; all the rest were from a 220-MHz nmr spectrometer at 98.5°C.

(four methyl groups). The six $\Delta\nu_{\frac{1}{2}}$ values in this category are independent of the temperature changes, as shown in Fig. 15. These $\Delta\nu_{\frac{1}{2}}$'s are slightly smaller at 85°C (~8 Hz) and then increase to about 10 Hz from 80° to 50°C (Figs. 14 and 15). These resonances all belong to the minor bases in the anticodon loop. The invariant $\Delta\nu_{\frac{1}{2}}$ values of Cm, Gm, and Y with respect to temperature changes are evident and indicate that the anticodon loop protrudes from the tRNA and does not associate with other parts of the molecule.

The second category contains the methyl resonances of m^2G, two m^5C's, T, and m^1A and the methylene resonances of two D's. The characteristic feature of this group is that the resonances are narrow at high temperature, become very broad at the transition temperature (75°C),

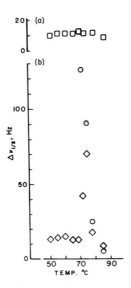

Fig. 15 The linewidths at half-height of methyl and methylene resonances versus temperature profile. (a) The average $\Delta\nu_{\frac{1}{2}}$ values of Cm, Gm, and Y (four methyl groups). (b) $\diamondsuit$ represents the average $\Delta\nu_{\frac{1}{2}}$ values of two D's, two m^5C's, m^2G, T, and m^1A; $\bigcirc$ represents the $\Delta\nu_{\frac{1}{2}}$ values of m_2^2G. Reprinted with permission from Kan *et al.* (1977), *Biochemistry* **16**, 3143. Copyright by the American Chemical Society.

and once again become narrow ($\sim$13 Hz) below the transition temperature. The entire process of variation in linewidth is completed within 10°C (Figs. 14 and 15). This phenomenon reflects the abrupt transition of tRNA^{Phe} in the presence of Mg^{2+}. The residues that contribute these resonances are most likely involved in the secondary and/or tertiary structures of tRNA^{Phe} in its native state. The broadening of these signals at the transition may represent the simultaneous existence of various conformational states.

The last category contains only one methyl resonance from m_2^2G. This resonance can be considered as an exception to the second category. The $\Delta\nu_{\frac{1}{2}}$ of m_2^2G is narrow at high temperature as well, broadening at the transition temperature. However, it becomes even more broadened and finally disappears at lower temperature (Figs. 14 and 15b), contrary to the resonances in the second category. Obviously, this residue is also involved in the secondary or tertiary structure of native tRNA^{Phe}. However, the broadening and disappearing of this resonance at low temperature may be caused by the immobility of the C_2—$N(CH_3)_2$ bond of the m_2^2G residue in the tRNA^{Phe} molecule in its native state. Therefore, the $\Delta\nu_{\frac{1}{2}}$ of m_2^2G may reflect the restriction of rotation of the C_2—$N(CH_3)_2$ bond.

The second point concerns the comparison between the observed and calculated chemical shifts of methyl/methylene resonances in yeast tRNA^{Phe}, which is related to the comparison between the structure of tRNA in aqueous solution and that in the crystalline state.

The calculated ring-current shielding/deshielding effects of the methyl or methylene groups from their neighboring bases in yeast tRNA^{Phe} are listed in Table IV. These calculations of shielding or de-shielding effects were based on the structure of yeast tRNA^{Phe} in or-thorhombic crystals as described using atomic coordinates (A. Rich, personal communication) and the graphic approach to the computation of the ring-current effects (Giessner-Prettre, et al., 1976). A comparison is made between the calculated ring-current effect (in terms of parts per million) and the observed shielding/deshielding effects, which are de-fined by the differences (in parts per million) in the chemical shift values from the intact tRNA versus those from the mononucleotides (mononucleosides) determined at 40°–50°C, that is, the lower plateau region with respect to the temperature perturbation (Table III).

The results in this comparison can be classified into four categories. The first category contains seven resonances, Cm_{32}, Gm_{34}, Y_{37} (all four methyl resonances), and m^5C_{40}. The observed and the calculated $\Delta\delta$ values of these seven resonances are in agreement with each other within 0.1 ppm. Since these four modified residues are either in antico-don loop or in anticodon stem, the results suggest no difference in the conformations of anticodon stem and anticodon loop of this tRNA in aqueous solution versus those in crystalline form that can be found by this approach. In addition, this category also contains one unusual case, which is $m_2^2G_{26}$. There are two ring-current effects predicted for the two methyl groups in $m_2^2G_{26}$ (0.82 and 0.13 ppm) based on a static structure of tRNA in crystal, but only one broad signal was seen on the ^1H nmr spectrum at ~65°C (Figs. 10 and 14), having an observed $\Delta\delta$ value of 0.46 ppm. This observed value of $\Delta\delta$ is very close to the average of the two calculated values. As expected, the $C_2-N(CH_3)_2$ bond in m_2^2G rotates with a sufficiently fast rate so that only one broad peak is observed. Therefore, it is possible that the average conformation of the two methyl groups from m_2^2G in aqueous solution may also be no different from the average positions from its crystalline form. Since $m_2^2G_{26}$ is located on top of anticodon stem, this reasoning again reinforces the preceding conclu-sion; that is, the conformation of the anticodon region of yeast tRNA^{Phe} in aqueous solution is similar to that in the crystalline state.

The second category contains m^2G_{10}, $D_{16,17}(C_5)$, $D_{16,17}(C_6)$, and m^5C_{49}. The observed $\Delta\delta$ values of these resonances in this category are much higher (more than 0.1 ppm) than the calculated $\Delta\delta$ values, indicating that these protons are *more* shielded than the predicted values based on

PAUL O. P. TS'O

TABLE IV

Ring-Current Anisotropic Shift Effect on the Methyl and Methylene Proton Resonances of a Yeast tRNA[Phe] Molecule in the Crystalline State[a]

		Ring-current effect (ppm)			
Modified base	Shielded by	Five-membered ring	Six-membered ring	Sum	Observed Δ (ppm)
m^2G_{10}	C_{11}		0.03		
	U_{12}		0.01		
	C_{25}		−0.01	0.03	0.37
$D_{16,17}$ (C_5)	None			0	0.27
$D_{16,17}$ (C_6)	None			0	0.53
$m_2^2G_{26}$ (α)	m^2G_{10}	0.68	0.21		
	C_{11}	0.02			
	A_{44}	−0.02	−0.07	0.82	
$m_2^2G_{26}$ (β)	m^2G_{10}	0.13	—		0.46
	C_{27}		0.04		
	A_{44}	−0.04	—	0.13	
Cm_{32}	A_{31}	—	0.07		
	Y_{37}	−0.01	−0.01		
	Adduct ring, −0.01[b]				
	A_{38}	−0.01	−0.08	−0.05	−0.08
Gm_{34}	U_{33}		0.01		
	A_{35}	−0.03	−0.03	−0.05	−0.13
Y_{37} (CCH_3)	A_{38}	—	~0		
	A_{39}	—	~0	0	0.08
(NCH_3)	A_{36}	—	0.1		
	A_{38}	0.05	—	0.15	0.17
$[COOCH_3\,(\alpha)]$	U_{33}		~0		
	A_{36}	−0.02	−0.03		
	A_{38}	—	+0.01	−0.04	−0.02
$[COOCH\,(\beta)]$	A_{36}	—	−0.03	−0.03	−0.01
m_5C_{40}	A_{29}	—	−0.02		
	G_{30}	−0.01	−0.02		
	A_{38}	0.10	0.16		
	ψ_{39}		0.12[c]	0.33	0.39
m^7G_{46}	A_9	—	0.09		
	A_{21}	0.11	0.02		
	G_{22}	−0.02	−0.01		
	G_{45}	0.04	—	0.21	<0.1
m^5C_{49}	U_6		0.02		
	U_7		0.10		
	U_{47}		~0		
	G_{65}	−0.02	−0.02		
	A_{67}		0.08	0.16	0.37

TABLE IV (Continued)

Modified base	Shielded by	Ring-current effect (ppm)			Observed Δ (ppm)
		Five-membered ring	Six-membered ring	Sum	
T_{54}	U_{52}		0.02		
	G_{53}	0.50	0.10		
	ψ_{55}		0.02^c		
	A_{58}	-0.02	-0.03		
	A_{62}	—	0.08	0.67	0.39 and 0.87
m^1A_{58}	G_{18}	0.23	0.05		
	T_{54}		-0.01		
	C_{60}		0.04	0.31	0.15

[a] The atomic coordinates of yeast tRNA^{Phe} were kindly provided by Dr. A. Rich (personal communication). The graphic analyses of the ring-current effect are from Giessner-Prettre et al. (1976). Reprinted with permission from Kan et al. (1977), Biochemistry **16**, 3143. Copyright by the American Chemical Society.

[b] By using the isoshielding contour of the five-membered ring of the guanine base.

[c] By using the isoshielding contour of the uracil base.

crystal structure and the ring-current effects. While there exists some doubt about the assignment of the $D_{16,17}(C_6)$ resonance in the native tRNA spectrum, this conclusion is most likely to be correct, since it is supported by the result on $D_{16,17}(C_5)$.

The third category contains m^7G_{46} and m^1A_{58}. The observed $\Delta\delta$ values of these two resonances are much lower (0.15–0.2 ppm) than the calculated $\Delta\delta$ values, indicating that these protons are *less* shielded than the predicted values based on crystal structure and the ring-current effects.

Finally, the fourth category contains the T residue. The methyl resonance from T has only one predicted ring-current effect value, but two $\Delta\delta$ values are observed (Table III). However, both experimental values are not in good agreement with the predicted value; interestingly, one was too high and the other was too low.

In summary, the comparison indicates that no differences between observed and calculated $\Delta\delta$ values from resonances in the anticodon stem and loop can be found, but differences in TΨC stem/loop and D stem/loop are uncovered. The nature of the results, with both agreement and disagreement involving too high and too low values, suggests that the difference may indeed be due to the difference in conformation. It should be noted that the regions showing the differences in comparison are the regions of the tertiary structure of the molecule, which are more readily influenced by packing, and the areas that are less defined by the

X-ray diffraction data. The segment involving the $m_2^2G_{26}$ residue, which connects the D stem and anticodon stem, may have the same conformation in aqueous solution as in the crystalline state except that the C_2—$N(CH_3)_2$ bond in m_2^2G is rotatable in aqueous solution but fixed in the solid state. There are no high-field nmr data concerning the acceptor stem region, except one preliminary result on the yeast tRNA^{Phe} with a phenylalanine attached at the CCA end through an amide linkage on A_{76}. It was found that the —CH_2— resonance on phenylalanine is not broadened at low temperature (L. S. Kan, P. O. P. Ts'o, M. Sprinzl, and F. Cramer, unpublished data). This indicates that the CCA end of yeast tRNA^{Phe} is not associated with any other part of the molecule in the native form. A similar approach is being used for the hydrogen-bonded NH—N resonances of yeast tRNA^{Phe} (Kan and Ts'o, 1977). The difference in this case is that the resonances have not been assigned with certainty.

Figure 16 shows the 360-MHz spectrum of the ^{1}H nmr resonances of the hydrogen-bonded NH from the yeast tRNA^{Phe} sample at 23°C. Under this condition, the spectrum is essentially insensitive to temperature variation within ±10°C and can be considered to be a reliable representation of the hydrogen-bonded NH resonances of yeast tRNA^{Phe} in the native conformation. This spectrum has a good signal-to-noise ratio and contains 15 well-resolved peaks plus a shoulder (k') in the

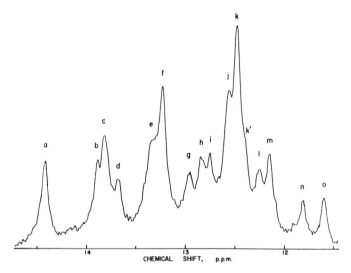

Fig. 16 A 360-MHz spectrum of yeast tRNA^{Phe} in the region of 11.5–14.5 ppm from DSS showing the hydrogen-bonded NH resonances at 23°C. The sample contained 25 mg/ml of tRNA^{Phe} dissolved in 0.01 M MgCl$_2$, 0.15 M NaCl, 0.002 M EDTA, and 0.01 M potassium phosphate buffer, pH 7.0. The letters a to o denote the resolved spectral positions of the resonances. From Kan and Ts'o (1977).

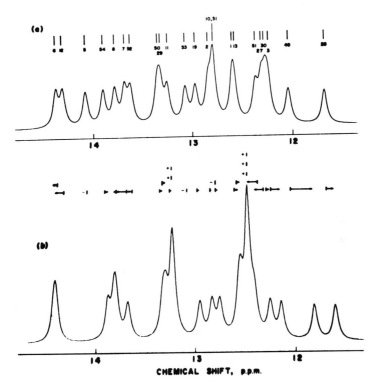

Fig. 17 (a) Computed spectrum of NH—N hydrogen-bonded proton resonances in yeast tRNA^{Phe} based on the calculated chemical shifts shown in Table I. The numbers of the NH resonances represent the following base pairs: 6, U_6A_{67}; 12, $U_{12}A_{23}$; 5, A_5U_{68}; 54, $T_{54}m^1A_{58}$; 7, U_7A_{66}; 8, U_8A_{14}; 52, $U_{52}A_{62}$; 50, $U_{50}A_{64}$; 29, $A_{29}U_{41}$; 11, $C_{11}G_{24}$; 53, $G_{53}C_{61}$; 19, $G_{19}C_{56}$; 2, C_2G_{71}; 10, $m^2G_{10}C_{25}$; 31, $A_{31}\psi_{39}$; 1, G_1C_{72}; 13, $C_{13}G_{22}$; 51, $G_{51}C_{63}$; 27, $C_{27}G_{43}$; 30, $G_{30}m^5C_{40}$; 3, G_3C_{70}; 49, $m^5C_{49}G_{65}$; 28, $C_{28}G_{42}$. Every computed peak has an equal linewidth at half-height, 36 Hz on a 360-MHz scale. (b) Simulated spectrum for the observed spectrum shown in Fig. 16. Every peak has a 36-Hz linewidth at half-height on a 360-MHz scale. The symbols between (a) and (b) represent the adjustments needed to transform the computed spectrum (a) to the simulated spectrum (b); (−) represents removal of resonance peaks, (+) represents the addition of resonance peaks, and (→ or ←) represents moving the chemical shifts to high field or low field, respectively. From Kan and Ts'o (1977).

region of 11 to 15 ppm from DSS (2,2-dimethyl-2-silapentane-5-sulfonate). This observed spectrum closely resembles the published and unpublished spectra of the same tRNA obtained under slightly different conditions (Kearns and Shulman, 1974; Kearns, 1976; Robillard *et al.*, 1976; B. Reid, personal communication).

Based on the experimental spectrum in Fig. 16 and the estimation that the spectrum contains 25 NH's, a simulated spectrum was constructed

(Fig. 17b). This simulated spectrum is considered to represent the spectrum of the hydrogen-bonded NH resonances of yeast tRNA[Phe] in the native state as observed in Fig. 16. This simulated spectrum is used subsequently in a comparison with the computed spectrum (Fig. 17 and Table V).

In the calculation of the chemical shifts of these hydrogen-bonded proton resonances, two parameters are needed: first, the shielding (or deshielding) effect on the resonance and, second, the intrinsic chemical shift of the resonance of this proton in different base pairs. For the NH—N resonance in base pairs of the Watson–Crick type, we adopted the values of Kearns and Shulman (1974), that is, 14.7, 13.4, and 13.6 ppm for $A \cdot U$, $A \cdot \Psi$, and $G \cdot C$ base pairs, respectively. As for the tertiary hydrogen bonds, there are many types, most of which are not Watson–Crick base pairs, such as $G_{15}C_{48}$. Kallenbach et al. (1976) have examined the $(U)N_3H$—$(A)N_7$ hydrogen-bonded proton resonance of reverse Hoogsteen type in a $U \cdot A \cdot U$ triple-stranded helix by ^{1}H nmr and recommended an intrinsic chemical shift of 14.1 ppm. From their data, together with a careful evaluation of the geometry and shielding effect, we estimated a value of 14.3 ppm as the intrinsic chemical shift of a reverse Hoogsteen $(U)N_3H$—$(A)N_7$ hydrogen-bonded proton resonance in an A'-RNA conformation (Arnott et al., 1973). As for the resonances of hydrogen-bonded NH—O in $G \cdot U$, $G \cdot \Psi$, or $G \cdot C$ base pairs, as well as that of NH—N in $G \cdot A$ base pairs, no intrinsic values of their chemical shifts have been determined experimentally. Therefore, no calculations of the resonances of these five base pairs were made (Table V).

The interatomic/intermolecular magnetic field experienced by the NH—N resonances in yeast tRNA[Phe] was then calculated based on the ring-current effect, as evaluated from the coordinates derived from the X-ray diffraction data (Kan and Ts'o, 1977). This result was plotted by the PDP-10 computer, as shown in Fig. 17a. The comparison of the computed spectrum, simulated spectrum, and observed spectrum of the hydrogen-bonded NH resonances of yeast tRNA[Phe] is given in Table V. In addition, the adjustments needed to transform the computed spectrum to the simulated experimental spectrum are described in Table V and shown in Figs. 17a and 17b.

As discussed above, this computed spectrum contains 23 resonances, 6 resonances less than the total recommended by the three-dimensional structure of tRNA determined by X-ray diffraction (Ladner et al., 1975b; Quigley et al., 1975).

Four general conclusions can be drawn from the above comparison. First, the simulated experimental spectrum and the adjustment of the computed spectrum to the simulated, experimental spectrum suggest

TABLE V

Comparison of the Computed, Simulated, and Observed Spectra of the Hydrogen-Bonded NH Resonances of Yeast tRNA[Phe] [a]

Computed spectrum		Simulated spectrum		Observed spectrum		
Base pairs[b]	Chemical shift (ppm)	Chemical shift (ppm)	Adjustment needed for transformation of the computed spectra to the simulated spectra[c] (ppm)	Chemical shift (ppm)	Peak desig- nation[d]	Approxi- mate proton number
6	14.38	14.42	0.04	14.41	a	2
12	14.31	14.40	0.09			
5	14.08	Excluded		Not observed		
54	13.90	13.875	−0.025	13.87	b	1
8	13.78	13.81	0.03	13.81	c	2
7	13.68		0.11			
52	13.62	13.67	0.05	13.67	d	1
50	13.35	13.32	−0.03	13.32	e	2
29	13.32	13.29	−0.03			
11	13.25	13.22	−0.03 (added two protons)	13.22	f	3
53	13.07	Shifted		Not observed		
19	12.97	12.95	−0.02	12.95	g	1
2	12.84	12.83	−0.01	12.83	h	1
31	12.79	12.75	−0.04	12.75	i	1
10	12.79	Shifted		Only one proton was observed in this region		
1	12.60	12.56	−0.04	12.55	j	2
13	12.58	12.54	−0.04			
51	12.37	12.47	0.10 (added three protons)	12.47	k	4
27	12.31	12.406	0.095	12.40	k′	1
30	12.27	12.25	−0.02	12.24	l	1
3	12.24	12.14	−0.10	12.14	m	1
49	12.04	11.81	−0.23	11.81	n	1
28	11.68	11.60	−0.08	11.60	o	1

[a] From Kan and Ts'o (1977).
[b] The designation numbers of the base pairs are described in the legend to Fig. 17a.
[c] Adjustments are shown in Fig. 17b.
[d] These peaks are shown in Fig. 16.

that there are 25 (±1) NH resonances in the spectrum. This number of hydrogen-bonded NH (to N or O) is considerably larger than needed in the cloverleaf structure (20) but is also definitely smaller than that exhibited by the crystalline structure (29). The data suggest that four (or three to five) NH's in the base pairs existing in the crystalline structure may exchange with water so rapidly that they cannot be detected by the nmr technique.

Second, within the tolerance range of ±0.1 ppm, only 4 resonances out of 25 are shown to be incorrectly calculated by the ring-current procedure (Table V). This result not only suggests that the ring-current procedure is indeed useful, but also that the intrinsic chemical shifts adopted are probably correct. In addition, the atomic coordinates adopted for the tRNA in solution may also be accurate for the most part, a conclusion that will be discussed further. Four computed resonances require more substantial adjustments. The first resonance is from the $m^5C_{49}G_{65}$ pair, which gives the only computed peak within a range close to that for the observed peak at 11.81 ppm. Therefore, the NH resonance of this $m^5C_{49}G_{65}$ pair is tentatively assigned to the n peak at 11.81 ppm, though an adjustment of -0.23 ppm is needed. The second resonance is from the A_5U_{68} pair, which is calculated to be at 14.08 ppm and is not within ±0.1 ppm from any observed peak. This A_5U_{68} pair is immediately below the putative G_4U_{69} pair recommended in the crystalline structure. Our current explanation is that this NH resonance is likely to be shifted upfield or that this NH may exchange rapidly in water because its adjacent putative G_4U_{69} base pair may not survive in water. The third resonance is from the $G_{53}C_{61}$ pair, which is calculated to be at 13.07 ppm, and cannot be fitted to the observed spectrum by adjustment within ±0.1 ppm. This base pair is located at the junction between the TΨC loop and TΨC stem. While this base pair most likely does exist, the NH—N resonance could be shifted elsewhere, such as to the f peak located 0.15 ppm downfield. The fourth resonance is either from the $m^2G_{10}C_{25}$ pair or the $A_{31}\Psi_{39}$ pair; both are calculated to be at 12.79 ppm. Since only one hydrogen-bonded NH resonance was observed at 12.75 ppm and, furthermore, the second one cannot be accommodated to this spectral position within the range of ±0.1 ppm, one of these two NH resonances must be removed or shifted elsewhere. The NH resonance in the $m^2G_{10}C_{25}$ pair, which is located at the beginning of the D stem and at the center of the folded region, is the likely candidate for removal or shifting. Since the number of the observed hydrogen-bonded NH resonances (25 ± 1) is less than the number suggested from the tRNA structure in the crystal, three or four proposed hydrogen-bonded NH's may exchange very rapidly in aqueous solution. No *definite information* is available indicating which base pair recommended in the crystal structure may not be detectable by nmr in aqueous solution, though the G_4U_{69} base pair is a good prospect. In addition, the spectral location of the NH resonance of the putative G·U base pair cannot be calculated since no information is available about the intrinsic chemical shift of this NH—O resonance.

Third, while the comparison between the calculated spectrum and the simulated experimental spectrum may not depend on the assignment of all the NH resonances, certain definitive information about the assignments can be derived from this investigation. For instance, due to the isolated position of these resonances in both the experimental spectrum and the calculated spectrum, as well as the close agreement between these two spectra about these resonances, the assignments of the NH resonances in U_6A_{67}, $U_{12}A_{23}$, and $C_{28}G_{42}$ are likely to be correct. In this case, the resonance observed at 11.6 ppm, now assigned to $C_{28}G_{42}$, is therefore unlikely to come from $C_{13}G_{22}$ (now assigned to 12.6 ppm) as recommended by Kearns and Shulman (1974). It is interesting to note that the NH resonance from the tertiary $G_{19}C_{56}$ Watson–Crick base pair (Quigley et al., 1975) was calculated to be at 12.97 ppm. Recently, the NH resonance of a tertiary hydrogen bond of the $G_{15}C_{49}$ reverse Watson–Crick base pair in *Escherichia coli* tRNA$_{fi}^{Met}$ has been unambiguously assigned by Daniel and Cohn (1975, 1976) at this chemical shift.

Calculations of hydrogen-bonded NH resonances have been made by others (Robillard et al., 1976; Kearns, 1977; Geerdes and Hilbers, 1977). A discussion has been made in comparison with these calculations (Kan and Ts'o, 1977). Special attention should be given to the interesting results by Geerdes and Hilbers (1977). Their work showed that the calculated nmr spectra are very sensitive to slight changes in structure.

In conclusion, despite some uncertainties in the theoretical treatment, this quantitative comparison between the simulated experimental spectrum and the calculated spectrum based on the atomic coordinates of the tRNA structure in the crystal and on ring-current effects clearly indicates that the native conformation of yeast tRNA^{Phe} in solution is fundamentally similar to that in the crystalline state. The minor difference is probably in the tertiary structure involving the folding of the TΨC loop and stem to the D loop and stem. This conclusion is reinforced by the ^1H nmr studies on the methyl/methylene resonances of the minor bases reported in the preceding section. In addition, some of the hydrogen-bonded base pairs existing in the tRNA in the crystal state may not be detectable in solution.

D. Investigation of the Backbone of the Nucleic Acids by ^{13}C nmr Studies

Carbon-13 nmr spectroscopy is particularly useful in studying the sugar–phosphate backbone conformation via the carbon–phosphorus coupling constants (Alderfer and Ts'o, 1977), since the backbone has not been very amenable to study by proton magnetic resonance. The early work of Smith and co-workers (Mantsch and Smith, 1972; Lapper et al.,

1972, 1973; Smith *et al.*, 1973) demonstrated that it was possible to obtain conformational information concerning the backbone of nucleic acids from the vicinal coupling constants, $^3J_{C-2',P-3'}$, $^3J_{C-4',P-3'}$, and $^3J_{C-4',P-5'}$. From these coupling-constant values, it is possible to obtain angular preferences of the bonds ϕ (C-5'—O-5') and ϕ' (C-3'—O-3'). Although rotational preferences of these bonds were apparent, no significant changes in rotamer preferences were demonstrated by varying the base moiety (e.g., UpU and ApA) or by varying the temperature (Schleich *et al.*, 1975, 1976). The conclusion from these investigations was that very little rotation, if any, of the ϕ and ϕ' bonds occurs in these nucleic acid backbones.

In this study, the dimers, UpU and ApA, have been reinvestigated, along with the polymers, poly(A), poly(U), poly(2'MeA), and poly(2'MeU). Our results led us to exchange the previous assignments of resonances for the C-2' and C-3' of poly(U) (Mantsch and Smith, 1972) and poly(A) (Smith *et al.*, 1975) and the C-2' and C-3' (Np portion) of UpU and ApA (Smith *et al.*, 1973). The data also indicate significant differences in the vicinal carbon–phosphorus coupling-constant values among these nucleic acids.

Although the arguments for the correct resonance assignment are essential, the details will not be presented here (see Alderfer and Ts'o, 1977). The results of this study are shown in Table VI.

The conventional analysis based on the distribution of population among three staggered rotamers of ϕ was adopted without apparent problems. From the equation of Govil and Smith (1973),

$$^3J_{C,P} \simeq 9.5 \cos^2 \theta - 0.6 \cos \theta \qquad (3)$$

where θ is the dihedral angle between the planes ^{13}CCO and $CO^{31}P$, the values of $^3J_{trans} = 10.1$ Hz and $^3J_{gauche} = 2.1$ Hz were calculated. It must be noted that subsequent calculations of rotamer distributions are dependent upon the applicability of Eq. (3), which has been used for both ϕ and ϕ'. For instance, since the electronegativity of the substituents of C-2' and C-4' may not be the same, the application of the same equation to the analysis of both $^3J_{C-2',P}$ and $^3J_{C-4',P}$ may lead to certain inaccuracies. Nevertheless, using these parameters, the distribution of rotamer populations for the 5' mononucleotides is approximately 85% *gg* and 15% (*gt* + *tg*). These rotamer populations are in reasonable agreement with those calculated by Davies and Danyluk (1975) (72–77% for *gg*, 28–32% for *gt* + *tg*) using proton–^{31}P coupling constants. The problem of employing this approach for the analyses of the coupling constants related to ϕ in the 3' mononucleotides is discussed in the following section.

TABLE VI

^{13}C—^{31}P Coupling Constants (Hz) of Monomers, Dimers, and Polymers[a]

Carbon	Residue	3'-AMP 35°C	5'-AMP 35°C	ApA 20°C	ApA 70°C	Poly(A) 35°C	Poly(A) 75°C	Poly(2'MeA) 60°C	3'-UMP 35°C	5'-UMP 35°C	UpU 27°C	UpU 60°C	Poly(U) 35°C	Poly(U) 80°C	Poly(2'MeU) 60°C
C-2'	Np-	4.9		3.7	4.6	2–3	4.0	3.7	4.2		4.2	4.7	4.8	4.6	4.1
	-pN														
C-3'	Np-	4.8		6.1	5.5	4–5	5.5	4.5	4.6		4.6	5.5	5.5	5.5	4.8
	-pN														
C-4'	Np-	3.9		5.5	4.3	4–5	5.3		4.6		4.9	4.8	4.0	4.3	
	-pN		8.5	9.8	8.5	8–9	8.1			8.5	8.9	8.5	8.5	8.5	
C-5'	Np-														
	-pN		4.9	4.6	4.5	—[b]	4.6			4.5	5.3	5.5	5.3	6.1	4.8

[a] Reprinted with permission from Alderfer and Ts'o (1977), Biochemistry 16, 2410. Copyright by the American Chemical Society.
[b] Not resolved.

In analyzing the rotamer distribution of bond angle ϕ', the calculations using the conventional approach of the three rotamers, that is, $\phi' = 60°$ (gg or trans), 180° (gt), and 300° (tg), the $^3J_{C,P}$ data, and Eq. (1) indicate the gg (or trans, $\phi' = 60°$) rotamer to be significantly populated to the extent of 30–50%. This result is not in agreement with the theoretical calculations (Olson and Flory, 1972a,b; Olson, 1973, 1975a,b,c; Perahia et al., 1974a,b; Yathindra and Sundaralingam, 1974; Tewari et al., 1974; Broyde et al., 1975) and the crystalline structures of 3' nucleotides (Sundaralingam, 1969) and dinucleoside monophosphates (Rubin et al., 1972; Sussman et al., 1972; Rosenberg et al., 1973; Hingerty et al., 1975) determined by X-ray diffraction. In these studies, the trans rotamers ($\phi' = 60°$) were found to be only slightly populated or nonexistent. This discrepancy has been noted by Lee et al. (1976) and by Cheng and Sarma (1977a). The conclusion based on theoretical investigations can be readily demonstrated with the manipulations of the space-filling Corey–Pauling–Koltun (CPK) models. The model of the ApA dimer shows the difficulty in achieving the base–base stacking interaction with ϕ' in the trans conformation and the extensive steric interference upon transferring from the gauche rotamers (either gt or tg) to the trans rotamer. Therefore, it appears unreasonable to adopt this three-rotamer distribution model for the analysis of ϕ' in the 3'-nucleotidyl unit, which affords the same high percentage of trans population (36 to 42%) for UpU, poly(U), ApA, and poly(A) in the 27°–60°C range (Table VI) while all of these compounds have vastly different extents in base–base stacking under these conditions.

A similar discrepancy was also encountered in the analyses of the $^3J_{H,P}$ data for ϕ' in the 3' ribonucleotides (Davies and Danyluk, 1975; Cheng and Sarma, 1977a). In recognition of this problem, Ezra et al. (1976) and Cheng and Sarma (1977a) proposed the following modification of the conventional three-rotamer model: (1) arbitrarily depopulate the trans rotamer ($\phi' = 60°$); (2) modify the classical gauche positions to $\phi' = 210°$ and 270°, at which the rotamers are either fixed or in rapid equilibrium.

We have attempted to analyze the $^3J_{C,P}$ data of ϕ with the same approach. Again, omitting details, the results of both approaches, that is, the "three-rotamer model" and the "two-rotamer model," are shown in Table VII. We concluded that the two-rotamer model is more applicable to the analysis of $^3J_{C,P}$ data, and those rotamers are likely to exist in rapid equilibrium between $\phi' \simeq 210°$ and 270°.

A conformational analysis of the furanose–phosphate backbone concerning the rotamer distribution of ϕ' can now be performed using an approach similar to that applied to the C-2'-endo $\rightleftharpoons$ C-3'-endo equilibrium states of the ribosyl–furanose ring system (Davies and Danyluk, 1974). This furanose conformational analysis calculated the distribution

TABLE VII

Calculated Rotamer Distributions of ϕ' in Uridylic and Adenylic Acids[a]

Compound	Temperature (°C)	Three-rotamer model[b]			Two-rotamer model[c]	
		f_{gt}	f_{tg}	f_{gg}	$f_{gt'}$	$f_{tg'}$
UpU	27	0.38	0.26	0.36	0.54	0.46
	60	0.33	0.28	0.39	0.52	0.48
Poly(U)	27	0.24	0.35	0.41	0.46	0.54
	60	0.25	0.33	0.42	0.46	0.54
ApA	20	0.43	0.20	0.37	0.60	0.40
	60	0.31	0.31	0.38	0.50	0.50
Poly(A)	75	0.40	0.24	0.36	0.58	0.42

[a] Reprinted with permission from Alderfer and Ts'o (1977), *Biochemistry* **16**, 2410. Copyright by the American Chemical Society.

[b] Calculated from Eq. (1) and

$J_{C-2',P-3'} = f_{gt}J_{gauche} + f_{tg}J_{trans} + f_{gg}J_{gauche}$;

$J_{C-4',P-3'} = f_{gt}J_{trans} + f_{tg}J_{gauche} + f_{gg}J_{gauche}$;

$f_{gt} + f_{tg} + f_{gg} = 1$.

[c] Calculated from Eqs. (1) and (5)–(7).

of C-2'-*endo* and C-3'-*endo* forms using $^3J_{H-1',H-2'}$, $^3J_{H-3',H-4'}$, and Eqs. (4)–(6):

$$f_{(C-3'-endo)} = {}^3J_{H-3',H-4'}/({}^3J_{H-1',H-2'} + {}^3J_{H-3',H-4'}) \qquad (4)$$

or

$$f_{(C-2'-endo)} = {}^3J_{H-1',H-2'}/({}^3J_{H-1',H-2'} + {}^3J_{H-3',H-4'}) \qquad (5)$$

and

$$f_{(C-3'-endo)} + f_{(C-2'-endo)} = 1.0 \qquad (6)$$

The furanose–phosphate backbone conformation can also be subjected to a similar analysis to determine the distribution of gt' and tg' rotamers, using $^3J_{C-2',P-3'}$, $^3J_{C-4',P-3'}$, and Eqs. (7)–(9):

$$f_{(gt)} = {}^3J_{C-4',P-3'}/({}^3J_{C-2',P-3'} + {}^3J_{C-4',P-3'}) \qquad (7)$$

or

$$f_{(tg')} = {}^3J_{C-2',P-3'}/({}^3J_{C-2',P-3'} + {}^3J_{C-4',P-3'}) \qquad (8)$$

and

$$f_{(gt')} + f_{(tg')} = 1.0 \qquad (9)$$

The proton data of ApA from Lee *et al.* (1976) were analyzed using Eqs. (4)–(6) and are listed in Table VIII along with the calculated $f_{gt'}$ populations using Eqs. (7)–(9). These calculations suggest that a relationship between the C-3'-*endo* and gt' conformations (and between the C-2'-*endo* and tg' conformations) probably exists for the angle ϕ' in the ribosyl–furanose–phosphate system. For the dimer data presented in Table VIII, a close correlation (actually a 1:1 relationship) is observed, whereas for the mononucleotides only a loose coupling between the sugar and sugar–phosphate conformations is found. It should be noted that these comparisons are only made with systems that have been shifted slightly from the 50/50 population distribution of the two conformations. Thus, correlations between the populations of C-3'-*endo* and gt' still remain to be determined under the conditions where large deviations from midpoint occur.

The coupling-constant values (Table VI), which semiquantitatively reflect the furanose–phosphate backbone conformation of the mononucleotides, dinucleoside monophosphates, and polynucleotides, and the calculated rotamer distributions (Tables VII and VIII) can be compared (1) for a given base with respect to temperature variation or (2) at a given temperature with respect to uracil and adenine. Such a comparative analysis taken in combination with the proton nmr data

TABLE VIII

Comparison of Ribosyl–Furanose and Sugar Backbone Conformations[a]

Nucleic acid		pH	Temperature (°C)	$f_{(C-3'-endo)}$	$f_{(gt')}$
ApA					
	Ap-	7.0	20	0.60	0.60
	Ap-	7.0	70	0.48	0.48
UpU					
	Up-	7.0	20	0.56	0.55
	Up-	7.0	60		0.53
	Up-	7.0	89	0.52	
3'-AMP		6.0	27	0.37	
		5.9	27		0.49
		7.0	27	0.38	
		6.7	27		0.57
3'-UMP		4.3	27	0.49	
		4.2	27		0.52
		8.3	27	0.57	
		8.4	27		0.69

[a] Reprinted with permission from Alderfer and Ts'o (1977), *Biochemistry* **16**, 2410. Copyright by the American Chemical Society.

concerning the dimers and polymers (Alderfer *et al.*, 1974; Kondo *et al.*, 1970) provides a greater understanding about the governing forces of nucleic acid conformation. The coupling constants of the mononucleotides as the monoanion at pH 4 to 5 (Table VI) are used as a point of reference in this comparison. Considering the uridylic acid series, comparison of $^3J_{C-2',P}$ and $^3J_{C-4',P}$ of monomer, dimer, and polymer indicates little change at low or high temperature. Thus, in terms of the dihedral angles of ϕ and ϕ', no significant alterations are detected. Proton nmr data (Ts'o *et al.*, 1969; Lee *et al.*, 1976) have shown that bases in UpU do not have a significant base-stacking interaction as judged from either dimerization shift or furanose conformation changes with temperature. The situation for the adenylic acid series is considerably different, where marked variations in both $^3J_{C-2',P}$ and $^3J_{C-4',P}$ are observed when poly(A) is compared with 3'-AMP (Table VI); the situation is also different for ApA (Fig. 18). It is well known that the adenyl compounds have extensive base-stacking properties. In the case of ApA, as the temperature increases, $^3J_{C-2',P-3'}$ increases, $^3J_{C-4',P-3'}$ decreases, and $^3J_{C-4',P-5'}$ decreases. Such temperature-dependent variations of $J_{C,P}$ values of ApA were not seen by others (Schleich *et al.*, 1975). The reasons for these differences are not understood. These variations in backbone conformation with temperature can be easily understood from variations in the base-stacking interactions with temperature (Kondo *et al.*, 1970). For ApA at low temperature, the larger fraction of gt' relative to tg' for bond ϕ' is not unexpected for a dimer that possesses an extensive stacking of bases, since the gt' rotamer facilitates a

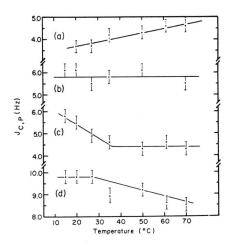

Fig. 18   The temperature dependence of (a) $^3J_{C-2',P}$, (b) $^2J_{C-3',P}$, (c) $^3J_{C-4',P-3'}$, and (d) $^3J_{C-4',P-5'}$ in ApA. Reprinted with permission from Alderfer and Ts'o (1977), *Biochemistry* **16**, 2410. Copyright by the American Chemical Society.

maximum amount of base–base overlap. Increasing the temperature reduces the amount of base stacking in ApA and the gt' rotamer is depopulated with a concomitant increase in the tg' population. At high temperature, the conformation of ϕ' in ApA is approaching that of the monomer 3'-AMP. A similar trend was observed for the furanose conformation of ApA (Kondo *et al.*, 1970, 1972) and poly(A) (Alderfer *et al.*, 1974) (monitored by $J_{\mathrm{II\ 1',II\ 2'}}$). At low temperatures, the furanoses of the dimer and polymer were mainly in the C-3'-*endo* conformation but reverted more to the C-2'-*endo* form at higher temperature, assuming the main conformational form of the mononucleotide. These observations support the discussion in the preceding section and suggest that the furanose conformation and ϕ' conformation are probably interrelated (Table VII). It should also be noted for ApA (Table VI and Fig. 18) that $^3J_{C-4',P-5'}$ increases with decreasing temperature to a value larger than that from 5'-AMP. This suggests an increased population of the gg rotamer of the ϕ bond (C-5'—O-5') as the result of increased base-stacking interactions. Included in Fig. 18 for comparative purposes is the effect of temperature on $^2J_{C-3',P}$ of ApA, which is negligible. It should be pointed out that, in the case of poly(A) at low temperatures, the linewidths of both the proton and carbon data preclude an accurate determination of the coupling constants; however, the data are highly suggestive of conformations that are predominantly, if not exclusively, C-3'-*endo* (furanose), gt' (C-3'—O-3'), and gg (C-5'—O-5'). These observed differences in the uridylic and adenylic acid series with respect to the furanose and furanose–phosphate backbone conformations clearly delineate the role and importance of base–base stacking as a major force in determining backbone conformations of RNA in aqueous solution.

III. EXPERIMENTAL EVALUATION OF ELECTROSTATIC INTERACTION AND STUDIES ON NONIONIC OLIGONUCLEOTIDE ANALOGUES

Since 1970, our laboratory has been engaged in a systematic study of the nonionic oligonucleotides, or oligonucleotides without electrostatic charges. This research program has two main objectives. First, this research should provide an experimental evaluation of the electrostatic interaction in a quantitative manner. Essentially, this goal has been achieved, although an interesting complication in stereochemistry has arisen in the modification of the oligonucleotide. The continuing research in this area appears to be straightforward and rewarding. Second, this research may open a new frontier permitting the study of the structure and function of nucleic acids *inside* the living cells. The results in this direction indicate that this is indeed very fascinating and promising.

A. Experimental Evaluation of the Electrostatic Interaction and Investigation of the Backbone of Nucleic Acids

In our first report in this series (Miller *et al.*, 1971), the synthesis and properties of the methyl and ethyl phosphotriester derivatives of TpT and dApdA were reported. Each derivative was prepared as a pair of diastereoisomers due to the asymmetry of the phosphorus atom after substitution (Fig. 19). The diastereoisomerism of these compounds has proved to be very interesting as a probe for nucleic acid conformation, a point that receives further attention in later paragraphs. Contrary to the prevailing myth at that time, the phosphotriesters, particularly the ethyl triester, of the deoxyoligonucleotides (as well as the 2'-O-methyl ribosyl nucleotides) are stable in neutral aqueous solutions. In addition, the alkyl phosphotriester analogues are totally resistant to hydrolysis by snake venom and spleen phosphodiesterase, micrococcal nuclease, fetal calf serum nuclease, and homogenates of Syrian hamster cells. As shown by pmr, circular dichroism (CD), and ultraviolet hypochromicity measurements, the conformations of the triesters in solution are quite similar to those of the parent diesters, although there is less base stacking in the triesters. The phosphate–alkyl groups serve as monitors for the interaction between the backbone and the bases. The pmr resonances of these groups provide valuable information about the dynamics of the dimer conformations. Thermal perturbation or denaturation by dimethyl sulfoxide (DMSO) causes a loss of the stacked conformation and an overall rotation of the base planes about the C—O—P

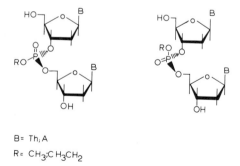

B= Th, A

R= CH_3; CH_3CH_2

Fig. 19 The diastereomeric dinucleoside alkyl phosphotriesters.

bonds of the pentose–phosphate backbone. The triesters of dApdA form complexes with poly(uridylic acid) in 0.01 M Mg^{2+} with a stoichiometry of 2U : 1A. The parent diester, dApdA, forms a complex with identical stoichiometry, and the secondary structures of all three complexes appear to be the same. The thermal stabilities of the 2 poly(U)·triester complexes are greater than that of the 2 poly(U)·dApdA complex due to decreased repulsion between the negatively charged phosphates of the poly(U) and the neutral phosphotriester backbone.

The association of the nonionic ethyl phosphotriester analogues of Tp(Et)T and dAp(Et)dA through hydrogen bonding in deuterochloroform was investigated by infrared spectroscopy (DeBoer et al., 1973). This is the first time that such studies have been done at the dimer level; because of solubility requirements, such experiments could be done only when nucleoside alkyl phosphotriesters became available. Association constants for both self-association and cross-association were determined by analysis of divergence from Beer's law of NH and NH_2 stretching bands. The complementary dimers associated with an intrinsic constant of 192 M^{-1}, as compared to a value of 92 M^{-1} for the corresponding monomers (Table IX). The relatively low value for the dimers indicates that the association process is not cooperative. The two residues of the dimers behave nearly independently in the association process; therefore, the restriction on the rotation about the bonds of the backbone must be small. This conclusion is discussed in relation to recent reports on unperturbed dimensions of polynucleotides in aqueous solution (see the discussion in DeBoer et al., 1973). The results reinforce the conclusion from the studies on polynucleotides that the restriction of the rotation of the backbone of a polynucleotide in solution cannot be due to steric hindrance alone. Our data support the suggestion that the electrostatic interaction between the charged groups of the polynucleotide backbone and between the charged groups and water could be the cause of the rigidity of the polynucleotide in a random-coil conformation observed in aqueous solution.

We have completed a comprehensive investigation of the duplex formation of a nonionic oligo(deoxythymidylate) analogue, heptadeoxythymidylyl (3'-5')deoxythymidine heptaethyl ester or d([Tp(Et)]$_7$T) = $d(T_8 \cdot Et)$ with poly(deoxyadenylate) or $(dA)_n$, as an experimental system for the evaluation of the electrostatic interaction (Pless and Ts'o, 1977). About 300 mg of $d(T_8 \cdot Et)$ were synthesized in good yield and the structure was verified by a variety of techniques including nmr.

Preliminary experiments established that in 0.15 M NaCl–0.04 M potassium phosphate (pH 6.9) an equimolar mixture of $(dA)_n$ and $d(T_8 \cdot Et)$ at a total concentration of 10^{-4} M exhibited a fairly broad transi-

TABLE IX

Intrinsic Association Constants in CdCl$_3$ at Room Temperature[a]

Compound	K_i (M^{-1})		Literature values
Self-association			
T		3.6	3.2[b]
Tp(Et)T		4.2	
A	Pair formation	3.1	3.1[c]
	Chain formation	2.2	1.4,[d] 4.8[d]
dAp(Et)dA	Pair formation	6.5	
	Chain formation	4.0	
Cross-association			
T + A		92	130[b]
Tp(Et)T + A		113	
T + dAp(Et)dA		72	
Tp(Et)T + dAp(Et)dA		192	

[a] From 22° to 24°C. Reprinted with permission from DeBoer *et al.* (1973), *Biochemistry* **12**, 720. Copyright by the American Chemical Society.

[b] Kyogoku *et al.* (1967b); constants at 25°C.

[c] Kyogoku *et al.* (1967a); value at 24°C.

[d] Nagel and Hanlon (1972); the first value is for pair formation and the second for higher associations, all taking place in a chain formation reaction.

tion with a midpoint in the proximity of 18°C. A mixing curve for this system is shown in Fig. 20.

Since the sample of d(T$_8$·Et) should consist of a large number of diastereoisomers, the annealing curve observed might result from the superimposition of a multitude of oligomer–polymer transitions with different stabilities. To address this possibility, an attempt was made to fractionate the octamer triester according to its ability to comigrate with the deoxyadenylate polymer on a Sephadex G-50 column at different temperatures. The eluant used was 0.001 M sodium cacodylate (pH 6.9).

A solution of 3.4 μmoles of (dA)$_n$ and 3.2 μmoles of d(T$_8$·Et) in 5.2 ml of 0.001 M sodium cacodylate, pH 6.9, was incubated at 4°C for 1 hour and then chromatographed on a Sephadex G-50 column (137 ml) equilibrated at 7°C. The elution profile is shown in Fig. 21a. The high molecular weight peak contains (dA)$_n$·d(T$_8$·Et) complexes. The second peak exhibits the same absorption spectrum as d(T$_8$·Et) and will be termed d(T$_8$·Et) (<7°C), that is, the d(T$_8$·Et) fraction that does not comigrate with (dA)$_n$ at 7°C; this fraction constitutes 13% of the total d(T$_8$·Et) introduced. The large molecular weight peak was lyophilized

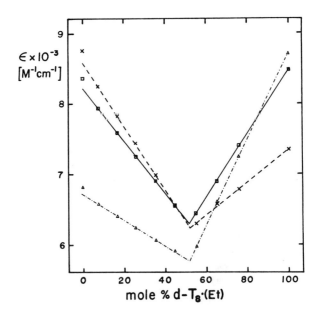

Fig. 20 Mixing curve for $(dA)_n$ + $d(T_8 \cdot Et)$ at 0°C in 0.15 M NaCl, 0.04 M potassium phosphate, pH 6.9. Total concentration, $10^{-4} M$. $\times$ --- $\times$, ϵ_{255nm}; $\square$——$\square$, ϵ_{260nm}; $\triangle$ -·- $\triangle$, ϵ_{265nm}. Reprinted with permission from Pless and Ts'o (1977), *Biochemistry* **16**, 1239. Copyright by the American Chemical Society.

and rechromatographed on Sephadex G-50 in 0.001 M sodium cacodylate at 20°C (Fig. 21b). The low molecular weight peak makes up 47% of the total $d(T_8 \cdot Et)$ and is termed $d(T_8 \cdot Et)$ (7°–20°C), that is, the $d(T_8 \cdot Et)$ fraction that comigrates with $(dA)_n$ at 7°C but does not comigrate at 20°C. Rechromatography of the high molecular weight fraction on Sephadex G-50 at 30°C (Fig. 21c) yielded $d(T_8 \cdot Et)$ (20°–30°C), 22% of the input $d(T_8 \cdot Et)$. The high molecular weight peak in this chromatography had an absorption spectrum of pure $(dA)_n$, and its rechromatography at 40°C did not liberate any more of the oligomer (Fig. 21d).

Annealing curves for equimolar mixtures of the various $d(T_8 \cdot Et)$ fractions and $(dA)_n$, at a total concentration of $10^{-4} M$ in 0.1 M sodium cacodylate, pH 6.9, are shown in Fig. 22. At temperatures below 10°C, the $(dA)_n$ + $d(T_8 \cdot Et)$ mixture exhibits a hypochromicity of 26.6% at 260 nm. The mixtures of $(dA)_n$ with $d(T_8 \cdot Et)$ (7°–20°C) and with $d(T_8 \cdot Et)$ (unfractionated) approach the same value of hypochromicity at temperatures below 0°C. The $(dA)_n$ + $d(T_8 \cdot Et)$ (<7°C) mixture is almost 20% hypochromic at −5°C, the lowest temperature examined. The temperatures at which these mixtures attain a hypochromicity of 10% ($T_{0.9}$) are 26.0° (+), 16.9° ($\square$), 19.0 ($\blacksquare$), and 2.6°C ($\bullet$), respectively, in Fig. 22. Thus, the transitions for $(dA)_n$ + $d(T_8 \cdot Et)$ (<7°C) and $(dA)_n$ + $d(T_8 \cdot Et)$ (20°–30°C) are separated by 23°C. A comparison of the transition profiles for $(dA)_n + d(T_8 \cdot Et)$ (7°–20°C) and $(dA)_n + d(T_8 \cdot Et)$ (unfractionated) shows that they have similar midpoint temperatures but different breadths. The transition for the $(dA)_n + d(T_8 \cdot Et)$ (unfractionated) mixture is considerably broader, reflecting the superimposition of oligomer–polymer transitions with more widely different stabilities.

The annealing profile of the $(dA)_n + d(T_8 \cdot Et)$ (<7°C) + d(T_8 \cdot Et) (20°–30°C) (2:1:1) mixture is bimodal [Fig. 22, due to the removal of the fraction of octamers with intermediate binding affinity, $d(T_8 \cdot Et)$ (7°–20°C), from the $d(T_8 \cdot Et)$ mixture]. The upper portion of the biphasic curve closely follows the course of the $(dA)_n + d(T_8 \cdot Et)$ (20°–30°C) (1:1) profile. Thus, the interaction of $d(T_8 \cdot Et)$ (20°–30°C) with $(dA)_n$ is not affected by the addition of $d(T_8 \cdot Et)$ (<7°C). This observation indicates that the different binding affinities of $d(T_8 \cdot Et)$ (<7°C) and of $d(T_8 \cdot Et)$ (20°–30°C) for $(dA)_n$ are not due to the presence of an inhibiting contaminant in the $d(T_8 \cdot Et)$ (<7°C) fraction. Reexamining the size of $d(T_8 \cdot Et)$ (<7°C) by cellulose thin-layer chromatography in solvent I', it was found to comigrate exactly with $d(T_8 \cdot Et)$ (unfractionated) and was clearly separated from $d(T_6 \cdot Et)$. In addition, the ultraviolet absorption spectra of $d(T_8 \cdot Et)$ (<7°C) and $d(T_8 \cdot Et)$ (20°–30°C) are identical. Therefore, we conclude that the observed differences in binding affinity are the result of the diastereoisomeric heterogeneity of the $d(T_8 \cdot Et)$ sample.

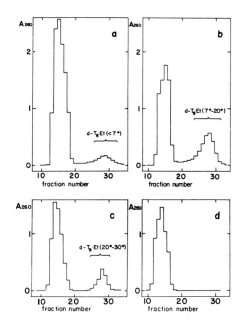

Fig. 21 Fractionation of $d(T_8 \cdot Et)$ by cochromatography with $(dA)_n$ on Sephadex G-50. Temperature (centigrade): a, 7°; b, 20°; c, 30°; d, 40°. Reprinted with permission from Pless and Ts'o (1977), *Biochemistry* **16**, 1239. Copyright by the American Chemical Society.

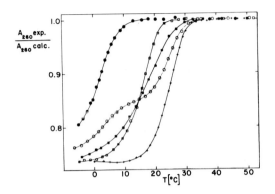

Fig. 22 Annealing of $(dA)_n$ with various fractions of $d(T_8 \cdot Et)$ in $0.1\,M$ sodium cacodylate, pH 6.9. Total concentration, $10^{-4}\,M$. ■, $(dA)_n + d(T_8 \cdot Et)$ (unfractionated) (1:1); ●, $(dA)_n + d(T_8 \cdot Et)$ (<7°C), (1:1); □, $(dA)_n + d(T_8 \cdot Et)$ (7°–20°C) (1:1); +, $(dA)_n + d(T_8 \cdot Et)$ (20°–30°C); ○, $(dA)_n + d(T_8 \cdot Et)$ (<7°C) + $d(T_8 \cdot Et)$ (20°–30°C) (2:1:1). Reprinted with permission from Pless and Ts'o (1977), *Biochemistry* **16**, 1239. Copyright by the American Chemical Society.

The octamer triester was also fractionated by cochromatography with poly(riboadenylic acid). Chromatography of an equimolar mixture of $(rA)_n$ and $d(T_8 \cdot Et)$ (unfractionated) on Sephadex G-50 in 0.001 M sodium cacodylate, pH 6.9, at 2°C, followed by rechromatography of the high molecular weight peak at 20°C yielded a small amount [7% of input $d(T_8 \cdot Et)$] of $d(T_8 \cdot Et)$ $(2°–20°C)_{rA}$. In $0.1\,M$ sodium cacodylate, pH 6.9, an equimolar mixture of $(dA)_n$ and $d(T_8 \cdot Et)$ $(2°–20°C)_{rA}$ at a total concentration of $10^{-4}\,M$ exhibited a $T_{0.9}$ of 28.1°C (data not shown), 2 degrees higher than the $T_{0.9}$ value for the $(dA)_n + d(T_8 \cdot Et)$ (20°–30°C) mixture.

For a determination of the enthalpy changes in the interaction of $(dA)_n$ with various octamer triester fractions in $0.1\,M$ sodium cacodylate, we followed the method developed by Damle (1970) and by Crothers (1971), which has previously been applied to the oligo(I) · poly(C) system in this laboratory (Tazawa *et al.*, 1972). This method calls for plotting the reciprocal of T_m (on the absolute scale) versus the logarithm of the free oligomer concentration at T_m and determining the enthalpy of binding from the slope via the equation $n\,\Delta H = 2.303\,R\,\Delta \log c_m / \Delta(1/T_m)$, where n is the degree of polymerization of the oligomer. To be consistent in the determination of ΔH values for this system and the $(dA)_n + d(T_8)$ system to be described below, we chose to obtain the binding enthalpy via $n\,\Delta H = 2.303\,R\,\Delta \log c_{0.9} / \Delta(1/T_{0.9})$ from a plot of $1/T_{0.9}$ (10% hypochromicity) versus $\log c_{0.9}$.

Figure 23 presents annealing curves for equimolar mixtures of $(dA)_n + d(T_8 \cdot Et)$ (20°–30°C) in $0.1\,M$ sodium cacodylate at five different total concentrations, from 10^{-5} to $5.3 \times 10^{-4}\,M$. The plot of $1/T_{0.9}$ versus

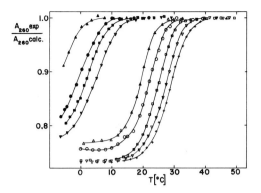

Fig. 23 Concentration dependence of the annealing of $(dA)_n + d(T_8 \cdot Et)$ ($<7°C$) (1:1) and of $(dA)_n + d(T_8 \cdot Et)$ (20°–30°C) (1:1) in 0.1 M sodium cacodylate, pH 6.9. $(dA)_n + d(T_8 \cdot Et)$ ($<7°C$): ▲, $0.072 \times 10^{-4}\ M$; ●, $0.359 \times 10^{-4}\ M$; ■, $1.01 \times 10^{-4}\ M$; ▼, $3.86 \times 10^{-4}\ M$. $(dA)_n + d(T_8 \cdot Et)$ (20°–30°C): △, $0.099 \times 10^{-4}\ M$; ○, $0.367 \times 10^{-4}\ M$; □, $1.02 \times 10^{-4}\ M$; ▽, $2.54 \times 10^{-4}\ M$; +, $5.35 \times 10^{-4}\ M$. Reprinted with permission from Pless and Ts'o (1977), *Biochemistry* **16**, 1239. Copyright by the American Chemical Society.

$\log (c_{0.9}/1\ M)$ (Fig. 24) gives a straight line yielding a value of $\Delta H = -8.6$ kcal/mole of base pairs. For the interaction of $d(T_8 \cdot Et)$ (2°–20°C)$_{rA}$ with $(dA)_n$ the concentration dependence of $T_{0.9}$ (determined for 1.00×10^{-4}, 0.29×10^{-4}, and $0.084 \times 10^{-4}\ M$ equimolar solutions; data not shown) gave a ΔH value of -8.6 kcal/mole of base pairs (Fig. 24). The determination of the enthalpy change for the interaction of $(dA)_n + d(T_8 \cdot Et)$ (7°C) was similarly undertaken, using concentrations of from 7.2×10^{-6}

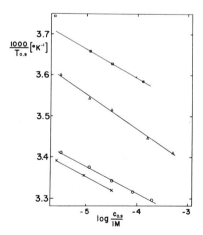

Fig. 24 Determination of ΔH values in 0.1 M sodium cacodylate, pH 6.9. □, $(dA)_n + d(T_8 \cdot Et)$ ($<7°C$); △, $(dA)_n + d(T_8)$; ○, $(dA)_n + d(T_8 \cdot Et)$ (20°–30°C); ×, $(dA)_n + d(T_8 \cdot Et)$ (2°–20°C)$_{rA}$. Reprinted with permission from Pless and Ts'o (1977), *Biochemistry* **16**, 1239. Copyright by the American Chemical Society.

to $3.9 \times 10^{-4}\,M$ (Fig. 23). In a plot of $1/T_{0.9}$ versus log $(c_{0.9}/1\,M)$, the point corresponding to $c_{total} = 7.2 \times 10^{-6}\,M$ falls off the straight line, which is well defined by the other three points (Fig. 24). Repeated measurements at this concentration gave identical results. A similar deviation to higher values of $1/T_m$ at low temperatures is seen in a $1/T_m$ versus log c plot published by Martin *et al.* (1971). Since this point was measured at $-5°C$, this phenomenon may reflect changes in the water structure at temperatures around $0°C$. The straight lines in Fig. 24 give ΔH values of -8.1 kcal/mole of base pairs.

For comparison, the interaction of the octamer diester $d(T_8)$, equal to $d[(Tp)_7T]$, with $(dA)_n$ in $0.1\,M$ sodium cacodylate, pH 6.9, was studied. Annealing curves for five equimolar mixtures, at total concentrations ranging from 9.7×10^{-6} to $1.6 \times 10^{-3}\,M$, were determined, and a plot of $1/T_{0.9}$ versus log $(c_{0.9}/1\,M)$ gave a ΔH value of -6.8 kcal/mole of base pairs (Fig. 24).

The salt dependence of the interaction of $d(T_8)$ with $(dA)_n$ was studied by annealing equimolar mixtures, at $10^{-4}\,M$ concentration, in a series of buffers containing $0.01\,M$ sodium cacodylate and varying concentrations of sodium chloride at pH 6.9. Figure 25 presents a plot of $T_{0.9}$ versus the decadic logarithm of the sodium ion activity for sodium ion concentrations of from 0.04 to 1.01 M. The straight line has a slope of $18.8°C$.

In contrast to the interaction between $(dA)_n$ and $d(T_8)$, the binding of the octamer triester to $(dA)_n$ is hardly affected by changes in ionic strength. As seen in Fig. 26, the annealing curves for $(dA)_n + d(T_8 \cdot Et)$ $(20°–30°C)$ $(1:1)$ at sodium ion concentrations of 0.29 and 0.10 M are superimposable $(T_{0.9} = 26.0°C)$, whereas at $0.013\,M$ sodium ion concentration the transition is slightly shifted to a higher temperature $(T_{0.9} = 26.4°C)$.

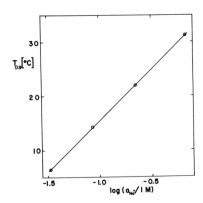

Fig. 25 Salt dependence of the annealing of $(dA)_n + d(T_8)(1:1)$. Reprinted with permission from Pless and Ts'o (1977), *Biochemistry* **16,** 1239. Copyright by the American Chemical Society.

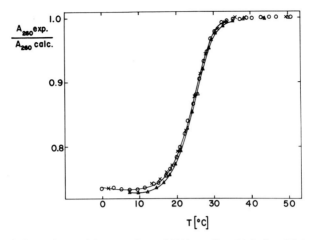

Fig. 26   Salt dependence of the annealing of $(dA)_n + d(T_8 \cdot Et)$ $(20°–30°C)$ $(1:1)$, $10^{-4} M$, pH 6.9. $\triangle$, 0.013 M sodium cacodylate; $\bigcirc$, 0.10 M sodium cacodylate; $\times$, 0.27 M NaCl, 0.02 M sodium cacodylate. Reprinted with permission from Pless and Ts'o (1977), *Biochemistry* **16**, 1239. Copyright by the American Chemical Society.

Comparing the binding of $d(T_8)$ and of $d(T_8 \cdot Et)$ with $(dA)_n$ to their interaction with the ribopolymer, $(rA)_n$, considerable differences were found. In 0.1 M sodium cacodylate, at $10^{-4} M$ total concentration, the equimolar mixtures $(rA)_n + d(T_8)$ and $(rA)_n + d(T_8 \cdot Et)$ (unfractionated) have $T_{0.9}$ values of 8.6° and ca. $-11°C$, respectively. Thus, the octamer triester has a much weaker interaction with the ribopolymer than does the diester.

It is important to note that, for two octamer triester fractions $[d(T_8 \cdot Et)$ $(<7°C)$ and $d(T_8 \cdot Et)$ $(20°–30°C)]$ with widely varying affinities for $(dA)_n$, Damle's (1970) method gives average ΔH values of -8.1 and -8.6 kcal/ mole of base pairs, respectively. This small variation in ΔH values can largely be attributed to the difference in the two transition temperatures (approx. 0° versus approx. 25°C) at which the measurements were made. Therefore, for the octamer triester fractions with their narrow temperature transitions, the ΔH values for the various constituent diastereoisomers should be nearly identical. The use of Damle's (1970) and Crothers' (1971) method is therefore warranted, provided that the relative proportion of the diastereoisomers and the input ratio of total $d(T_8 \cdot Et)$ to $(dA)_n$ are invariant for all concentration variation experiments.

We have examined the salt dependence of duplex formation according to the method of Manning (1972). Together with our circular dichroism data, which indicate that $(dA)_n \cdot d(T_8 \cdot Et)$ has a B-type conformation, we conclude that the average charge spacing in the $(dA)_n \cdot d(T_8 \cdot Et)$ duplex is

3.3 Å (not 1.65 Å, since the T strand is electroneutral), in single-stranded $(dA)_n$ at 26°C is also 3.3 Å, and in single-stranded T strand is 4.9 Å. These results together with the thermodynamic measurements of ΔH and T_m allow us to evaluate the entropy changes. After a proper correction of the ΔS_{cratic} term, we were able to derive a difference in ΔS_{elec} of about 2 eu (cal/mole of base pair)$^{-1}$ degree^{-1} in 0.1 M salt in favor of reaction $(dA)_n + d(T_8 \cdot Et)$ ($\Delta S_{elec} = 0$ for this process) over reaction $(dA)_n + d(T_8)$ and a difference of ΔS_{conf} of 7 e.u. in favor of reaction $(dA)_n + d(T_8)$ over $(dA)_n + d(T_8 \cdot Et)$.

In conclusion, the electrostatic interaction of double-helical nucleic acids has been the focus of considerable interest (Schildkraut and Lifson, 1965; Delisi and Crothers, 1971; Krakauer, 1974; Record et al., 1976). In the present investigation, a novel approach to this problem was adopted: transformation of one negatively charged strand [$d(T_8)$ in this case] into a nonionic strand. Measurements of the enthalpy changes by the concentration variation method indicate that in 0.1 M salt, at 11°C, there is a difference of 1.6 ± 0.4 kcal/mole of base pairs between duplex formation from two charged strands [$(dA)_n + d(T_8)$, $\Delta H = -6.8$ kcal] and duplex formation from one charged strand and one nonionic strand [$(dA)_n + d(T_8 \cdot Et)$, $\Delta H = -8.4$ kcal]. This value of 1.6 kcal is a measure of the electrostatic enthalpy attendant on formation of the double helix in B conformation. It should consist of a term for the contraction of one of the two charged strands [$d(T_8)$] from a linear phosphate spacing of about 4.9 to 3.3 Å upon duplex formation and a term for the interpenetration of the two polyelectrolytes, since the electrostatic effects of these two processes were eliminated on substituting the octamer triester for the diester. The term corresponding to the contraction of the second strand is not included. In the specific instance of formation of $(dA)_n \cdot d(T_8)$ this term should be negligible, since the single-stranded $(dA)_n$ is already fully contracted (linear phosphate spacing of 3.3 Å) in 0.1 M salt at 26°C. In duplex formation from two extended single strands, however, the contraction of the second strand is expected to contribute a further term to the electrostatic enthalpy change, and ΔH_{elec} could be higher than 1.6 kcal/mole of base pairs.

The stability of the duplex containing the phosphotriester oligomer, $(dA)_n \cdot d(T_8 \cdot Et)$, is lowered by a special entropic effect, which is not present in the interaction of $(dA)_n$ with $d(T_8)$. The ΔS value characterizing this effect was estimated to be about -7 cal (mole of base pairs)$^{-1}$ degree^{-1}. The study of molecular models suggests that this phenomenon is due to a loss of rotational freedom of the phosphate ethyl groups upon duplex formation and that this effect is more pronounced in the formation of the double helix in the A than in the B form.

B. Biochemical Studies of the Interaction and the Effects of the Nonionic Analogues of Oligonucleotides

Oligodeoxyribonucleotide ethyl phosphotriesters, d[Tp(Et)Gp(Et)G] and d[Tp(Et)Tp(Et)Cp(Et)A], were synthesized by a stepwise chemical procedure. These triesters are complementary, respectively, to 3'-CpCpA-OH (the 3'-amino acid-accepting terminus) and to —UpGpApA— (the anticodon region) of phenylalanine tRNA from yeast and *E. coli*. The association constants for binding of the triesters to their complementary regions on tRNA were measured by equilibrium dialysis and were compared with those of oligodeoxyribonucleotides and oligoribonucleotides of the same sequences. In 1 M NaCl–10 mM MgCl$_2$ at 0°C, the association constants of the oligomers with both tRNA$_{yeast}^{Phe}$ and tRNA$_{coli}^{Phe}$ are very similar (Table X). The association constants of the ribooligonucleotides are 8 to 20 times larger than those of the corresponding deoxyribooligonucleotides, whereas the deoxyribooligonucleotide triesters exhibit binding constants slightly higher than those of the deoxyribooligonucleotides. These differences are due to the differences in conformations of the various oligomers. At low salt concentration (0.1 M NaCl, 1 mM EDTA), the oligonucleotide triesters have the same binding constants as at high salt concentration, whereas the corresponding deoxyribo- and ribooligonucleotides show a four- to six-fold decrease in their binding constants. This reflects the removal of the charge repulsion between the neutral triesters and the tRNA. The binding of oligomers to modified tRNA$_{yeast}^{Phe}$ was also examined. Removal of the Y base decreased the binding of anticodon-complementary oligomers 6-fold, whereas removal of the 3'-CpA residues decreased the binding of the 3'-CpCpA-OH complementary oligomers 6- to 20-fold. This study provides the chemical and physicochemical basis for the investigation of the biochemical effects of these triesters on the aminoacylation of tRNA.

The effect of d[Tp(Et)Gp(Et)G] and d[Tp(Et)Tp(Et)Cp(Et)A] on the aminoacyl-tRNA synthetase-catalyzed aminoacylation of tRNA was examined. At 0°C both triesters inhibit the formation of phenylalanyl-tRNA^{Phe} by approximately 50–60%. The inhibition decreases with increasing temperature. A Lineweaver–Burk analysis at 0°C shows that the inhibition by both triesters is competitive. The results suggest that the inhibition is a consequence of the formation of complexes between the triesters and the tRNA.

As expected, d[Tp(Et)Gp(Et)G] inhibits the aminoacylation of all five tRNA's examined (phenylalanine, lysine, tyrosine, proline, and leucine) by 55–80%, whereas the inhibition by d[Tp(Et)Tp(Et)Cp(Et)A] is more

56   PAUL O. P. TS'O

TABLE X

Effect of Salt Concentration on the Binding of Oligomers to tRNA$_{\text{yeast}}^{\text{Phe}}$ at 0°C[a]

	$K\ (M^{-1})$	
	1 M NaCl, 10 mM MgCl$_2$	0.1 M NaCl, 1 mM EDTA
d[Tp(Et)Tp(Et)Cp(Et)A]	3,000	3,400
d[Tp(Et)Cp(Et)C]	2,400	2,500
d(TpTpCpA)	2,200	0
UpUpCpA	48,000	14,800
UpGpG	4,800	800

[a] Each buffer contained 10 mM Tris, pH 7.5. The tRNA concentration was 40 μM. Reprinted from P. Miller, J. C. Barrett, and P. O. P. Ts'o (1974), *Biochemistry* **13**, 4887. Copyright by the American Chemical Society.

specific and is greatest for tRNA^{Phe}. The effect of the triesters on various related reactions catalyzed by the aminoacyl-tRNA synthetases was investigated. d[Tp(Et)Gp(Et)G] inhibits the enzymatic deacylation of phenylalanyl-tRNA^{Phe} at 0°C by 15%, but has no effect on the synthetase-catalyzed ATP–pyrophosphate exchange reaction. d[Tp(Et)-Tp(Et)Cp(Et)A] does inhibit pyrophosphate exchange; however, the inhibition is not specific for any amino acid. The transfer of an activated amino acid from a preformed aminoacyl adenylate–synthetase complex to tRNA at 0°C is inhibited by d[Tp(Et)Gp(Et)G] for three amino acids (phenylalanine, tyrosine, and leucine), whereas d[Tp(Et)-Tp(Et)Cp(Et)A] only inhibits the transfer of phenylalanine.

The effect of d(TpGpG), d(GpGpT), d(TpTpCpA), UpGpG, GpGpU, and UpUpCpA on aminoacylation of tRNA^{Phe} was examined. At 0°C, the inhibitory activities of these oligonucleotides directly parallel the magnitudes of their association constants with tRNA^{Phe} as determined by equilibrium dialysis.

These findings demonstrate that the triesters can inhibit tRNA aminoacylation by specifically masking complementary regions of the tRNA through complex formation.

C. Studies on the Effect of the Nonionic Oligonucleotide Analogues on Cells

In view of the higher association constants of the ribosyl oligomers to their complementary polynucleotides as shown in the above section, we have prepared the nonionic 2'-O-methyl ribotrinucleotide ethyl phosphotriester, G^{m}p(Et)G^{m}p(Et)U (Miller *et al.*, 1977). This nonionic trimer is complementary to the . . . ApCpC . . . sequence found in the amino acid-accepting stem of most tRNA's and the anticodon region of tRNA^{Gly} and to the threonine codon of mRNA. G^{m}p(Et)G^{m}p(Et)U forms hydrogen-bonded complexes with the amino acid-accepting stem of tRNA$^{Phe}_{yeast}$ and unfractionated tRNA$_{coli}$ under physiological salt conditions at 37°C as determined by equilibrium dialysis. The extent of phenylalanine aminoacylation of tRNA$^{Phe}_{coli}$ is inhibited 39% by G^{m}p(Et)G^{m}p(Et)U at 37°C in solution. The triester is resistant to hydrolysis by serum nucleases and cell lysates. The triester is readily taken up by transformed Syrian hamster fibroblasts growing in monolayer. Within the cell, the triester is deethylated to give the trinucleotide species G^{m}p(Et)G^{m}pU, G^{m}pG^{m}p(Et)U, and G^{m}pG^{m}pU and is also hydrolyzed to dimeric and monomeric units.

While both the trimer–diester, G^{m}pG^{m}pU, and the trimer–triester, G^{m}p(Et)G^{m}p(Et)U, are equally active in inhibiting the aminoacylation reaction with the *E. coli* tRNA^{Phe} at 37°C (39% inhibition with 50 μM oligomer, 2 μM tRNA), only the trimer–triesters are active in inhibiting the protein synthesis (Table XI) or the colony formation of the hamster fibroblast in culture (Table XII). This result suggests that only the trimer–triester can effectively enter the living cells.

Treatment of transformed fibroblasts in monolayer with 25 μM G^{m}p(Et)G^{m}p(Et)U results in a 40% inhibition of cellular protein synthesis with a concurrent slight increase in cellular RNA synthesis during the first 4 hours (Fig. 27). After 4 hours, the rate of cellular protein

TABLE XI

Effect of Oligomers on Protein Synthesis in Syrian Hamster
Transformed Fibroblast Cells[a]

	Oligomer concentration (μM)	Inhibition[b] (%)
G^{m}p(Et)G^{m}p(Et)U	200	73
	100	53
	50	39
G^{m}pG^{m}pU	200	8
	100	4
	50	12

[a] Reprinted with permission from Miller *et al.* (1977), *Biochemistry* **16**, 1988. Copyright by the American Chemical Society.

[b] Percentage inhibition of incorporation of [³H]leucine into protein at 37°C. Hamster fibroblasts in monolayer were preincubated with oligomer for 1 hour before adding [³H]leucine.

TABLE XII

**Effect of Oligomers on Colony Formation by Transformed
Hamster Fibroblast Cells[a]**

	Oligomer concentration (μM)	Control[b] (%)
G^{m}p(Et)G^{m}p(Et)U	25	50
	10	61
	5	89
G^{m}pG^{m}pU	25	89
	10	94
	5	94

[a] Reprinted with permission from Miller *et al.* (1977), *Biochemistry* **16**, 1988. Copyright by the American Chemical Society.

[b] Average result of two experiments. The control contained 18 colonies after 5 days.

synthesis begins to recover, while RNA synthesis returns to that of the control. Our biochemical studies suggest that inhibition of cellular protein synthesis might be expected if G^{m}p(Et)G^{m}p(Et)U, G^{m}p(Et)G^{m}pU, G^{m}pG^{m}p(Et)U, and G^{m}pG^{m}pU, which have been taken up by or formed within the cell, physically bind to tRNA and mRNA and inhibit the function of these nucleic acids. The reversible inhibition of protein synthesis may be a consequence of further degradation of the trinucleotide

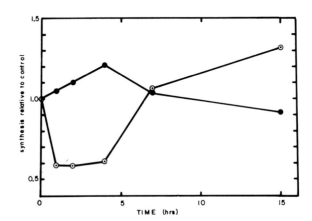

Fig. 27 Effect of 25 μM G^{m}p(Et)G^{m}p(Et)U on protein synthesis (○) and RNA synthesis (●) in transformed hamster fibroblasts growing in monolayer. Cells were preincubated with triester for the indicated time periods. L-[³H]Amino acids or [¹⁴C]adenosine were added and the amount of radioactivity incorporated into protein and RNA was determined after 40 minutes. Reprinted with permission from Miller *et al.* (1977), *Biochemistry* **16**, 1988. Copyright by the American Chemical Society.

species within the cell as well as of an increase in supply of RNA molecules involved in protein synthesis. The growth of the transformed fibroblasts is inhibited during the first 24 hours of incubation with 25 μM $G^mp(Et)G^mp(Et)U$, after which growth proceeds at a normal rate. In cloning experiments, the number and size of colonies formed by the transformed fibroblasts after 5 days exposure to 25 μM triester are decreased by 50% relative to untreated controls (Table XII). The temporary inhibition of cell growth may reflect the transitory inhibition of cellular protein synthesis caused by the triester. The more permanent effect on colony formation may result from perturbation of the ability of single cells to attach and form colonies during a critical time period.

IV. FUTURE PROSPECTS

In the physicochemical studies of the duplex formation between the oligo(T) phosphotriester $[d(T_8 \cdot Et)]$ with $(dA)_n$ or $(rA)_n$, the steric hindrance of the ethyl group was uncovered. In fact, the loss of entropy in this hindrance $(\Delta\delta_{conf})$ was quantitatively evaluated to be about 7 entropy units less favorable than the duplex formation between the diester $d(T_8)$ with $(dA)_n$. The steric hindrance of the ethyl group is even more serious for the interaction between $d(T_8 \cdot Et)$ and $(rA)_n$.

In the study of the uptake and the biological effects of the 2'-O-methyl ribooligonucleotide ethyl phosphotriester, it was also found that these phosphotriesters are degraded by an unknown mechanism, since these oligonucleotide alkyl phosphotriesters were found to be completely resistant to spleen and venom nucleases, micrococcal nuclease, serum nuclease, and even the cell lysate.

To avoid the steric hindrance in the complex formation and to increase the resistance of the oligonucleotide to hydrolysis, we have prepared a type of nonionic analogue, namely, the deoxyribodinucleoside methyl phosphonates, $NO(O)P(CH_3)ON$ or $[Np(CH_3)N]$ (Miller et $al.$, 1977). The stacking of $d[Ap(CH_3)A]$ and $d[Ap(CH_3)Ap(CH_3)]$ in aqueous solution is shown by nmr data and by hypochromicity (13% and 11.3%, respectively, at room temperature). In 10 mM Tris, 10 mM MgCl$_2$, 5 mM NaCl, the T_m's of complexes of $d[Ap(CH_3)A]$ and $d[Ap(CH_3)Ap(CH_3)]$ with poly(U) in a stoichiometry of 2U : 1A are 14.4° and 14°C, respectively, whereas that of dApdA : 2 poly(U) is only 7°C. We were able to separate the diastereoisomers of $d[Ap(CH_3)Ap(CH_3)]$. The T_m of the poly(U) complex with one isomer (I) is 13.5°C and the T_m of the poly(U) complex with the other isomer (II) is 17.4°C. The investigation of the absolute configuration of these isomers is in progress. Neither $Tp(CH_3)T$ nor $d[Ap(CH_3)T]$ can be hydrolyzed by serum nucleases but both are taken up intact by the Syrian hamster fibroblasts. The biological effects of this type of nonionic oligonucleotide analogue are currently under active investigation.

Two other areas of future investigation should now be mentioned. We are interested in utilizing this approach of nonionic oligonucleotide analogues of the desired sequence as a means to control the function of viral nucleic acid inside the living cells in a sequence-specific manner. This approach has the potential to suppress specifically the function of any single-stranded, unpaired, or open region of nucleic acid, either in a biochemical reaction or inside the living cells. This potential will be fully investigated.

The research in this direction brings us back to the nmr. Since these oligonucleotide analogues can permeate to the interior of the living cells, can survive nuclease attack there, and can form complexes with complementary regions of nucleic acids inside the cell, this system may offer a unique opportunity to study some of the properties and structure of nucleic acids inside the living cell by the ^{31}P nmr technique. It was further revealed from our preliminary studies that the chemical shifts of the phosphorus resonance in the triester form (about 1 ppm) and in the phosphonate form (about 40 ppm) are sufficiently different from the chemical shifts of the phosphorus in monoester or diester forms. Therefore, it is relatively easy to locate the phosphorus triester and particularly the phosphorus–phosphonate resonances in a cellular mixture if there is sufficient sensitivity in the nmr measurement. The development of the Fourier transform technique and magnet with a wide bore accommodating a 20-mm nmr sample tube provides the possibility of such an approach. We are certainly inspired by the pioneering experiments in ^{31}P nmr studies about the intracellular phosphate species inside the Ehrlich ascites tumor cells reported by Navon et al. (1977).

ACKNOWLEDGMENTS

This research was supported in part by grants from the National Science Foundation and from the National Institute of General Sciences as well as by various fellowships granted to the individual investigators over the past 10 years. The valuable assistance of Cathryn Alden and Christine Dreon in the preparation of the manuscript is gratefully acknowledged.

REFERENCES

Alderfer, J. L., and Ts'o, P. O. P. (1977). *Biochemistry* **16**, 2410–2416.
Alderfer, J. L., Tazawa, I., Tazawa, S., and Ts'o, P. O. P. (1974). *Biochemistry* **13**, 1615–1622.
Arnott, S., and Hukins, D. W. L. (1972). *Biochem. Biophys. Res. Commun.* **47**, 1504.
Arnott, S., and Hukins, D. W. L. (1973). *J. Mol. Biol.* **81**, 93.

Arnott, S., Hukins, D. W. L., Dover, S. D., Fuller, W., and Hodgson, A. R. (1973). *J. Mol. Biol.* **81,** 107.

Arter, D. B., Walker, G. C., Uhlenbeck, O. C., and Schmidt, P. G. (1974). *Biochem. Biophys. Res. Commun.* **61,** 1089.

Borer, P. N., Kan, L. S., and Ts'o, P. O. P. (1975). *Biochemistry* **14,** 4847–4863.

Broyde, S. G., Wartell, R. M., Stellman, S. D., Hingerty, B., and Langride, R. (1975). *Biopolymers* **14,** 1597–1613.

Cheng, D. M., and Sarma, R. H. (1977a). *Biopolymers* **16,** 1687–1711.

Cheng, D. M., and Sarma, R. H. (1977b). *Am. Chem. Soc.* **99,** 7333–7348.

Crothers, D. M. (1971). *Biopolymers* **10,** 2147.

Crothers, D. M., Cole, P. E., Hilbers, C. W., and Shulman, R. G. (1974). *J. Mol. Biol.* **87,** 63.

Damle, V. N. (1970). *Biopolymers* **9,** 353.

Daniel, W. E., Jr., and Cohn, M. (1975). *Proc. Natl. Acad. Sci. U.S.A.* **72,** 2582–2586.

Daniel, W. E., Jr., and Cohn, M. (1976). *Biochemistry* **15,** 3917–3924.

Davies, D. B., and Danyluk, S. S. (1974). *Biochemistry* **13,** 4417–4434.

Davies, D. B., and Danyluk, S. S. (1975). *Biochemistry* **14,** 543.

DeBoer, G., Miller, P. S., and Ts'o, P. O. P. (1973). *Biochemistry* **12,** 720–726.

Delisi, C., and Crothers, D. M. (1971). *Biopolymers* **10,** 2323.

Early, T. A., Kearns, D. R., Burd, J. F., Larson, J. E., and Wells, R. D. (1977). *Biochemistry* **16,** 541–551.

Ezra, F. S., Lee, C. H., Kondo, N. S., Danyluk, S. S., and Sarma, R. H. (1976). *Biochemistry* **16,** 1977–1987.

Geerdes, H. A. M., and Hilbers, C. W. (1977). *Nucleic Acids Res.* **4,** 207–222.

Giessner-Prettre, C., and Pullman, B. (1970). *J. Theor. Biol.* **27,** 87.

Giessner-Prettre, C., Pullman, B., Borer, P. N., Kan, L. S., and Ts'o, P. O. P. (1976). *Biopolymers* **15,** 2277.

Giessner-Prettre, C., Pullman, B., and Caillet, J. (1977). *Nucleic Acids Res.* **4,** 99–116.

Govil, G., and Smith, I. C. P. (1973). *Biopolymers* **12,** 2589–2598.

Hingerty, B., Subramanian, E., Stellman, S. D., Broyde, S. B., Sato, T., and Langridge, R. (1975). *Biopolymers* **14,** 227–236.

Kallenbach, N. R., Daniel, W. E., Jr., and Kaminker, M. A. (1976). *Biochemistry* **15,** 1218–1224.

Kan, L. S., and Ts'o, P. O. P. (1977). *Nucleic Acids Res.* **4,** 1633–1648.

Kan, L. S., Ts'o, P. O. P., von der Haar, F., Sprinzl, M., and Cramer, F. (1974). *Biochem. Biophys. Res. Commun.* **59,** 22–29.

Kan, L. S., Borer, P. N., and Ts'o, P. O. P. (1975a). *Biochemistry* **14,** 4864–4869.

Kan, L. S., Borer, P. N., and Ts'o, P. O. P. (1975b). *Abstr., 169th Natl. Meet. Am. Chem. Soc.* No. MEDI-29.

Kan, L. S., Ts'o, P. O. P., Sprinzl, M., von der Haar, F., and Cramer, F. (1977). *Biochemistry* **16,** 3143–3154.

Kastrup, R. V., and Schmidt, P. G. (1975). *Biochemistry* **14,** 3612.

Kearns, D. R. (1976). *Prog. Nucleic Acids Res. Mol. Biol.* **18,** 91–149.

Kearns, D. R., and Shulman, R. G. (1974). *Acc. Chem. Res.* **7,** 33–39.

Kim, S. H., Suddath, F. L., Quigley, G. J., McPherson, A., Sussman, J. L., Wang, A. H.-J., Seeman, N. C., and Rich, A. (1974). *Science* **185,** 435–440.

Kondo, N. S., Holmes, H. M., Stempel, L. M., and Ts'o, P. O. P. (1970). *Biochemistry* **9,** 3479–3498.

Kondo, N. S., Fang, K. N., Miller, P. S., and Ts'o, P. O. P. (1972). *Biochemistry* **11,** 1991–2003.

Krakauer, H. (1974). *Biochemistry* **13,** 2579.

Kyogoku, Y., Lord, R. C., and Rich, A. (1967a). *J. Am. Chem. Soc.* **89,** 496–504.

Kyogoku, Y., Lord, R. C., and Rich, A. (1967b). *Proc. Natl. Acad. Sci. U.S.A.* **57,** 250–257.

Ladner, J. E., Jack, A., Robertus, J. D., Brown, R. S., Rhodes, D., Clark, B. F. C., and Klug, A. (1975a). *Proc. Natl. Acad. Sci. U.S.A.* **72,** 4414–4418.

Ladner, J. E., Jack, A., Robertus, J. D., Brown, R. S., Rhodes, D., Clark, B. F. C., and Klug, A. (1975b). *Nucleic Acids Res.* **2,** 1629–1637.

Lapper, R. D., Mantsch, H. H., and Smith, I. C. P. (1972). *J. Am. Chem. Soc.* **94,** 6243–6244.

Lapper, R. D., Mantsch, H. H., and Smith, I. C. P. (1973). *J. Am. Chem. Soc.* **95,** 2878–2880.

Lee, C. H., Ezra, F. S., Kondo, N. S., Sarma, R. H., and Danyluk, S. S. (1976). *Biochemistry* **15,** 3627–3629.

Manning, G. S. (1972). *Biopolymers* **11,** 937.

Mantsch, H. H., and Smith, I. C. P. (1972). *Biochem. Biophys. Res. Commun.* **46,** 808–815.

Martin, F. H., Uhlenbeck, O. C., and Doty, P. (1971). *J. Mol. Biol.* **57,** 201.

Miller, P. S., Fang, K. N., Kondo, N. S., and Ts'o, P. O. P. (1971). *J. Am. Chem. Soc.* **93,** 6657–6665.

Miller, P. S., Braiterman, L. T., and Ts'o, P. O. P. (1977). *Biochemistry* **16,** 1988–1996.

Nagel, G. M., and Hanlon, S. (1972). *Biochemistry* **11,** 816.

Navon, G., Ogawa, S., Shulman, R. G., and Yamane, T. (1977). *Proc. Natl. Acad. Sci. U.S.A.* **74,** 87–91.

Olson, W. K. (1973). *Biopolymers* **12,** 1787–1814.

Olson, W. K. (1975a). *Biopolymers* **14,** 1775–1795.

Olson, W. K. (1975b). *Biopolymers* **14,** 1797–1810.

Olson, W. K. (1975c). *Nucleic Acids Res.* **2,** 2055–2068.

Olson, W. K., and Flory, P. J. (1972a). *Biopolymers* **11,** 1–24.

Olson, W. K., and Flory, P. J. (1972b). *Biopolymers* **11,** 25–56.

Patel, D. J. (1976). *Biopolymers* **15,** 533–558.

Patel, D. J., and Tonelli, A. (1974). *Biopolymers* **13,** 1943.

Patel, D. J., and Tonelli, A. E. (1975). *Biochemistry* **14,** 3990–3996.

Perahia, D., Pullman, B., and Saran, A. (1974a). *Biochim. Biophys. Acta* **340,** 299–313.

Perahia, D., Pullman, B., and Saran, A. (1974b). *Biochim. Biophys. Acta* **353,** 16–27.

Pless, R. C., and Ts'o, P. O. P. (1977). *Biochemistry* **16,** 1239–1250.

Pople, J. A., Schneider, W. G., and Bernstein, H. J. (1959). "High-Resolution Nuclear Magnetic Resonance," Chapter 10. McGraw-Hill, New York.

Porschke, D., Uhlenbeck, O. C., and Martin, F. H. (1973). *Biopolymers* **12,** 1313.

Quigley, G. J., Seeman, N. C., Wang, A. H.-J., Suddath, F. L., and Rich, A. (1975). *Nucleic Acids Res.* **2,** 2329–2341.

Ravetch, J., Gralla, J., and Crothers, D. M. (1974). *Nucleic Acids Res.* **1,** 109.

Record, M. T., Woodbury, C. P., and Lohman, T. M. (1976). *Biopolymers* **15,** 893.

Reid, B. R., and Robillard, G. (1975). *Nature (London)* **257,** 287–291.

Rich, A., and RajBhandary, U. L. (1976). *Annu. Rev. Biochem.* **45,** 805.

Robillard, G. T., Tarr, C. E., Vosman, F., and Berendsen, H. J. C. (1976). *Nature (London)* **262,** 363–369.

Rosenberg, J. M., Seeman, N. C., Kim, J. J. P., Suddath, F. L., Nicholas, H. B., and Rich, A. (1973). *Nature (London)* **243,** 150–154.

Rubin, J., Brennan, T., and Sundaralingam, M. (1972). *Biochemistry* **11,** 3112–3128.

Schildkraut, C., and Lifson, S. (1965). *Biopolymers* **3,** 195.

Schleich, T., Cross, B. P., Blackburn, B. J., and Smith, I. C. P. (1975). In "Structure and Conformation of Nucleic Acids and Protein-Nucleic Acid Interactions" (M. Sundaralingam and S. T. Rao, eds.), p. 223. Univ. Park Press, Baltimore, Maryland.

Schleich, T., Cross, B. P., and Smith, I. C. P. (1976). *Nucleic Acids Res.* **3,** 355–370.

Smith, I. C. P., Mantsch, H. H., Lapper, R. D., Delauriers, R., and Schleich, T. (1973). In "Conformations of Biological Molecules and Polymers" (E. Bergmann and B. Pullman, eds.), p. 381. Academic Press, New York.

Smith, I. C. P., Jennings, H. J., and Delauriers, R. (1975). *Acc. Chem. Res.* **8,** 306–313.
Stout, C. D., Mizuno, H., Rubin, J., Brennan, T., Rao, S. T., and Sundaralingam, M. (1976). *Nucleic Acids Res.* **3,** 1111–1123.
Sundaralingam, M. (1969). *Biopolymers* **7,** 821–860.
Sussman, J. L., and Kim, S.-H. (1976). *Biochem. Biophys. Res. Commun.* **68,** 89–96.
Sussman, J. L., Seeman, N. C., Kim, S. H., and Berman, H. M. (1972). *J. Mol. Biol.* **66,** 403–421.
Tazawa, I., Tazawa, S., and Ts'o, P. O. P. (1972). *J. Mol. Biol.* **66,** 115.
Tewari, R., Nanda, R. K., and Govil, G. (1974). *Biopolymers* **13,** 2015–2035.
Ts'o, P. O. P. (1974). *In* "Basic Principles in Nucleic Acid Chemistry" (P. O. P. Ts'o, ed.), Vol. 1, p. 453. Academic Press, New York.
Ts'o, P. O. P., Kondo, N. S., Schweizer, M. P., and Hollis, D. P. (1969). *Biochemistry* **8,** 997–1029.
Yathindra, N., and Sundaralingam, M. (1974). *Proc. Natl. Acad. Sci. U.S.A.* **71,** 3325–3329.

2

The Structure of Transfer RNA Molecules in Solution

DAVID R. KEARNS, THOMAS EARLY, AND PHILIP H. BOLTON

Department of Chemistry, University of California–San Diego, La Jolla, California

I. INTRODUCTION

High-resolution nuclear magnetic resonance (nmr) has become a very useful tool in the investigation of the base pairing and other structural features of polynucleotides in solution (Kearns, 1976). In this chapter, we limit our considerations to nmr studies of the transfer RNA molecules. We begin by presenting experimental evidence that most tRNA's (probably all Class I tRNA's) have nearly identical tertiary structures in solution and then describe a set of experimental criteria that allow us to distinguish between those resonances that arise from common tertiary interactions and those due to secondary structure Watson–Crick base pairs. We then present a comprehensive analysis of the low-field nmr spectra of a number of tRNA's in which we use a modified ring-current shift calculation for nucleic acid bases first described by Giessner-Prettre and Pullman (1970) and a specific model for the tRNA structure. The results of these calculations are then compared with experimental spectra and with calculations based on the three crystal models for yeast tRNA[Phe]. From this analysis it will be evident that nmr data, in conjunction with ring-current calculations, may be useful in aiding the refinement of the X-ray diffraction data on yeast tRNA[Phe]. Finally, we present nmr data on the role that the 2'-OH group plays in stabilizing the conformation of RNA's.

65

FRONTIERS IN PHYSICOCHEMICAL BIOLOGY

II. TERTIARY STRUCTURE OF tRNA IN SOLUTION

The majority of nmr studies of tRNA have been concerned with resonances in the low-field region of the spectrum (10–15 ppm downfield)
from sodium 4,4-dimethyl-4-silapentane-1-sulfonate (DSS) (see Fig. 1),
which arise from imino protons hydrogen bonded to the ring nitrogen
atoms of other bases. Most resonances in this spectral region are due to
secondary structure Watson–Crick base pairs (one from the imino proton in each A · U or G · C base pair), but a number (~four) of tertiary
interactions can also contribute resonances to this region (Kearns and
Shulman, 1974; Wong and Kearns, 1974; Reid *et al.*, 1975; Hilbers and
Shulman, 1974; Bolton and Kearns, 1975; Wong *et al.*, 1975). Resonances
from nonexchangeable (or slowly exchanging) aromatic C—H protons of
the bases (typically about 90) and from exchangeable amino protons
(Ts'o, 1973, 1974; Ts'o *et al.*, 1969) are located between 7 and 9 ppm. The
ribose C—H proton resonances occur between 4 and 6 ppm, and most
methyl and methylene resonances are between 1 and 3 ppm (Koehler

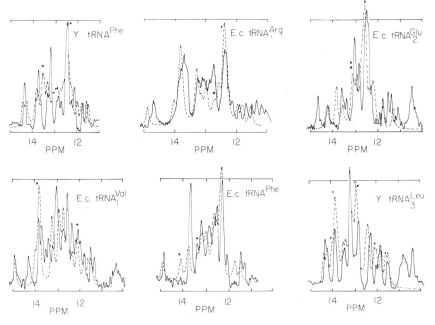

Fig. 1 An example of the low-field nmr spectra of different tRNA's. The solid lines are
the observed spectra and the dashed lines are spectra computed using ring-current shift
theory and a model for the tRNA structure as described in the text. Y, yeast; E.c., *E. coli*;
asterisks, predicted resonances from tertiary interactions.

and Schmidt, 1973). In this section we concentrate on the low-field spectral region, although the methyl resonances will also be considered.

Each A · U and G · C secondary structure base pair and some tertiary interactions contribute resonances to the low-field spectrum (Kearns and Shulman, 1974; Bolton and Kearns, 1975; Reid *et al.*, 1975). Therefore, by integrating the low-field nmr spectrum, the total number of secondary structure base pairs and the number of tertiary interactions (minimum number) can be determined. Accurate internal calibration methods have shown that most Class I tRNA's exhibit 23–24 resonances between 11.5 and 15 ppm when spectra are obtained at sufficiently low temperatures and in the presence of adequate levels of magnesium (Bolton *et al.*, 1976). Reid and Robillard (1975) obtained slightly higher total values (26 ± 2) using a different approach but, within the accuracy of the methods (~10%), it is clear there are four or five more resonances between 11.5 and 15 ppm than predicted by the cloverleaf secondary structure alone. Therefore, the integrations demonstrate that tertiary interactions are contributing at least four or five extra resonances to the low-field spectrum of most tRNA. X-Ray diffraction studies (Quigley *et al.*, 1975; Sussman and Kim, 1976; Ladner *et al.*, 1975; Stout *et al.*, 1976) have shown that a number of the bases involved in tertiary interactions in yeast tRNA^{Phe} are common to other tRNA's (see Fig. 2). If there is a set of tertiary interactions that are common to most tRNA's, it is reasonable to expect that these might give rise to resonances that would appear as *common* resolved resonances in the low-field nmr spectra of a mixture of tRNA's. As anticipated, the spectrum of *Escherichia coli* tRNA^{Mixed} exhibits a number of moderately well-resolved resonances located at 14.8, 13.8, 13.0, 11.5, and 10.5 ppm (Fig. 3). The following experimental criteria have been used to show that resonances located at these positions, and another one at 11.8 ppm, are due to tertiary interactions common to most tRNA.

A. The Resolved Resonances Show Early Melting Behavior

One characteristic that permits resonances from secondary structure and tertiary interactions to be distinguished is that high salt or divalent metal ions are required to stabilize tertiary interactions (Römer and Hach, 1975; Rialdi *et al.*, 1972; Schreier and Schimmel, 1974). Hence, when magnesium is absent tertiary interactions are preferentially destabilized. The 44°C spectra of mixed *E. coli* tRNA in the presence and absence of magnesium are compared in Fig. 3. It is evident that the losses of intensity occur selectively at the positions where the resolved resonances appear. Since the net loss in intensity corresponds to only

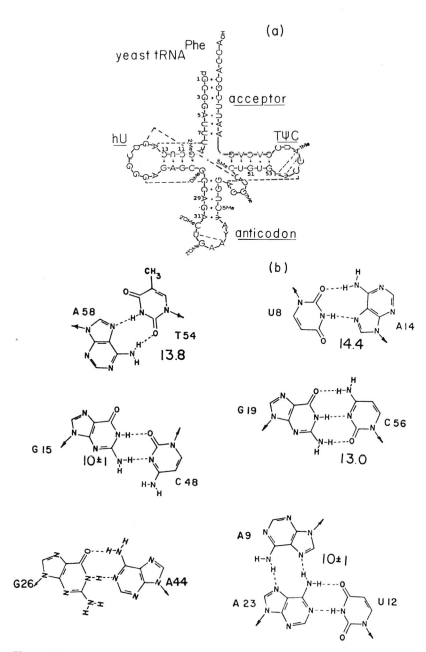

Fig. 2 (a) The cloverleaf model for the secondary structure of yeast tRNA[Phe] showing some of the tertiary interactions that give rise to resonances in the low-field nmr spectrum. (b) The hydrogen-bonding schemes for some of the tertiary interactions seen in yeast tRNA[Phe] (Quigley *et al.*, 1975; Sussman and Kim, 1976; Ladner *et al.*, 1975). The bold values are the estimated chemical shifts of resonances from some hydrogen-bonded protons.

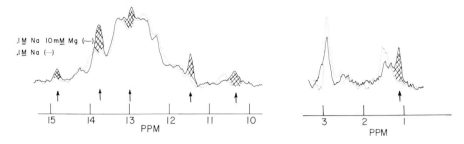

Fig. 3 The nmr spectra of *E. coli* tRNA demonstrating the presence of resolved resonances and their sensitivity to magnesium. Dotted line, no magnesium; solid line, with magnesium. Shaded areas indicate losses of intensity in the absence of magnesium and the arrows indicate the positions of the common resonances in the low-field region.

four or five protons per molecule (Bolton and Kearns, 1975), this correlates well with the approximately four extra resonances that were inferred on the basis of the total intensity measurements (Bolton *et al.*, 1976; Reid and Robillard, 1975).

B. A Resolved Resonance Is Observed from the Methyl Group of the Common T_{54} Residue

Since we assign one of the common resonances to the $A_{58} \cdot T_{54}$ tertiary interaction, it should also be possible to observe the methyl resonance of the common T_{54} as a resolved peak. This is confirmed by the spectrum shown in Fig. 3 (Chao and Kearns, 1977), where a resolved resonance at 1.1 ppm [assigned to T_{54} in the spectra of several individual tRNA's (Koehler and Schmidt, 1973; Kastrup and Schmidt, 1975; Smith *et al.*, 1969; Kan *et al.*, 1975)] is observed. In T, TMP, or poly(dT), the methyl resonance is located at ~1.9, so in the native tRNA it has been shifted upfield by about 0.7–0.8 ppm from its intrinsic unshifted position. In spite of this sensitivity to the conformation of the TΨC loop, the methyl resonance appears as a well-resolved single peak in the spectra of mixed tRNA's. This proves that it is possible to observe a *resolved* resonance in mixed tRNA's from a base involved in a common tertiary interaction despite variations in the sequence of bases located elsewhere in the molecule.

C. The Resolved Resonances Are Broadened by Low Levels of Paramagnetic Metal Ions

Since divalent metal ions are intimately involved in stabilizing tertiary interactions, it is likely that some strong metal binding sites are located close to the tertiary interactions they stabilize. To test this possibility, trace amounts of paramagnetic Mn^{2+} ions were added to tRNA solutions containing high levels of Mg^{2+}, and we found (Chao and Kearns, 1977) that small amounts of Mn^{2+} (1 per 100 tRNA's) caused substantial broadening of several of the common resonances. It is important to note that this broadening of resonances assigned to tertiary interactions occurs before most other resonances are affected (Chao and Kearns, 1977) and, hence, provides another way to discriminate between resonances from tertiary and secondary structure interactions.

Various tertiary interactions have been proposed on the basis of the crystal structure studies of yeast $tRNA^{Phe}$, but only six or seven can give rise to low-field resonances (see Fig. 2). The 14.8-ppm resonance in E. coli (14.4 in yeast tRNA) is assigned to the $s^4U_8 \cdot A_{14}$ reversed Hoogsteen base pair on the basis of chemical modification studies showing that the s^4U_8 base is involved (Wong and Kearns, 1974a; Wong et al., 1975; Reid et al., 1975). A resonance at 13.8 ppm, one of the sharpest features in the E. coli $tRNA^{mixed}$ spectrum, is assigned to the reversed Hoogsteen $A_{58} \cdot T_{54}$ tertiary interaction because (1) it is common to all E. coli tRNA's, (2) the 13.8-ppm peak is less prominent in yeast, where A_{58} is often modified (Bolton and Kearns, 1975), and (3) studies of AU_2 (Kallenbach et al., 1976) are consistent with such an assignment. A resonance at 13.0 ppm is assigned to the highly conserved $G_{19} \cdot C_{56}$ tertiary interaction (Bolton and Kearns, 1975) because this is in the region expected for a Watson–Crick $G \cdot C$ base pair, and $G_{19} \cdot C_{56}$ is the only homology that occurs at a high enough frequency to account for the observed intensity. The 11.5-ppm resonance is assigned to a tertiary interaction involving a hydrogen bond between N_3—H of U_{33} and either the phosphate or the ribose of A_{36} (Kim, 1977; Quigley and Rich, 1976) based on the observation of an analogous "extra" resonance in the anticodon hairpin fragments of several tRNA's (Wong and Kearns, 1974b; B. F. Rordorf and D. R. Kearns, unpublished results; Kearns, 1976). Since the crystal studies indicate that there may be a similar hydrogen bonding between N_3—H of Ψ_{55} and the phosphate of A_{58} (Quigley et al., 1975; Sussman and Kim, 1976; Ladner et al., 1975), a second resonance may be expected in this same spectral region. As will be noted, an extra resonance is invariably found in the spectra of individual tRNA's in the 11.5- to 12.0-ppm region, which is not accounted for by the ring-current shift calculations. This may be due to Ψ_{55} or possibly to the $G_{15} \cdot C_{48}$ interaction.

The above observations are important in at least two respects. First, they permit us to distinguish experimentally between those resonances in the low-field nmr spectra that arise from common tertiary interactions and other resonances that are due to secondary structure base pairs. This is crucial, of course, to any interpretation of low-field nmr spectra. Second, and perhaps more important, the fact that resolved resonances from tertiary interactions can be observed in the spectra of mixed tRNA's demonstrates that, in solution, most tRNA's are folded into the same overall tertiary structure. Since the chemical shifts of some of these resonances (e.g., from the methyl protons of T) are particularly sensitive to molecular conformation, the tRNA molecules must have a high degree of similarity in many of the local structural features as well. This conclusion that most tRNA's have the same structure in solution receives support from other chemical and biochemical studies (Kim, 1977; Rich and RajBhandary, 1976), but the nmr results offer more detailed and comprehensive evidence for this.

Now that we have discussed some of the experimental observations concerning the locations and assignments of resonances from the common tertiary interactions, we turn to an analysis of the remaining resonances in the low-field spectra arising from secondary structure base pairs.

III. RING-CURRENT SHIFT THEORY ANALYSIS OF LOW-FIELD nmr SPECTRA

The above analysis permitted us to identify those resonances in the low-field spectra that arise from common tertiary interactions. The remaining ~20 resonances must therefore be attributed to secondary structure $A \cdot U(\Psi)$ and $G \cdot C$ base pairs, and an examination of the spectra of individual tRNA's indicates (see Fig. 1) that these resonances are subjected to a wide range of chemical shifts (Shulman et al., 1973; Kearns and Shulman, 1974; Kearns et al., 1973). Therefore, to analyze these spectra we must know the (hypothetical) chemical shifts of the imino proton resonance in isolated $A \cdot U$ and $G \cdot C$ base pairs in which there are no chemical shifts due to neighboring bases. From a consideration of various model systems and tRNA spectra, we have concluded that the intrinsic positions are 14.5 ppm for $A \cdot U$ and 13.6 ppm for $G \cdot C$ (Kearns et al., 1973; Kearns, 1976; Kearns and Shulman, 1974; Patel and Tonelli, 1974). The wide range of chemical shifts observed in native tRNA spectra and in various model systems is then attributed to ring-current shifts from neighboring bases (Kearns et al., 1973; Shulman et al., 1973). Therefore, to compute the low-field nmr spectrum of a tRNA, the following information must be known: (1) the molecular structure of the tRNA and (2) a theory for computing ring-current shifts from neighboring bases. In both respects we are faced with uncertainties. On one hand, we have atomic coordinates (crystalline state) for only one tRNA, and as we shall see these are not adequate at this time to serve our purposes. On the other hand, we are also uncertain as to the reliability of the ring-current shift theory as applied to nucleic acid bases. In many respects the problem is analogous to the dilemma encountered in attempts to predict the circular dichroism (CD) spectra of even simple polynucleotide systems, where there are uncertainties both in the theoretical model used and in the molecular coordinates of RNA double helices (Cech and Tinoco, 1977). If the theory is assumed to be correct, then the experimental data could be used to determine structures and vice versa. However, when calculations and experiment disagree, it could be due to deficiencies in the theory and/or incorrect assumptions regarding the molecular structure in solution. The approach we have taken is to assume a specific but plausible model for the structure of all tRNA's and then use ring-current shift parameters (scaled to fit benzene) to predict the spectra of a number of different tRNA molecules (T. Early and D. R. Kearns, unpublished results). In this way, systematic deviations can reveal any major inadequacies in the basic assumptions. We first discuss in detail the ring-current shift calculations and present the model we use to analyze tRNA spectra. The predictions are then compared with the experimental observations and, as will be demonstrated, the approach does seem to be successful. This suggests that our model and the shift calculations are basically sound.

A. Ring-Current Shift Calculations

Many authors (Robillard *et al.*, 1976; Arter and Schmidt, 1976; Geerdes and Hilbers, 1977) have used a rather complicated expression involving elliptical integrals of the first and second kinds to compute ring-current shifts, $\Delta\delta$, arising from the nucleic acid bases (Giessner-Prettre and Pullman, 1970).

For distances greater than $\sim$3 Å from the bases, we find (Early *et al.*, 1977) that the results given by these more complicated expressions can be accurately reproduced by a simple dipolar sum of the form

$$\Delta\delta = \sum_i K_i(3 \cos^2 \theta - 1) \, r_i^{-3}$$

where dipoles of magnitude K_i are placed at the center of the pyrimidine ring, or at the centers of the six- and five-membered rings in the case of purine, with the principal axis oriented perpendicular to the plane of the ring. In the case of the pyrimidines, K was then adjusted to agree with the results of Giessner-Prettre and Pullman (1970) at $\theta = 0$ and $r = 3.4$ Å

TABLE I

K Parameters Used in the Dipolar Approximation to the Ring-Current Shift Expressions

| Base | K (ppm Å^{-3}) | |
	Six-membered ring	Five-membered ring
A	19.654	13.2665
G	7.8616	11.0062
C	6.2893	—
U	2.555	—

(i.e., directly above the center of the ring). For purines, two dipoles, centered in the middle of the six- and five-membered rings, were adjusted in magnitude so that the point of maximal shift in a plane (3.4 Å from the base) agreed in magnitude and position with the calculations of Giessner-Prettre and Pullman. In the worst case, the values differed by only 0.07 ppm (total shift of 1.3 ppm), but for most regions the agreement was within 0.03 ppm. The set of K values obtained by this procedure is listed in Table I. Further refinements may be required at some future time, but until the fundamental validity of the approach is demonstrated and more accurate molecular coordinates are available there is no need to use the more complicated theoretical expressions for the calculations.

B. A Model for tRNA Structure

To compute the spectra of various tRNA's, the following specific model for the tRNA structure and set of assumptions have been adopted.

1. All secondary structure base pairs are in regular RNA-11 helices (Arnott, 1970).
2. Bases in single-stranded regions are stacked on adjacent bases as though part of regular RNA helices. The one exception is that A_{14} is not stacked on the terminal $G \cdot C_{13}$ base pair of the hU stem because it is paired with U_8.
3. Ring-current shifts on the hydrogen-bonded imino protons are computed as described above for the RNA-11 geometry.
4. The intrinsic positions for the Watson–Crick base pairs are taken to be $A \cdot U = 14.5$, $G \cdot C = 13.6$, and $A \cdot \Psi = 13.5$ ppm (Kearns *et al.*, 1973; Kearns, 1976).

5. Resonances from $A \cdot U$ base pairs adjacent to nonterminal $G \cdot U$ pairs have been omitted for reasons discussed elsewhere (Kearns *et al.*, 1973).
6. Resonances from tertiary interactions are located at 14.9 ($s^4U_8 \cdot A_{14}$ in *E. coli*), 14.4 ($U_8 \cdot A_{14}$ in yeast), 13.75 ($A_{58} \cdot T_{54}$, or 13.45 if using m^1A_{58}), 12.95 ($G_{19} \cdot C_{56}$), 11.8 (Ψ_{55} or $G_{15} \cdot C_{48}$), and 11.5 ppm (U_{33}) (Bolton and Kearns, 1975, 1976, 1977; Kearns, 1976).

C. Comparison of Predicted and Observed Spectra

Using the assumptions listed above, the low-field spectra of a number of tRNA's were calculated (see Fig. 1). For a number of tRNA's the agreement between the observed and predicted spectra is quite good, although the position of the resonances from the seventh base pair of the amino acid acceptor stem, and in many cases the position of the resonance from the interior terminal base pair of the TΨC stem, is incorrectly predicted. This suggests that the TΨC and amino acid acceptor stems are not stacked on each other in the manner assumed in the model. If we make special assumptions about the stacking of these two helices, then an almost perfect fit of most of the tRNA spectra can be obtained. For several tRNA's (yeast $tRNA^{Trp}$, $tRNA^{Tyr}$, *E. coli* $tRNA_f^{Met}$), the fits are reasonable, but in each case there are small discrepancies throughout the spectrum. Finally, the fit of yeast $tRNA_1^{Val}$ is so poor that there must be some important error that may be related to the fact that the hU stem of this molecule formally contains only two good Watson–Crick base pairs (both $A \cdot U$) and may be melted. In view of the reasonably good agreement between the predicted and observed spectra for a number of tRNA, we tentatively conclude that our approach is reasonable. Taken literally, it suggests that the secondary structure helices of the tRNA in solution are more regular than is suggested by the present X-ray models for yeast $tRNA^{Phe}$. It will be interesting to see if the helices become more regular after further refinements.

Clearly, an important unsolved problem in nmr studies of tRNA is the experimental determination of the assignments of specific resonances to specific secondary structure base pairs. If our predictions are confirmed, then the validity of both the ring-current calculations and the model would be established. Once this is accomplished the ring-current analysis of the spectra will become a more powerful structural tool for studying the structure of other RNA's and DNA's in solution.

It is of interest to compare our results obtained using standard RNA helices with those obtained using published crystal coordinates [Quigley *et al.*, 1975 (MIT); Sussman and Kim, 1976 (DUK); Ladner *et al.*, 1975

(MRC); Stout *et al.*, 1976 (SHM)]. The models are labeled in the same manner as given by Stout *et al.* (1976). The same set of intrinsic positions for A · U, G · C, and A · Ψ base pairs and the same ring-current shift theory are used, with the results shown in Fig. 4. Resonances from the tertiary interactions were also included using the experimentally determined positions. While there are some similarities between the predicted spectra, we find that the different models give quite different predictions with regard to the chemical shifts of resonances from specific base pairs in the molecule. For example, in the MRC model, the resonance from AU_7 is predicted to be the lowest field resonance in the spectrum, whereas the DUK and MIT models predict that this is one of the more highly shifted AU resonances in the spectrum. Similar differences are noted for many of the other resonances (e.g., GC_{28}, GC_{13}, and AU_{29}). In light of this and the fact that there are important discrepancies between the predicted and observed spectra, we *cannot* conclude on the basis of ring-current calculations that the structure of tRNA^{Phe} is the same in the crystal as in solution. Rather, it is evident that the three different models lead to quite different predictions. None gives as good a fit as our model, which employs standard RNA helices.

Robillard *et al.* (1976) arrived at a somewhat different conclusion on the basis of a similar set of calculations using the DUK coordinates

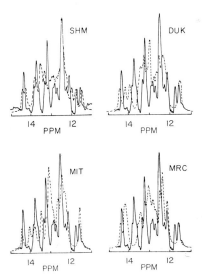

Fig. 4 A comparison of the yeast tRNA^{Phe} low-field nmr spectrum calculated using the standard helix model (SHM) for tRNA and the three crystal models that have been proposed [Sussman and Kim, 1976 (DUK); Ladner *et al.*, 1975 (MRC); Quigley *et al.*, 1975 (MIT); Stout *et al.*, 1976 (SHM)].

(Sussman and Kim, 1976). However, we note that they assume that the DUK structure gives the correct description of the solution state structure of the molecule and then adjust the AU and GC intrinsic positions (two parameters), four parameters that go into the ring-current shift theory, and the positions of a number of tertiary interactions (different from the ones we deduced on the basis of experimental data) to obtain a good fit of the experimental data. With 10 adjustable parameters to fit approximately 25 resonance positions, it is not surprising that an excellent fit was obtained. Unfortunately, there is no reason to believe that the DUK structure is better than the three other structures at that stage in the refinement of the diffraction data. Furthermore, the approach of Robillard et al. requires reassignments and reinterpretations of spectra, which are either unjustified or demonstrably wrong. Specifically, the position of resonances from the $A_{58} \cdot T_{54}$ base pair is assigned at 14.4 ppm in yeast tRNA^{Phe} and at 14.2 ppm in E. coli tRNA$_1^{Val}$ (Robillard et al., 1976). Interestingly, E. coli tRNA^{Phe} has no resonance between 14.7 and 13.8 ppm and yet it has a TΨC loop identical to that of tRNA$_1^{Val}$ (Reid et al., 1975). Therefore, the Robillard et al. assignment of the $A_{58} \cdot T_{54}$ resonance cannot be correct. Ethidium bromide binding data also rule out the Robillard et al. assignment of the $A_{58} \cdot T_{54}$ resonance in E. coli tRNA$_1^{Val}$ (Jones et al., 1978). Their choice of 14.35 ppm for the intrinsic position of A · U is also in disagreement with model system data (Early et al., 1977; Patel and Tonelli, 1974; Patel and Canuel, 1976).

IV. HYDROGEN BINDING OF THE RIBOSE 2'-OH GROUP IN RNA

Since the 2'-OH group is the only structural difference between RNA and DNA, this group is, in some way, responsible for the differences found in the physical properties of DNA and RNA. The role of the 2'-OH has not been determined because there are relatively few experimental techniques that offer direct information about the role of the 2'-OH in the conformation of RNA. Investigations of the circular dichroism and proton magnetic resonance (pmr) of dinucleotide monophosphates (Ts'o, 1974) and the ^{31}P nmr of 3'-mononucleotides (Cozzone and Jardetzky, 1976) indicate that an interaction between the 2'-OH and 3'-phosphate (but not the 5'-phosphate) gives rise to the differences between RNA and DNA (Ts'o, 1974). However, observations on 2'-O-methyl-substituted RNA indicate that the 2'-OMe RNA and RNA have similar spectral properties (nmr and CD) and that they differ in essentially the same manner from DNA (Ts'o, 1974); this apparently rules out a direct interaction between the 2'-OH proton and 3'-phosphate. X-Ray studies of RNA fibers (Arnott, 1970) and tRNA crystals (Quigley and Rich, 1976) ascribe no role to the 2'-OH in the secondary structure. We now present experimental evidence that the 2'-OH in RNA is hydrogen bonded to a water molecule that is simultaneously hydrogen bonded to a 3'-phosphate (Bolton and Kearns, 1978).

During the examination of the nmr spectra of various synthetic RNA's at low temperatures, we invariably observed an "extra" resonance at 6.8 ± 0.2 ppm in H_2O but not in D_2O solutions. An example is shown in Fig. 5a, where the spectrum of poly(C) is presented. An extra resonance at ~6.8 ppm is also observed in the spectra of poly(U), poly(A), and poly(I)-poly(C), but in no case can this resonance be assigned to any of the imino or amino protons (e.g., U contains no amino group). Since the extra ~6.8-ppm resonance appears to be common to most of the RNA systems, it is not associated with the bases but rather arises from backbone protons. We therefore expect to find a similar resonance in tRNA, and the spectrum of E. coli tRNA[mixed] confirms this expectation (see Fig. 5b).

The behavior of the 6.8-ppm resonance is very sensitive to RNA conformation. Upon heating poly(C) from 13° to 36°C, the 6.8-ppm resonance broadens and is reduced to about one-half the intensity, whereas with poly(U) over one-half the resonance intensity is lost before 20°C. The thermal stability of poly(C) is greater than that of poly(U) (Ts'o, 1974), and the nmr results show that the presence of the 6.8-ppm resonance correlates with structural ordering of the RNA. Gadolinium

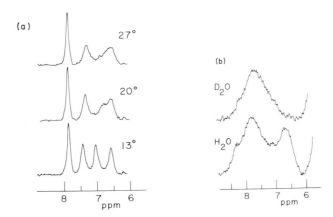

Fig. 5 (a) The 300-MHz nmr spectrum of poly(C) illustrating the sensitivity of the 2'-OH resonance to temperature. (b) The nmr spectrum of E. coli mixed tRNA at 20°C in D_2O and H_2O illustrating the presence of the ribose 2'-OH peak at ~6.7 ppm.

broadening experiments indicate that the proton responsible for the 6.8-ppm resonance is about one-half the distance from the phosphate binding site, as are the H_6 and $H_{1'}$ protons. [Similar results were obtained for poly(U) and poly(I)-poly(C) and for poly(C) using manganese as a paramagnetic probe.]

On the basis of these and other observations, the "extra" resonance observed in the nmr spectra of RNA is attributed to the 2'-OH proton since (1) this is the only exchangeable proton per nucleotide unit otherwise unaccounted for, (2) it is common to all RNA's, and (3) it is located about the proper distance away from the phosphate to agree with the paramagnetic metal binding studies. The fact that the 2'-OH resonance is only observed below the optical T_m indicates that stabilization of the 2'-OH is correlated with RNA helix formation.

Ordinarily, the exchange of the 2'-OH proton of free ribose in water is too fast for this resonance to be observed, but in dimethyl sulfoxide (DMSO) the exchange is slow and the resonance is observed at 5.4 ppm. By adding water to DMSO solutions, the behavior of this resonance can be followed as a function of water content. As water is added, this resonance in cytidine shifts downfield and finally broadens beyond detection. By extrapolation of the data, we infer a chemical shift of ~6.6 ppm for the 2'-OH proton of ribose under conditions where it is hydrogen bonded to water but slowly exchanging. We therefore conclude that hydrogen bonding to water accounts for most of the chemical shift of the ribose 2'-OH observed in the polynucleotides; however, to account for the much reduced rate of exchange of the 2'-OH proton with bulk water, we further suggest that the water molecule to which it is hydrogen bonded is slowly exchanging with the bulk water. The gadolinium broadening experiments indicate that the 2'-OH is located ~6 Å from the phosphate group (judging from the broadening of the 2'-OH resonance relative to other resonances in the molecule), and this places the associated water molecule close enough to be simultaneously hydrogen bonded to the phosphate oxygen atom.

It is apparent from an examination of models that the formation of simultaneous hydrogen bonds in which the bound water functions as a proton donor to the phosphate and as a proton acceptor for the 2'-OH group places considerable restrictions on the allowed conformations of the RNA backbone; these restrictions are not present in DNA. We further note that a similar hydrogen-bonding scheme could be present in 2'-OMe-substituted RNA, provided the water molecule acts as the proton donor to both the 2'-OMe and phosphate groups. This hydrogen bonding in RNA, not found in DNA, may contribute to the relative conformational rigidity and the slightly greater thermal stability of

RNA. Furthermore, since the hydrogen bonding occurs in both single-stranded and double-stranded RNA's, it suggests that the backbone conformations in helical single- and double-stranded RNA's may be very similar.

X-Ray diffraction data on RNA fibers (Arnott, 1970) and tRNA crystals (Quigley and Rich, 1976) offer no positive evidence for the hydrogen bonding of the 2'-OH proposed here, although the X-ray results are consistent with our proposal. It is unclear whether or not fiber diffraction studies can offer the resolution needed to observe the 2'-OH hydrogen bonding, and there is the further complication in RNA fibers that some of the 2'-OH may be involved in intermolecular interactions (Arnott, 1970).

V. SUMMARY

The nmr studies summarized in this chapter provide concrete experimental evidence that most tRNA molecules in solution have virtually identical tertiary structures. A number of resonances arising from these common tertiary interactions have been identified in low-field nmr spectra, and tentative assignments to tertiary structure base pairs seen in the crystalline form of yeast tRNA^{Phe} have been proposed. The chemical shifts of resonances from secondary structure Watson–Crick base pairs have been predicted using the ring-current shift theory and a model for tRNA that assumes that all arms of the cloverleaf of the tRNA are standard RNA-11 helices. The agreement between the calculated and observed spectra supports the validity of the approach, but this must be tested further. At the present stage of refinement, the crystal structure data on yeast tRNA^{Phe} are not accurate enough to be used for predicting the nmr spectra. As a final topic we discussed evidence for the participation of the 2'-OH group in hydrogen-bonding interactions that stabilize the RNA-11 conformation.

REFERENCES

Arnott, S. (1970). *Prog. Biophys. Mol. Biol.* **21**, 265–319.
Arter, D. B., and Schmidt, P. G. (1976). *Nucleic Acids Res.* **3**, 1437–1447.
Bolton, P. H., and Kearns, D. R. (1975). *Nature (London)* **255**, 347–349.
Bolton, P. H., and Kearns, D. R. (1976). *Nature (London)* **262**, 423–424.
Bolton, P. H., and Kearns, D. R. (1977). *Biochim. Biophys. Acta* **477**, 10–19.
Bolton, P. H., and Kearns, D. R. (1978). *Biochim. Biophys. Acta* **517**, 329–337.
Bolton, P. H., Jones, C. R., Bastedo-Lerner, D., Wong, K. L., and Kearns, D. R. (1976). *Biochemistry* **15**, 4370–4376.

Cech, C. L., and Tinoco, I. (1977). *Biopolymers* **16,** 43–66.
Chao, Y.-Y. H., and Kearns, D. R. (1977). *Biochim. Biophys. Acta* **477,** 20–27.
Cozzone, P., and Jardetzky, O. (1976). *Biochemistry* **15,** 4583–4859.
Early, T., Kearns, D. R., Burd, J. F., Larson, J. E., and Wells, R. D. (1977). *Biochemistry* **16,** 541–555.
Geerdes, H. A. M., and Hilbers, C. W. (1977). *Nucleic Acids Res.* **4,** 207–221.
Giessner-Prettre, C., and Pullman, B. (1970). *J. Theor. Biol.* **27,** 87–95.
Hilbers, C. W., and Shulman, R. G. (1974). *Proc. Natl. Acad. Sci. U.S.A.* **71,** 3239–3242
James, T. L. (1975). "Nuclear Magnetic Resonance in Biochemistry." Academic Press, New York.
Jones, C. R., Bolton, P. H., and Kearns, D. R. (1978). *Biochemistry* **17,** 601–607.
Kallenbach, N. R., Daniel, W. E., Jr., and Kaminker, M. A. (1976). *Biochemistry* **15,** 1218–1224.
Kan, L. S., Ts'o, P. O. P., von der Haar, F., Sprinzl, M., and Cramer, F. (1975). *Biochemistry* **14,** 3278–3291.
Kastrup, R. V., and Schmidt, P. G. (1975). *Biochemistry* **14,** 3612–3618.
Kearns, D. R. (1976). *Prog. Nucleic Acids Res.* **18,** 92–149.
Kearns, D. R., and Shulman, R. G. (1974). *Acc. Chem. Res.* **7,** 33–39.
Kearns, D. R., Lightfoot, D. R., Wong, K. L., Wong, Y. P., Reid, B. R., Cary, L., and Shulman, R. G. (1973). *Ann. N.Y. Acad. Sci.* **222,** 324–336.
Kim, S. H. (1977). In "Transfer RNA" (S. Altman, ed.). MIT Press, Cambridge, Massachusetts.
Koehler, K. M., and Schmidt, P. G. (1973). *Biochem. Biophys. Res. Commun.* **50,** 370–376.
Ladner, J. E., Jack, A., Robertus, J. D., Brown, R. S., Rhodes, D., Clark, B. F. C., and Klug, A. (1975). *Nucleic Acids Res.* **2,** 1629–1637.
Patel, D. J., and Canuel, L. (1976). *Proc. Natl. Acad. Sci. U.S.A.* **73,** 674.
Patel, D. J., and Tonelli, A. E. (1974). *Biopolymers* **13,** 1943–1964.
Quigley, G. J., and Rich, A. (1976). *Science* **194,** 796–806.
Quigley, G. J., Seeman, N. C., Wang, A. H.-J., Suddath, F. L., and Rich, A. (1975). *Nucleic Acids Res.* **2,** 2329–2341.
Reid, B. R., and Robillard, G. T. (1975). *Nature (London)* **257,** 287–291.
Reid, B. R., Ribeiro, N. S., Gould, G., Robillard, G., Hilbers, C. W., and Shulman, R. G. (1975). *Proc. Natl. Acad. Sci. U.S.A.* **72,** 2049–2053.
Rialdi, G., Levy, J., and Biltonen, R. (1972). *Biochemistry* **11,** 2472–2479.
Rich, A., and RajBhandary, U. L. (1976). *Annu. Rev. Biochem.* **45,** 805.
Robillard, G. T., Tarr, C. E., Vosman, F., and Berendsen, H. J. C. (1976). *Nature (London)* **262,** 363–369.
Römer, R., and Hach, R. (1975). *Eur. J. Biochem.* **55,** 271–284.
Schreier, A. A., and Schimmel, P. R. (1974). *Biochemistry* **13,** 1841–1852.
Shulman, R. G., Hilbers, C. W., Kearns, D. R., Reid, B. R., and Wong, Y. P. (1973). *J. Mol. Biol.* **78,** 57–69.
Smith, I. C. P., Yamane, T., and Shulman, R. G. (1969). *Can. J. Biochem.* **47,** 480–484.
Stout, C. D., Mizuno, H., Rubin, J., Brennan, R., Rao, S. R., and Sundaralingam, M. (1976). *Nucleic Acids Res.* **3,** 1111–1123.
Sussman, J. L., and Kim, S. H. (1976). *Biochem. Biophys. Res. Commun.* **68,** 89–96.
Ts'o, P. O. P. (1973). *Biol. Macromol.* **4,** 49–190.
Ts'o, P. O. P. (1974). "Basic Principles in Nucleic Acid Chemistry," vol. 1, pp. 453–583. Academic Press, New York.
Ts'o, P. O. P., Schweizer, M. P., and Hollis, D. P. (1969). *Ann. N.Y. Acad. Sci.* **158,** 256–297.
Wong, K. L., and Kearns, D. R. (1974a). *Nature (London)* **252,** 738–739.
Wong, K. L., and Kearns, D. R. (1974b). *Biopolymers* **13,** 371–380.
Wong, K. L., Bolton, P. H., and Kearns, D. R. (1975). *Biochim. Biophys. Acta* **383,** 446–451.

3

The Chemistry of Membrane-Active Peptides and Proteins

YU. A. OVCHINNIKOV

Shemyakin Institute of Bioorganic Chemistry, U.S.S.R. Academy of Sciences, Moscow, U.S.S.R.

Among the vast class of peptides, the biological functions of which extend from hormones and antibiotics to toxins and releasing factors, there is one group of compounds whose structure–function relationships are so unique, and which have proved to be such important tools in membrane study, that they have become a separate domain of bioorganic chemistry.

These are the membrane-active complexones, substances capable of interacting with metal ions and mediating their transport across artificial and biological membranes. Highly varied in chemical structure, but united in their function, they have even transcended the domain of peptides to include representatives from other chemical classes in their group. They vary in molecular weight and in the nature of their functional groups, they can be linear or, especially, ring compounds, and they comprise both naturally occurring and synthetic substances. Their ion-complexing capacity, in most cases associated with the inclusion of the ion in a molecular cavity where it is held by ion–dipole forces, as well as their structure and properties have been treated in detail elsewhere (Ovchinnikov *et al.*, 1974a); here I will focus on the work that is not found in this monograph.

Interesting as they are by themselves, the complexones are significant largely as stepping stones to the more complicated molecules actually

FRONTIERS IN PHYSICOCHEMICAL BIOLOGY

engaged in the naturally occurring transport process in cellular membranes. Recently, several binding proteins have been discovered in the periplasmatic space; these proteins apparently participate in the transmembrane transport process, at least in the initial stages, and much information is being accumulated on another type of protein, bacterial rhodopsins, which participate in proton transfer across the bacterial membrane. It is only fitting, therefore, to conclude this chapter with a brief description of our work on these proteins, for it is actually the peptide complexones that have provided us with the methodology and insight to study the transport mechanism of the proteins.

With regard to their transmembrane ion-transporting properties, the complexones can be divided into two groups: those acting as carriers and those forming ion-permeable pores or channels. Simple models depicting these two types of actions are shown in Fig. 1. Among the ion carriers (ionophores) a preeminent place is occupied by macrocyclic depsipeptides of the groups of valinomycin and the enniatins, whereas the best-studied representatives of the channel type of ion-mediating agents are gramicidins A, B, and C. These substances together with the antitoxin antamanide and their numerous synthetic analogues form the largest and most thoroughly studied group of peptide complexones.

Valinomycin consists of three identical DVal-LLac-LVal-DHyIv fragments (HyIv, hydroxyisovaleric acid; Lac, lactic acid) (Fig. 2). Its 36-membered ring contains 6 amide and 6 ester bonds; the side chains are hydrophobic alkyl radicals. Valinomycin is capable of binding numerous cations including those of the alkali, alkaline-earth, and transition metals (Table I).

An outstanding property of this ionophore is its remarkably high K/Na-complexing selectivity, the stability of the K^+ complex being three or four orders of magnitude higher than that of the Na^+ complex.

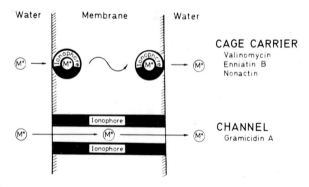

Fig. 1   Principle of ionophore action on membranes.

TABLE I

Stability Constants of Equimolar Valinomycin Complexes with Various Cations at 25°C (liters mole^{-1})

Solvent	Li$^+$	Na$^+$	K$^+$	Rb$^+$	Cs$^+$	NH$_4^+$	Ag$^+$	Tl$^+$	Mg^{2+}	Ca^{2+}	Sr^{2+}	Ba^{2+}	References
Methanol	<5	4.7 12[a] 14[c]	80,000 27,000[b] 8,000[a]	180,000	26,000	47	8,000	5,400	<5	500	170	2,200	Grell et al. (1972)
Methanol: water (9:1)	—	—	4,700	—	—	—	—	—	—	—	—	—	Grell et al. (1975)
Methanol: water (7:3)	—	—	93	—	—	—	—	—	—	—	—	—	Grell et al. (1975)
Methanol: water (1:1)	—	—	10.5	—	—	—	—	—	—	—	—	—	Grell et al. (1975)
Ethanol	—	<50	2,600,000	2,900,000	650,000	—	—	—	—	—	—	—	Andreev et al. (1971); Shemyakin et al. (1969)
99.5% aqueous ethanol	—	—	1,200,000	—	—	—	—	—	—	—	—	—	Möschler et al. (1971)
Water	—	—	2.3	5.9	0.12	—	—	—	—	—	—	—	Feinstein and Felsenfeld (1971)

[a] Wipf et al. (1968).
[b] Andreev et al. (1971).
[c] Haynes et al. (1971).

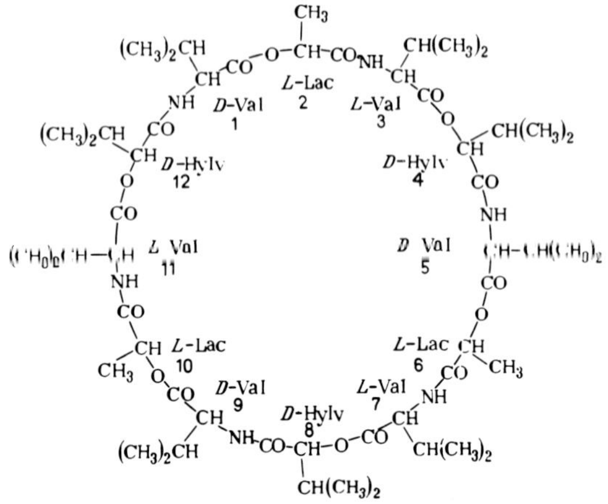

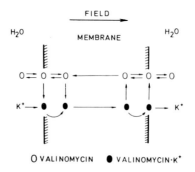

Fig. 2   Structure of valinomycin. Hylv, hydroxyisovaleric acid; Lac, lactic acid.

Nowhere in nature are any substances known that can at least approach valinomycin in this respect. Valinomycin is quite universal in its action on membranes, inducing K^+ conductivity in any type of membrane system with almost no exception, and its action is manifested at low concentrations (10^{-8} M and lower).

Studies on bilayers, which very satisfactorily simulate the lipid regions of biological membranes, have made it possible to create a model of the ion transport, which is shown in Fig. 3 (Ovchinnikov *et al.*, 1974a). By heterogeneous complexation, the K^+ ion becomes attached to a valinomycin molecule situated on the membrane surface. From this it passes to a second ionophore molecule moving within the membrane. The complexed ion thus migrates along the potential gradient to the opposite side of the membrane, where the above reactions are now reversed. From this, one may draw a number of important conclusions.

Fig. 3   Model of valinomycin-mediated potassium transport across a lipid membrane.

To function efficiently as an ionophore, the valinomycin molecule must be capable of forming sufficiently stable K^+ complexes. Free valinomycin must be capable of partitioning between the surface and the interior of the membrane; that is, it must have both lipophilic and surface-active properties. The K^+ complex must be highly lipophilic in order to pass through the interior of the membrane, and the transfer of a K^+ ion from one ionophore molecule to another must occur at a sufficiently high rate. Interestingly, nature seems to have made the valinomycin molecule so perfect that all stages of migration of the ion are of a similar activation energy, which means that no bottlenecks occur in the energy profile of potassium transport.

Physicochemical studies of valinomycin revealed that such variegated and sometimes even contradictory requirements could be fulfilled only by virtue of conformational rearrangements of the valinomycin molecule, that is, by appropriate changes in its three-dimensional structure. We were thus interested in the conformational dynamics, and to study it we made use of a composite approach whereby the conformational states of peptide–protein molecules in solution were determined by a wide variety of physicochemical methods (proton magnetic resonance, optical rotatory dispersion, circular dichroism, infrared and ultraviolet spectra, dipole moments, etc.) together with theoretical conformational analysis. While these methods were inferior to X-ray analysis in resolving power, they permitted measurements to be made over a wide range of environmental conditions, which aided the characterization of the dynamic aspects of the molecular conformation.

In this way it was found (Ivanov et al., 1971a; Ovchinnikov and Ivanov, 1975) that valinomycin undergoes a variety of conformational changes depending on the environmental conditions. In nonpolar solvents it assumes a compact "braceletlike" structure (Fig. 4a) stabilized by six intramolecular hydrogen bonds between the amide CO and NH groups, with the ester carbonyls pointing away from the symmetry axis and all (twelve) alkyl radicals on the outer boundary of the molecule. In general, the crystalline and solution conformations of valinomycin are quite similar, although in the former two hydrogen bonds are formed by ester carbonyls (Duax et al., 1972; Smith et al., 1975; Karle, 1975). The arrows in Fig. 5 show the atomic displacements that accompany the transition from the crystalline to the "bracelet" conformation. Curiously, the so-called meso-valinomycin, a close analogue of valinomycin, cyclo[-(DVal-LHyIv-LVal-DHyIv)$_3$-], differing from the parent compound in having α-hydroxyisovaleryl isopropyl instead of lactyl methyl side chains, forms a pseudo C$_3$-symmetric and true centrosymmetric bracelet structure in the crystal, which is shown in Fig. 6 (Pletnev et al., 1977).

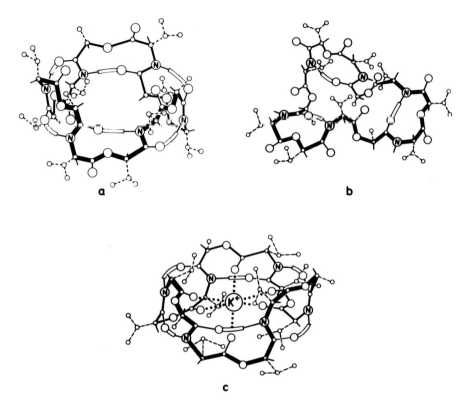

Fig. 4 Conformation of valinomycin in nonpolar solvents (a), in solvents of medium polarity (b), and in the K$^+$ complex (c).

In solvents of medium polarity the more planar "propeller" conformation (form B with three hydrogen bonds) is predominant. In its center is a hydrophobic nucleus consisting of three D-valyl isopropyl groups and three lactyl methyl groups; this is surrounded by the depsipeptide chain with its polar amide and ester groups. Apparently valinomycin is in form A in the inner hydrophobic regions of the membrane, whereas at the membrane–water interface it is in form B.

In polar (including aqueous) media, valinomycin B exists as an equilibrium mixture of a large number of conformers devoid of intramolecular hydrogen bonds.

In its complex with K$^+$, valinomycin assumes in all solvents and in the crystal (Neupert-Laves and Dobler, 1975) a rigid conformation with the six hydrogen bonds, as in structure A, but the ester carbonyls are now pointing inward, forming six ion–dipole bonds with the unsolvated cation in the center of the molecular cavity (Fig. 4c). The cation is effec-

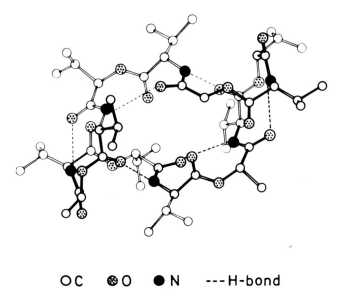

OC ⊗O ●N ---H-bond

Fig. 5 Conformation of crystalline valinomycin.

tively screened from interacting with the solvent and the anion by the depsipeptide skeleton and the pendent side chains. The exterior of the complex consists mainly of isopropyl and methyl groups, the source of its hydrophobicity. The diameter of the internal cavity (2.7–2.9 Å) corresponds to that of potassium ($r = 1.33$ Å) and rubidium ($r = 1.49$ Å) ions, but steric hindrances occur with a Cs^+ ion ($r = 1.65$ Å) giving rise to weaker complexes (Table I).

The outstandingly high K/Na-complexing selectivity of valinomycin is ascribed to the rigidity of the bracelet conformation, the molecular cavity being incapable of contracting sufficiently to optimize the ion–dipole interaction with a Na^+ ion and thereby being incapable of compensating for the larger desolvation energy of this ion relative to K^+.

If details are disregarded, a certain analogy between the action of valinomycin and that of enzymes can be discerned. Indeed, both enzymes and valinomycin manifest activity at very low concentrations; for example, valinomycin increases the K^+ conductivity of lipid bilayers at a concentration of 10^{-9} M. Both enzymes and valinomycin are regenerated after each act, so they can repeat the process. The function of enzymes is to catalyze a chemical reaction; that is, enzymes lower the energy barrier. Valinomycin lowers the energy barrier for the K^+ ion so that it can cross easily from one side of the membrane to the other. Enzymatic action is selective: Enzymes interact only with their characteristic sub-

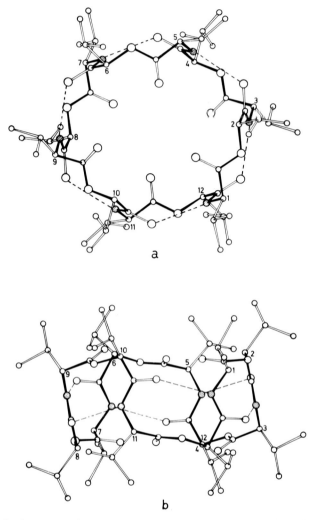

Fig. 6 Conformation of crystalline mesovalinomycin. (a) View along the C_3 axis; (b) side view. Shaded atoms are nitrogen.

strates. Valinomycin interacts selectively with K^+ ions. Finally, the characteristic ability of enzymes to adapt to the size and shape of the substrates is well known. As we have seen, valinomycin shapes its active center, the molecular cavity, to conform to the size and properties of the substrate, that is, the K^+ ion. Thus, valinomycin and other ionophores might be considered to be permeases catalyzing transmembrane potassium ion transport.

The investigation of valinomycin has not only resulted in an explanation of its K/Na efficiency, but has also led to the directed synthesis of analogues with properties lacking in the parent ionophore. Several dozen valinomycin analogues have been synthesized that differ in either the nature of the side chains, the configuration of the residues, the nature of the ligands, or the size of the ring (or in several of these characteristics). Of these analogues, cyclo[-(DVal-LLac-LVal-DHyIv)$_4$-], comprising four instead of three valinomycin monomeric fragments, binds such bulky cations as trimethyl ammonium and protonated acetylcholine (Eisenman et al., 1976) and transports them across membranes. A number of analogues also exist for which the K$^+$ complexes are more stable than the complex of valinomycin (Ovchinnikov et al., 1974a).

It has been shown that under certain conditions (e.g., excess ionophore at high absolute concentration) valinomycin forms 2:1 macrocycle:cation "sandwich" complexes as well as the usual 1:1 complex (Ivanov et al., 1975; Ivanov, 1975). For such rather unstable complexes, a hypothetical structure (Fig. 7) has been proposed in which the ion interacts with six ester carbonyls—three from each valinomycin molecule. The conformation of the valinomycin molecule is close to that of the equimolar complexes. Much more stable "sandwiches" are formed by the valinomycin analogues, cyclo[-(DVal-LPro-LVal-DHyIv)$_3$-] and cyclo-[-(DVal-LLac-LVal-DPro)$_3$-], the hydroxy acid residues (Lac or HyIv) of which are substituted by proline residues of the same configuration. In this case the ligands are most likely the amide carbonyls of the Val-Pro bond (Fig. 8) (Fonina et al., 1976).

The results obtained pose the question as to whether sandwich complexes might not take part in the transmembrane transport of K$^+$ ions. Indeed, K$^+$ ion transition from the surface complex to the ionophore in the membrane interior (see Fig. 3) can be treated in terms of preliminary "sandwiching." Such a mode of ion transport agrees with Grell's pro-

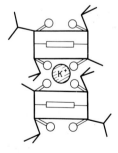

Fig. 7 Hypothetical sandwich-type complex of valinomycin.

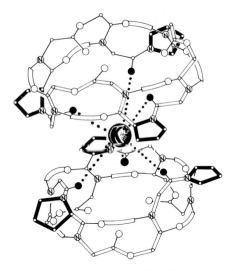

Fig. 8 Proposed conformation for the cyclo[-(DVal-LPro-LVal-DHyIv)₃-]₂·K⁺ complex.

Fig. 9 Valinomycin "floating" in a dimyristoyl lecithin bilayer.

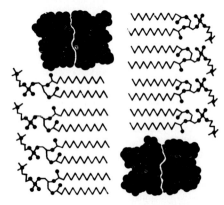

Fig. 10 Valinomycin "incorporated" into the dimyristoyl lecithin bilayer.

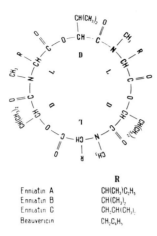

	R
Enniatin A	CH(CH₃)C₂H₅
Enniatin B	CH(CH₃)₂
Enniatin C	CH₂CH(CH₃)₂
Beauvericin	CH₂C₆H₅

Fig. 11   Enniatin antibiotics.

posal (Grell *et al.*, 1975) that valinomycin could hardly simply float about as such in the membrane (Fig. 9), but rather forms dimeric or larger clusters (Fig. 10), a circumstance that would facilitate the formation of double complexes.

The enniatin ring (Fig. 11) is half the size of the valinomycin ring and, moreover, enniatin lacks the NH groups that through hydrogen binding endow rigidity to the valinomycin complexes. Enniatin is inferior to valinomycin in K^+ complex stability, K/Na selectivity, and membrane-affecting efficiency (the ionophore concentration must be at least ~10^{-6} *M* to affect the membrane permeability). However, enniatins are interesting as wide-range complexones capable of binding and transporting ions of diverse sizes and valencies (Table II). The equimolar enniatin complexes (Fig. 12a) resemble a disk with a lipophilic brim and a central cation. As in the valinomycin complexes, the cation is held in the molecular cavity by ion–dipole interactions with six carbonyl ligands. The absence of hydrogen bonds enables enniatins to adapt their cavity

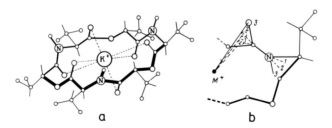

Fig. 12   (a) Conformation of the (enniatin B) · K^+ complex. (b) Schematic presentation of the conformational rearrangement accompanying the transition from the Li^+ complex (1) to complexes with K^+ (2) and Cs^+ (3).

TABLE II

Stability Constants of Equimolar Complexes of Enniatin Antibiotics with Various Cations at 25°C (liters mole⁻¹)

Compound and solvent	Li^+	Na^+	K^+	Rb^+	Cs^+	NH_4^+	Tl^+	Mg^{2+}	Ca^{2+}	Sr^{2+}	Ba^{2+}	Mn^{2+}	References
Enniatin A													
Methanol	—	—	1200	—	—	—	—	—	—	—	—	—	Wipf et al. (1968)
Ethanol	—	2900	9800	—	—	—	—	—	—	—	—	—	Shemyakin et al. (1967)
96% aqueous ethanol	100	550	3200	2400	1280	—	—	—	—	—	—	—	Ovchinnikov et al. (1974b)
Enniatin B													
Methanol	19 / 30[a]	260 / 240[b]	830 / 840[b]	550	220	83	520	16	900	450	850	4	Grell et al. (1972)
Methanol:water (9:1)	—	—	280	—	—	—	—	—	—	—	—	—	Grell et al. (1975)
Methanol:water (7:3)	—	—	44	—	—	—	—	—	—	—	—	—	Grell et al. (1975)
Methanol:water (1:1)	—	—	14.5	—	—	—	—	—	—	—	—	—	Grell et al. (1975)

											Reference
Ethanol	—	1300	3700	4000	2200	—	—	—	—	—	—	Andreev et al. (1971); Shemyakin et al. (1967, 1969)
96% aqueous ethanol	50	340	2100	3000	740	—	330	13	120	250	2600	Ovchinnikov et al. (1974b)
Enniatin C												
Ethanol	100	2900 2500[b]	4900 5500[b]	7500	—	—	—	—	—	—	—	Ovchinnikov et al. (1974b)
96%aqueous ethanol	<50	240	1500	1270	1400 4200[b]	—	—	—	—	—	—	Ovchinnikov et al. (1974b)
Beauvericin												
Ethanol	100	300	3100	3500	3500	—	—	—	—	—	—	Ovchinnikov et al. (1974b)

[a] Shemyakin et al. (1967).
[b] Wipf et al. (1968).

93

94   YU. A. OVCHINNIKOV

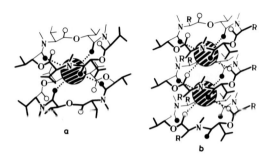

Fig. 13 Conformation of the sandwich complex of enniatin B (a) and of the beauveri-cin complex with a 3:2 macrocycle:cation ratio (R = CH₂C₆H₅) (b).

size to the complexing ion (Fig. 12b), which explains the low complex-ing selectivity.

If the occurrence in solution and particularly in the transmembrane transport process of other than 1:1 macrocycle:cation valinomycin complexes is as yet in the realm of conjecture, for enniatins the forma-tion of such complexes in solution, in two-phase systems, and in bilayers can be considered to rest on a fully sound basis (Ivanov, 1975; Ovchinnikov *et al.*, 1974b; Ivanov *et al.*, 1973a). In this case not only 2:1 "sandwiches" (Fig. 13a) but also 3:2 "stacks" have been found (Fig. 13b).

While such complexes are less stable than the equimolar complexes, they provide a better screening of the cation from the solvent and anion. The macrocycle:cation ratio has a tendency to increase on passing from Na⁺ to K⁺ and increases further on passing to Cs⁺. Accordingly, the enniatin-mediated transmembrane transport of Na⁺, K⁺, and Cs⁺ pro-ceeds largely by way of 1:1, 2:1, and 3:2 complexes, respectively (Fig. 14).

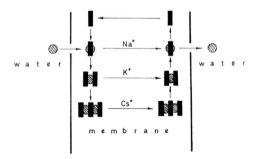

Fig. 14 Alkali ion transport model through lipid bilayers in the presence of enniatin ionophores.

Interestingly, the bis derivatives of enniatin B, where two depsipeptide cycles are covalently bound through a long and sufficiently flexible "bridge" (Fig. 15a), form extremely stable "intramolecular sandwiches" in solution [stability constant of $(1-2) \times 10^4$ M^{-1} in 96% aqueous alcohol, R = H] (Fig. 15b) (Sumskaya et al., 1977), and it is in this form that potassium is transported across artificial membrane systems by this complexone (Ivanov et al., 1973a; Sumskaya et al., 1977).

With sufficiently large valinomycin clusters or enniatin "stacks" one can picture a "channel" mechanism to explain the transport phenomenon. In this case the ionophore stacks span the membrane and the ion can migrate along the electrical potential or concentration gradient across the membrane, passing from one ionophore molecule to another without significantly displacing them. In fact, as we have seen, the mechanism does not assume such an ultimate form with the depsipeptide antibiotics; their mode of action, depicted in Figs. 3 and 14, is intermediate between the individual (single) carrier model, shown in Fig. 1, and the channel mechanism. Of the ionophores investigated the nactin macrotetrolides (Szabo et al., 1969; Hladky, 1975; Benz and Stark, 1975) obey most clearly the former mechanism and the gramicidin group of peptide antibiotics obeys the latter mechanism (see Ovchinnikov et al., 1974a, and references therein as well as Neher, 1975).

Gramicidin A, as follows from the formula, is a 15-membered linear peptide built up of hydrophobic amino acid residues with strictly alternating asymmetric center configurations (with the exception of optically inactive Gly[2]). Gramicidins B and C differ only in the nature of the aromatic residue at position 11. The gramicidins induce ion conductivity of alkali metal ions, thallium ions, and protons in artificial and biological membranes at extremely low concentrations (10^{-10} M and lower) (see Ovchinnikov et al., 1974a, Chapter 6). At the same time they show no definite signs of complexing the metal ions in solution and manifest only a very weak ability to extract potassium and barium pic-

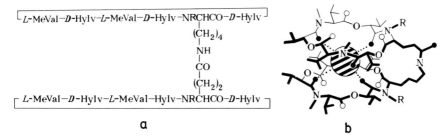

L-MeVal—D-Hylv—L-MeVal—D-Hylv—NRCHCO—D-Hylv
(CH$_2$)$_4$
NH
CO
(CH$_2$)$_2$
L-MeVal—D-Hylv—L-MeVal—Hylv—NRCHCO—D-Hylv

a

b

Fig. 15 Primary structure of bisenniatins B (a) and conformation of their K$^+$ complexes (b). R = H or CH$_3$.

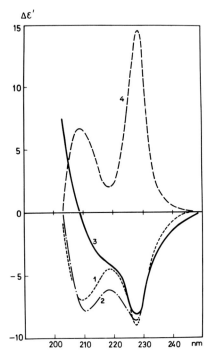

Fig. 16 CD curves of the individual gramicidin A dimeric forms in dioxane (Veatch *et al.*, 1974). The numbers on the curves designate the corresponding species.

rates from water to methylene chloride (Byrn, 1974). We also know that gramicidin forms at least four quite stable slowly interconverting dimers in solution (species 1, 2, 3, and 4) with distinct spectral parameters (Fig. 16) (Veatch *et al.*, 1974; Veatch and Blout, 1974). These dimeric forms are responsible for the induced ion conductivity (Veatch *et al.*, 1975).

Two elegant hypotheses have been proposed to explain the accumulated facts. Urry (1971) has suggested that two gramicidin molecules, in a novel, hitherto undescribed helical conformation (the π_{LD}-helix), associate "head to head" by forming four to eight intermolecular hydrogen bonds via the CO and NH groups (see also Urry *et al.*, 1971). Such a dimer with six interhelical H bonds is shown in Fig. 17a.

According to the model of Veatch and Blout (1974), gramicidins form double helices in which all hydrogen bonds are intermolecular, whereas the chain directions can be parallel or antiparallel. Within each of these groups the helices may differ in the number of residues (six or seven) per turn and in the direction of rotation (right- or left-hand helices). An example of such a double helix is shown in Fig. 18a. Aside from the

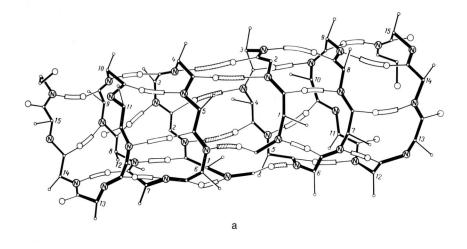

a

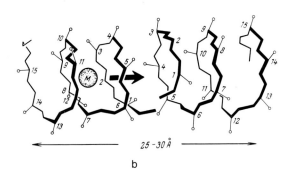

25 – 30 Å

b

Fig. 17 Schematic representation of the π^6_{LD}-helix of gramicidin A in solution (a) and functioning in the membrane as a pore (b). The side chains are omitted for clarity.

significant differences between these two types of structures, they possess a number of common features: (a) The lengths range from 30 to 50 Å (i.e., the size is commensurate with the thickness of the hydrophobic zones of the lipid bilayer); (b) the periphery of the helices comprises the hydrophobic side chains of the amino acids, which are capable of readily penetrating the lipid zones of a membrane; (c) all the helices possess an axial cavity 1.4–6.0 Å in diameter, which can thus accommodate unsolvated metal ions [from the smallest (Li$^+$, 1.4 Å) up to the largest (Cs$^+$, 3.4 Å)]; (d) all amide carbonyls are situated on the surface of the molecular cavity, thereby facilitating the entry of metal ions into the cavity. Thus, whatever the concrete structure of the dimer, the mode of

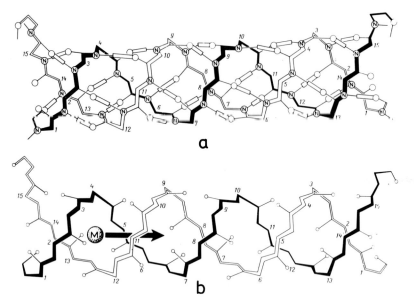

Fig. 18 Right-handed antiparallel $\pi\pi^6$-helix of gramicidin A in solution (a) and functioning in the membrane as a pore (b). The side chains are omitted for clarity.

action of gramicidins apparently reduces to the formation of "tubes" leading from the solution on one side of the membrane to the solution on the other side, and these tubes are capable of passing metal ions (Figs. 17b and 18b). Movement of the cation along the helical axis can be accompanied by slight displacements of the O atoms with the formation of "instantaneous" complexes. Further study is necessary to elucidate the details of this process and to determine the solution conformations of gramicidin A, its structure in the membrane systems, and the change induced by interactions with metal ions.

As a contribution toward filling this gap, we have synthesized a number of gramicidin A analogues (see Scheme 1) (Shepel *et al.*, 1976) that differ from the parent antibiotic in chain length (compounds II–VIII, XI, and XII), in the L and D sequence of the amino acid residues (compounds X and XI), and in some of the residues themselves (compound IX). The analogues were then investigated with respect to their associative capacity by determining the concentration dependence of the circular dichroism (CD) curves and the fluorescence polarization over the maximum possible concentration range (10^{-2}–10^{-5} M) in ethanol and dioxane (Table III).

The concentration dependence of the CD spectra of some of the analogues is shown in Figs. 19 and 20, for which the spectra were

$$
\begin{array}{cccccc}
& \text{D} & \text{D} & \text{D} & \text{D} & \text{D} & \text{D}
\end{array}
$$

I   HCO-Val-Gly-Ala-Leu-Ala-Val-Val-Val-Trp-Leu-Trp-Leu-Trp-Leu-Trp-NH(CH$_2$)$_2$OH

G r a m i c i d i n A

II   HCO-Val-Gly-Ala-Leu-Ala-Val-(- -)-Trp-Leu-Trp-Leu-Trp-Leu-Trp-NH(CH$_2$)$_2$OH

III   HCO-Val-Gly-Ala-Leu-Ala-Val-Val-Val-(- -)-Trp-Leu-Trp-Leu-Trp-NH(CH$_2$)$_2$OH

IV   HCO-Val-Gly-(- - - - -)-Val-Trp-Leu-Trp-Leu-Trp-Leu-Trp-NH(CH$_2$)$_2$OH

V   HCO-Val-Gly-Ala-Leu-Ala-(- - - - -)-Leu-Trp-Leu-Trp-Leu-Trp-NH(CH$_2$)$_2$OH

VI   HCO-Val-Gly-Ala-Leu-Ala-Val-Val-(- - - - -)-Trp-Leu-Trp-Leu-Trp-NH(CH$_2$)$_2$OH

VII   HCO-Val-Gly-(- - - - - - -)-Trp-Leu-Trp-Leu-Trp-Leu-Trp-NH(CH$_2$)$_2$OH

VIII   HCO-Val-Gly-Ala-Leu-Ala-(- - - - - - -)-Leu-Trp-Leu-Trp-Trp-NH(CH$_2$)$_2$OH

IX   HCO-Val-Gly-Val-Leu-Val-Leu-Val-Leu-Trp-Leu-Trp-Leu-Trp-Leu-Trp-NH(CH$_2$)$_2$OH

X   HCO-Val-Gly-Ala-Leu-Ala-Val-Val-Val-Gly-Trp-Leu-Trp-Leu-Trp-Leu-Trp-NH(CH$_2$)$_2$OH

Gly

XI   HCO-Val-Gly-Ala-Leu-Ala-Val-Val-Val Trp-Leu-Trp-Leu-Trp-Leu-Trp-NH(CH$_2$)$_2$OH

XII   HCO-Leu-Val-Gly-Ala-Leu-Ala-Val-Val-Val-Trp-Leu-Trp-Leu-Trp-Leu-Trp-NH(CH$_2$)$_2$OH

Scheme 1

TABLE III

Stability Constants of Gramicidin A Analogue Dimers

Number	Solvent	Dimerization constant, K (M^{-1})	
		Circular dichroism	Fluorescence polarization
III[a]	Ethanol	$(3.0 \pm 0.4) \times 10^2$	$(3.5 \pm 1) \times 10^2$
	Dioxane	—	$(2.0 \pm 0.5) \times 10^3$
VI[a]	Ethanol	$(1.4 \pm 0.4) \times 10^2$	$(2 \pm 4) \times 10^2$
	Dioxane	$(9.6 \pm 0.5) \times 10^4$	$\sim 10^5$
IV[b]	Ethanol	$(3 \pm 1) \times 10^2$	$\sim 10^3$
	Dioxane	$\sim 10^4$	$\sim 10^4$
VII[b]	Ethanol	$(1 \pm 0.5) \times 10^2$	$\sim 10^2$
	Dioxane	$(3.5 \pm 2) \times 10^3$	$(2.6 \pm 2) \times 10^3$
IX[b]	Ethanol	$(3 \pm 2) \times 10^4$	$(9 \pm 4) \times 10^3$
	Dioxane	$\sim 10^5$	$(4.0 \pm 1.5) \times 10^4$
II[c]	Ethanol	$(2.5 \pm 0.5) \times 10^2$	$(3 \pm 1) \times 10^2$
	Dioxane	—	$(5.5 \pm 0.5) \times 10^2$
XII[c]	Ethanol	$(3.2 \pm 0.7) \times 10^2$	$(3.8 \pm 0.5) \times 10^2$
	Dioxane	—	$(2 \pm 1) \times 10^2$
V[c]	Ethanol	$< 10^2$	$\leq 10^2$
	Dioxane	$\sim 10^3$	$\sim 3 \times 10^3$
VIII[c]	Ethanol	$\leq 10^2$	$\leq 10^2$
	Dioxane	—	—
XI[c]	Ethanol	—	$< 10^3$
	Dioxane	—	—

[a] Species 3.

[b] Species 4.

[c] See the text.

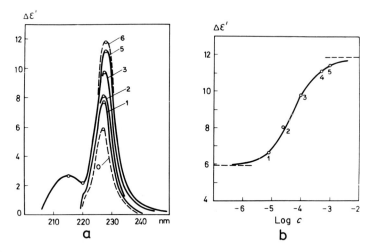

Fig. 19 Concentration dependence of (a) the CD curves and (b) $\Delta\epsilon'$ for $\lambda = 228$ nm of analogue VI in dioxane. The dashed lines are for the monomer (0) and the dimer (6).

obtained under equilibrium conditions. These figures also show a plot of the differential dichroic absorption at $\lambda = 228$ nm versus $\log c$, where c is the overall concentration of the antibiotic.

The CD parameters provide a clue to both the monomer and the dimer conformations. Generally speaking, a clear-cut relationship between the

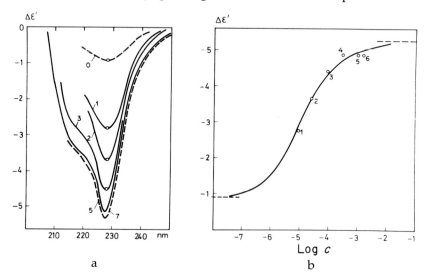

Fig. 20 Concentration dependence of (a) the CD curves and (b) $\Delta\epsilon'$ at $\lambda = 228$ nm for analogue IV in dioxane. The dashed lines are for the monomer (0) and the dimer (7).

CD curves and the chain length or the nature of the displaced residue can be observed only for some, not all, the analogues. A comparison of the CD spectra of the analogue dimers with those of the individual gramicidin A dimeric species showed that in certain analogues the conformational equilibrium is shifted toward one or the other of these species. In this respect the analogues can be divided into four groups (Table III). In one group, consisting of analogues III and VI, the CD spectra are close to that of species 3 (cf. Figs. 16 and 19). Of particular interest here is analogue VI since its CD spectrum at total dimerization concentrations in dioxane can be described in terms of a combination of all four individual species, with species 3 amounting to 65% of the total. A second group, the CD spectra of which closely resemble that of gramicidin A species 4, comprises analogues IV, VII, and IX, the most interesting being analogue IX, which possesses the species 4 conformation to the extent of about 80%. Analogues II and XII comprise a third group, with CD spectra resembling those of species 1 and 2 but with about half their amplitude. These spectra cannot be described by a linear combination of the spectra of the four gramicidin A species. The last group contains analogues V, VIII, and XI, the CD amplitudes of which are less than 0.2 of those characteristic of the gramicidin A dimers.

A comparison of the equilibrium shift in the analogue dimers with the structure of the monomers uncovers a number of interesting facts. A decrease in the relative Trp content is paralleled by augmentation of the species 3 content. In the group in which species 4 conformations are predominant, the Trp content is higher than those for the other analogues. Moreover, in analogue IX, Ala has been replaced by Val and Val by Leu; that is, the less bulky hydrophobic residues have been replaced by more bulky ones. This all demonstrates that an essential part in stabilization of species 4 is played not only by H bonds, but also by hydrophobic interactions.

In the light of the data presented here, let us now consider the effect of gramicidin A and its synthesized analogues on the ion conductance of lipid bilayers. It is well known that in the presence of very small amounts of gramicidin A the membrane current at constant voltage undergoes jumps, indicating the "opening" and "closing" of the conductance channels. At the same time, the channels have definite conductance values, some of which are realized more often than others. We found that, in egg lecithin membranes, gramicidin A forms four basic types of channels that differ in both conductance and dwell times (Fig. 21). It therefore seems logical to relate these types with the four dimer species discussed above. Substantiation for this may be found in the

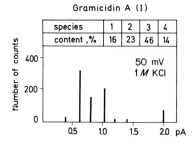

Gramicidin A (I)

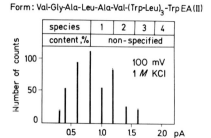

Form: Val-Gly-Ala-Leu-Ala-Val-(Trp-Leu)₃-Trp EA (II)

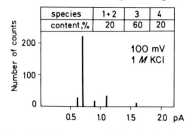

Form: Val-Gly-Ala-Leu-Ala-(Val)₃-(Trp-Leu)₂-Trp EA (III)

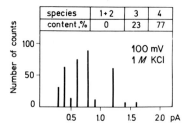

Form: Val-Gly-(Val-Leu)₃-(Trp-Leu)₃-Trp EA (IX)

Fig. 21 Histograms of the single-channel conductances induced by gramicidin A (I) and its analogues II, III, and IX. EA, ethanol amide.

fact that with the thirteen-membered analogue III, which in solution exists largely in the form of a single type of dimer, we have observed mainly a single type of channel with a conductance of 7 pmho. At the same time, with another thirteen-membered analogue (II), whose dimeric forms in solution differ from those of the gramicidin A species, the histogram characterizing the conductance distribution of the channel displays a wide range of values. The fifteen-membered analogue IX was also found to be active in bilayers, but its individual channels are characterized by a much larger range of conductances than might have been expected since its solution contains predominantly a single dimer species. While there are several different conducting types of this analogue in the bilayer, the differences between them are much less than in the case of the naturally occurring gramicidin A species, and we are inclined to ascribe them to small conformational differences that could have escaped detection in solution.

Other indirect support of the proposal advanced here may be seen in the fact that the CD curves of analogue XI, which displays no conductance channels in bilayer membranes, are of low intensity, evidence of its lack of preferable conformations in solution. The fact that in this analogue the LD alternation of the asymmetric centers is violated shows that such ordering must be required for the formation of the gramicidin A type of channel.

Thus, the correlation that apparently exists between the number of different solution dimers and the number of different types of conductance channels serves as an argument in favor of the concept that the antibiotic's conductance channels are formed by its dimers; this correlation also allows one to penetrate deeper into the conductance mechanism. The dimer species does not seem to be crucial, but this requires further study.

Antamanide (Fig. 22) is a cyclodecapeptide isolated from the extracts of the poisonous mushroom *Amanita phalloides*. It is an inhibitor of phallotoxins, strong poisons produced by the mushroom (Wieland *et al.*, 1968). Phallotoxins cause irreversible pathological changes in mammalian liver (Wieland, 1972; Wieland and Govindan, 1974; Frimmer *et al.*, 1974). Wieland *et al.* (1972a; Faulstich *et al.*, 1974) ascribe the antitoxic effect of antamanide to its specific ability to "tighten" the cellular membranes of the liver, thereby lowering their permeability to the toxin molecules.

In an independent series of experiments it has been shown that antamanide and several of its synthetic analogues are capable of binding alkali and alkaline earth metal ions in solution, the most stable complexes involving the smaller ions (Na^+ and Ca^{2+}) (Table IV). A comparison of the complexing and antitoxic properties of antamanide and its analogues led to the conclusion that the ability to complex Na^+ (or Ca^{2+}) ions is a necessary but insufficient condition for the manifestation of the antitoxic action. The reason for this is still unclear.

Considerable effort was spent in elucidating the spatial structure of antamanide and its complexes in the crystalline state (Karle *et al.*, 1973;

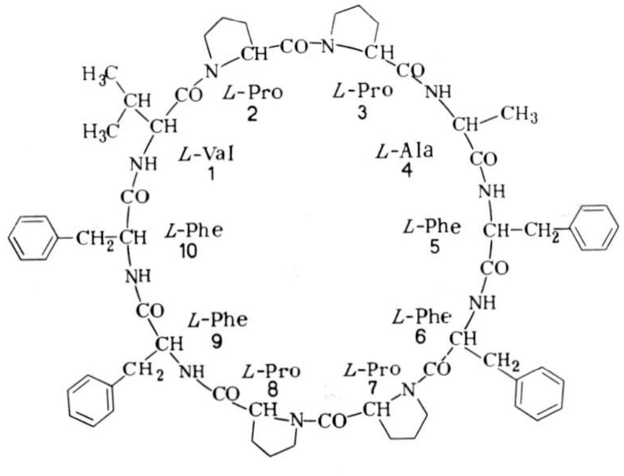

Fig. 22   Structure of antamanide.

TABLE IV

Stability Constants of Equimolar Antamanide Complexes with
Various Cations (liters mole^{-1})

Solvent	Li$^+$	Na$^+$	K$^+$	Tl$^+$	Ca^{2+}	References
Ethanol	—	2,800	270	—	—	Andreev et al. (1971); Miroshnikov et al. (1973); Wieland et al. (1970)
Ethanol : water (96:4)	—	2,000 1,300	180	—	—	Wieland et al. (1972c) Miroshnikov et al. (1973)
Methanol	10	500	10	190	30	Wieland et al. (1972c)
Acetonitrile	—	30,000	290	—	100,000	Wieland et al. (1972c)
Acetonitrile : water (96:4)	1,300	2,600	20	—	—	Wieland et al. (1972c)
Acetonitrile : water (92:8)	—	1,200	280	—	—	Wieland et al. (1972c)

Karle, 1974a,b) and in solution (Ivanov et al., 1971b, 1973b; Patel, 1973, 1974). It has been established that, under all conditions, equimolar complexes of antamanide with Li$^+$, Na$^+$, K$^+$, and Ca^{2+} assume a common type of saddlelike conformation, the cation interacting with four amide carbonyls situated approximately in the apices of a square. An example of this is illustrated in Fig. 23 using the structure of crystalline Li$^+$-antamanide. The reasons for the Na/K selectivity of antamanide are, evidently, similar to those of the K/Cs selectivity of valinomycin (see above). The accommodation of the more bulky cation is associated with sterically disadvantageous conformational changes. Besides, it is quite possible that the selectivity is also aided by the greater tendency of amide groups than of ester groups to interact with small ions (Eisenman and Krasne, 1973).

A distinguishing feature of the equimolar complexes of antamanide in comparison with those of enniatin or valinomycin is that in the

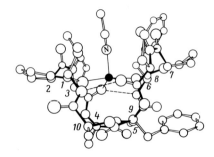

Fig. 23 Conformation of the antamanide·Li$^+$ complex (Shepel et al., 1976). Side chains are partially omitted. The dashed lines indicate the CO····HN hydrogen bonds.

former only part of the ion solvate sheath is supplanted by ligands of the macrocycle. For instance, Li^+ bound to antamanide retains a solvate molecule of acetonitrile (Fig. 23), and Na^+ complexed with [Phe^4, Val^6]-antamanide is in contact with a molecule of ethanol (Karle et al., 1973; Karle, 1974a,b).

The relative accessibility of the cation to the environment in antamanide complexes is also reflected in their ability to form ion pairs with the counterion (picrate) and double complexes, a possible structure of which is shown in Fig. 24 (Ivanov et al., 1975; Ivanov, 1975).

From the above data, it can be assumed that the antitoxic action of antamanide stems from its interaction with Na^+ or Ca^{2+} ions sorbed on the membrane surface by protein or lipid components (Fig. 25, position a or b, respectively). As a result, the bound antamanide covers quite considerable areas of the latter (250–300 Å^2 according to molecular models), thereby modifying its properties, including its permeability to the A. phalloides toxins. Since the antitoxic action of antamanide is stereospecific [enantiomeric antamanide is less active by an order of magnitude than the natural antitoxin (Wieland et al., 1972b)], the protein components of the membrane apparently also participate in the binding process. The most probable type of interaction is stacking of the aromatic groups of the protein with the phenyl groups of antamanide, inasmuch as substitution of the latter by cyclohexyls, while practically without effect on the complex stability, causes complete loss of the antitoxic potency (Wieland et al., 1972c). It is interesting that the perhydroantamanide formed by such a substitution is now endowed with well-defined ionophore properties, which are lacking for antamanide itself (Ovchinnikov et al., 1972).

These results show that it is sometimes very easy to pass over the threshold dividing two such different classes of compounds as ionophores and antitoxins. It stands to reason that the principles of the selective interaction of peptide–protein systems with alkali and alkaline earth metals are not confined only to such relatively simple examples.

Fig. 24   Proposed structure for the antamanide sandwich complexes.

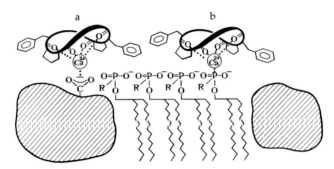

Fig. 25 Model of antamanide interaction with a biomembrane (shaded areas are protein globules).

Their significance lies in the fact that they bring to light the action of these principles in greater detail and more fundamentally than do the other presently known biological systems.

The protein molecules mentioned in the beginning of this chapter, like the peptide ionophores, participate in transmembrane transport; however, in contrast to the former, which only affect the membrane permeability, the protein molecules are actually responsible for the membrane permeability *per se*, that is, for processes that occur naturally in the membrane.

Bacterial rhodopsin, a protein of the purple membrane of *Halobacterium halobium*, participates in the process of light-dependent proton transfer across the membrane. To elucidate the molecular mechanism of action of rhodopsin as a proton pump, it is necessary to know the mutual spatial orientation of its component parts within the membrane and their properties. For this we have undertaken a study of the action of various proteases on purple membranes [see Abdulaev *et al.* (1976) for the first report in this series]. Trypsin and chymotrypsin exert no apparent action on this protein, as can be seen from the results of N-terminal and C-terminal analysis and of electrophoresis in polyacrylamide gels in the presence of sodium dodecyl sulfate.

Carboxypeptidase A splits off one threonine, one serine, and two alanine residues. Kinetic analysis and Akabori hydrazinolysis yielded the following C-terminal sequence of the protein: Ala-Ala-Thr-Ser.

Brief digestion of the membranes with papain at a low enzyme : substrate ratio leads to cleavage of the final 17 amino acids of the C terminus (Asp, 1; Thr, 1; Ser, 2; Pro, 2; Glu, 4; Ala, 5; Gly, 2); consequently, this C-terminal fragment is acidic. After this action of papain at a low concentration, the remainder of the protein has Gly at the C terminus and an apparent MW of 20,500, whereas the molecular weight of the original protein was 22,500.

At a higher papain concentration and with a longer reaction time, cleavage in the vicinity of the N-terminal region and of 17 C-terminal amino acids is observed. As a result, two more fragments of MW 16,000 and 12,000 with Gly and Ile as the N-terminal amino acids are formed, since kinetic analysis has shown that bacteriorhodopsin is initially converted into the 20,500 fragment and that this yields the MW 16,000 fragment; the MW 12,000 fragment evidently results from the further degradation of the MW 16,000 fragment. Analysis showed that each of these three rhodopsin fragments has the same C-terminal amino acid, namely, Gly. It is interesting that under the same conditions, but in the presence of 2.3 M NaCl, papain attacks only the C-terminal part of bacteriorhodopsin.

Apparently, papain attacks the protein at the water-exposed site connecting the hydrophobic α-helical stretches discovered by Henderson and Unwin (1975), whereas the hydrophilic, acidic C-terminal region is localized in the aqueous phase and is therefore accessible to the action of various proteases. Indeed, under the experimental conditions, thermolysine also acts on the C terminus. Treatment with this enzyme leaves a fragment of MW 21,500 in the membrane. The dansyl method for determining the N-terminal amino acid is invalid for this fragment and for rhodopsin itself, indicating that both the native protein and this remaining fragment do not contain a free amino acid group. The papain, thermolysine, and carboxypeptidase A cleavage of the C-terminal region of bacterial rhodopsin does not cause any change in its spectral parameters, whereas cleavage of the N-terminal fragment shifts its long-wave adsorption peak from 560 to 600 nm.

The modified membranes (i.e., the preparations containing the polypeptides of MW 21,500, 20,500, and 16,000 + 12,000) are all capable of participating in the photochemical cycle, functioning as a proton pump (Barsky et al., 1975).

We hope to determine the minimal fragment that retains the functional properties of bacteriorhodopsin. Following this, we shall embark on the same type of physicochemical investigations that have been so fruitful in the study of peptide antibiotic ionophores.

As stated in the introduction, a number of proteins that specifically bind low molecular weight compounds, such as amino acids, sugars, coenzymes, and inorganic ions, have been isolated from the periplasmatic space of bacterial cells. The interest in these proteins is due to just this property, which serves as grounds for the conjecture that they are involved in active transport of metabolites across the cellular membrane, at least in the first, receptor stage of the process (Oxender, 1972, 1975).

It has been shown in a number of laboratories that, in *Escherichia coli* cells, branched chain amino acid transport is effected by a multiple system. By the method of cold osmotic shock, proteins have been isolated, one of which, the LIV protein, complexes Leu, Ile, and Val, and two others specifically bind only Leu (LS protein) and Ile (I protein) (Piperno and Oxender, 1966; Anraku, 1968; Furlong and Weiner, 1970; Furlong *et al.*, 1973). The selective binding mechanism of these proteins is as yet unknown, as is their mode of action within the cell. Therefore, of considerable interest is the determination of the primary structures of the binding proteins and a comparison of the amino acid sequences of members of a single group that display different substrate specificities.

The complete amino acid sequence of LIV protein, the first representative of the periplasmatic binding proteins, has been established (Ovchinnikov *et al.*, 1972a, b). The approach used in this study was to split the protein main chain into large fragments at the Met and Arg residues and at the aspartyl–propyl bonds and to determine the sequence of these fragments by degradation on an automated sequencer.

LIV protein (MW, 36,700) comprises 344 amino acid residues with one disulfide bridge (Fig. 26, residues 57–78). Interestingly, all the Arg residues of this protein are localized in its central region and the Trp residues occur in its C-terminal region.

The amino acid composition of LS protein (MW, 37,000) closely resembles that of LIV protein (the only substantial difference being nine instead of five Met residues). In the N-terminal sequences of both proteins as determined by automated Edman degradation (Fig. 27), the first 14 amino acids, with the exception of the terminus itself, are identical, whereas among the following 26 amino acids there are differences in five positions. The cysteine-containing tryptic peptides of both proteins are structurally quite similar, and the cysteine residues seem to occupy the same positions.

A preliminary analysis of the fragments obtained in the cyanogen bromide cleavage of LS protein shows that at least four Met residues occupy the same positions as in LIV protein, and the fragments have partial N-terminal sequences similar to those of the corresponding LIV-protein fragments. Thus, the data obtained indicate a considerable structural similarity between the LIV and LS proteins.

To locate functionally important groups in the LIV-protein molecule, the effect of the chemical modification of some of the amino acid residues on ligand binding was studied. It was found that the binding of leucine was accompanied by some changes in the CD of the aromatic long-wave spectral region of LIV protein. From these data we assumed that the aromatic amino acid residues (in particular, tyrosine) partici-

Glu-Asp-Ile-Lys-Val-Ala-Val-Val-Gly-Ala-Met-Ser-Gly-Pro-Val-Ala-Gln-Tyr-Gly-Asp-Gln-Glu-Phe-Thr-Gly- 25

Ala-Glu-Gln-Ala-Val-Ala-Asp-Ile-Asn-Ala-Lys-Gly-Gly-Ile-Lys-Gly-Asn-Lys-Leu-Gln-Ile-Val-Lys-Tyr-Asp- 50

Asp-Ala-Cys-Asp-Pro-Lys-Gln-Ala-Val-Ala-Val-Ala-Asn-Lys-Val-Val-Asn-Asp-Gly-Ile-Lys-Tyr-Val-Ile-Gly- 75

His-Leu-Cys-Ser-Ser-Thr-Gln-Pro-Ala-Ser-Asp-Ile-Tyr-Glu-Asp-Glu-Gly-Ile-Leu-Met-Ile-Thr-Pro-Ala- 100

Ala-Thr-Ala-Pro-Glu-Leu-Thr-Ala-Arg-Gly-Tyr-Gln-Leu-Ile-Leu-Arg-Thr-Gly-Leu-Asp-Ser-Asp-Gln-Gly- 125

Pro-Thr-Ala-Ala-Lys-Tyr-Ile-Leu-Glu-Lys-Val-Lys-Pro-Gln-Arg-Ile-Ala-Ile-Val-His-Asp-Lys-Gln-Gln-Tyr- 150

Gly-Glu-Gly-Leu-Ala-Arg-Ala-Val-Gln-Asp-Gly-Leu-Lys-Lys-Gly-Asn-Ala-Asn-Val-Phe-Phe-Asp-Gly-Ile- 175

Thr-Ala-Gly-Glu-Lys-Asp-Phe-Ser-Thr-Leu-Val-Ala-Arg-Leu-Lys-Lys-Glu-Asn-Ile-Asp-Phe-Val-Tyr-Tyr-Gly- 200

Gly-Tyr-His-Pro-Glu-Met-Gly-Gln-Ile-Leu-Arg-Gln-Ala-Arg-Ala-Ala-Gly-Leu-Lys-Thr-Gln-Phe-Met-Gly-Pro- 225

Glu-Gly-Val-Ala-Asn-Val-Ser-Leu-Ser-Asn-Ile-Ala-Gly-Glu-Ser-Ala-Glu-Gly-Leu-Leu-Val-Thr-Lys-Pro-Lys- 250

Asn-Tyr-Asp-Gln-Val-Pro-Ala-Asn-Lys-Pro-Ile-Val-Ala-Asp-Ile-Lys-Ala-Lys-Gly-Lys-Asp-Pro-Ser-Gly-Ala- 275

Phe-Val-Trp-Thr-Thr-Tyr-Ala-Ala-Leu-Gln-Ser-Leu-Gln-Ala-Gly-Leu-Asn-Gln-Ser-Asp-Asp-Pro-Ala-Glu-Ile- 300

Ala-Lys-Tyr-Leu-Lys-Ala-Asn-Ser-Val-Asp-Thr-Val-Met-Gly-Pro-Leu-Thr-Trp-Asp-Glu-Lys-Gly-Asp-Leu-Lys- 325

Gly-Phe-Glu-Phe-Gly-Val-Phe-Asp-Trp-His-Ala-Asn-Gly-Thr-Ala-Thr-Ala-Asp-Lys

Fig. 26 Amino acid sequence of LIV protein.

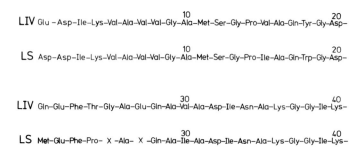

Fig. 27 N-Terminal sequences of the LIV and LS proteins.

pate in the complexing process. Nitration of native LIV protein led to a modification of one Tyr residue (out of a total of thirteen) with retention of 70% of the binding capacity. The ligand leucine did not protect this Tyr residue from attack. Peptide maps of the chymotryptic hydrolysate showed that nitration affects more than one Tyr residue of this protein. Nitration in urea solutions results in the introduction of a large number of nitro groups; exhaustive nitration occurs in 4 M detergent. The enzyme activity after removal of the urea is decreased proportionally to the number of nitro groups introduced.

Thus, it appears that Tyr does not directly participate in the complexing reaction, but is apparently important for retention of the spatial structure of the protein. Changes in the spatial structure drastically affect the activity (Gavrilova and Antonov, 1977).

Reaction of LIV protein with acetyl acetone leads to a modification of 10 Lys residues (out of a total of 29) and of 2 (out of 7) Arg residues with

TABLE V

Binding of L-[^{14}C]Leucine by LIV Protein

Protein concentration (μM)	L-[^{14}C]Leucine concentration range (μM)	Number of points	$K_{d(app)}$ (μM)	Maximal saturation value (r)	Apparent Hill coefficient (n)[a]
0.32	0.07–0.80	12	0.1	1.02	0.43
	0.75–28.6	19	5.5	5–6	0.43
6.4	0.07–1.0	16	0.3	0.45	0.57
	1.0–28.2	16	4.0	1.16	0.57
44.8	0.05–0.90	16	3.3	0.3	1.09
	1.3–31.5	10	3.3	0.3	1.09

[a] Calculated by the method of Silonova et al. (1969).

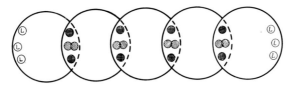

Fig. 28   Aggregation of LIV protein.

complete loss of binding activity. Protein inactivation takes place when over 6 eq of the modifying agent are added. Selective regeneration of the Lys residues by treatment with hydroxylamine, during which the Arg residues remain in their modified state, leads to complete restoration of the binding activity. Thus, three or four Lys residues play an essential part in the complexing process; their identification is currently under way.

Complexing studies of LIV protein within a wide range of protein and leucine concentrations show that the binding activity, namely, the apparent number of active sites (r), and the binding cooperativity (apparent Hill coefficient) depend on both protein and ligand concentrations (Table V) (Antonov *et al.*, 1975, 1976).

Such behavior can be explained by reversible linear aggregation of the protein. As a result of this aggregation, active binding sites become screened, leaving only the terminal sites exposed (see Fig. 28). With an increase in protein concentration, the number of exposed sites per monomer diminishes. With an increase in the ligand amino acid concentration, the extent of aggregation diminishes with a concomitant increase in the number of active sites per monomer. Aggregation of the protein has been followed by gel filtration on Sepharose 6B (maximum MW, 300,000), ultracentrifugation, and polarization fluorescence. In principle, this self-aggregating property could serve as a means for maintaining the necessary level of Leu, Val, and Ile in the *E. coli* cell.

REFERENCES

Abdulaev, N. G., Kiselev, A. P., and Ovchinnikov, Yu. A. (1976). *Bioorg. Khim.* **2**, 1148.

Andreev, I. M., Malenkov, G. G., Shkrob, A. M., and Shemyakin, M. M. (1971). *Mol. Biol. (Moscow)* **5**, 614.

Anraku, Y. (1968). *J. Biol. Chem.* **243**, 3116.

Antonov, V. K., Vorotyntseva, T. I., Alexandrov, S. L., Gavrilova, N. A., and Arsenyeva, E. L. (1975). *Dokl. Akad. Nauk SSSR* **221**, 1215.

Antonov, V. K., Alexandrov, S. L., and Vorotyntseva, T. I. (1976). *Adv. Enzyme Regul.* **14**, 269.

Barsky, E. A., Drachev, L. A., Kaulen, A. D., Kondrashin, A. A., Liberman, E. A., Ostroumov, C. A., Samuilov, V. D., Semenov, A. Yu., Sculachev, V. P., and Yasaitis, A. A. (1975). *Bioorg. Khim.* **1**, 113.

Benz, R., and Stark, G. (1975). *Biochim. Biophys. Acta* **382,** 27.

Byrn, S. R. (1974). *Biochemistry* **13,** 5186.

Duax, W. L., Hauptman, H., Weeks, C. M., and Norton, D. A. (1972). *Science* **176,** 911.

Eisenman, G., and Krasne, S. L. (1973). In "MTP International Review of Science, Biochemistry Series One" (F. Fox, ed.), Vol. 2, p. 28. Butterworths, London.

Eisenman, G., Krasne, S., and Ciani, S. (1976). In "Ion Selective Electrodes and Enzyme Electrodes in Medicine and in Biology" (M. Kessler *et al.,* eds.), p. 3. Urban & Schwarzenberg, Vienna.

Faulstich, H., Wieland, T., Walli, A., and Birkmann, K. (1974). *Hoppe-Seyler's Z. Physiol. Chem.* **355,** 1162.

Feinstein, M. B., and Felsenfeld, H. (1971). *Proc. Natl. Acad. Sci. U.S.A.* **68,** 2037.

Fonina, L. A., Savelov, I. S., Avotina, G. Ya., Ivanov, V. T., and Ovchinnikov, Yu. A. (1976). In "Peptide" (A. Loffet, ed.), p. 635. Edition de l'Université de Bruxelles, Bruxelles.

Frimmer, M. R., Kroker, R., and Postendörfer, J. (1974). *Naunyn-Schmiedebergs Arch. Pharmacol.* **284,** 395.

Furlong, C. E., and Weiner, J. H. (1970). *Biochem. Biophys. Res. Commun.* **38,** 1076.

Furlong, C. E., Cirakoglu, C., Willis, R. C., and Santy, P. A. (1973). *Anal. Biochem.* **51,** 297.

Gavrilova, N. A., and Antonov, V. K. (1977). *Bioorg. Khim.* **3,** 768.

Grell, E., Eggers, F., and Funk, T. (1972). In "Molecular Mechanisms of Antibiotic Action on Protein Biosynthesis and Membranes" (E. Muñoz, F. Garcia-Ferrandiz, and D. Vazquez, eds.), p. 646. Elsevier, Amsterdam.

Grell, E., Funck, T., and Eggers, F. (1975). In "Membranes" (G. Eisenman, ed.), Vol. 3, p. 2. Dekker, New York.

Haynes, D. H., Pressman, B. C., and Kowalsky, A. (1971). *Biochemistry* **10,** 852.

Henderson, R., and Unwin, R. N. T. (1975). *Nature (London)* **257,** 28.

Hladky, S. B. (1975). *Biochim. Biophys. Acta* **375,** 350.

Ivanov, V. T. (1975). *Ann. N.Y. Acad. Sci.* **264,** 221.

Ivanov, V. T., Laine, I. A., Abdullaev, N. D., Pletnev, V. Z., Lipkind, G. M., Arkhipova, S. F., Senyavina, L. B., Meshcheryakova, E. A., Popov, E. M., Bystrov, V. F., and Ovchinnikov, Yu. A. (1971a). *Khim. Prir. Soedin.* p. 221.

Ivanov, V. T., Miroshnikov, A. I., Abdullaev, N. D., Senyavina, L. B., Arkhipova, S. F., Uvarova, N. N., Khalilulina, K. Kh., Bystrov, V. F., and Ovchinnikov, Yu. A. (1971b). *Biochem. Biophys. Res. Commun.* **42,** 654.

Ivanov, V. T., Evstratov, A. V., Sumskaya, L. V., Melnik, E. I., Chumburidze, T. S., Portnova, S. L., Balashova, T., and Ovchinnikov, Yu. A. (1973a). *FEBS Lett.* **36,** 65.

Ivanov, V. T., Miroshnikov, A. I., Kozmin, S. A., Meshcheryakova, E. A., Senyavina, L. B., Uvarova, N. N., Khalilulina, K. Kh., Bystrov, V. F., and Ovchinnikov, Yu. A. (1973b). *Khim. Prir. Soedin.* p. 378.

Ivanov, V. T., Fonina, L. A., Uvarova, N. N., Kozmin, S. A., Kropotnitskaya, T. B., Chekhlayeva, N. M., Balashova, T. A., Bystrov, V. F., and Ovchinnikov, Yu. A. (1975). In "Peptides Chemistry, Structure and Biology" (R. Walter and J. Meienhofer, eds.), p. 195. Ann Arbor Sci. Publ., Ann Arbor, Michigan.

Karle, I. L. (1974a). *J. Am. Chem. Soc.* **96,** 4000.

Karle, I. L. (1974b). *Biochemistry* **13,** 2155.

Karle, I. L. (1975). *J. Am. Chem. Soc.* **97,** 4379.

Karle, I. L., Karle, J., Wieland, T., Bürgermeister, W., Faulstich, H., and Witkop, B. (1973). *Proc. Natl. Acad. Sci. U.S.A.* **70,** 1836.

Miroshnikov, A. I., Khalilulina, K. Kh., Uvarova, N. N., Ivanov, V. T., and Ovchinnikov, Yu. A. (1973). *Khim. Prir. Soedin.* p. 214.

Möschler, H. J., Weder, H.-G., and Schwyzer, R. (1971). *Helv. Chim. Acta* **54,** 1437.

Neher, E. (1975). *Biochim. Biophys. Acta* **401,** 540.

Neupert-Laves, K., and Dobler, M. (1975). *Helv. Chim. Acta* **58,** 432.

Ovchinnikov, Yu. A., and Ivanov, V. T. (1975). *Tetrahedron* **31**, 2177.
Ovchinnikov, Yu. A., Ivanov, V. T., Barsukov, L. I., Melnik, E. I., Oreshnikova, N. I., Bogolyubova, N. D., Ryabova, I. D., Miroshnikov, A. I., and Rimskaya, V. A. (1972). *Experientia* **28**, 399.
Ovchinnikov, Yu. A., Ivanov, V. T., and Shkrob, A. M. (1974a). "Membrane Active Complexones." Elsevier, Amsterdam.
Ovchinnikov, Yu. A., Ivanov, V. T., Evstratov, A. V., Mikhaleva, I. I., Bystrov, V. F., Portnova, S. L., Balashova, T. A., Meshcheryakova, E. A., and Tulchinsky, V. M. (1974b). *Int. J. Pept. Protein Res.* **6**, 465.
Ovchinnikov, Yu. A., Aldanova, N. A., Grinkevich, V. A., Arzamazova, N. M., Moroz, I. N., and Nazimov, I. V. (1977a). *Bioorg. Khim.* **3**, 564.
Ovchinnikov, Yu. A., Aldanova, N. A., Grinkevich, V. A., Arzamazova, N. M., and Moroz, I. N. (1977b). *FEBS Lett.* **78**, 313.
Oxender, D. L. (1972). *Annu. Rev. Biochem.* **41**, 777.
Oxender, D. L. (1975). *Ann. N.Y. Acad. Sci.* **264**, 358.
Patel, D. J. (1973). *Biochemistry* **12**, 677.
Patel, D. J., and Tonelli, A. E. (1974). *Biochemistry* **13**, 788.
Piperno, J. R., and Oxender, D. L. (1966). *J. Biol. Chem.* **241**, 5732.
Pletnev, V. Z., Galitsky, N. M., Ivanov, V. T., and Ovchinnikov, Yu. A. (1977). *Bioorg. Khim.* **3**, 1427.
Shemyakin, M. M., Ovchinnikov, Yu. A., Ivanov, V. T., Antonov, V. K., Shkrob, A. M., Mikhaleva, I. I., Evstratov, A. V., and Malenkov, G. G. (1967). *Biochem. Biophys. Res. Commun.* **29**, 834.
Shemyakin, M. M., Ovchinnikov, Yu. A., Ivanov, V. T., Antonov, V. K., Vinogradova, E. I., Shkrob, A. M., Malenkov, G. G., Evstratov, A. V., Ryabova, I. D., Laine, I. A., and Melnik, E. I. (1969). *J. Membr. Biol.* **1**, 402.
Shepel, E. N., Iordanov, S., Ryabova, I. D., Miroshnikov, A. I., Ivanov, V. T., and Ovchinnikov, Yu. A. (1976). *Bioorg. Khim.* **2**, 581.
Silonova, G. V., Livanova, N. B., and Kurganov, B. J. (1969). *Mol. Biol.* **3**, 768–784.
Smith, G. D., Duax, W. L., Langs, D. A., de Titta, G. T., Edmonds, J. W., Rohrer, D. C., and Weeks, C. M. (1975). *J. Am. Chem. Soc.* **97**, 7242.
Sumskaya, L. V., Balashova, T. A., Mikhaleva, I. I., Chumburidze, T. S., Melnik, E. I., Ivanov, V. T., and Ovchinnikov, Yu. A. (1977). *Bioorg. Khim.* **3**, 5.
Szabo, G., Eisenman, G., and Ciani, S. (1969). *J. Membr. Biol.* **1**, 346.
Urry, D. W. (1971). *Proc. Natl. Acad. Sci. U.S.A.* **68**, 672.
Urry, D. W., Goodall, M. C., Glickson, J. D., and Myers, D. F. (1971). *Proc. Natl. Acad. Sci. U.S.A.* **68**, 1907.
Veatch, W. R., and Blout, E. R. (1974). *Biochemistry* **13**, 5257.
Veatch, W. R., Fossel, E. T., and Blout, E. R. (1974). *Biochemistry* **13**, 5249.
Veatch, W. R., Mathies, R., Eisenberg, M., and Stryer, L. (1975). *J. Mol. Biol.* **99**, 75.
Wieland, T. (1972). *Naturwissenschaften* **59**, 225.
Wieland, T., and Govindan, V. M. (1974). *FEBS Lett.* **46**, 531.
Wieland, T., Lüben, G., Ottenheym, H., Fassel, J., de Vries, J. X., Konz, W., Prox, A., and Schmid, J. (1968). *Angew. Chem.* **80**, 209.
Wieland, T., Faulstich, H., Bürgermeister, W., Otting, W., Mohle, W., Shemyakin, M. M., Ovchinnikov, Yu. A., Ivanov, V. T., and Malenkov, G. G. (1970). *FEBS Lett.* **9**, 89.
Wieland, T., Faulstich, H., Jahn, W., Govindan, V., Puchinger, H., Kopitar, Z., Schmaus, H., and Schmitz, A., (1972a). *Hoppe-Seyler's Z. Physiol. Chem.* **353**, 1337.
Wieland, T., Birr, C., Bürgermeister, W., Trietsch, P., and Rehr, G. (1972b). *Justus Liebig's Ann. Chem.* **759**, 71.
Wieland, T., Faulstich, H., and Bürgermeister, W. (1972c). *Biochem. Biophys. Res. Commun.* **47**, 984.
Wipf, H.-K., Pioda, L. A. R., Štefanac, Z., and Simon, W. (1968). *Helv. Chim. Acta* **51**, 377.

4

Conformational Analysis of Polypeptides: Application to Homologous Proteins

MARY K. SWENSON, ANTONY W. BURGESS, AND HAROLD A. SCHERAGA

Department of Biophysics, Weizmann Institute of Science, Rehovoth, Israel, and Department of Chemistry, Cornell University, Ithaca, New York

I. INTRODUCTION

Procedures for empirical energy calculations have been developed to provide an understanding of how the interactions in polypeptides and proteins determine their stable conformations and their reactivity with other molecules. This field has been reviewed on several occasions, and the articles by Pullman and Pullman (1974), Ramachandran (1974), Scheraga (1974b), and Nemethy and Scheraga (1977) may be cited as examples.

The assumption that the conformation of a protein is a thermodynamically stable one underlies all computational procedures that attempt to predict this conformation from the amino acid sequence. Therefore, to make a prediction, it is necessary to adopt a reasonable set of bond lengths and bond angles, to determine the conformational energy minima, and to calculate the statistical weights (which include the conformational entropy) at each minimum within a reasonable range of energy of the minimum corresponding to the experimental structure.

115

FRONTIERS IN PHYSICOCHEMICAL BIOLOGY

Details of this approach have been discussed elsewhere (Scheraga, 1971).

In this chapter, we summarize the various *types* of problems in the field of polypeptide and protein structures that have been subjected to this kind of conformational analysis and give, as an example, the details of the application of this method to the computation of the three-dimensional structures of homologous proteins, namely, several neurotoxin inhibitors that have amino acid sequence homologies with bovine pancreatic trypsin inhibitor.

II. STANDARD GEOMETRY

Ideally, calculations on polypeptides should allow variation of all degrees of freedom. However, such a procedure makes excessive demands on the potential functions used; that is, it is necessary to have reliable parameters for the many components of the potential function. A procedure often adopted to circumvent this difficulty partially is to maintain rigid geometry (fixed bond lengths, bond angles, and planar *trans*-peptide groups) and to use internal variables (dihedral angles for rotation about single bonds). Efforts are currently being made to incorporate force constants for varying bond lengths and bond angles; in evaluating such procedures, care should be taken to see that the force constants were selected on a rational basis.

The use of rigid geometry can also be justified on other grounds. In polymeric systems, near-neighbor interactions (i.e., those between the side chain and backbone of the same amino acid residue) tend to dominate in determining geometry and conformation. Thus, if the geometry of the monomeric units is selected properly in advance (i.e., each monomeric unit having its own characteristic geometry), this "standard" geometry can be maintained in the polymer. Arnott and Scott (1972) and Momany *et al.* (1975) demonstrated this for carbohydrates and polypeptides, respectively. For amino acids, Momany *et al.* (1975) surveyed the X-ray and neutron diffraction literature and found that the $\tau(NC^{\alpha}C')$ bond angles for a given type of amino acid, say alanine, differ from those of, say, tyrosine, but that all alanines (and correspondingly all tyrosines) have essentially the same bond angle in various (unstrained) peptides. Therefore, different values of $\tau(NC^{\alpha}C')$ are used for each amino acid (see Table I of Momany *et al.*, 1975), and these are maintained constant in calculations on polypeptides. The other backbone bond angles and bond lengths show less variation from one type of amino acid crystal to another, and a uniform set has been selected for all amino acids except proline, for which a different geometry is used (Momany *et al.*, 1975). The dominance of near-neighbor interactions also manifests itself in the tendency of methyl groups to be staggered in small-molecule crystals, even though the barrier to internal rotation is only about 3 kcal; that is, long-range packing interactions adapt to allow the attainment of the best short-range conformations.

From a similar survey of crystal structures of peptides, Momany *et al.* (1975) concluded that the peptide group could be taken in the planar *trans* conformation, except for X-Pro peptide bonds. In the latter case, the possibility of a *cis* conformation has to be taken into account. Recently, Zimmerman and Scheraga (1976) identified the interactions that favor the *trans* conformation in Gly-Gly and that allow for a small percentage of the *cis* conformation in Gly-Pro. They also showed that, while deviations from planarity of 1–3° can occur in Gly-Gly and as much as 10° in Gly-Pro, the conformation-dependent properties of blocked dipeptides (i.e., those with *N*-acetyl and *N'*-methyl end groups) can be represented adequately by keeping the peptide groups planar. Therefore, while both *cis* and *trans* (planar) conformations are allowed for X-Pro peptide bonds, all other peptide groups are taken in the planar *trans* conformation. It is recognized that deviations from planarity are sometimes reported for well-refined X-ray structures of proteins. However, considering (1) that protein structures are determined at best at ~1.5-Å resolution, whereas the structures of small peptides are determined at higher resolutions, and (2) that the principle of the dominance of near-neighbor interactions in polymers is well established, it seems reasonable to assume that amino acids and peptides should maintain similar geometries in unstrained monomeric and polymeric structures. Therefore, at the present stage of conformational energy calculations on polypeptides, we have chosen to maintain the geometry rigid and the peptide groups in the planar *trans* conformation. The final results of such computations certainly can be carried further by introducing flexible geometry, if it is warranted by the acquisition of reliable force constants and if compelling evidence accumulates from X-ray studies to demonstrate that bond lengths and bond angles of the amino acid residues of proteins differ from the accurately determined values for their corresponding small molecules.

III. COMPUTATION PROCEDURES

When a standard geometry is adopted, any conformation of a polypeptide chain can be generated with matrix methods by varying the backbone and side-chain dihedral angles. The conformational energy of any initial conformation can then be minimized. A variety of empirical potential energy functions are currently in use, and Hagler *et al.* (1974) have provided an evaluation of some of them.

The potential functions currently in use in this laboratory (Momany *et al.*, 1975) were based on, among other things, calculations on crystal and gas-phase structures of small molecules. A documented computer program (ECEPP, Empirical Conformational Energy Program for Peptides) using these potentials (together with a procedure for generating the polypeptide chain) is available from the Quantum Chemistry Program Exchange (see Footnote 60 of Momany *et al.*, 1975). The effect of solvation is introduced by a separate algorithm (Gibson and Scheraga, 1967; Hopfinger, 1971).

Parenthetically, it is worth mentioning that, while ECEPP provides results that are in agreement with experimental results in many cases, nevertheless efforts continue to be made to improve the nature of the potential functions. In this connection, we refer to a new approach developed recently and termed EPEN (Empirical Potential Using Electrons and Nuclei) (Shipman *et al.*, 1975; Burgess *et al.*, 1975b, 1976). This new potential function is currently being used in Monte Carlo calculations to try to improve the treatment of hydration (Owicki *et al.*, 1978).

While several minimization procedures are available (see Section V), empirical conformational energy calculations suffer from the existence of the multiple-minimum problem (Scheraga, 1974a), that is, the existence of many local minima in the multidimensional energy surface. The procedures being used to circumvent this problem are discussed in Section IV.

Finally, the computation of the statistical weights has been discussed elsewhere (Scheraga, 1971), and it has been applied in the conformational analysis of the twenty naturally occurring amino acid residues (Zimmerman *et al.*, 1977a) and of bends in dipeptides (Zimmerman *et al.*, 1977b).

IV. APPROACHES TO THE MULTIPLE-MINIMUM PROBLEM

For convenience, we may divide polypeptides into three categories: (1) small open-chain and cyclic peptides, (2) synthetic analogues of fibrous proteins, and (3) globular proteins. It has been possible to surmount the multiple-minimum problem in the first two categories because the number of variables is small, and the conformational space can be covered adequately to locate the minimum corresponding to the experimental structure. Thus, conformational analyses have been carried out for single residues, di-, tri-, tetra-, and larger oligopeptides, cyclic peptides, random coils, preferred type (e.g., α or ω) and right- or left-handedness of helical polypeptides, and thermodynamic parameters for helix-coil transitions in poly(amino acids) and for helix–helix transitions [e.g., the interconversion between forms I and II of poly(L-proline)] (Scheraga, 1974b). As recent examples in the first two categories, we cite the computations of the conformations of thyrotropic releasing factor, pGlu-His-Pro-NH$_2$ (Burgess et al., 1975a), enkephalin, Tyr-Gly-Gly-Phe-Met (Isogai et al., 1977), luteinizing hormone-releasing hormone (Momany, 1976a,b), gramicidin S (Dygert et al., 1975), and the synthetic collagenlike polymer (Gly-Pro-Pro)$_n$ (Miller and Scheraga, 1976).

The multiple-minimum problem has not been solved yet for globular proteins, and much research is now being directed toward the solution of this problem. These efforts have been reviewed by Nemethy and Scheraga (1977). Two types of problems, involving globular proteins, are already amenable to solution, namely, the energetic relief of high-energy overlaps in X-ray structures of proteins and the computation of the structures of homologous proteins. In these cases, the multiple-minimum problem is alleviated because a starting structure is available, thereby avoiding the necessity to search other regions of conformational space. Both of these problems are considered in more detail in Section V.

Finally, adequate search procedures have been developed to treat enzyme–substrate interactions, namely, the interactions of α-chymotrypsin with amides and oligopeptides (Platzer et al., 1972) and of hen egg white lysozyme with oligosaccharides (Pincus et al., 1976, 1977).

V. COMPUTATION OF MODELS FOR THE THREE-DIMENSIONAL STRUCTURES OF THREE SNAKE VENOM INHIBITORS

As an illustration of the techniques mentioned briefly in Sections II–IV, we present here a description of the computation of the structures of homologous proteins. There is considerable evidence that indicates that proteins with homologous amino acid sequences have similar three-dimensional structures. On the basis of this assumption, the three-dimensional structure of α-lactalbumin has been computed from that of lysozyme (Warme et al., 1974) and (in less detail) that of thrombin from those of chymotrypsin, trypsin, and elastase (Endres et al.,

1975; Nishikawa and Ooi, 1974). In this section, we use a computer modeling procedure (Warme et al., 1974; Warme and Scheraga, 1973) to compute three-dimensional structures of three snake venom inhibitors, Russell's viper venom inhibitor II (*Vipera* Russelli, RV II) (Takahashi et al., 1972, 1974) and the black mamba venom inhibitors I and K (*Dendroapsis polylepis*, BMI and BMK) (Strydom, 1973), which are homologous to bovine pancreatic trypsin inhibitor (BPTI).

The detailed atomic coordinates of BPTI have been determined by single-crystal X-ray diffraction (W. Steigemann, personal communication; Deisenhofer and Steigemann, 1975). These coordinates have been used as a basis for the computer modeling of the three-dimensional structures of these snake venom inhibitors. The computer modeling (see also Warme et al., 1972; Warme and Scheraga, 1973, 1974, 1975) is based on the structural and conformational properties of small molecules (Momany et al., 1975) and, as indicated in Section II, relies in part on the constancy of specific bond lengths and bond angles observed in small molecules; that is, cognizance is taken of the assumption that the geometry of the amino acid residues and peptide groups in a protein is very likely to be similar to that found from single-crystal studies on amino acids and their derivatives. Thus, bond lengths and bond angles can be maintained fixed, and only the dihedral angles, ϕ, ψ, and χ, are varied when changing the atomic positions of a protein to produce a structure with a standard geometry (Momany et al., 1975) for each amino acid residue. The values of ϕ, ψ, and χ were adjusted, first, to obtain a best least-squares fit between the computer-generated atomic coordinates (using standard geometry and the experimental values of ϕ, ψ, and χ) for BPTI and those obtained from the X-ray crystallographic data and then to minimize the conformational energy (Momany et al., 1975) of the protein. In this work, two sets of atomic coordinates of BPTI were used to obtain a model of this protein having standard geometry; one set of coordinates (designated X-ray 1) was obtained directly from the Kendrew wire model (W. Steigemann, personal communication), and the other (designated X-ray 2) was a larger data set than X-ray 1 and was obtained by Deisenhofer and Steigemann (1975) using the refinement procedures of Diamond (1966, 1971) (W. Steigemann, personal communication). These techniques tend to place the atomic coordinates so that the bond lengths and bond angles are near the standard values for small molecules. However, errors in measurement of the initial wire model change the atomic coordinates so that the standard geometry of the wire model is not preserved; also, the Diamond refinement procedure adjusts ϕ's, ψ's, and χ's and the bond angle $\tau(NC^{\alpha}C')$ (while retaining standard geometry for the other backbone atomic positions) to improve the agreement between the calculated coordinates and the electron density maps obtained from the observed structure factors and calculated phases. It is interesting to note, from the results reported here, how closely the experimental structure for BPTI can be modeled by a set of coordinates with standard geometry, in which the interatomic interactions are optimized to yield a low-energy conformation.

In the spirit of starting with the original coordinates, rather than with those improved by the Diamond refinement, we used the initial set of atomic coordinates (X-ray 1) to obtain a standard-geometry model of BPTI and, from this, to compute low-energy conformations for three proteins homologous to BPTI, namely, RV II, BMI, and BMK. The amino acid sequences of RV II, BMI, and BMK have been summarized by Creighton (1975). They have a high degree of homology with BPTI, and no insertions or deletions are required to align the six equivalent half-cystine residues. Two of these proteins (RV II and BMK) inhibit trypsin (Strydom, 1973; Takahashi *et al.*, 1972, 1974), whereas the third (BMI) inhibits chymotrypsin; this difference arises from the presence of tyrosine instead of lysine in the active site of BMI (Strydom, 1973). The results of these calculations demonstrate the conservation of three-dimensional homology, particularly in the active-site and long-range interactions. It must be emphasized that, as in the case of the computation of the structure of α-lactalbumin from that of lysozyme (Warme *et al.*, 1974), the procedure is based on the *assumption* that sequence homology implies structural homology. Thus, while structural conclusions are drawn from the results of the computations, no inferences should be made about the relative efficiencies of the various homologues with respect to their ability to inhibit the action of trypsin or chymotrypsin.

A. Generation of Standard-Geometry Models for BPTI and Three Homologous Proteins

The set of backbone and side-chain bond lengths and bond angles for this "standard-geometry" model is that intended for use in conformational energy calculations on proteins (Momany *et al.*, 1975). The backbone geometry differs slightly for each type of amino acid, primarily in the bond angle $\tau(NC^\alpha C')$. In the calculations of the models for BPTI and its homologues, a modification of the procedure of Warme *et al.* (1972) was used to generate the atomic coordinates. In the stage I procedure of Warme *et al.* (1972), the positions of the backbone atoms N, C^α, C^β, C', and O and the side chains of 20-residue segments are fit as closely as possible to the X-ray coordinates by least squares, maintaining standard geometry. Although this procedure, when used carefully, is capable of producing a close fit to the experimental coordinates, it is possible to introduce systematic artifacts when the calculated coordinates deviate considerably from the experimental ones. These artifacts manifest themselves primarily as a rotated peptide group because of a compensation of errors in ψ_1 and ϕ_2, which may arise as follows. Because the N, C^α, and C^β atoms of residue 1 are held in a fixed conformation (overlapping the X-ray positions as closely as possible), the first dihedral angle that is allowed to vary in order to optimize the positions of the backbone atoms of the first 20 residues is ψ_1. The initial deviation of the calculated coordinates from the X-ray values for residue 20 may be quite large because of possible cumulative errors that arise when a standard-geometry chain is generated from approximate values of ϕ and ψ (keeping ω fixed at 180°). Variation of ψ_1 moves residue 20 on a cone without bringing it closer to the X-ray position, so that a close fit at residue 20 during the least-squares minimization is obtained by large changes in ϕ_2; that is, if ψ_1 undergoes a large change to reduce the initial deviation, then ϕ_2 may undergo a large compensating change during minimization to provide a good fit from the C^α—C' bond of residue 2 up to residue 20, but leaving the oxygen atom of the planar peptide group between residues 1 and 2 (and the C^β of residue 2) considerably altered from its X-ray position. This possible source of difficulty was eliminated here by initially fitting only 5 instead of 20 residues and then continuing the fitting by one or two additional residues at a time. After 10–15 residues had been fitted by the standard-geometry chain, the fitting procedure was restarted at the fourth residue from the C terminus of that segment, and again one or two residues were added at a time and their positions were optimized. For example, after the first segment attained a length of 15 residues, the new segment was started at residue 12, and residues 13, 14, and 15, plus residue 16, were the new portion to be fit to the X-ray coordinates; then residues 17, 18, etc., were added successively. After the atomic coordinates for the whole chain had been generated, the procedure was repeated, using 20-residue segments (i.e., 1–20, 17–36, etc.) and finally the entire polypeptide chain, to improve the fit. The use of 20-residue segments at this point was possible because the computed atomic positions were already close to the X-ray coordinates. Fletcher's (1970) variable metric optimization algorithm was used in this geometric fitting procedure.

Other procedures have also been used to surmount the difficulty mentioned above. For example, the original fit can be made to residue 2 (the coordinates of which may be more accurate than those of residue 1), and then the fitting to the rest of the chain can be carried out in both directions (i.e., toward the N and C termini) (Rasse et al., 1974).

Once the backbone atoms had been positioned, the standard-geometry side-chain atoms beyond C^β (Momany et al., 1975) were fit to the X-ray coordinates by varying the side-chain dihedral angles. The optimization algorithm of Powell (1964) was used in this calculation.

The positions of the C, N, O, and S atoms in the model for BPTI corresponded to a standard-geometry structure, but the positions of the hydrogen atoms were not yet defined. The nonpolar hydrogen atoms were included in the backbone and side chains by treating them as "united atoms" together with the carbons to which they are attached; that is, appropriate radii were selected for the aliphatic and aromatic carbons, as in the procedure of Gibson and Scheraga (1967), to reflect the presence of the nonpolar hydrogen atoms. Polar hydrogens (those bound to nitrogens and oxygens) were attached using standard bond lengths and bond angles (Momany et al., 1975); for those with rotational freedom, the dihedral angles that involved motion of OH or NH_2 groups were varied until the conformational energy of interaction of the OH or NH_2 group with the rest of the BPTI molecule was at a minimum. The program used for this purpose is designated here as FOLD; it includes nonbonded, torsional, electrostatic, hydrogen-bond, and disulfide loop-closing energies and allows conformational energy calculations to be carried out on structures with any geometry.

The energy of the resulting complete chain with standard geometry (and hydrogens properly accounted for) was then minimized by the stage II procedure of Warme and Scheraga (1973) using the energy parameters, Fourier treatment of 1–4 interactions, disulfide loop-closing potential, and fitting potential algorithm described therein. The minimization in stage II (Warme and Scheraga, 1973) was carried out by varying the backbone dihedral angles ϕ and ψ and all side-chain dihedral angles except those involving OH and NH_2 groups (however, the energies of interaction of OH and NH_2 hydrogens were included), using the Fletcher–Powell (1963) modification of the Davidon (1959) minimization procedure. In this minimization, a weighting factor (W) was used to keep the atomic positions of the standard-geometry structure close to the X-ray positions (Warme and Scheraga, 1973). Initially, energy minimization was performed on 10-residue segments (in contact with the rest of the molecule), maintaining an overlap of 4 residues between segments, and W was set to 50 kcal/$Å^2$. After this initial minimization of segments, the energy of the entire protein molecule (including the disulfide closing potential) was minimized. As the conformational energy of the whole molecule decreased, W was reduced in steps gradually and, by the final cycle of minimization, $W = 0.05$ kcal/Å^2 for BPTI, with other values (referred to later in the chapter) for the homologous proteins. After reaching a minimum, the backbone atoms and most of the side-chain atoms were maintained in fixed positions, whereas the dihedral angles that led to movement of polar hydrogen atoms and atoms connected to disulfide bonds were optimized using FOLD. The hydrogen atoms attached to the polar groups of serine, threonine, lysine, asparagine, glutamine, glutamic acid, and aspartic acid were rotated independently until their energy of interaction with the rest of the molecule was at a minimum. FOLD was also used to compute the total conformational energy of the resulting structure.

The reported X-ray coordinates for BPTI have a varying geometry at the C$^\alpha$ atoms and different dihedral angles around the peptide bonds. The strain energy associated with bond angle bending ($\Delta\tau$) at the C$^\alpha$ atom and rotation of the peptide group ($\Delta\omega$) away from planarity was computed for the X-ray 1 and X-ray 2 structures using the methods of Ramachandran and Venkatachalam (1968) and Bixon and Lifson (1967), respectively. The contributions to the strain energy of the standard-geometry model structures were assumed to be zero, except in the case of the disulfide bonds, where this is represented in part by the loop-closing potential.

The backbone atoms of the proteins, RV II, BMI, and BMK, are homologous to those of BPTI and were first fitted to the *calculated* standard-geometry structure of BPTI obtained by our modification of the stage I procedure. However, to fit the structures as closely as possible to the original X-ray coordinates (i.e., the X-ray 1 data) of BPTI, the same modification of the stage I procedure was then used to optimize the fit of these initial standard-geometry structures for RV II, BMI, and BMK to the X-ray 1 data of BPTI. This refitting step was carried out because changes in sequence (of the homologous proteins) introduced changes in $\tau(NC^\alpha C')$, with possible large deviations from the X-ray coordinates of BPTI. By first fitting to the calculated structure of BPTI, the ultimate fit to the X-ray data could be carried out more rapidly. In this refitting step, the dihedral angles changed very little [e.g., in RV II, one dihedral angle changed by $\sim 7°$; four changed by $\sim 5°$; ϕ_{13}, ψ_{13}, and ϕ_{14} changed by $17°$, $-17°$, and $17°$, respectively (residue 13 is Pro in BPTI, but Arg in RV II); all other dihedral angles changed by less than $2°$].

When generating the side chains for the homologous proteins, those that were identical to the corresponding residues of BPTI (considering, e.g., Phe and Tyr to be "identical" residues) were fit to the X-ray 1 coordinates by varying the side-chain dihedral angles to optimize the atomic positions. For nonidentical residues, the initial side-chain dihedral angles were taken as the lowest-energy ones (for the given values of ϕ and ψ) for the N-acetyl- and N'-methylamides of single amino acids (Lewis *et al.*, 1973). After generation of a standard-geometry polypeptide chain, which fit the X-ray 1 coordinates as closely as possible, the positions of those side chains involved in side chain–side chain or side chain–backbone overlaps were optimized with FOLD. At the same time, the positions of the polar hydrogens were optimized with FOLD. Then the stage II procedure was applied; this was followed again by FOLD to optimize the positions of the polar hydrogens, and the disulfide bonds, and to calculate the total conformational energy of the molecule.

B. Standard-Geometry Model for Three-Dimensional Structure of BPTI

To compare the total conformational energies of the X-ray and the computed structures for BPTI, the polar hydrogens were added to the X-ray structures, using standard geometry, and their rotational positions were optimized with FOLD by the procedure described in Section V, A. It should be noted that the *comparisons* between the calculated and observed structures could not include residues 56–58 since these residues were not located in the electron density maps computed in the X-ray analysis (W. Steigemann, personal communication; Deisenhofer and Steigemann, 1975). The results of calculations for only residues 1–55 are reported here. Table I shows the conformational energies of X-ray 1 (initial wire model coordinates) and X-ray 2 (Diamond-refined coordinates), as well as those of the energy-minimized structures deduced from them.

The conformations of both of the experimental structures, X-ray 1 and X-ray 2, contain very few overlapping atoms and the strain energy due to bond angle bending is only a small part ($\sim 10\%$) of the computed energy (Table I). The electrostatic energy term for the experimental structures is attractive and similar to the final value for the lowest energy conformations (Table I).

The X-ray 1 structure (with polar hydrogens added) had a total (conformational plus geometric strain) energy of 1416 kcal/mole; this positive energy results mainly from nonbonded interactions between overlapping atoms of residue pairs $Pro_2 \cdots Phe_4$, $Ala_{16} \cdots Gly_{36}$, and $Asn_{24} \cdots Gln_{31}$ and the C^γ atom of Asp_{50} with its own peptide N atom. Initially, the standard-geometry polypeptide chain generated by the

TABLE I
rms Deviations and Energies of Computed Structures

	X-Ray 1	Low-energy conformation (X-ray 1)	X-Ray 2	Low-energy conformation (X-ray 2)	Low-energy standard-geometry conformations		
					RV II	BMI	BMK
rms Deviations (Å)							
Backbone atoms	0.7[a]	0.8[b]	—	0.6[a]	0.8[b]	1.3[b]	0.9[b]
Side-chain atoms	1.7[a]	1.5[b]	—	0.9[a]	1.4[b]	1.7[b]	1.4[b]
Total[c]	1.2[a]	1.1[b]	—	0.7[a]	1.0[b]	1.4[b]	1.1[b]
Energy contributions (kcal/mole)							
Torsional	456	303	302	297	306	315	295
Disulfide	27	19	7	12	18	19	25
Electrostatic	-898	-937	-965	-994	-995	-969	-931
Nonbonded	1683	-137	159	-177	-86	-104	-98
Strain	148	0	71	0	0	0	0
Total[c]	1416	-752	-426	-862	-757	-739	-709

[a] rms deviation from the refined X-ray coordinates (X-ray 2) obtained by Deisenhofer and Steigemann (1975).
[b] rms deviation from X-ray 1.
[c] These rms deviations and energies pertain only to residues 1–55 because accurate X-ray data were not available for residues 56–58.

stage I procedure to fit the X-ray 1 coordinates of BPTI also had several high-energy interactions (introduced by the imposition of standard geometry); the largest involved an interaction between the side-chain atoms of Phe_4 and Tyr_{35}. No high-energy interactions between backbone atoms were observed. The side-chain bond angle $\tau(C^\alpha C^\beta C^\gamma)$ of Phe_4 in the X-ray 1 structure was reported to be 128.5°, a deviation of 14.5° from the standard-geometry bond angle (114°) used for the Phe residue (Momany et al., 1975). When fitting the side chain of the standard-geometry structure of BPTI, the position of the C^β atom of Phe_4 is fixed (along with the backbone), so that the large value of $\Delta\tau(C^\alpha C^\beta C^\gamma) = 14.5°$ makes it impossible to position the phenyl ring in the same orientation as it is in the X-ray 1 structure. Even after energy minimization, the average deviation between the standard-geometry structure and the X-ray 1 coordinates of the Phe_4 side chain was 1.6 Å. In the Diamond-refined coordinates for BPTI (X-ray 2), $\tau(C^\alpha C^\beta C^\gamma)$ for Phe_4 was reported as 114.8° (Deisenhofer and Steigemann, 1975). The average deviation between the X-ray 2 and the X-ray 1 coordinates for the Phe_4 side chain was 1.3 Å; that is, the Diamond refinement also involved considerable movement of the Phe_4 side chain in going from the X-ray 1 structure to the X-ray 2 structure. A similar situation occurred for Tyr_{35}, where the X-ray 1 coordinates corresponded to unusual bond angles around the C^α and C^β atoms, and it was difficult to obtain close agreement between the standard-geometry structure and the X-ray 1 structure for the side chain of this residue.

Using FOLD, it was possible to relieve the high-energy interactions between side-chain atoms by varying the side-chain dihedral angles to minimize the energy of interaction with the rest of the protein. The total conformational energy of the standard-geometry model of the BPTI structure decreased to −752 kcal/mole (Table I) after energy minimization; although most of the energy decrease was due to the removal of atomic overlaps, all of the energy terms improved (Table I).

Figure 1 shows the values of the deviations $\Delta\omega$ and $\Delta\tau(NC^\alpha C')$ of the X-ray 1 structure from standard geometry, together with the corresponding strain energies. The rms deviations of atomic positions between the X-ray 1 structure and the low-energy standard-geometry conformation deduced from it, and the conformational energies of each residue, are also shown in Fig. 1. The dihedral angles for the low-energy conformation of this standard-geometry model for BPTI are given in Table II.

The total (conformational plus strain) energy of the X-ray 2 structure of BPTI (with polar hydrogens added) was −426 kcal/mole. Unfavorable interactions between the side chains of residues $Asn_{24} \cdots Gln_{31}$, Phe_4

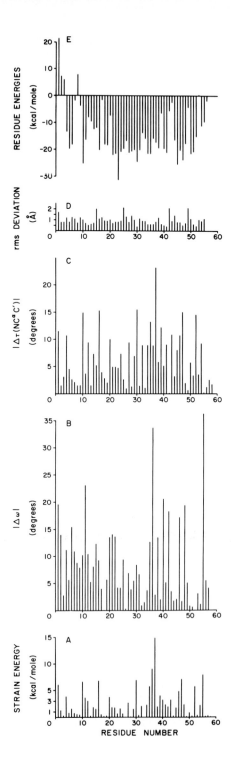

TABLE II.   DIHEDRAL ANGLES OF LOW-ENERGY CONFORMATION OF BPTI COMPUTED FROM XRAY1 COORDINATES[a]

Residue	ϕ	ψ	X's						
1 ARG	-159.0	171.4	-73.7	105.2	75.7	-172.2	-167.5	5.2	0.7
2 PRO	-75.0	-17.3							
3 ASP	83.8	-40.3	-52.7	-69.8	-10.2				
4 PHE	-55.4	-23.3	98.4	84.9					
5 CYS	-73.5	-17.2	-90.0						
6 LEU	-95.4	23.6	-61.9	-179.2					
7 GLU	-87.4	165.0	-75.0	-174.1	104.7	8.7			
8 PRO	-75.0	165.0							
9 PRO	-75.0	127.5							
10 TYR	-123.8	91.4	-165.1	67.8	5.4				
11 THR	-65.9	-36.1	-173.1	85.1					
12 GLY	104.7	172.2							
13 PRO	-75.0	-41.8							
14 CYS	-57.1	154.2	-68.1	88.1	72.1	-167.3	64.2		
15 LYS	-128.6	55.3	-109.0						
16 ALA	-92.2	-168.8							
17 ARG	-114.0	58.7	-110.3	-167.4	-78.0	-175.8	-168.2	5.1	0.7
18 ILE	-101.6	113.9	-52.1	-72.5					
19 ILE	-88.9	116.0	-78.5	86.3					
20 ILE	-125.8	172.9	-73.8	-110.0	-167.7	175.2	-178.2	5.0	0.7
21 TYR	-117.7	146.0	-98.7	93.9					
22 PHE	-146.8	147.7	86.1	101.7					
23 TYR	-75.7	134.8	-171.5	-108.9	6.1				
24 ASN	-90.6	102.5	-161.6	-35.3	4.9				
25 ALA	-71.6	61.5							
26 LYS	-168.8	-56.8	-173.0	177.7	-179.2	178.4	-55.7		
27 ALA	-73.5	-38.4							
28 GLY	100.2	19.3							
29 LEU	-137.6	174.3	47.4	83.3					
30 CYS	-99.3	141.0	-66.8						
31 GLN	-131.7	165.4	-95.7	-179.1	-85.6	4.9			
32 THR	-79.2	148.8	-14.7	-61.0					
33 PHE	-153.5	134.6	110.5	89.5					
34 VAL	-70.3	124.0	-60.3						
35 TYR	-107.7	136.8	170.7	49.3	-8.8				
36 GLY	-59.5	-68.3							
37 GLY	126.3	47.6							
38 CYS	175.5	141.1	63.1						
39 ARG	41.1	76.4	-56.3	174.0	-162.1	-177.7	-168.2	5.1	0.7
40 ALA	-102.4	177.4							
41 LYS	-113.9	174.1	-89.8	166.9	175.8	178.3	-55.0		
42 ARG	-68.2	-35.1	-168.6	70.4	-165.4	78.0	-164.4	5.2	0.7
43 ASN	-70.5	63.0	-141.7	-25.1					
44 ASN	-165.2	139.1	115.2	3.9					
45 PHE	-143.0	162.3	-72.2	88.1					
46 LYS	-84.9	-45.8	-82.5	168.8	166.6	-176.9	-55.9		
47 SER	-110.0	147.8	63.4	-85.4					
48 ALA	-68.9	-41.8							
49 GLU	-49.7	-45.5	61.0	-180.0	88.6	-0.3			
50 ASP	-65.4	-50.6	-159.5	150.7	-174.7				
51 CYS	-51.9	-52.4							
52 MET	-63.8	-45.7	172.8	142.4	-175.6				
53 ARG	-53.1	-46.1	178.4	-175.0	-176.7	179.4	-168.0	5.1	0.7
54 THR	-75.5	-51.0	151.2	-105.2					
55 CYS	-55.7	-52.8	46.0	-71.0					

[a] $\omega = 180$ DEGREES FOR ALL RESIDUES. ALL OTHER DIHEDRAL ANGLES ARE IN DEGREES.

··· Glu_7, and Asn_{43} ··· Glu_7 were responsible for 455 kcal/mole. The high-energy interactions in these Asn and Gln residues arise from the interactions between aliphatic CH_2 groups of the side chains. After minimization, the standard-geometry polypeptide chain generated to fit the X-ray 2 coordinates had a conformational energy of -862 kcal/mole (see Table I). The dihedral angles for this low-energy conformation of BPTI are given in Table III, and this structure is compared to the X-ray 2 structure in Fig. 2.

Figures 1 and 2 indicate the degree to which an optimized standard-geometry structure can fit the X-ray 1 and X-ray 2 data, respectively. Since ω and each value of τ were constant for a given residue in the computed structures, a comparison of the values of $|\Delta\tau(NC^\alpha C')|$ and $|\Delta\omega|$ in Figs. 1 and 2 provides a direct measure of the improvement in these parameters that was achieved by Deisenhofer and Steigemann (1975) in

Fig. 1 Data for X-ray 1 structure and the low-energy structure deduced from them. (A) Strain energy per residue in X-ray 1 structure due to deviations in $\tau(NC^\alpha C')$ and ω from standard geometry. (B) Departure of ω from 180° in X-ray 1 structure. (C) Departure of $\tau(NC^\alpha C')$ of X-ray 1 structure from standard geometry. (D) Total rms deviation per residue of low-energy structure from observed X-ray 1 structure. (E) Total residue energy of low-energy structure.

TABLE III.   DIHEDRAL ANGLES OF LOW-ENERGY CONFORMATION OF BPTI COMPUTED FROM XRAY2 COORDINATES[a]

Residue	ϕ	ψ	χ's						
1 ARG	-57.7	170.9	77.7	-163.3	51.9	89.8	-157.3	-174.9	-165.9
2 PRO	-75.0	149.0							
3 ASP	-70.2	-36.2	-71.7	-5.2	-175.0				
4 PHE	-68.6	2.6	77.8	64.6					
5 CYS	-104.3	-8.1	-79.5						
6 LEU	-88.5	-26.8	-63.7	-176.3					
7 GLU	-57.9	150.5	50.7	173.7	76.0	119.1			
8 PRO	-75.0	169.7							
9 PRO	-75.0	131.3							
10 TYR	-108.0	108.1	-170.8	72.6	5.0				
11 THR	-61.2	-57.4	-77.9	-53.4					
12 GLY	113.2	171.2							
13 PRO	-75.0	-27.7							
14 CYS	-59.9	157.0	-75.5						
15 LYS	119.7	30.1	90.5	100.3	140.4	92.0	148.3		
16 ALA	-87.6	168.7							
17 ARG	-116.2	77.1	-75.3	161.2	-125.0	135.8	-159.0	-175.0	-165.5
18 ILE	-111.4	120.5	-43.8	-144.4					
19 ILE	-79.9	113.1	-54.8	-62.2					
20 ARG	-119.8	173.0	-78.7	-51.9	168.9	80.2	-162.5	-175.0	-166.0
21 TYR	-115.9	135.3	-86.7	90.2	3.0				
22 PHE	-137.9	133.6	87.1	92.3					
23 TYR	-73.0	143.1	-178.0	-103.9	5.3				
24 ASN	-117.7	101.7	-161.7	-26.6	-174.0				
25 ALA	-53.6	-29.8							
26 LYS	-68.9	-69.4	-97.7	-73.6	-169.6	177.1	178.4		
27 ALA	-52.9	-40.3							
28 GLY	97.6	9.8							
29 LEU	-142.1	-178.7	40.7	72.1					
30 CYS	-97.2	144.0	-72.4						
31 GLN	-131.6	167.8	-60.1	-163.7	140.5	157.9			
32 THR	-91.0	144.3	57.1	152.9					
33 PHE	-155.3	141.1	87.1	98.1					
34 VAL	-66.2	128.3	-178.6						
35 TYR	-111.0	131.9	49.9	-11.4					
36 GLY	-46.5	-50.5							
37 GLY	115.9	9.2							
38 CYS	-158.0	151.4	61.9						
39 ARG	55.5	56.1	-42.7	-54.1	-175.3	168.1	-157.5	-175.0	-165.9
40 ALA	-79.9	149.0							
41 LYS	-166.5	56.5	-82.8	-147.3	-179.0	143.2	177.8		
42 ARG	-92.3	-46.1	-96.6	-97.5	161.1	170.9	-158.8	-175.0	-165.8
43 ASN	-78.1	77.2	-155.3	-1.4	-174.6				
44 ASN	-166.8	112.5	178.8	-22.6	-175.0				
45 PHE	-122.6	171.5	-62.8	87.0					
46 LYS	-98.2	-18.3	-71.4	-164.0	-98.4	-51.4	173.8		
47 SER	-143.1	158.0	91.8	-170.3					
48 ALA	-77.0	-35.3							
49 GLU	-57.4	-45.9	-76.0	95.6	85.5	-174.5			
50 ASP	-75.7	-31.7	-96.8	45.2	-175.7				
51 CYS	-64.1	-54.2	-177.6						
52 MET	-67.4	-28.2	-77.3	-58.8	-85.4				
53 ARG	-66.2	-37.7	174.1	153.1	32.9	-106.9	-160.8	-175.0	-165.0
54 THR	-96.9	-55.6	-73.2	174.3					
55 CYS	-68.3	-56.1	-90.0						

[a] ω = 180 DEGREES FOR ALL RESIDUES. ALL OTHER DIHEDRAL ANGLES ARE IN DEGREES.

their Diamond refinement of the BPTI structure. In the X-ray 1 structure, six residues have a value of $|\Delta\tau| \geq 15°$, whereas only one residue in the X-ray 2 structure has a value of $|\Delta\tau| \geq 15°$. Furthermore, only one of the six residues with $|\Delta\tau| \geq 10°$ corresponds to any of the thirteen values of $|\Delta\tau| \geq 10°$ in the X-ray 1 structure. In the X-ray 1 structure, twenty of the residues are reported to have values of $|\Delta\omega| \geq 10°$. After the Diamond refinement (Deisenhofer and Steigemann, 1975), only seven of the values of ω deviate from planarity by more than 10°, but only three of these correspond to the distorted peptide groups in the X-ray 1 structure. The X-ray 2 structure of Deisenhofer and Steigemann is a highly accurate one; if it is possible to collect data at higher resolutions (and refine these by Diamond's procedure), it will be interesting to see whether these deviations in τ and ω from the standard values adopted here represent a real distortion of the molecular geometry by interactions within the protein molecule.

Although the two computed low-energy structures for BPTI were constrained to remain close to the X-ray 1 and X-ray 2 structures, respectively, from which they were derived, even with the low value of the fitting parameter W used in the later stages of energy minimization they represent local-minimum low-energy conformations. The one obtained from the X-ray 2 data has a lower conformational energy. The

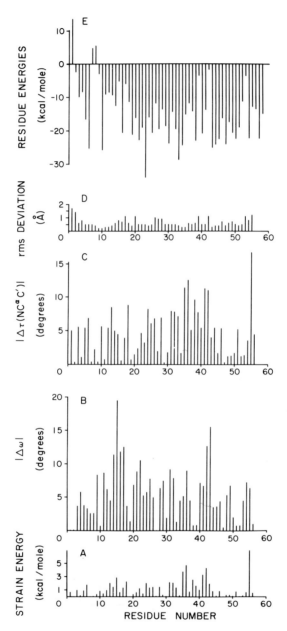

Fig. 2 Same as Fig. 1, but for the X-ray 2 structure and the low-energy structure deduced from the data.

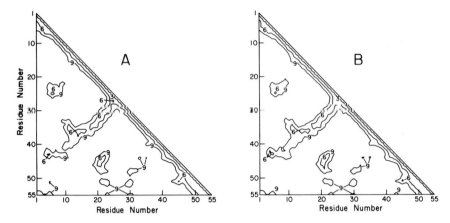

Fig. 3 Triangular contour map of $C_i^\alpha \cdots C_j^\alpha$ distances (in angstroms) for the standard-geometry models of BPTI based on (A) the X-ray 1 coordinates and (B) the X-ray 2 coordinates. The numbers on the contours of the similar maps in Figs. 1–4 of Burgess and Scheraga (1975) should all be augmented by 2 Å.

similarity between the two standard-geometry models for BPTI is illustrated in Fig. 3.

C. Three-Dimensional Structures for Homologous Inhibitor Proteins

The regions of homology between the amino acid sequences of RV II, BMI, BMK, and BPTI are shown in Fig. 4. Using the definition of homology given by Takahashi et al. (1974), the overall homologies between BPTI and RV II, BMI, and BMK are 50, 47, and 52%, respectively. The calculations reported here show that this degree of similarity is sufficient to allow the chains to fold into almost identical three-dimensional conformations. This suggests that there must be considerable redundancy in both the short- and long-range conformational information present in a given amino acid sequence.

1. Russell's Viper Venom Inhibitor II

After the initial geometric fit of the RV II sequence to the X-ray 1 structure, several unfavorable side chain–side chain or side chain–backbone interactions were found. The residues involved were $Pro_2 \cdots Thr_3$, backbone–side chain; $Phe_4 \cdots Ala_{42}$, side chain–backbone; $Asn_6 \cdots Leu_{25}$, side chain–side chain; $Ile_{21} \cdots Thr_{47}$, side chain–backbone; $Tyr_{22} \cdots Lys_{31}$, side chain–side chain; and $Lys_{29} \cdots Arg_{48}$, side chain–

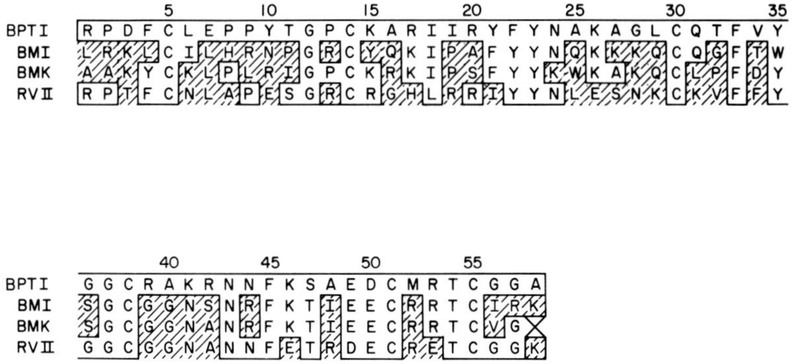

Fig. 4 Homologies of RV II, BMI, and BMK with BPTI (Creighton, 1975), expressed by the single-letter code in the numbering system of BPTI. Residues are considered to be homologous if they are identical or similar (e.g., Phe and Tyr). Open area indicates that the residues are homologous; shaded regions are nonhomologous; × represents a deletion. (Insertions prior to residue 1 of BPTI are ignored.)

side chain. None of the interacting residues is homologous with the parent BPTI molecule. Polar hydrogens were added and their positions were optimized with FOLD. All of the high-energy interactions involving the side chains could be relieved by conformational energy minimization with respect to the side-chain dihedral angles using FOLD, resulting in a total conformational energy of 680 kcal/mole.

The stage II procedure was then used to vary the dihedral angles ϕ, ψ, and χ to obtain a low-energy structure for RV II. All of the backbone atoms and those of the side chains that were homologous to BPTI were constrained to the X-ray 1 coordinates using a fitting parameter W of 50 kcal/Å² for the first 20 cycles of minimization. This constraint was then reduced stepwise to 0.10 kcal/Å² (a total of 260 cycles of minimization was carried out, with 195 variables, or about 1 cycle per variable). The final conformational energy of the RV II molecule was −757 kcal/mole, and the rms deviations between homologous atoms of BPTI (from the X-ray 1 structure) and RV II were 0.8 and 1.4 Å for the backbone and side chains, respectively (Table I). This model for the RV II molecule has all residues after the first three in low-energy conformations and favorable conformations for all of the disulfide bonds. Inspection of the triangular map for the $C_i^\alpha \cdots C_j^\alpha$ distances (Fig. 5A) shows that it is similar to those of Fig. 3. Table IV lists the final dihedral angles for RV II, which can be used in conjunction with the standard-geometry program (Momany et al., 1975) to generate all of the coordinates for the RV II molecule.

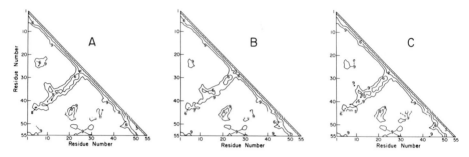

Fig. 5   Same as Fig. 3, but for (A) RV II, (B) BMI, and (C) BMK.

2. Black Mamba Toxin I

After the generation of the standard-geometry polypeptide chain, BMI had two unfavorable backbone–backbone interactions: between the O of Cys_{14} and the C^β of Ser_{36} (residue 36 in BPTI is Gly), and between the C^β of Ala_{20} and the N and C^α of Asn_{24}. Some other regions of the chain also contained atomic overlaps: $Ile_6 \cdots Gln_{25}$, backbone–side chain; $Arg_9 \cdots Phe_{33}$, side chain–side chain; $Arg_9 \cdots Thr_{34}$, side chain–backbone; $Pro_{19} \cdots Arg_{44}$, backbone–side chain; $Ala_{20} \cdots$

TABLE IV.   DIHEDRAL ANGLES OF LOW-ENERGY CONFORMATION OF RVII[a]

Residue	ϕ	ψ			χ's				
1 ARG	-158.9	169.0	-77.8	90.9	84.3	-81.7	159.9	4.6	0.7
2 PRO	-75.0	-1.0							
3 THR	69.6	-47.1	87.0	153.8					
4 PHE	-53.6	-20.9	101.3	97.1					
5 CYS	-68.5	-15.0	-101.5						
6 ASN	-96.0	31.4	-58.3	22.0	3.8				
7 LEU	-93.1	157.3	-105.1	-63.2					
8 ALA	-82.9	165.3							
9 PRO	-75.0	129.1							
10 GLU	-125.5	88.8	176.4	-173.2	-28.9	-23.8			
11 SER	-55.1	-48.1	57.6	176.6					
12 GLY	118.3	-175.0							
13 ARG	-92.3	-31.1	-172.6	179.7	173.4	179.7	-168.4	5.2	0.7
14 CYS	-67.0	150.8	-63.4						
15 ARG	-105.6	51.1	-155.2	179.6	-179.3	-80.0	179.9	5.0	0.7
16 GLY	-79.9	-171.2							
17 HIS	-133.8	56.1	-56.6	-69.2					
18 LEU	-99.7	117.5	-75.5	179.8					
19 ARG	-91.7	114.0	167.6	-170.1	168.9	172.2	-167.3	5.1	0.7
20 ARG	-122.9	-174.7	-61.0	-97.5	-171.1	176.4	-174.2	5.1	0.7
21 ILE	-139.9	154.2	77.7	133.2					
22 TYR	-150.2	154.3	93.4	108.8	2.9				
23 TYR	-73.6	129.5	-179.8	-104.7	4.7				
24 ASN	-96.9	114.4	-167.0	-14.3	5.0				
25 LEU	-72.4	57.3	-177.1	68.5					
26 GLU	-156.4	-55.3	-85.1	175.4	92.0	13.0			
27 SER	-82.7	-14.5	90.1	101.1					
28 ASN	76.8	17.7	-142.2	91.8	7.2				
29 LYS	-138.5	177.1	-110.8	178.1	177.5	-172.8	-55.1		
30 CYS	-96.2	144.8	-63.5						
31 LYS	-127.4	163.5	99.6	177.6	-172.5	-173.3	0.0		
32 VAL	-62.2	151.3	70.9						
33 PHE	-154.0	129.9	151.5	74.1					
34 PHE	-83.1	125.2	-166.4	76.6					
35 TYR	-114.9	136.6	-175.6	50.0	-29.3				
36 GLY	-67.4	-49.4							
37 GLY	114.0	42.4							
38 CYS	177.8	143.0	84.7						
39 GLY	38.2	67.8							
40 ASN	-107.5	-170.2							
41 ASN	-108.0	163.5	-143.8	-85.3	3.3				
42 ALA	-73.9	-30.3							
43 ASN	-71.0	60.4	-151.3	-179.0	38.9				
44 ASN	-156.0	136.1	102.3	87.1	3.4				
45 PHE	-141.8	158.3	61.2	92.0					
46 GLU	-86.4	-43.9	-169.4	172.0	81.6	1.6			
47 THR	-103.5	146.9	173.2	89.8					
48 ARG	-72.1	43.0	-149.7	173.5	-173.2	-178.1	179.3	5.9	0.7
49 ASP	-47.7	-57.5	-175.0	-4.8					
50 GLU	-53.0	-56.8	-173.2	176.7	79.2	-1.0			
51 CYS	-49.1	-54.5	-173.5						
52 ARG	-54.9	-47.8	-164.1	165.5	-177.9	179.4	-173.1	5.1	0.7
53 GLU	-58.3	-55.8	-105.4	51.0	28.6				
54 THR	-58.8	-59.9	46.5	-6.1					
55 CYS	-59.3	-52.7	-70.1						

[a] ω = 180 DEGREES FOR ALL RESIDUES.   ALL OTHER DIHEDRAL ANGLES ARE IN DEGREES.

Arg_{44}, backbone–side chain; Ala_{20} $\cdots$ Tyr_{23}, backbone–side chain; Phe_{21} $\cdots$ Ile_{48}, side chain–side chain; Trp_{35} $\cdots$ Gly_{37}, side chain–backbone; and Trp_{35} $\cdots$ Cys_{38}, side chain–backbone. The side-chain dihedral angles of the residues mentioned above were varied to remove most of these overlaps, and the conformational energy was minimized using FOLD.

The stage II program was then used for 180 cycles, and the initial fitting parameter W (50 kcal/Å^2 for homologous atoms) was reduced in steps to 0.01 kcal/Å^2, which was the value used in the last 20 cycles. The final conformational energy was -739 kcal/mole (see Table I for the energy contributions and rms deviations from the X-ray 1 structure). The similarity of the chain conformations of the BPTI and BMI molecules is obvious from a comparison of the $C_i^\alpha \cdots C_j^\alpha$ distance maps (Figs. 3 and 5B). Again, all of the disulfide bonds were able to form favorable conformations, and the side chains formed a close packed nucleus for the BMI molecule. Table V contains a list of the final dihedral angles used for the generation of the BMI structure.

3. Black Mamba Toxin K

The three-dimensional model for the BMK structure was generated in a manner similar to that for RV II and BMI. After the standard-geometry

TABLE V.   DIHEDRAL ANGLES OF LOW-ENERGY CONFORMATION OF BMI^a

Residue	ϕ	ψ	X's						
1 LEU	-159.6	166.0							
2 ARG	-64.1	-25.2	36.8	89.6	179.1	-179.0	-169.6	5.0	7.3
3 LYS	88.3	-24.0	-164.0	171.8	-179.8	84.1	-57.2		
4 LEU	-70.1	-10.8	-172.6	-178.3					
5 CYS	-76.5	-14.9	-93.7	69.2					
6 ILE	-98.3	24.5	-58.7	164.6					
7 LEU	-76.1	150.4	87.3	62.4					
8 HIS	-72.7	152.0	-107.5	-58.6					
9 ARG	-71.3	122.7	-70.8	118.4	86.6	-163.7		5.4	8.6
10 ASN	-117.9	118.7	149.1	-164.0	4.5	99.6			
11 PRO	-75.0	-56.3	-62.6						
12 GLY	113.1	179.3							
13 ARG	-102.7	-39.2	-168.8	171.6	179.0	-180.0	-168.1	5.0	7.7
14 CYS	-60.7	176.1	-73.2						
15 TYR	-140.6	61.1	-65.6	-74.7	6.5				
16 GLN	-95.9	-174.5	-82.8	-176.9	-86.8	5.1			
17 LYS	-133.8	82.5	-168.5	174.0	177.8	179.3	-55.8		
18 ILE	-100.9	112.3	-47.1	-92.6					
19 PRO	-75.0	108.2							
20 ALA	-157.3	-179.4							
21 PHE	-134.9	165.4	-52.1	107.8	3.2				
22 TYR	-142.4	154.5	89.2	105.7	3.5				
23 TYR	-85.3	132.5	-172.3	-100.9					
24 ASN	-100.7	112.5	-163.2	-19.5	7.4				
25 GLN	-71.5	57.9	-77.7	-96.9	162.5	-30.0	-55.7		
26 LYS	-157.4	-55.5	-176.1	-180.0	-179.6	179.3	-56.0		
27 LYS	-73.8	-33.7	-177.0	177.4	176.3	-179.7	-55.8		
28 LYS	100.6	-11.4	-162.9	-175.4	-179.0	-179.6			
29 GLN	-113.6	176.9	-76.3	-174.4	86.1	4.5			
30 CYS	-100.4	142.5	-51.4						
31 GLN	-132.5	168.2	-35.5	-178.2	-82.4	4.0			
32 GLY	-74.4	155.6							
33 PHE	-167.3	122.5	162.2	62.5					
34 THR	-71.2	128.0	-172.9	85.3					
35 TRP	-113.1	135.7	171.8	51.1					
36 SER	-47.3	-42.1	78.0	-75.9					
37 GLY	101.8	26.3							
38 CYS	-146.9	144.1	-78.3						
39 GLY	24.0	82.6							
40 GLY	-105.8	-164.0							
41 ASN	-119.2	-175.5	-89.3	175.5	41.1				
42 SER	-91.4	5.1	-44.6	-62.7					
43 ASN	-90.7	53.6	-153.9	-177.8	37.9				
44 ARG	-161.9	139.3	-146.5	168.0	-176.5	-106.3	-172.6	5.3	7.5
45 PHE	-140.8	151.0	-66.3	87.3					
46 LYS	-72.0	-43.4	-90.0	173.3	162.7	-173.3	-56.1		
47 THR	-115.4	160.8	178.6	146.8					
48 ILE	-73.2	-54.6	-75.9	-53.3					
49 GLU	-40.2	-48.9	34.8	-121.7	14.2	75.0			
50 GLU	-60.7	-43.4	-169.1	175.1	80.4	-1.2			
51 CYS	-62.4	-52.8	174.0						
52 ARG	-60.0	-34.1	-167.6	170.9	179.5	180.0	-168.4	5.3	7.1
53 ARG	-39.5	-49.0	153.7	-173.4	-160.6	175.0	-168.6	5.2	7.5
54 THR	-121.8	-66.5	143.0	88.5					
55 CYS	-57.3	-50.1	-89.3						

$^a \omega = 180$ DEGREES FOR ALL RESIDUES. ALL OTHER DIHEDRAL ANGLES ARE IN DEGREES.

polypeptide chain had been generated, there was only one high-energy backbone–backbone interaction (between the carbonyl oxygen of Cys_{14} and the C^β of Ser_{36}). The other high-energy interactions were between $Tyr_4 \cdots Ala_{42}$, side chain–backbone; $Tyr_4 \cdots Phe_{45}$, side chain–side chain; $Cys_5 \cdots Trp_{25}$, backbone–side chain; $Pro_{19} \cdots Arg_{44}$, backbone–side chain; $Ser_{20} \cdots Arg_{44}$, backbone–side chain; $Tyr_{22} \cdots Leu_{31}$, side chain–side chain; $Tyr_{23} \cdots Tyr_{22}$, side chain–side chain; and $Glu_{50} \cdots Thr_{54}$, backbone–side chain. The side chain of Trp_{25} (which is Ala in BPTI) contributed most to the high energy of interaction, not only with Cys_5, but also with Lys_6 and Tyr_{23}. These high energies undoubtedly arose from the fact that the initial values for the Trp side-chain dihedral angles had been chosen from a table of low-energy conformations (Lewis et al., 1973) with no consideration given to the juxtaposition of this side chain with the rest of the polypeptide chain. The high-energy regions involving side chains were relieved by conformational energy minimization by varying the side-chain dihedral angles of the residues mentioned above (and the positions of the polar hydrogens were optimized) using FOLD.

The stage II energy minimization procedure was then used to vary the dihedral angles ϕ, ψ, and χ to obtain a low-energy structure. The initial fitting parameter W was 50 $kcal/Å^2$ for homologous atoms. This was reduced stepwise to 0.5 $kcal/Å^2$, and a total of 220 cycles of minimization was applied. The final conformational energy was -709 kcal/mole (see Table I). Table I gives the energy contributions and the rms deviations of the homologous atoms from the parent X-ray 1 structure. Table VI lists the final dihedral angles for BMK. The triangular contour map of Fig. 5C is similar to those of Fig. 3.

The dihedral angles in Tables II–VI have been rounded to the nearest tenth of a degree. It should be emphasized that the precise atomic positions for these models should be generated from more accurate values of these dihedral angles and only with the geometry of Momany et al. (1975). Figures showing the residue energies and rms deviations of the backbone atoms of the low-energy structures of RV II, BMI, and BMK from the corresponding X-ray 1 coordinates of BPTI, and the computed Cartesian coordinates for all of the standard-geometry models of BPTI and the homologous inhibitors, may be requested from the ASIS National Auxiliary Publication Service.*

* These figures are available as document No. NAPS-03019 of the ASIS National Auxiliary Publication Service, % Microfiche Publications, P.O. Box 3513, Grand Central Station, New York, New York 10017. A copy may be secured by citing the document number and by remitting $3.00 for microfiche or $11.25 for photocopies. Outside the United States and Canada, postage is $1.00 for a microfiche or $3.00 for a photocopy. Advance payment is required. Make checks or money orders payable to Microfiche Publications.

TABLE VI.   DIHEDRAL ANGLES OF LOW-ENERGY CONFORMATION OF BMK[a]

Residue		ϕ	ψ	χ's					
1	ALA	-158.9	174.3						
2	ALA	-69.6	-20.6						
3	LYS	82.3	-44.0	-65.0	-55.0	-178.6	178.6	-55.3	
4	TYR	-53.2	-23.8	105.5	80.2	-12.2			
5	CYS	-62.9	-17.5	-64.0					
6	LYS	-91.5	24.6	-51.7	167.0	59.9	-174.8	-52.9	
7	LEU	-94.3	163.1	-74.2	175.7				
8	PRO	-75.0	159.2						
9	LEU	-76.2	135.5	-90.1	164.3				
10	ARG	-117.3	103.0	-154.0	-176.9	-176.5	-78.8	179.6	5.0	0.7
11	ILE	-68.6	-67.5	-62.8	140.4				
12	GLY	115.9	171.3						
13	PRO	-75.0	-35.0						
14	CYS	-50.9	166.3	-86.1					
15	LYS	-140.0	48.0	-124.9	94.7	78.9	-168.9	63.6	
16	ARG	-96.7	165.6	-146.6	-179.7	163.9	-178.4	-168.9	5.1	0.7
17	LYS	-106.2	75.3	-170.3	177.9	176.4	-179.2	0.0	
18	ILE	-101.3	112.6	-60.2	-85.0				
19	PRO	-75.0	106.9						
20	SER	-128.7	175.3	24.7	64.9				
21	PHE	-135.4	153.6	-79.0	82.4				
22	TYR	-143.7	145.9	129.4	102.0	3.0			
23	TYR	-78.1	124.2	179.6	-111.3	3.8			
24	LYS	-94.7	111.4	-177.0	-177.0	172.9	-175.1	66.6	
25	TRP	-71.2	62.9	-147.1	77.3				
26	LYS	-166.5	-52.1	-172.1	-171.8	169.6	-140.4	-55.3	
27	ALA	-75.7	-24.1						
28	LYS	91.5	-12.2	-166.3	178.3	179.5	179.7	-55.7	
29	GLN	-123.0	170.7	-143.3	48.0	-109.5	6.0		
30	CYS	-91.5	142.6	-59.9					
31	LEU	-118.0	166.1	57.1	62.2				
32	PRO	-75.0	147.1						
33	PHE	-151.6	119.7	159.9	69.9				
34	ASP	-67.8	123.0	-178.5	32.9	11.9			
35	TYR	-93.1	124.9	-173.1	40.1	-26.8			
36	SER	-55.8	-36.3	89.1	-97.6				
37	GLY	98.4	39.1						
38	CYS	-141.9	150.7	-105.3					
39	GLY	-33.3	98.5						
40	GLY	-105.7	-153.0						
41	ASN	-100.6	178.5	-136.0	-101.7	4.4			
42	ALA	-84.1	-8.5						
43	ASN	-78.2	58.1	-154.2	-149.2	17.2			
44	ARG	-163.7	138.4	70.9	178.8	-152.1	-110.9	-165.8	5.3	0.7
45	PHE	-136.4	157.0	-70.3	86.5				
46	LYS	-86.6	-29.1	-83.1	168.8	163.1	-174.3	-56.0	
47	THR	-123.7	157.2	139.6	87.5				
48	ILE	-73.4	-57.5	-56.1	138.4				
49	GLU	-45.7	-46.2	48.6	-109.2	131.3	-30.8		
50	GLU	-60.0	-39.3	-174.0	178.6	79.2	-0.8		
51	CYS	-70.1	-53.8	177.0					
52	ARG	-58.7	-36.0	-167.3	173.9	178.3	179.8	-166.6	5.2	0.7
53	ARG	-42.1	-43.4	152.9	-176.6	-165.2	175.2	-167.5	5.2	0.7
54	THR	-118.2	-62.4	150.4	87.6				
55	CYS	-61.3	-54.3	-86.8					

[a] ω = 180 DEGREES FOR ALL RESIDUES.   ALL OTHER DIHEDRAL ANGLES ARE IN DEGREES.

D. Discussion

The standard-geometry atomic coordinates obtained from both sets of X-ray data for BPTI were generally free of atomic overlaps. The geometry of the amino acid residues in the experimental structures for BPTI (X-ray 1 and X-ray 2) was similar to the standard geometry (including planar *trans*-peptide groups) obtained from amino acids and their derivatives (Momany *et al.*, 1975), as indicated by their relatively small strain energies (Table I). Consequently, there were only small differences in ϕ's, ψ's, and χ's between the experimental and computed structures for BPTI. Thus, it follows that procedures that refine protein structures such as BPTI by energy minimization, with the aim of improving the geometry and the conformational energy (Levitt and Lifson, 1969; Hermans and McQueen, 1974), will not be dominated by the constraints to fixed bond lengths and bond angles.

The procedures used here have led to conformations for BPTI with acceptable bond lengths and bond angles, and which have lower conformational energies than those of the original X-ray structures without large deviations from the X-ray coordinates. The torsional energy contribution is a measure of both the displacement of some of the rotations of groups around single bonds, from their minimum positions, and the interaction of dipoles on either side of such single bonds (the so-called 1–4 electrostatic interaction). The low values of the nonbonded energies indicate that all of the computed structures have fewer atomic overlaps than the original X-ray structures.

Although the X-ray conformations for BPTI had low conformational energies initially, the imposition of standard geometry in the modified stage I had introduced several unacceptable atomic overlaps. These atomic overlaps could usually be removed by optimization of the side-chain dihedral angles.

In adjusting side-chain positions of the standard-geometry models, difficulties were sometimes encountered in the modified stage I procedure when the side chains were large, when a reported bond angle in that residue differed by more than 5° from its standard value (Momany *et al.*, 1975), or when the value reported for ω differed by more than 10° from its standard value of 180°. For example, in the X-ray 2 data for BPTI, there are such large deviations in some bond angles and/or ω from standard values in the sequence of residues 14–17 (Cys-Lys-Ala-Arg) with large side chains, and the resulting standard-geometry fit was not as good as it was in sequences with small side chains (e.g., residues 36–38, Gly-Gly-Cys) despite large deviations in τ's and ω's from standard geometry among the latter residues. The values of the bond lengths and bond angles calculated from the X-ray 2 coordinates are very close to the standard values used here, although ω and $\tau(NC^\alpha C')$ were allowed to vary in the Diamond (1966, 1971) refinement that was used to obtain the X-ray 2 coordinates. [Our experience has shown that, when ω is allowed to vary, it remains close to 180° in peptides. Thus, the approximation in which the peptide group is held fixed in the planar *trans* conformation is likely to be a reasonable model of the real structure (Zimmerman and Scheraga, 1976).]

As a result of the conformational energy calculations reported here, the rms deviation of the side chains increased by only 0.3 Å for the X-ray 2 structure and actually decreased by 0.2 Å for the X-ray 1 conformation (Table I). The average rms deviation does not fully reflect the closeness of the fit between the X-ray and the standard-geometry structures. There appears to be a systematic variation in the rms deviation along the amino acid sequence of BPTI (see Figs. 1D and 2D). Those areas where the X-ray structure is less precise (W. Steigemann, personal communication; Deisenhofer and Steigemann, 1975) are the areas where the rms deviation of the computed structure is greatest (e.g., residue 1). Of the twelve residues showing large deviations (Figs. 1D and 2D), eight are Arg (in X-ray 2 all but one are Arg) and two of the others are Lys and Glu. These residues all have long side chains on the surface of BPTI and are poorly resolved by the X-ray crystallographic electron densities (Deisenhofer and Steigemann, 1975). It is interesting to compare the values of $|\Delta\omega|$ for each of the standard-geometry structures (Figs. 1B and 2B). Since ω is set to 180° for the standard-geometry low-energy conformations, these figures reflect the fact that the ω's in the X-ray 1 data set have large deviations from 180°, whereas those in the X-ray 2 data set are closer to 180°. The conformational energy of almost all of the residues is a stabilizing influence in both structures (Figs. 1E and 2E). The only residues with positive conformational energies are at the beginning of the chain, where there is also considerable uncertainty in the experimentally reported positions for these atoms.

The disulfide energy term in Table I reflects the geometry of the disulfide bonds; it includes deviation of the rotation around the $S^{\gamma_i}S^{\gamma_j}$ bond from its minimum-energy position and deviations in the bond angles at $C^{\beta_i}S^{\gamma_i}S^{\gamma_j}$ and $C^{\beta_j}S^{\gamma_j}S^{\gamma_i}$ and in the bond length $S^{\gamma_i}S^{\gamma_j}$ from their standard values. The three disulfide bridges in the X-ray 2 structure appear to have geometries and conformations close to the standard values reported by Momany et $al.$ (1975). Energy minimization directed the "chain folding" so that all of these disulfide bridges had acceptable geometries and conformations.

On using the procedures described in this paper, the conformational energy of the standard-geometry model of the X-ray 2 coordinates was ~400 kcal/mole lower than that of the X-ray 2 structure from which the calculation was started. This minimum-energy conformation has all of the polar protons positioned and should be useful for the computer simulation of the interaction of BPTI with trypsin.

The low-energy conformations obtained for the three homologues of BPTI had coordinates of their homologous backbones and side-chain atoms that were similar to those of BPTI. In all cases the side chains packed to form a compact interior without severe atomic overlaps. Residues 1–3 of RV II have positive conformational energies, probably due to the strain of accommodating a proline residue in position 2 of the BPTI sequence. The deviations of the RV II and BMK backbone atoms from the X-ray 1 coordinates were always less than 1 Å, and the side-chain atoms had rms deviations of less than 2 Å for all of the homologous proteins. The nonbonded energy contributed to the stability of the low-energy conformations for RV II, BMI, and BMK. The disulfide loops had reasonable geometries in all of the homologous proteins. Only three residues in BMK had positive conformational energies (Leu_{31}, Glu_{49}, and Arg_{53}), and all energies were below 10 kcal/mole. Several residues in BMI had positive conformational energies but only Glu_{49} had a conformational energy greater than 10 kcal/mole. The slightly larger values for the rms deviations for BMI were probably due to the use of fewer cycles of energy minimization in this calculation.

Our calculations support the concept that, within the limits of error detected experimentally, proteins have standard bond lengths and bond angles, with peptide groups in the planar *trans* conformation. Differences between the computed and experimental structures for BPTI occurred in regions of the chain that were less well defined by the X-ray data (i.e., those regions that changed significantly in going from the X-ray 1 to the X-ray 2 coordinates). It is important to note that the use of conformational energy calculations to develop models for the structures of homologous proteins is more likely to succeed if all of the residues of the starting conformation are in their correct local energy minimum. This information is generally available from most high-resolution X-ray crystallographic studies on proteins. By imposing the initial standard geometry on a residue by residue basis, as was done here, it is possible to avoid potential artifacts involving compensating errors in successive value of ϕ and ψ.

VI. CONCLUDING REMARKS

A variety of problems (indicated in Section IV) have been subjected to conformational analysis using computational procedures involving empirical potential energy functions. Some of these have been illustrated in Section V.

The multiple-minimum problem has been surmounted for small open-chain and cyclic peptides and for synthetic analogues of fibrous proteins. To solve this problem for globular proteins, so that we may understand how they fold to their native conformations, simplified procedures are being used to combine algorithms based on short-range interactions with those in which long-range interactions are introduced (Burgess and Scheraga, 1975; Tanaka and Scheraga, 1975, 1977); these procedures have been reviewed elsewhere (Nemethy and Scheraga, 1977).

As one type of simplification, use is made of rigid geometry, for the reasons outlined in Section II. This is in no way meant to explain away as artifacts the departures from standard geometry and from planarity of peptide groups that are observed in the X-ray structures of proteins. Rather, it is an attempt to determine how well the observations can be reconciled with the geometry observed in crystals of small molecules. As the quality of spectroscopic force constants improves, deviations from standard geometry can be assessed—to determine whether long-range interactions in a protein do indeed induce strain in bond angles and in peptide groups.

As far as the computations illustrated in Section V are concerned, low-energy structures of several homologues of BPTI have been computed. Even though their energies were initially very high, primarily because of overlaps among nonhomologous residues, these overlaps were relieved by energy minimization. Thus, we have available (in the coordinates deposited with Microfiche Publications) well-defined structures of three inhibitors that may possibly be of use to experimentalists in the interpretation of their data on these proteins in terms of the predicted specific locations of the atoms of the structures.

ACKNOWLEDGMENTS

This work was supported by research grants from the National Science Foundation (PCM75-08691) and from the National Institute of General Medical Sciences, National Institutes of Health, U.S. Public Health Service (GM-14312).

We are indebted to Dr. W. Steigemann for sending us the X-ray 1 and X-ray 2 coordinates, to Dr. Susan Fitzwater and Dr. Paul K. Warme for helpful comments on the manuscript, and to Zachary Hodes, Shirley Rumsey, and Dorothy Mei for checking some of the computations.

REFERENCES

Arnott, S., and Scott, W. E. (1972). *J. Chem. Soc., Perkin Trans.* 2 pp. 324–335.
Bixon, M., and Lifson, S. (1967). *Tetrahedron* **23,** 769–784.
Burgess, A. W., and Scheraga, H. A. (1975). *Proc. Natl. Acad. Sci. U.S.A.* **72,** 1221–1225.
Burgess, A. W., Momany, F. A., and Scheraga, H. A. (1975a). *Biopolymers* **14,** 2645–2647.
Burgess, A. W., Shipman, L. L., and Scheraga, H. A. (1975b). *Proc. Natl. Acad. Sci. U.S.A.* **72,** 854–858.
Burgess, A. W., Shipman, L. L., Nemenoff, R. A., and Scheraga, H. A. (1976). *J. Am. Chem. Soc.* **98,** 23–29.
Creighton, T. E. (1975). *Nature (London)* **255,** 743–745.
Davidon, W. C. (1959). *U.S. A. E. C. Res. Dev. Rep.* **ANL-5990.**
Deisenhofer, J., and Steigemann, W. (1975). *Acta Crystallogr., Sect. B* **31,** 238–250.
Diamond, R. (1966). *Acta Crystallogr.* **21,** 253–266.
Diamond, R. (1971). *Acta Crystallogr., Sect. A* **27,** 436–452.
Dygert, M., Go, N., and Scheraga, H. A. (1975). *Macromolecules* **8,** 750–761.
Endres, G. F., Swenson, M. K., and Scheraga, H. A. (1975). *Arch. Biochem. Biophys.* **168,** 180–187.
Fletcher, R. (1970). *Comput. J.* **13,** 317–322.
Fletcher, R., and Powell, M. J. D. (1963). *Comput. J.* **6,** 163–168.
Gibson, K. D., and Scheraga, H. A. (1967). *Proc. Natl. Acad. Sci. U.S.A.* **58,** 420–427.
Hagler, A. T., Lifson, S., and Huler, E. (1974). In "Peptides, Polypeptides, and Proteins" (E. R. Blout *et al.,* eds.), pp. 35–48. Wiley, New York.

Hermans, J., Jr., and McQueen, J. E., Jr. (1974). *Acta Crystallogr., Sect. A* **30,** 730–739.

Hopfinger, A. J. (1971). *Macromolecules* **4,** 731–737.

Isogai, Y., Nemethy, G., and Scheraga, H. A. (1977). *Proc. Natl. Acad. Sci. U.S.A.* **74,** 414–418.

Levitt, M., and Lifson, S. (1969). *J. Mol. Biol.* **46,** 269–279.

Lewis, P. N., Momany, F. A., and Scheraga, H. A. (1973). *Isr. J. Chem.* **11,** 121–152.

Miller, M. H., and Scheraga, H. A. (1976). *J. Polym. Sci., Polym. Symp.* **54,** 171–200.

Momany, F. A. (1976a). *J. Am. Chem. Soc.* **98,** 2990–2996.

Momany, F. A. (1976b). *J. Am. Chem. Soc.* **98,** 2996–3000.

Momany, F. A., McGuire, R. F., Burgess, A. W., and Scheraga, H. A. (1975). *J. Phys. Chem.* **79,** 2361–2381.

Nemethy, G., and Scheraga, H. A. (1977). *Q. Rev. Biophys.* **10,** 239–352.

Nishikawa, K., and Ooi, T. (1974). *J. Theor. Biol.* **43,** 351–374.

Owicki, J. C., Nemenoff, R. A., Snir, J., and Scheraga, H. A. (1978). In preparation.

Pincus, M. R., Zimmerman, S. S., and Scheraga, H. A. (1976). *Proc. Natl. Acad. Sci. U.S.A.* **73,** 4261–4265.

Pincus, M. R., Zimmerman, S. S., and Scheraga, H. A. (1977). *Proc. Natl. Acad. Sci. U.S.A.* **74,** 2629–2633.

Platzer, K. E. B., Momany, F. A., and Scheraga, H. A. (1972). *Int. J. Pept. Protein Res.* **4,** 201–219.

Powell, M. J. D. (1964). *Comput. J.* **7,** 155–162.

Pullman, B., and Pullman, A. (1974). *Adv. Protein Chem.* **28,** 347–526.

Ramachandran, G. N. (1974). In "Peptides, Polypeptides, and Proteins" (E. R. Blout *et al.,* eds.), pp. 14–34. Wiley, New York.

Ramachandran, G. N., and Venkatachalam, C. M. (1968). *Biopolymers* **6,** 1255–1262.

Rasse, D., Warme, P. K., and Scheraga, H. A. (1974). *Proc. Natl. Acad. Sci. U.S.A.* **71,** 3736–3740.

Scheraga, H. A. (1971). *Chem. Rev.* **71,** 195–217.

Scheraga, H. A. (1974a). In "Current Topics in Biochemistry, 1973" (C. B. Anfinsen and A. N. Schechter, eds.), pp. 1–42. Academic Press, New York.

Scheraga, H. A. (1974b). In "Peptides, Polypeptides, and Proteins" (E. R. Blout *et al.,* eds.), pp. 49–70. Wiley, New York.

Shipman, L. L., Burgess, A. W., and Scheraga, H. A. (1975). *Proc. Natl. Acad. Sci. U.S.A.* **72,** 543–547.

Strydom, D. J. (1973). *Nature (London), New Biol.* **243,** 88–89.

Takahashi, H., Iwanaga, S., and Suzuki, T. (1972). *FEBS Lett.* **27,** 207–210.

Takahashi, H., Iwanaga, S., Hokama, Y., Suzuki, T., and Kitagawa, T. (1974), *FEBS Lett.* **38,** 217–221.

Tanaka, S., and Scheraga, H. A. (1975). *Proc. Natl. Acad. Sci. U.S.A.* **72,** 3802–3806.

Tanaka, S., and Scheraga, H. A. (1977). *Proc. Natl. Acad. Sci. U.S.A.* **74,** 1320–1323.

Warme, P. K., and Scheraga, H. A. (1973). *J. Comput. Phys.* **12,** 49–64.

Warme, P. K., and Scheraga, H. A. (1974). *Biochemistry* **13,** 757–767.

Warme, P. K., and Scheraga, H. A. (1975). *Biochemistry* **14,** 3509–3517.

Warme, P. K., Go, N., and Scheraga, H. A. (1972). *J. Comput. Phys.* **9,** 303–317.

Warme, P. K., Momany, F. A., Rumball, S. V., Tuttle, R. W., and Scheraga, H. A. (1974). *Biochemistry* **13,** 768–782.

Zimmerman, S. S., and Scheraga, H. A. (1976). *Macromolecules* **9,** 408–416.

Zimmerman, S. S., Pottle, M. S., Nemethy, G., and Scheraga, H. A. (1977a). *Macromolecules* **10,** 1–9.

Zimmerman, S. S., Shipman, L. L., and Scheraga, H. A. (1977b). *J. Phys. Chem.* **81,** 614–622.

5

Aspects of Biomolecules in Their Surroundings: Hydration and Cation Binding

BERNARD PULLMAN

Institut de Biologie Physico-Chimique, Laboratoire de Biochimie Théorique, Associé au Centre National de la Recherche Scientifique, Paris, France

I. INTRODUCTION

The problem of the influence of the environment on the properties of biomolecules has been exemplified using the structure of tRNA in solution as compared to its structure in the crystal, that is, the different perturbation effects produced by two different types of milieu (see Chapter 2). However, there is a more fundamental way of looking at the problem, which is rather naturally that of the theoreticians. This approach distinguishes between the *intrinsic properties* of the free, isolated molecule, such as it would be at first approximation in the gas phase, and the effect that different environmental conditions produce upon these properties.

Quantum-mechanical studies were for a long time concentrated essentially, for technical reasons, on free molecules and their intrinsic electronic or conformational properties. The indisputable success obtained in the interpretation or prediction of a large number of experimental phenomena and the practical success of theories based on this type of research indicate that these intrinsic properties are largely conserved or only slightly modified under the influence of the molecular

FRONTIERS IN PHYSICOCHEMICAL BIOLOGY

environment. This is evidently not always the case and the determination of the role of the environment is then primordial.

From that point of view, one of the key problems is *water*. Biologists and, in particular, biophysicists have been fascinated for a long time by the possibility of the existence of *structured water* around biological molecules, in particular, polymers. The following questions are raised most frequently in connection with this possibility: Are there preferential binding sites of water molecules on these substances? Is there a formation of *shells* of organized, bound water and, if so, how many? What are the energies of interaction? What is the effect of bound water upon the properties of the biological substrate and, inversely, what is the effect of binding to the substrate on the properties of bound water?

In spite of the efforts by numerous groups, very little seems to be known experimentally about this fundamental subject. It is therefore a particularly tempting and exciting problem for theoreticians, the more so because recent developments in the concepts and methods of quantum mechanics and in the computational facilities enable it to be treated more fully. There is also the prospect of obtaining a precision far beyond the possibilities of current experimentation.

On the other hand, while water undoubtedly represents the essential element of the constant environment of biomolecules, it is not the only one. Ions and, in particular, metal cations (but also anions, which frequently accompany them) are nearly omnipresent in the biophase, and their interactions with a large number of biological substrates apparently are important, as judged by the consequences. Although the experimental data in this field (for example, on the binding sites) are somewhat more abundant than those for water, we are still far from knowing precisely the specificity and the essential features of this interaction.

Since 1974 our laboratory has been engaged in a systematic and detailed exploration of these two problems: hydration and cation binding of biomolecules. Without presenting the technical details of these investigations, I would like to indicate that they were carried out using the most refined methods of quantum chemistry, in particular, the *ab initio* self-consistent field method (see, e.g., Pullman and Pullman, 1974; Pullman and Saran, 1976), which avoids the use of empirical parametrizations. Numerous results have been obtained, too numerous to be described here in detail. I will just present a few representative examples, which will illustrate the essence and the scope of the results obtained.

II. HYDRATION

From a general point of view, there are two ways to investigate the effect of water the properties of biomolecules (see, e.g., Pullman, 1976). The traditional approach is to describe the bulk effect of the solvent by a *continuum* model, which considers it as a dielectric polarizable milieu around a cavity containing the solute. This *macroscopic* model may perhaps account correctly for certain effects of hydration but obviously does not provide information about the intimate interactions between the solvent and solute molecules.

The second procedure is based upon a *microscopic* model, which we call the *supermolecule* model and which consists of treating the solute and the surrounding water molecules as a single, unique polyelectronic entity. The usual methodology proceeds by successive inclusion of water molecules, in increasing number, up to the completion of the hydration shell or shells. This method is particularly well adapted to the requirements of our problem. In particular, it provides the answer to the question of the possible existence and structure of organized water around biomolecules. This second procedure has been used extensively in our work (Pullman and Pullman, 1975), and I will now present a few of the results obtained with it.

The examples will concern essentially the nucleic acids. These substances have, a priori, a number of hydration sites distributed on their constituent units: the purine and pyrimidine bases, the phosphodiester group, and the sugars. The overall strategy of the theoretical approach involves the following three steps.

1. The first step consists of a detailed exploration of *the hypersurface of interaction of the biological substrate with a single water molecule* approaching it from various directions, turning it around the solute and about its local axis.

Such a search yields information concerning the basic characteristics of the solute–solvent interaction at the microscopic level: The *minima* on the energy hypersurface indicate the most stable positions for a single solvent molecule, the value of the binding energy, the distance of equilibrium, and relative orientations of the solute and solvent. Moreover, when the spanning is done in sufficient detail, the *lability characteristics* of the interaction are obtained, which relate the extent to which it is possible to depart from the most stable positions without altering the binding appreciably.

Figure 1 presents the results of such an exploration for one of the nucleic acid bases, guanine [see Port and Pullman (1973) for the results

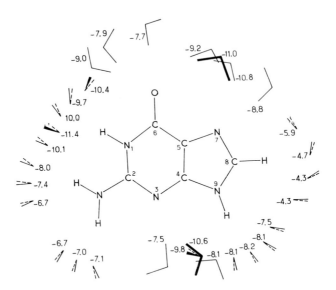

Fig. 1 Hydration sites in guanine. Energies are expressed as kilocalories per mole. Heavy lines, preferred hydration sites; full lines, coplanar arrangement of water and the base; half-dashed, perpendicular arrangement of water with respect to the plane of the base.

for the other bases]. At the principal binding sites to this planar heteroaromatic molecule, the oxygen atom of the water molecule is in the plane of the base. The preferred hydration sites are indicated by heavy lines. For the sake of comparison with the energy of water–water interactions, it is useful to indicate that this amounts in the same type of calculation to about 6.5 kcal/mole. Thus, the absolute values of the interactions are somewhat overevaluated. Our major interest resides, however, in the relative values, which are conserved.

Another, more complex example of such an exploration is offered by the phosphate, or rather the phosphodiester, group (represented in Fig. 2). This group has a tridimensional structure, which makes the exploration and description of its hydration scheme more delicate.

The results related to the possibilities of hydration by one water molecule are best visualized by first examining the situation in the various OPO planes of the phosphate. They are presented in Fig. 3 with ad hoc notations (Pullman *et al.*, 1975a,b; Perahia *et al.*, 1975) for the potential energy minima, which are the most stable sites of binding and are the only ones represented. It is obvious that many nearly equivalent binding sites exist. The plane O_1PO_3 contains the site corresponding to the *global minimum* (-28.6 kcal/mole) on the energy hypersurface: This occurs in two equivalent positions when water is bound to one of the

O3
O1
P
O4
C4
O2
C2

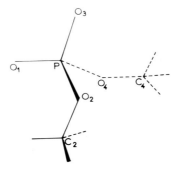

Fig. 2 Phosphodiester group in *gg* conformation.

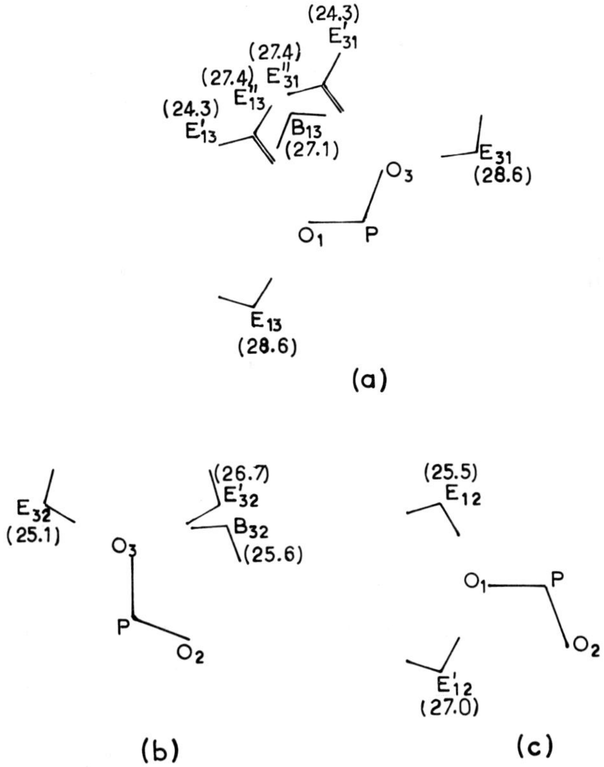

(24.3)
E'31
(27.4)
(27.4)
E''31
E''13
(24.3)
E'13
B13
(27.1)
O3
E31
(28.6)
O1
P
E13
(28.6)
(a)
E32
(25.1)
O3
P
O2
(26.7)
E'32
B32
(25.6)
(b)
(25.5)
E12
O1
P
O2
E'12
(27.0)
(c)

Fig. 3 Preferred hydration sites in the different OPO planes ($-E$ is expressed as kilocalories per mole). (a) Plane O_1PO_3; (b) plane O_2PO_3; (c) plane O_1PO_2.

anionic oxygens, O_1 or O_3, by one H bond at the exterior of the OPO angle (positions E_{13} and E_{31}). In fact, on the whole, the distribution of the favorable hydration sites in the neighborhood of the anionic oxygens points to the existence of a whole zone of attraction for water about these two atoms. This is confirmed by studying the variations in the binding energy obtained by rotating the plane of the water molecule about the PO_1 axis from its most stable position, E_{13}, in such a way that $O_1 \cdots$ HO describes a cone about this axis (see Fig. 4).

2. The second step in the strategy of hydration studies involves the *determination of the existence and structure of hydration shell or shells* around the biological substrate.

A particularly striking example of this is offered by our work on the phosphate group. Taking the results for monohydration as a basis, a simultaneous attachment of water molecules in progressively increasing numbers is accomplished, the structure of the supermolecule being optimized in each case.

The finding under step 1 of a large number of nearly equivalent positions for one molecule of water around the phosphate oxygens is a strong indication that there are many possible schemes for polyhydration. The results obtained for simultaneous fixation of a number of water molecules to the phosphate ion are given in Table I, from which it can be seen that many possibilities appear. *Up to six water molecules may be accommodated in the first hydration shell.* Intermediate hydration involving two to five water molecules may, of course, take place, with a number of energetically nearly equivalent possibilities occurring for a given number of water molecules (three for $n = 2$ to 4; two for $n = 5$). The average energy of water attachment is approximately -23 kcal/mole for $n = 2$ and 3 and decreases when n increases further, down to -17.3 for $n = 6$. The stable positions for water attachment in the polyhydrates are close enough to those established for the interaction of the phosphate with a single water molecule to enable us to conserve the same notations. Figure 5 gives a pictorial representation of the structure of the hexahydrate.

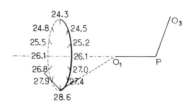

Fig. 4   Variation in $-E$ upon rotation of water out of position E_{13} ($O_1 \cdots$ HOH sweeping a cone about the PO_1 axis).

TABLE I

Polyhydration of DMP⁻ with n Water Molecules
in the First Shell

n	Occupied positions	$-E_{total/n}$ (kcal/mole)
2	E_{13}, E_{31}	27.1
	B_{14}, E'_{12}	23.6
	B_{14}, B_{12}	20.4
3	B_{13}, B_{32}, B_{14}	23.0
	B_{13}, E'_{32}, E'_{14}	23.5
	B_{13}, E_{31}, E_{13}	24.7
4	E_{13}, E_{31}, E_{14}, E_{32}	21.9
	E'_{12}, E'_{34}, E'_{14}, E'_{32}	21.7
	E_{14}, E_{32}, E'_{14}, E'_{32}	21.3
5	B_{13}, B_{32}, B_{14}, E'_{12}, E'_{34}	20.3
	B_{13}, E'_{32}, E'_{14}, E'_{12}, E'_{34}	20.3
6	E_{31}, E_{32}, E_{34}, E_{12}, E_{13}, E_{14}	17.3

The study of the hydration shells may be extended to a search for a second and further such shells. The relatively strong energies of interaction of the water molecules in the first hydration shell of the phosphate group suggest the possibility that a second hydration shell organized around the first one exists. Explicit computations (Pullman *et al.*, 1975a,b) confirm that this is indeed the case for this ionic biosubstrate and show that organized "bound" water is most probably essentially limited to two hydration layers. With nonionic substrates, such as the nucleic acid bases, only one hydration shell is expected.

3. Finally, the third stage of the strategy involves *the exploration of the effect of hydration upon the molecular properties of the species involved* and implies, therefore, two aspects, which are discussed below.

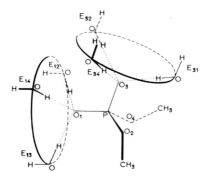

Fig. 5 Representation of the hexahydrate of DMP⁻.

a. *The effect of hydration upon the properties of the biological substrate.*
The case of the phosphodiester linkage offers again a good example of
this aspect. It is well established (Pullman and Saran, 1976) that the
phosphodiester linkage has an intrinsic preference for a *gauche–gauche*
(gg) conformation about the P—O$_{ester}$ bonds and that this preference is
largely conserved in the crystal structure of nucleic acids,
polynucleotides, and their components. The question may be raised
about the possible effect of hydration on this situation. The answer to
this question is given in Table II, which shows that progressive hydration
decreases the difference in stability between the different conformers to
the point that when the first hydration shell is completed the *tg (gt)* form
is practically equal in energy to the *gg* form. The *tt* form is always the
least stable, but draws closer to the others as hydration increases. Nuclear magnetic resonance (nmr) studies on the $^2J_{P,C-5'}$ spin–spin coupling in mononucleotides confirm an elongation of the mean value of
the P—O$_{5'}$ torsion angles in water solution (±120 to ±150) with respect
to the value in the crystal (±60 to ±09) (Giessner-Prettre and Pullman,
1974). Similarly, nmr studies on solution conformation of dinucleoside
monophosphates clearly show that the effect of solvation increases the
importance of *trans* conformations with respect to rotations about one or
the other of the P—O$_{ester}$ bonds: While these type of compounds are
predominantly g^-g^- in the crystals, the total population of their *gg* conformers is only in the range of 10–40% in water solution, with the
combined population of the *tg* or *gt* conformers making up the remainder (Sarma and Danyluk, 1977).

TABLE II

**Effect of Hydration on the
Conformation of DMP$^-$** [a]

	gg	*gt*	*tt*
DMP$^-$	0	3.4	8.0
H$_2$O			
1		2.4–3.9	
2		2.9	
3		2.6	
4		1.2	
5		1.0$_5$	
6		0.0	1.7$_5$

[a] Energies are expressed as kilocalories per mole with respect to the
gg form taken as zero energy.

b. The second aspect of the effect of hydration on the properties of the species involved is concerned with *the modification brought about in the electronic structure of the water molecules by their binding to the biological substrate*. This perturbation can be visualized, for example, by the distribution of electronic charges obtained by a Mulliken population analysis. It must be kept in mind that the "charges" are not "observable" in the quantum-mechanical sense and represent only a conventional convenient partition of the exact electron-density distribution into atom-centered fractions. Moreover, the numerical values of the charges depend on the basis set used in the computation. Nevertheless, the relative values as well as their behavior under a given perturbation are sufficiently independent of the basis set to provide general information about this problem.

Figure 6 shows the distribution of the net electronic charges obtained with an STO 3G basis set (Pullman *et al.*, 1975b; Pullman, 1976) in (a) free dimethyl phosphate anion (DMP⁻), considered as a model of a phosphodiester, (b) free water, and two monoadducts: (c) at the B_{13} site and (d) at the E_{31} site.

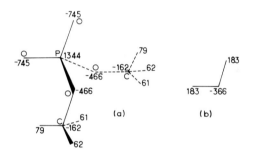

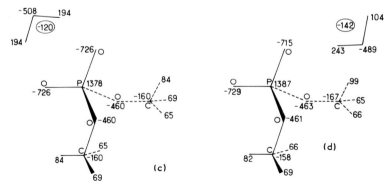

Fig. 6   Net charges in (a) DMP⁻, (b) H_2O, and (c, d) two monohydrates of DMP⁻.

It is seen that the structure of the bound water is significantly perturbed with respect to that of free water: Its oxygen atom has become appreciably more negative and its hydrogen(s) engaged in the hydrogen bond(s) with DMP⁻ have become slightly but significantly more positive. These data show that the binding of water to DMP⁻ produces a partial transfer of electrons from the latter to the former. Thus, the bound water carries an excess of electronic density from 0.120 to 0.142 e which may be compared to the case of water dimers in which there is a small charge transfer (0.034 e) from the proton acceptor to the proton donor molecule. This enhancement of the charge transfer is a result of the anionic nature of the solute, which transmits a *small partial anionic character to the bound water* molecule of the first layer, thereby enhancing its attraction toward other water molecules.

Polyhydration in the first shell indicates that the presence of a number of water molecules reduces, in each of them, the amount of charge transfer found for a single molecule, which becomes about 0.1 e in the hexahydrate.

The perturbation is reduced when water molecules are added in the second shell. This is shown in Fig. 7, which indicates the charges in a heptahydrate of DMP⁻. The seventh water molecule is added in the second shell, to the E_{31} molecule of the first shell. It is seen that these two molecules share (with a slight increase) the charge transferred to the E_{31} water molecule in the hexahydrate.

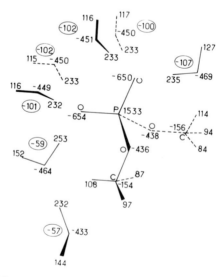

Fig. 7 Net charges in a heptahydrate of DMP⁻.

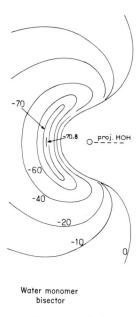

Water monomer
bisector

Fig. 8 Electrostatic potential of a water molecule.

The possible structure of water around the solute can also be considered with respect to the modifications of the molecular electrostatic potential of the bound water molecules compared to that of isolated water or of water bound to itself. This potential is defined (Scrocco and Tomasi, 1973) at point P by

$$V(P) = \sum_\alpha \frac{Z_\alpha}{r_{\alpha P}} - \int \frac{\rho(i)}{r_{\rho_i}} \, d\tau_i$$

where ρ_i is the electron density distribution corresponding to the wave function and α designates the nuclei. The interaction energy between the molecular distribution and an external point charge q is qV. The general representation of the potential is through isopotential curves or maps (Pullman, 1974). Figures 8 and 9 give the molecular electrostatic potential of a free water molecule and of a water dimer, respectively. Figure 10 presents the same result for the O_1PO_3 plane of the hexahydrate of DMP^-. A comparison of these figures shows the strong enhancement of the attractive character for a positive charge of the DMP^--bound water with respect to that of the molecules of the bulk solution. The perturbation of the potential of the molecules of the second shell has been studied in the same fashion. The results confirm that conclusion based on energy computations that bound perturbed water around

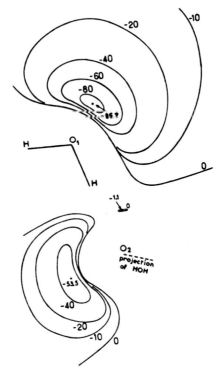

Fig. 9   Electrostatic potential of the most stable water dimer.

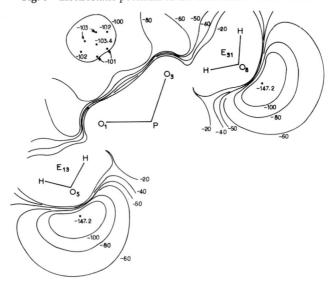

Fig. 10   Electrostatic potential in the O_1PO_3 plane of the hexahydrate of DMP^-.

DMP^- is essentially restricted to the first two hydration shells. Experimental information about the hydration of the phosphate group in nucleic acids and in phospholipids, although it does not lead to a unique scheme and does not fix precisely the preferred sites of hydration, indicates a number of bound water molecules and hydration shells; this is comparable to that suggested by the theoretical studies (Pullman, 1976).

III. CATION BINDING

Cation binding to the constituents of the nucleic acids may be studied within the same supermolecule approach. The essential target is represented, of course, by the anionic phosphate group. Detailed exploration (Pullman et al., 1975a,b, 1977) leads to the determination of the prinicipal binding sites, which are shown in Fig. 11. As given in Table III, the strongest binding corresponds to the B site. Studies on a series of alkali and alkali earth cations show that Na^+ and Mg^{2+} bind more strongly than K^+ and Ca^{2+}, respectively; the divalent cations bind much more strongly than the monovalent ones. The order of decreasing binding is $Mg^{2+} > Ca^{2+} > Na^+ > K^+$ as predicted by general theoretical considerations (A. Pulman et al., 1976).

Within the scheme of our general approach, the next problem to be considered is the effect of cation binding upon the properties of the

TABLE III

Cation Binding to DMP⁻

Cation site	Cation	Energy of cation binding[a]				ΔE between conformers[b]		
		gg	gt	tg	tt	gt–gg	tg–gg	tt–gg
Free DMP⁻						2.3	2.3	5.7
B	Na^+	−150.3	−148.3	−148.3	−146.0	4.3	4.3	10.0
	Mg^{2+}	−340.1	−335.2	−335.2	−330.0	7.2	7.2	15.9
	K^+	−128.3	−126.3	−126.3	−124.1	4.3	4.3	9.9
	Ca^{2+}	−285.0	−280.3	−280.3	−275.2	7.0	7.0	15.4
E	Na^+	−118.6	−115.8	−125.6	−122.7	5.1	−4.7	1.6
	Mg^{2+}	−227.8	−271.5	−289.9	−283.5	8.5	−9.8	−0.04
	K^+	−99.5	−96.9	−107.4	−104.6	4.9	−5.6	0.5
	Ca^{2+}	−227.0	−220.3	−240.0	−233.7	8.5	−10.8	−1.3

[a] ΔE with respect to DMP⁻ and cation at infinite separation.
[b] The energy differences between the gg, gt (tg), and tt forms in free DMP⁻ differ slightly from those of Table II because different basis sets were used.

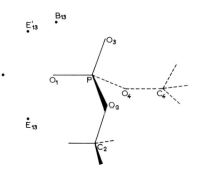

Fig. 11   Preferred cation binding sites on DMP⁻

phosphodiester group. The influences on two important properties have
been explored.

   1. *Effect on conformation.* The results are summarized in the right-
hand part of Table III. It is seen that the fixation of the cations at the B
site does not influence the order of conformational preferences with
respect to the torsion about the P—O_{ester} bonds existing in free DMP⁻:
The *gg* conformation remains the most stable one, followed by the *gt (tg)*
form, which in turn is followed by the *tt* form. In fact, the cation binding in its preferred B site increases even the relative stability of the *gg*
form with respect to the two other forms. However, the binding of the
cations to the external E site produces modifications in the previous
order of conformational stabilities, the perturbations being stronger
with divalent Ca^{2+} and Mg^{2+} than with monovalent Na^{+} and K^{+}; with
the monovalent cations, the *tg* form becomes the most stable one, followed, in decreasing order of stability, by *gg, tt,* and *gt.* With the divalent cation, both the *tg* and the *tt* conformers become more stable than
gg, and *gt* becomes less stable than it. The conclusion drawn from these
results is that, according to the site of binding, the cation may or may
not perturb the intrinsic conformational preferences of DMP⁻. The results also suggest useful working hypotheses. Thus, the preeminence of
the *tg⁻* conformation in the presence of divalent cations at the E site may
be of particular significance in connection with Sundaralingam's model
for the configuration of the tRNA–mRNA complex on the ribosome
during peptide bond synthesis (Sundaralingam *et al.,* 1974). The
stereochemical requirements for the interaction of messenger RNA with
the aminoacyl and peptidyl transfer RNA's are best satisfied when the
phosphodiester group bridging the adjacent codon triplets of mRNA is
in the extended *tg⁻* conformation, which could be produced by the
binding of Mg^{2+}, the presence of which is known to be necessary for the
process to occur.

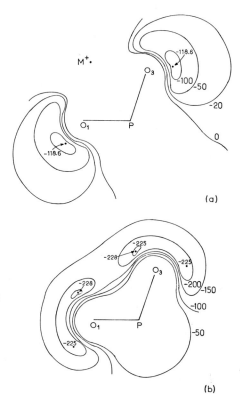

Fig. 12 Electrostatic potential of (a) DMP$^-$ in the presence of a Na$^+$ cation at site B and (b) DMP$^-$.

2. *Screening effect on the electrostatic molecular potential.* This is one of the major effects of neutralizing the negative charge of the phosphate anion by the positive charge of the cation. The quantitative aspects of the phenomenon are illustrated in Figs. 12 and 13, which indicate the screening in plane O_1PO_3 with monovalent and divalent cations (Figs. 12 and 13, respectively) fixed in the B position (Pullman and Berthod, 1976). A strong decrease in the value of the potential well and a strong reduction in the dimensions of the attractive zone of the phosphate are seen in Fig. 12 as a result of the binding of the monovalent cation. In the presence of the divalent cation (Fig. 13), the attractive isoenergy lines practically disappear completely. There remains only a very narrow zone of negative potential with a minimum of -12.8 kcal/mole, in the region of the external minimum of the isolated DMP$^-$, whereas the whole O_1PO_3 plane is now repulsive. In other words, a magnesium ion placed at 2 Å in the bridge position between O_1 and O_3 totally screens

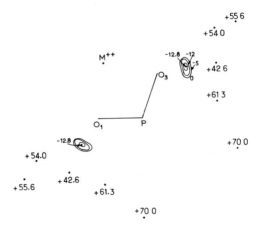

Fig. 13 Electrostatic potential of DMP⁻ in the presence of a Mg²⁺ cation in site B.

the electrostatic potential of DMP⁻ and in fact replaces its attractive character by a repulsive one.

This situation may be related to the observation that the value of the magnesium or calcium to phosphate ratio in studies of the effect of cation concentration on the binding of this cation to DNA reaches a maximum of 0.48 and remains constant up to a concentration of 50 mM of the cation (Chang and Carr, 1968).

At this point it may be useful to add that, although the phosphate group represents the principal center of cation binding in nucleic acids, the purine and pyrimidine bases constitute important secondary sites for such binding as evidenced by our supermolecule studies of the interaction of Na⁺ and Mg²⁺ with these bases (Perahia *et al.*, 1976, 1977) and with the ribose ring and substantiated by a number of X-ray crystallographic results (summarized in Perahia *et al.*, 1976, 1977). As an illustration, Fig. 14 shows the most stable Na⁺ binding sites in the four bases of the nucleic acids. Particularly interesting information about the nature of the forces involved in this interaction is obtained by comparing these results with those of the electrostatic potential energy maps and thus with proton affinities (Pullman, 1974). The two particularly outstanding features of cation binding to the bases are as follows.

a. The first is the increased significance of the individual oxygen binding sites in compounds containing both oxygen and nitrogen. For example, in cytosine, the interaction energies at sites 2 and 4, representing individual binding of the cation to the carbon carbonyl oxygens, are greater than at site 3, representing individual binding to a nitrogen atom. The reverse is true for proton interaction energies. A similar situa-

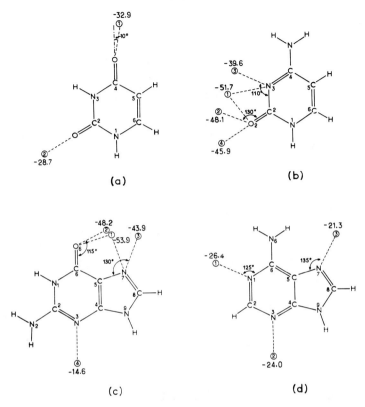

Fig. 14 Preferred interaction sites of Na$^+$ with nucleic acid bases and the corresponding interaction energies (kilocalories per mole). (a) Uracil; (b) cytosine; (c) guanine; (d) adenine.

tion concerns binding at sites 2 and 3 in guanine. This inversion in the intrinsic binding ability for a proton and for Na$^+$ is due to the exchange repulsive component of the binding energy, which, for the same distance, is larger for nitrogen than for oxygen.

b. The second point is the appearance in cytosine and guanine of a "bridged" site 1, involving simultaneously a nitrogen and a carbonyl oxygen, as the preferred binding position. In the electrostatic potential energy maps of these compounds these N and O atoms were generally associated with separate minima, and an intermediate bridged position for a proton was less favorable because the equilibrium distance for a proton is much smaller than that for a sodium ion, so the proton can move into the potential minima located close to the heteroatoms. These X-ray crystallographic results seem to confirm this general trend.

IV. THE TRIPLE SYSTEM:
SUBSTRATE–WATER–CATION OR THE PROBLEM
OF THE THROUGH-WATER CATION BINDING

In the two previous sections we have studied succesively the interaction of biological substrates with water and their interaction with cations. In the biological milieu, both water and cations are present, and the question arises of the possibilities and modalities of simultaneous interactions.

We have investigated this problem using the phosphate–water–Na^+ system (Berthod and Pullman, 1977). Let us go back to Fig. 10, which shows the isoenergy contours of the molecular electrostatic potential in the O_1PO_3 plane of the hexahydrate of DMP^-. The anionic character acquired by the water molecules is perceptible in the neighborhood of the in-plane external water molecules E_{13} and E_{31}, where the potential minimum appears to be strongly enhanced (−147.2 kcal/mole) with respect to that of an isolated water molecule (minimum, −71 kcal/mole; Fig. 8) or of a water molecule bound to the oxygen of another neutral water (minimum, −86.2 kcal/mole; Fig. 9). Moreover, the value of the potential inside the O_1PO_3 angle indicates that the attractive character for a positive ion found in this region for the phosphate itself remains quite large in the hydrated species. The disposition of the inner water molecules of hydration E_{14}, E_{12}, E_{34}, and E_{32} is such (Fig. 5) that they act pairwise so as to reinforce their attractive potential mutually. This can be seen in the map of Fig. 15 drawn in the bisector plane of water E_{34}, where the potential minimum reaches −176.8 kcal/mole and thus repre-

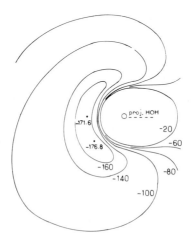

Fig. 15 Electrostatic potential in the bisector plane of the water molecule at site E_{34} in the DMP^-–$(H_2O)_6$ complex.

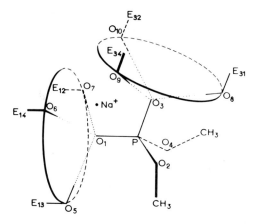

Fig. 16 Na$^+$ association with the DMP$^-$ hexahydrate. The distances from Na$^+$ are 2.1 Å to O$_9$ and O$_6$, 4.87 Å to O$_{10}$ and O$_7$, 3.96 Å to O$_1$ and O$_3$, 5.94 Å to O$_4$, 4.67 Å to O$_2$, and 4.42 Å to P.

sents an enhancement of about 30 kcal/mole with respect to the corresponding minimum for the in-plane molecules E$_{31}$, the enhancement being brought about by the effect of the water molecule in site E$_{14}$. The same effect occurs for the pair E$_{12}$/E$_{32}$ on the other side of the O$_1$PO$_3$ plane.

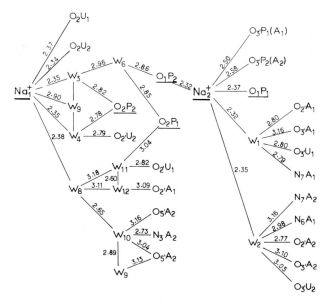

Fig. 17 Part of the interactions between water, Na$^+$ ions, and ApU in the crystal structure of soldium ApU hexahydrate; data are from Seeman *et al.* (1976).

This situation and the geometrical disposition of the water molecules point to the possibility that a cation may find a favorable binding position between two water molecules, say E_{34} and E_{14}, in the region of maximum attraction along the intersection of the above-considered planes. A search in this region was thus carried out by explicit computation on the $DMP^--(H_2O)_6-Na^+$ system. An equilibrium position was indeed found for a location of Na^+, as shown in Fig. 16, at a distance of 2.1 Å from O_9 and O_6, with a binding energy of -139 kcal/mole with respect to $DMP^--(H_2O)_6$. (The energy of direct binding of Na^+ to DMP^- in its most favorable position, which is a bridge position in the O_1PO_3 plane, is -168 kcal/mole using the same basis set.) An equivalent position exists on the other side of the O_1PO_3 plane due to the conjugate attraction of the water molecules E_{32} and E_{12}.

This result indicates that a sodium cation can bind to the phosphate anion through an intermediate molecule of water, which may thus be considered to serve as a water of hydration of both the anion and cation. This mode of binding of Na^+ allows it to remain surrounded by the water molecules of its hydration shell. Since the same is true of the anion, two of the molecules of water may then be shared. This situation points to the multiplicity of interactions possible in such a triple system. Figure 17 illustrates part of the interactions observed in the crystal structure of sodium adenylyl-3',5'-uridine hexahydrate, which substantiates the reality of diversified interactions (Seeman *et al.*, 1976).

REFERENCES

Berthod, H., and Pullman, A. (1977). *Chem. Phys. Lett.* **46,** 249–252.

Chang, K. Y., and Carr, C. W. (1968). *Biochim. Biophys. Acta* **157,** 127–138.

Giessner-Prettre, C., and Pullman, B. (1974). *J. Theor. Biol.* **48,** 425–443.

Perahia, D., Pullman, A., and Berthod, H. (1975). *Theor. Chim. Acta* **40,** 47–60.

Perahia, D., Pullman, A., and Pullman, B. (1976). *Theor. Chim. Acta* **42,** 23–31.

Perahia, D., Pullman, A., and Pullman, B. (1977). *Theor. Chim. Acta* **43,** 207–214.

Port, G. N. J., and Pullman, A. (1973). *FEBS Lett.* **31,** 70–74.

Pullman, A. (1974). *Jerusalem Symp. Quantum Chem. Biochem.* **6,** 1–14.

Pullman, A. (1976). *In* "The New World of Quantum Chemistry" (B. Pullman and R. Parr, eds.), pp. 148–187. Reidel Publ., Dordrecht, Netherlands.

Pullman, A., and Berthod, H. (1976). *Chem. Phys. Lett.* **41,** 205–209.

Pullman, A., Berthod, H., and Gresh, N. (1976). *Int. J. Quantum. Chem., Symp.* **10,** 59–76.

Pullman, A., and Pullman, B. (1975). *Q. Rev. Biophys.* **7,** 505–566.

Pullman, B., and Pullman, A. (1974). *Adv. Protein Chem.* **28,** 347–526.

Pullman, B., and Saran, A. (1976). *Prog. Nucleic Acid Res. Mol. Biol.* **18,** 215–322.

Pullman, B., Gresh, N., and Berthod, H. (1975a). *Theor. Chim. Acta* **40,** 71–74.

Pullman, B., Pullman, A., Berthod, H., and Gresh, N. (1975b). *Theor. Chim. Acta* **40,** 93–111.

Pullman, B., Gresh, N., Berthod, H., and Pullman, A. (1977). *Theor. Chim. Acta* **44,** 151–163.

Sarma, R. H., and Danyluk, S. S. (1977). *Int. J. Quantum Chem., Quant. Chem. Symp.* **4,** 269–276.

Scrocco, E., and Tomasi, J. (1973). *Topics Current Chem.* **42,** 95–170.

Seeman, N. C., Rosenberg, J. M., Suddath, F. L., Kim, J. J. P., and Rich, A. (1976). *J. Mol. Biol.* **104,** 109–144.

Sundaralingam, M., Brennan, T., Yathindra, N., and Ichikawa, T. (1974). *In* "Structure and Conformation of Nucleic Acids and Protein–Nucleic Acids Interaction" (M. Sundaralingam and S. T. Rao, eds.), pp. 101–115. University Park Press, Baltimore.

Part II

PHYSICOCHEMICAL ASPECTS
OF THE MECHANISMS OF
GENETIC EXPRESSION

CHAIRMAN: E. P. GEIDUSCHEK
Chapters 6 and 7

CHAIRMAN: B. CLARK
Chapters 8, 9, and 10

6

SITE-DIRECTED MUTAGENESIS AS A TOOL IN GENETICS

C. WEISSMANN, T. TANIGUCHI, E. DOMINGO,
D. SABO, AND R. A. FLAVELL
Institut für Molekularbiologie I, Universität Zürich, Zürich, Switzerland

The advent of hybrid DNA technology portends a breakthrough in our understanding of the structure and function of the eukaryotic genome. It will undoubtedly soon be possible to integrate any gene of interest with its neighboring regions into an appropriate replicon *in vitro*, clone and amplify it, and reintroduce it into eukaryotic cells. This constitutes the basis for what may be designated as "reversed genetics," an approach in which DNA regions are modified at predetermined positions *in vitro* and the effects of these interventions are scored *in vivo* or *in vitro*. This is in contrast to classical genetics, where deviant phenotypes are first isolated and then the lesion giving rise to them is identified.

Gross site-directed modifications of DNA, such as deletions or insertions of DNA segments, can be generated by the techniques of *in vitro* recombination; point mutations or mutations limited to a short, localized region can be produced by site-directed mutagenesis. In this chapter we will summarize some experiments in site-directed mutagenesis as applied to the RNA phage Qβ and outline how the technique may be applied to hybrid plasmids.

167

FRONTIERS IN PHYSICOCHEMICAL BIOLOGY
Copyright © 1978 by Academic Press, Inc.
All rights of reproduction in any form reserved.
ISBN 0-12-566960-7

I. PHAGE Qβ AND ITS REPLICATION*

Phage Qβ, a small spherical virus, contains an RNA molecule of about 4500 nucleotides that serves as both genome and messenger RNA. As shown in Fig. 1, Qβ RNA consists of three translatable and four untranslatable (extracistronic) segments. While the regions immediately preceding the cistrons are involved in the initiation and regulation of protein synthesis, the function of the longer untranslatable segments at the ends of the genome is not known. It has been speculated that the precise conservation of these sequences is essential for the viability of RNA phages (Min Jou *et al.*, 1972).

After penetrating its host, the viral RNA first serves as messenger RNA. As shown *in vitro*, ribosomes initially bind exclusively at the initiation site of the coat cistron; binding at the replicase cistron occurs only after coat translation has begun, and the A_2 cistron is probably only translated off nascent RNA strands. RNA replication begins after Qβ replicase has been assembled from the phage-coded subunit and three host-specific components.

Purified Qβ replicase, in conjunction with a host factor, replicates Qβ RNA *in vitro*, yielding infectious progeny RNA in large excess over the input template. Qβ replicase shows little affinity for the 3' end of Qβ RNA, where RNA synthesis begins, but binds tightly to at least two internal sites of Qβ RNA, one of which (S site) partly overlaps the coat cistron ribosome binding site (cf. Fig. 1). Presumably this interaction places the 3' terminus of the RNA into the initiation site of the polymerase. The product of the first step of synthesis, a single-stranded Qβ minus strand, is noninfectious but serves as an excellent template for the synthesis of infectious Qβ RNA (Feix *et al.*, 1968). The stability *in vitro* of the replication complexes at all stages allows manipulations to be carried out with little loss of enzymatic activity.

II. SITE-DIRECTED MUTATIONS IN Qβ RNA

As pointed out above, little is known about the functions of the untranslated regions in RNA phages. We have therefore generated Qβ RNA's with point mutations in the 3'-terminal extracistronic region and studied their viability *in vivo*. In addition, we have prepared Qβ RNA's with mutations located in the coat cistron initiator region and tested their ribosome binding capacity.

---

* For the original references, see the reviews by Weissmann *et al.* (1973), Weissmann (1974), and Kamen (1975).

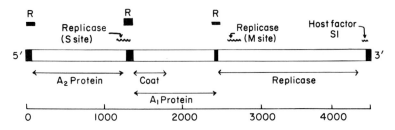

Fig. 1 Map of Qβ RNA. Untranslated areas are black; the cistrons are indicated by double-headed arrows. The ribosome binding sites are marked R; the binding sites for Qβ replicase, host factor, and protein S1 (a component of Qβ replicase) are indicated by wavy lines. Based on Weissmann *et al.* (1973).

Our procedure is based on the following principle. A mutagenic nucleotide analogue is inserted into a specific position of the nucleic acid chain. The nucleotide analogue, when utilized as a template, can direct the insertion of either of two nucleotides. For example, in the case of N^4-hydroxyCTP, it is believed that one of the tautomeric forms (cf. Fig. 2) can hydrogen-bond with G, whereas the other interacts with A. Therefore, if a nucleic acid strand containing hydroxyCMP in a certain position is used as a template by a replication enzyme, the newly synthesized strand contains either an A or a G at the position opposite the analogue, depending on the tautomeric form of the template nucleotide at the instant of incorporation. Since hydroxyCMP can be introduced into a nucleic acid in lieu of either CTP or UTP, it is clear that it can be used to generate $G \rightarrow A$ or $A \rightarrow G$ transitions. To insert an analogue into a precise, predetermined position, we utilized the principle of stepwise, substrate-limited RNA synthesis described below.

A. Qβ RNA with an A → G Substitution in Position 40 from the 5′ Terminus*

1. Stepwise Synthesis of Qβ Minus Strands with N^4-HydroxyCMP in Position 39

As shown schematically in Fig. 3, incubation of Qβ replicase with host factor, Qβ RNA, and with GTP, ATP, and CTP as the only substrates led to the synthesis of a 23-nucleotide-long minus strand segment extending to the position where UTP would be required for further elongation. After removal of the substrates by Sephadex chromatography, the nascent minus strand was elongated up to position 37 with

---

* The details are given in Domingo *et al.* (1976).

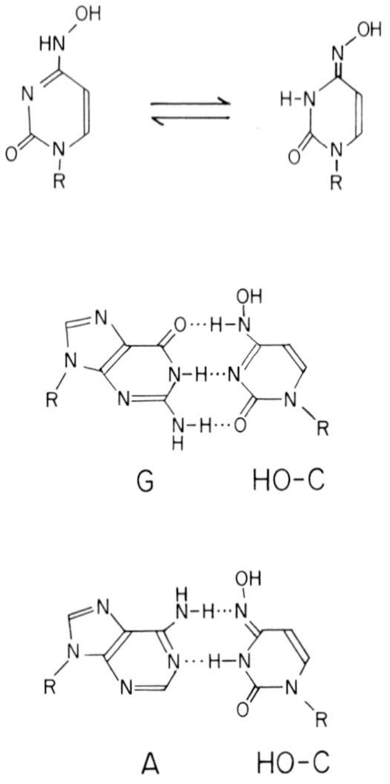

Fig. 2 Tautomeric forms and base-pairing capability of N^4-hydroxyCMP.

ATP, CTP, and UTP. The complex was again purified and residue 38, GMP, was added. The substrate was removed and the complex was incubated with N^4-hydroxyCTP (in lieu of UTP) and [α-^{32}P]ATP to add the sequence pHOCpApApA$_{OH}$, which could subsequently be detected by nucleotide analysis. The complex was reisolated, and the minus strands were completed with the four standard triphosphates and purified free of plus strands. Digestion of a sample with RNase T_1 yielded HOC-pApApGp as the main radioactive oligonucleotide (70% of the total radioactivity), showing that the nucleotide analogue had efficiently replaced UMP in position 39.

2. Synthesis of Qβ Plus Strands Using Minus Strands as Template

The purified, substituted minus strands were used as template for the synthesis of ^{32}P-labeled plus strands by Qβ replicase. As shown in Fig. 4, this is expected to yield two types of plus strands: one containing a GMP residue in position 40 (the mutant), and the other with an AMP residue in that position (the wild type). It may also be seen from the

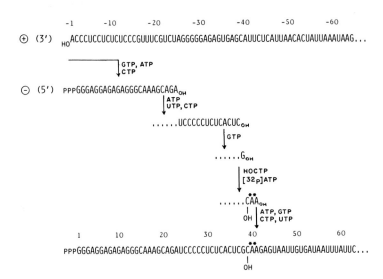

Fig. 3 Scheme for the stepwise synthesis of Qβ minus strands with the introduction of HOCMP in position 39 from the 5' terminus.

figure that digestion with RNase T$_1$, an enzyme that specifically cleaves at the 3' side of GMP residues, is expected to yield two different digestion products from the region of interest. Since these oligonucleotides, designated T1* in the case of the mutant RNA and T1 in the case of the wild-type RNA, have different mobilities, they can be distinguished by two-dimensional polyacrylamide gel electrophoresis. It was indeed found that Qβ RNA synthesized on substituted minus strands yielded a

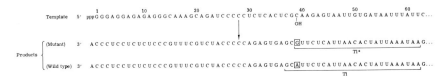

Fig. 4 Use of minus strands with a N^4-hydroxyCMP residue in position 39 instead of UMP as template for Qβ replicase *in vitro* yields two types of plus strands containing, in position −40 (i.e., position 40 from the 3' end), a GMP residue (the mutant strand) and an AMP residue (the wild-type strand). Since the synthesis of the minus strands of RNA phages starts at the penultimate nucleotide of the plus strand (Kamen, 1969; Rensing and August, 1969; Weber and Weissmann, 1970), nucleotide 40 from the 3' end of the plus strand is complementary to position 39 from the 5' end of the minus strand, that is, the position into which N^4-hydroxyCMP had been introduced.

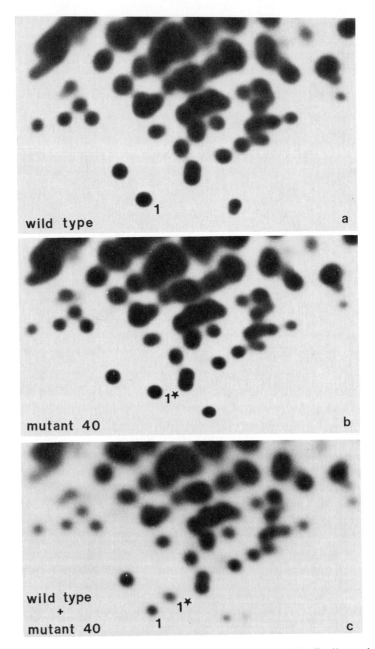

Fig. 5 Two-dimensional polyacrylamide gel electrophoresis of the T_1 oligonucleotides of uniformly ^{32}P-labeled wild-type RNA and mutant ($A_{-40} \rightarrow G$) RNA prepared from cloned phage. (a) Wild-type RNA; (b) mutant ($A_{-40} \rightarrow G$) RNA; (c) a mixture of wild-type and mutant ($A_{-40} \rightarrow G$) RNA. From Domingo *et al.* (1976).

new large oligonucleotide with the structure shown for T1*, whereas the amount of the wild-type oligonucleotide T1 was diminished. Figure 5 shows the positions of these two oligonucleotides in the fingerprint.

3. Infectivity of Plus Strands Containing an A → G Substitution in Position 40 from the 3' End

It has been known for a long time that pure phage RNA, when introduced into *Escherichia coli* spheroplasts, initiates an infectious cycle that results in a burst of complete phage particles.

To examine whether plus strands containing the A → G substitution in position 40 were infectious we transfected *E. coli* spheroplasts with a Qβ RNA preparation synthesized *in vitro*, consisting of about 23% of chains with the A → G substitution and the rest wild-type RNA. The infected spheroplasts were plated on a bacterial lawn, each of the resulting plaques were punched out of the agar plate, and the phages were eluted from it. The ^{32}P-labeled RNA's of 18 such clones were examined by T_1 fingerprinting. Of these 18 preparations, 4 showed T1*, diagnostic for the mutant A_{-40} → G, and no significant amounts of T1 (Fig. 5); 14 RNA's gave rise to T1 but not to T1*. This proportion of mutant phage, 22%, reflects rather accurately the proportion of mutant RNA, 25%, determined chemically in the preparation used for transfection. The mutant RNA is thus as infectious to spheroplasts as the wild type.

4. Competitive Growth in Vivo between Mutant A_{-40} → G and Wild-Type Phage

To test if wild-type phage had a selective advantage over mutant 40 *in vivo*, *E. coli* Q13 was infected with a 1:1 mixture of cloned mutant and wild-type phage at an overall multiplicity of infection of about 20 phages per cell, to ensure intracellular competition. The resulting lysate was then used to infect a fresh culture at the same multiplicity. A total of ten such cycles of infection was carried out. Cultures infected with mutant 40 alone were propagated in parallel. As shown in Fig. 6, after four cycles of infection the mutant content was only about 2% and no mutant was detected after ten cycles. Propagation of carefully recloned mutant phage A_{-40} → G in the absence of added wild type reproducibly resulted in the appearance of an increasing proportion of wild-type phage after a few cycles, showing that revertants arose at a substantial rate and outgrew the mutant. From the data of Fig. 6, we have estimated a reversion rate of 10^{-4} per doubling and a growth rate of 0.25 of the mutant relative to wild type under competitive conditions (Batschelet *et al.*, 1976).

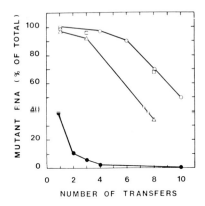

Fig. 6 Competitive growth of mutant Qβ (A_{-40} → G) and wild-type Qβ phage *in vivo*. Wild-type and mutant phage were isolated from single plaques. *E. coli* were infected with either mutant phage (three independent experiments, open symbols) or a 1:1 mixture of mutant and wild-type phage (closed symbols) at a multiplicity of 15–25. The resulting lysate was used to infect a fresh bacterial culture at the same multiplicity; this procedure was reported for a total of 10 transfers. The content of mutant was determined by T_1 fingerprint analysis of ^{32}P-labeled RNA. From Domingo *et al.* (1976).

5. Competitive Replication in Vitro of Wild-Type and Mutant (A_{-40} → G) Qβ RNA

One possible explanation for the reduced propagation rate of mutant (A_{-40} → G) phage is that its RNA is replicated less effectively by the replicase under competitive conditions. To test the relative replication rates of wild-type and mutant RNA, a mixture of the two was used as template for Qβ replicase in a serial transfer synthesis, in which each transfer corresponded to a doubling of the input RNA. The mutant RNA content decreased from 52% after the first transfer to 30% after eighteen transfers (Domingo *et al.*, 1976). Since one cycle of infection *in vivo* involves the equivalent of about ten to fifteen doublings of the input RNA, the disadvantage of the mutant phage *in vivo* could be roughly accounted for by the slower replication of its RNA, as measured *in vitro*.

It is of interest that the nucleotide in position −40 is part of a sequence (−63 to −38) that binds both host factor I and S1 protein (Senear and Steitz, 1976). Host factor I is required by Qβ replicase for initiation on plus strands (Franze de Fernandez *et al.*, 1972); its interaction with the 3′-terminal region of Qβ RNA, as well as with a further sequence (Senear and Steitz, 1976) located in the middle of the RNA, may be required to bring the RNA into the proper conformation for initiation. S1 is a ribosomal protein (Inouye *et al.*, 1974; Wahba *et al.*, 1974; Hermoso and Szer, 1974), which, after infection, is recruited as the

α-subunit of Qβ replicase (Kamen, 1970; Kondo *et al.*, 1970). S. Goelz and J. A. Steitz (personal communication) have recently found that the mutant oligonucleotide T1* is bound less efficiently by protein S1 than its wild-type counterpart, T1, suggesting that the reduced efficiency of RNA replication could be due to weaker binding of replicase and/or host factor to the mutated binding site.

B. Qβ RNA with a G → A Substitution in Position 16 from the 3′ Terminus

*Properties of Mutant (G$_{-16}$ → A) Qβ RNA**

The synthesis of Qβ RNA with a G → A substitution in position 16 from the 3′ terminus was carried out by the same general approach as described above (cf. Flavell *et al.*, 1974, 1975). Again a mixture of mutant and wild-type RNA was obtained.

To determine whether Qβ RNA with the extracistronic G$_{-16}$ → A mutation was infectious, spheroplasts were infected with plus strands synthesized on minus strands with the HOCMP substitution in position -15. Seventy-eight clones derived from spheroplasts infected with RNA from a one-round synthesis and forty-two clones from infections with RNA from multiple rounds of synthesis were analyzed by T$_1$ fingerprinting: None contained a nucleotide substitution in position -16. Since the RNA preparations used for transfection contained 30 and 60%, respectively, of the mutant (G$_{-16}$ → A) RNA and no mutant clones were found, plaque formation efficiency of mutant relative to wild-type RNA is less than about 0.03. The failure to find mutant clones cannot be ascribed to the manipulations to which the RNA had been subjected, since the specific infectivity of wild-type RNA synthesized *in vitro* on the substituted minus strand was about equal to that of natural Qβ RNA. Moreover, the results obtained with the mutant (A$_{-40}$ → G) RNA described above showed that the procedures used can in fact yield viable mutants.

To determine the efficiency of replication of the mutant (G$_{-16}$ → A) RNA by Qβ replicase, a serial transfer experiment was initiated with a mixture of mutant and wild-type RNA, and the ratio of the two was determined by T$_1$ fingerprinting. Contrary to the experiment with the mutant (A$_{-40}$ → G) RNA, the proportion of mutant RNA increased with successive transfers and reached a value of about 80% after extensive replication (Flavell *et al.*, 1975). Thus, the deleterious effect of the

---

* The details are given in Flavell *et al.* (1974, 1975) and Sabo *et al.* (1977).

C. WEISSMANN *ET AL.*

($G_{-16} \rightarrow$ A) transition on the infectivity of the RNA does not appear to be due to impaired RNA replication.

C. Qβ RNA with Mutations in the Coat Cistron Initiator Region*

Under initiation conditions, *E. coli* ribisomes bind to Qβ RNA almost exclusively at the coat cistron initiation site. To determine to what extent the AUG triplet is required for 70 S complex formation, we prepared Qβ RNA with G → A transitions of the third and fourth nucleotides of the coat cistron, that is, with modifications in the third position of the A-U-G codon and in the nucleotide following it (cf. Fig. 7).

1. Synthesis of Qβ Minus Strands with $\overline{HOCMP}$ Substitutions in the Positions Complementary to the Third and Fourth Nucleotides of the Coat Cistron

To carry out stepwise synthesis in the required region, we synchronized minus strand synthesis at a ribosome attached to the coat initiation site (Kolakofsky *et al.*, 1973). As outlined in Fig. 8, the 70 S Qβ RNA ribosome complex was used as template for Qβ replicase. Synthesis proceeded up to the position corresponding to the sixteenth nucleotide of the coat cistron (Kolakofsky *et al.*, 1973). The ribosome was then dislodged by treatment with EDTA and stepwise synthesis was carried out as shown in Fig. 8B, leading to insertion of $\overline{HOCMP}$ into the positions complementary to the fourth and third nucleotides of the coat cistron. Analysis of the appropriately labeled minus strands showed that in the region of interest 29% of the RNA had the sequence . . .$\overline{HOC}$-$\overline{HOC}$-A-U-G. . . , 15% had . . .C-$\overline{HOC}$-A-U-G. . . , 29% . . .$\overline{HOC}$-C-A-U-G. . . , and 27% . . .C-C-A-U-G. . . ; the presence of CMP is due to traces of CTP, which were difficult to remove in this experiment and competed effectively with $\overline{HOCTP}$.

2. Characterization of Plus Strands Synthesized with the Substituted Minus Strands as Template

Qβ plus strands were synthesized on substituted minus strands and replicated *in vitro*. The ratio of wild type : mutant ($G_{C_3} \rightarrow$ G) : mutant ($G_{C_4} \rightarrow$ A) : mutant ($G_{C_3,C_4} \rightarrow$ A) of Fig. 7 was found to be 1 : 1.8 : 1.6 : 4.3 (average of three experiments). We have not yet determined whether any of the mutant RNA's are infectious.

---

* Details to be given in Taniguchi and Weissmann (1978).

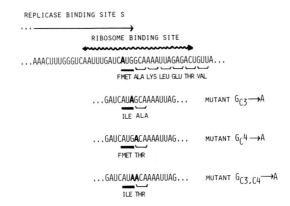

Fig. 7 Nucleotide sequence around the ribosome binding site of the coat cistron of wild-type Qβ RNA and mutants generated by site-directed mutagenesis at C_3 and C_4. The wild-type nucleotide sequence was established by Hindley and Staples (1969) and Weber *et al.* (1972).

3. Ribosome Binding Capacity of Qβ RNA with G → A Transitions in Positions C_3 and/or C_4

A preparation of ^{32}P-labeled Qβ RNA consisting of the species described above was bound to ribosomes under initiation conditions, the 70 S complex was treated with RNase A, and the RNA fragment retained by the ribosome was isolated (Steitz, 1969; Hindley and Staples, 1969) and analyzed with regard to the C_3/C_4 region. All RNA fragments recovered were derived from the coat initiation region.

The composition of the ribosome-bound fragments with regard to this region (average of three experiments) was wild type : mutant $(G_{C_3} \rightarrow A)$: mutant $(G_{C_4} \rightarrow A)$: mutant $(G_{C_3,C_4} \rightarrow A) = 1 : <0.1 : 4.5 : 1.5$. Taking into consideration the original composition of the RNA preparation, we estimate the relative binding efficiencies to be $1 : <0.1 : 2.8 : 0.35$.

Both mutant RNA's lacking the A-U-G triplet have a reduced ribosome binding capacity, but the double mutant RNA still bound with considerable efficiency. This suggests that the tRNA$_f^{Met}$–AUG interaction contributes substantially to the formation and/or stabilization of the 70 S ribosome complex.

It is striking that ribosomes are bound more efficiently to mutant $(G_{C4} \rightarrow A)$ RNA than to wild-type RNA and to mutant $(G_{C3,C4} \rightarrow A)$ more efficiently than to mutant $(G_{C3} \rightarrow A)$ RNA. Perhaps the nucleotides flanking the codon and the anticodon contribute to the stability of the interaction with tRNA$_f^{Met}$. As shown in Scheme 1, the mutant RNA's

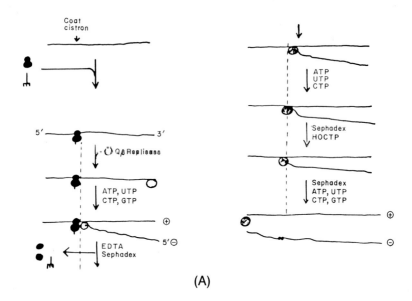

(A)

Insertion of N^4-hydroxyCMP opposite position 3 and 4
of the coat cistron

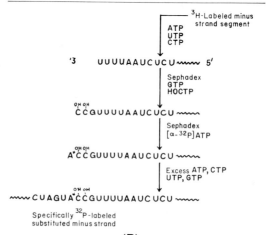

(B)

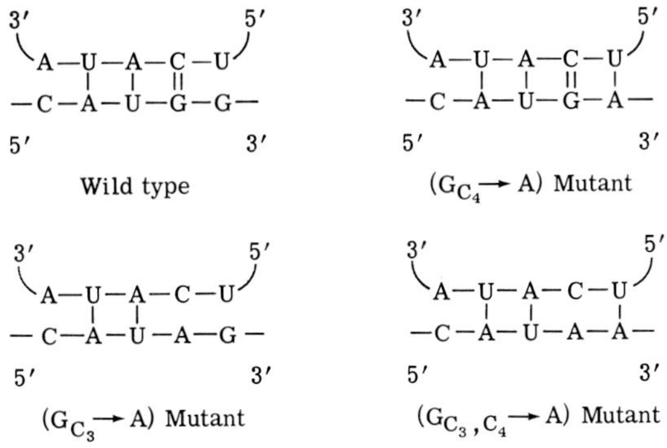

```
3'                    5'              3'                    5'
  \                  /                  \                  /
   A—U—A—C—U                            A—U—A—C—U
   |   |   |   ‖                        |   |   ‖   |
  —C—A—U—G—G—                          —C—A—U—G—A—
5'                    3'              5'                    3'

      Wild type                        (G_C4 → A) Mutant


3'                    5'              3'                    5'
  \                  /                  \                  /
   A—U—A—C—U                            A—U—A—C—U
   |   |   |                            |   |   |
  —C—A—U—A—G—                          —C—A—U—A—A—
5'                    3'              5'                    3'

   (G_C3 → A) Mutant                  (G_C3 , C4 → A) Mutant
```

Scheme 1

$(G_{C_4} \to A)$ and $(G_{C_3,C_4} \to A)$ can form an additional A-U base pair as compared to wild-type RNA and mutant $(G_{C_3} \to A)$ RNA, respectively.

III. PERSPECTIVES FOR SITE-DIRECTED MUTAGENESIS

A. Further Work on Qβ RNA

The work described above has demonstrated the feasibility of generating point mutations in Qβ RNA using the principle of substrate-controlled RNA synthesis and introduction of a mutagenic nucleotide, $\overline{HO}$CMP, in place of either CMP or UMP in a predetermined position. Other nucleotide analogues, substituting for purine nucleotides, should also become available. This procedure yields a mixture of wild-type and mutant RNA, which can be used as such for some studies: If the mutant phage RNA is viable it can be cloned *in vivo* and obtained in a pure form.

We have synthesized Qβ-specific DNA and inserted it into plasmids (Mekler *et al.*, 1977). So far we have not prepared a complete Qβ DNA transcript. However, several approaches are planned in this direction. It should become possible to clone in DNA form each of the mutant species generated in experiments such as those described above, syn-

---

Fig. 8 Scheme for the synchronization of Qβ minus strand synthesis at the coat ribosome binding site (A) and stepwise synthesis with introduction of $\overline{HO}$CMP at positions complementary to the third and fourth nucleotides of the coat cistron (B). See the text for an explanation. The black spheres represent ribosomes, the open circles Qβ replicase, and the forklike symbols tRNA$_f^{Met}$.

thesize RNA from it with RNA polymerase, and investigate the biological properties of the pure mutant RNA.

B. Site-Directed Mutagenesis in DNA

We believe that the application of our methodology to DNA, in particular to hybrid plasmids, will prove to be feasible and useful. Figure 9 shows an approach to site-directed DNA mutagenesis currently being pursued in our laboratory by W. Müller and H. Weber. Access to the region of interest is obtained with use of specific restriction enzymes. In principle, a single-stranded, purified restriction fragment is hybridized to denatured, circular DNA and used as primer for DNA polymerase.

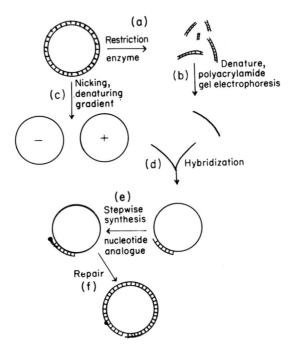

Fig. 9 Scheme for site-directed mutagenesis of a DNA plasmid. (a) The plasmid is cleaved by an appropriate restriction enzyme. (b) Restriction fragments are denatured and fractionated by agarose gel electrophoresis, and the appropriate single-stranded restriction fragment is isolated. (c) Plasmid is nicked and denatured, and the intact circles are isolated by centrifugation through a denaturing gradient. (d) The single-stranded restriction fragment is annealed with the circular DNA to yield a primer template, which is (e) elongated in a substrate-controlled, stepwise fashion with DNA polymerase. After insertion of the nucleotide analogue in the appropriate position, standard triphosphates are used to reconstitute the double-stranded circular DNA (f). This DNA is cloned and mutant DNA is identified by sequence analysis, starting in from the appropriate restriction site.

Substrate-controlled synthesis, insertion of mutagenic nucleotide analogues at a single position or in a more extended region, and completion of the DNA chain are then carried out, in principle, as described in the case of $Q\beta$ RNA. The resulting plasmids are cloned and the mutants are identified by sequence analysis. It is obvious that, with the use of hybrid plasmids, mutations that would be deleterious or lethal in a eukaryotic cell can be generated, cloned, and investigated *in vitro* or, when appropriate vectors become available, *in vivo* in eukaryotic cells.

ACKNOWLEDGMENT

This work was supported by the Schweizerische Nationalfonds (Grant No. 3.475-0.75).

REFERENCES

Batschelet, E., Domingo, E., and Weissmann, C. (1976). *Gene* **1**, 27–32.

Domingo, E., Flavell, R. A., and Weissmann, C. (1976). *Gene* **1**, 3–25.

Feix, G., Pollet, R., and Weissmann, C. (1968). *Proc. Natl. Acad. Sci. U.S.A.* **59**, 145–152.

Flavell, R. A., Sabo, D. L., Bandle, E. F., and Weissmann, C. (1974). *J. Mol. Biol.* **89**, 255–272.

Flavell, R. A., Sabo, D. L. O., Bandle, E. F., and Weissmann, C. (1975). *Proc. Natl. Acad. Sci. U.S.A.* **72**, 367–371.

Franze de Fernandez, M. T. Hayward, W. S., and August, J. T. (1972). *J. Biol. Chem.* **247**, 824–831.

Hermoso, J. M., and Szer, W. (1974). *Proc. Natl. Acad. Sci. U.S.A.* **71**, 4708–4712.

Hindley, J., and Staples, D. H. (1969). *Nature (London)* **224**, 964–967.

Inouye, H., Pollack, Y., and Petre, I. (1974). *Eur. J. Biochem.* **45**, 109–117.

Kamen, R. (1969). *Nature (London)* **221**, 321–325.

Kamen, R. I. (1970). *Nature (London)* **228**, 527–533.

Kamen, R. I. (1975). *In* "RNA Phages" (N. D. Zinder, ed.), pp. 203–234. Cold Spring Harbor Lab., Cold Spring Harbor, New York.

Kolakofsky, D., Billeter, M. A., Weber, H., and Weissmann, C. (1973). *J. Mol. Biol.* **76**, 271–284.

Kondo, M., Gallerani, R., and Weissmann, C. (1970). *Nature (London)* **228**, 525–527.

Mekler, P., van den Berg, J., Billeter, M. A., and Weissmann, C. (1977). *Experientia* **33**, 824.

Min Jou, W., Haegeman, G., Ysebaert, M., and Fiers, W. (1972). *Nature (London)* **237**, 82–88.

Rensing, U., and August, J. T. (1969). *Nature (London)* **224**, 853–856.

Sabo, D. L., Domingo, E., Bandle, E. F., Flavell, R. A., and Weissmann, C. (1977). *J. Mol. Biol.* **112**, 235–252.

Senear, A. W., and Steitz, J. A. (1976). *J. Biol. Chem.* **251**, 1902–1912.

Steitz, J. A. (1969). *Nature (London)* **224**, 957–964.

Taniguchi, T., and Weissmann, C. (1978). *J. Mol. Biol.* **118**, 533–565.

Wahba, A. J., Miller, M. J., Niveleau, A., Landers, T. A., Carmichael, G. G., Weber, K., Hawley, D. A., and Slobin, L. I. (1974). *J. Biol. Chem.* **249**, 3314–3316.

Weber, H., and Weissmann, C. (1970). *J. Mol. Biol.* **51**, 215–224.

Weber, H., Billeter, M., Kahane, S., Hindley, J., Porter, A., and Weissmann, C. (1972). *Nature (London), New Biol.* **237,** 166–170.
Weissmann, C. (1974). *FEBS Lett.* **40,** S10–S18.
Weissmann, C., Billeter, M. A., Goodman, H. M., Hindley, J., and Weber, H. (1973). *Annu. Rev. Biochem.* **42,** 303–328.

7

Gene Control by the
λ Phage Repressor

MARK PTASHNE

The Biological Laboratories, Harvard University, Cambridge, Massachusetts

The DNA of coliphage λ encodes about 50 genes. In the lysogenic state, the phage DNA molecule is integrated into the host chromosome, and most of the phage genes are turned off by the product of the phage cI gene, the so-called λ phage repressor. This repressor binds to two regions of λ DNA, the left and right operators (O_L and O_R) (see Fig. 1). Repressor bound at O_L prevents transcription of gene N, whereas repressor bound at O_R prevents transcription of a gene variously called cro or *tof*. Partly because the protein of gene N is a positive regulator that is required to turn on transcription of other phage genes, these genes are not expressed in lysogens. The repressor has evolved an elaborate mechanism for self-regulation. This ensures, among other things, that the enzyme required for prophage excision is never expressed during the growth of an uninduced lysogen. Various agents including ultraviolet light and other activated carcinogens cause induction by initiating a process that results in repressor cleavage by a protease (Roberts and Roberts, 1975); this aspect of λ's life will not be discussed further here.

Work of the early 1970s has resulted in a rather detailed understanding of the molecular basis of gene regulation by the λ repressor. I shall consider explicitly here three examples of gene regulation involving the repressor. For a more complete discussion of these matters as well as for a list of the pertinent references, the reader is referred to Ptashne *et al.* (1976).

FRONTIERS IN PHYSICOCHEMICAL BIOLOGY

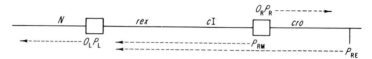

Fig. 1 Schematic representation of transcriptional patterns in a portion of the λ genome. The arrows show the directions of transcription of genes N, cro, cI, and rex. Genes cI and rex are transcribed either from the promoter P_{RM} in lysogens or from P_{RE} after phage infection of nonlysogens. $O_L P_L$ represents the "leftward" and $O_R P_R$ the "rightward" operator promoter.

    1. As mentioned above, repressor bound to O_L and O_R blocks transcription of the two adjacent genes. I shall describe the mechanisms of this negative control.

    2. The cI gene is transcribed in two modes. In the lysogenetic state, transcription begins near the right end of cI near O_R at the promoter called P_{RM} (promoter for repressor maintenance). This transcription is itself subject to both positive and negative control by the repressor. Thus, the amount of repressor in lysogens is carefully regulated. I will describe our currect understanding of the molecular mechanisms of this autogenous control.

    3. Upon infection of a nonlysogenic cell, cI transcription begins about 1000 bases to the right of cI and hence well to the right of O_R, at a promoter called P_{RE} (promoter for repressor establishment) (see Fig. 1). (Transcription beginning at P_{RE} requires the positive regulatory factors coded by the phage cII and cIII genes. Once repression has been established, transcription of cII and cIII is turned off by the repressor, and cI is no longer transcribed from P_{RE}. The mechanism of action of cII and cIII is not understood and is not considered further here.) P_{RE} directs the synthesis of five- to tenfold more repressor, per genome, than does P_{RM} and provides the large burst of repressor necessary for the establishment of lysogeny. P_{RE} directs the synthesis of more repressor than does P_{RM} by utilizing a novel form of posttranscriptional control, as I shall describe.

    Before considering these three issues, I will describe our understanding of the structures of the λ operators, promoters, and repressor.

I. OPERATOR STRUCTURE

    The most striking aspect of the λ operators is that each contains three repressor binding sites ($O_L 1$, 2, and 3; $O_R 1$, 2, and 3). The sequences specifically recognized are seventeen base pairs long and are separated

Fig. 2   DNA, RNA, and protein sequences in and around the two control regions of phage λ. The repressor binding sites O_L1, 2, and 3 in O_L (left operator) and O_R1, 2, and 3 in O_R (right operator) are set off in horizontal brackets. The start points of transcription of genes N, cro, and cI are indicated. Also shown are amino-terminal residues of the repressor. Six bases on O_R3, presumed to code for a strong ribosome binding site for cI, are marked with asterisks. The O_L has been reversed from its orientation in Fig. 1.

by "spacers" rich in A (adenine) and T (thymine) that are three to seven base pairs long. The terminal binding sites O_L1 and O_R1, which are adjacent to the controlled genes N and cro, bind repressor with a higher affinity than do the remaining sites. The complete nucleotide sequences of the λ operators are shown in Fig. 2. Figure 3 shows a cartoon of these sequences that emphasizes several important features. The evidence for the preceding statements is summarized in Ptashne et al. (1976; see also Walz et al., 1976).

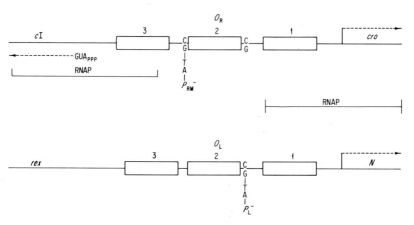

Fig. 3   Cartoon of the λ operators and a portion of genes N, cro, cI, and rex. The boxes show the positions of the 17-base pair repressor binding sites. The start points of cro, N, and P_{RM}-directed cI transcription are indicated. The approximate positions of the RNA polymerase binding sites at P_L, P_R, and P_{RM}, defined as the DNA protected from deoxyribonuclease digestion by polymerase, are shown. The extents of the fragments protected (solid or dashed lines) depend on the particular deoxyribonuclease used in the protection experiment (see the text). Two promoter mutations are shown.

II. PROMOTER STRUCTURE

A promoter is defined as a DNA sequence necessary for recognition and binding of RNA polymerase and for initiation of transcription. The promoters P_L, P_R, and P_{RM} denote, respectively, promoters for the transcription of genes N, cro, and cI. Figure 2 shows the regions of DNA protected from deoxyribonuclease digestion by RNA polymerase bound at these promoters. As shown first with P_R, the protected fragment is about 45 base pairs long when pancreatic deoxyribonuclease is used, and transcription begins roughly in the middle of this protected sequence. A similar relationship between polymerase-protected fragments and transcription start points has been found in other cases where pancreatic deoxyribonuclease has been used. Barbara Meyer (unpublished results) has recently repeated these polymerase protection experiments at P_R, using λ exonuclease and the single-stranded specific nuclease S1 in place of pancreatic deoxyribonuclease. Under these conditions the protected piece is roughly 65 base pairs long, and its approximate extent is indicated in Fig. 2. I will comment below on the fact that polymerase-protected fragments overlap repressor binding sites in the operators.

Two promoter mutations have been sequenced. One, located in the spacer between O_L1 and O_L2, damages P_L. The other, located in the spacer between O_R2 and O_R3, damages P_{RM}. The former is 31 and the latter 33 base pairs from the respective start points of transcription, and each changes the sole G · C in a spacer to A · T. We also know that a promoter mutation occurs in P_R within a few base pairs of the position analogous to that of the P_L mutation, but the exact base change has not been determined. As indicated in Fig. 2, the RNA polymerase-protected fragment generated by pancreatic deoxyribonuclease digestion does not include the regions in which these promoter mutations occur. It is not surprising, therefore, that these fragments do not bind polymerase. In contrast, the larger protected pieces obtained by λ exonuclease and nuclease S1 treatment include these regions, and these fragments bind polymerase and direct transcription. We do not know why digestion of polymerase–DNA complexes with different nucleases yields fragments of different sizes.

III. λ REPRESSOR

The λ repressor is an acidic protein, the monomer of which has a MW of about 26,000. These monomers are in concentration-dependent equilibrium with dimers and tetramers. The repressor binds tightly to DNA as an oligomer, but it is not known whether dimers or tetramers bind to the sites within the operators.

Repressor is produced in small amounts in ordinary lysogens, about 200 monomers per cell. We have constructed *in vitro* a recombinant DNA molecule that contains the *cI* gene read from two *lac* promoters. This recombinant is incorporated in a plasmid, and bacteria carrying this plasmid (pKB252) overproduce repressor some 50- to 100-fold. Sufficient quantities of repressor have been isolated to permit complete sequence analysis. A striking feature of the amino terminus of repressor is the strong clustering of basic residues. Although arginine and lysine constitute about 10% of the total residues, they account for 33% of the 27 amino-terminal residues. It has been suggested that amino-terminal residues of repressor make specific contacts with operator DNA, as has also been suggested for the *lac* repressor.

The three examples of gene regulation mentioned earlier will be discussed below.

IV. REPRESSOR CONTROL OF N AND *cro*

The two terminal repressor binding sites in O_R (O_R1 and O_R2) and in O_L (O_L1 and O_L2) mediate repression of *cro* and *N*, respectively. This was deduced from the fact that mutations that render *cro* transcription constitutive—that is, mutations that reduce repression at that operator—have been found in O_R1 and O_R2; mutations with a similar effect of *N* transcription have been found in O_L1 and O_L2. At each operator, mutation of two sites has a more dramatic effect on repression than does mutation of either site alone. RNA polymerase and repressor binding sites overlap in each operator, and the repressor excludes binding of the polymerase. Apparently repressor bound to two sites excludes polymerase more efficiently than does repressor bound to a single site. The repressor blocks transcription only if added to the template before RNA polymerase.

V. SELF-REGULATION OF *cI*

A. Negative Control

Repressor bound to O_R3 turns off transcription of *cI*. This was deduced as follows. RNA was transcribed *in vitro* from a DNA fragment bearing P_{RM} and a portion of the *cI* gene, and the *cI* transcript was identified by several criteria. Most important is that this message was not produced if the template bore a mutation that prevents transcription from P_{RM} *in vivo*. The sequence of the *c*I transcript corresponds to the DNA sequence, as shown in Figs. 1 and 2. Relatively high concentrations of repressor turn off transcription of this message *in vitro* and *in vivo* (see below). Mutations in O_R1 and O_R2 do not drastically affect this repression as measured *in vitro*. As at P_R, the repressor blocks transcription only if added to the template before RNA polymerase.

B. Positive Control

Repressor bound to O_R1 enhances transcription of *c*I. The efficiency with which *c*I is transcribed *in vitro* can be increased five- to tenfold by the addition of repressor. This effect requires an intact O_R1. The mechanism of this positive effect is unknown—in particular, we do not know the role of O_R2—but two possibilities are as follows.

1. RNA polymerase bound to P_R prevents, by steric inhibition, other polymerase molecules from binding to P_{RM}; repressor bound to O_R1 prevents polymerase binding to P_R, but does not block access of polymerase to P_{RM}.

2. Repressor bound to O_R1 directly enhances polymerase binding at P_{RM}, either by providing a protein–protein contact or by subtly altering DNA structure. We note that the distance from the center of O_R1 to the start point of transcription of *c*I is about the same as that between the center of the CAP (catabolic gene activator protein) binding site and the start point of transcription of the *lac* operon. The CAP enhances transcription of the *lac* gene, but the mechanism is not understood.

We favor some version of the second model, but we feel that the evidence is not conclusive.

VI. TRANSLATIONAL CONTROL OF *c*I

Why does transcription of *c*I initiated at P_{RE} produce more repressor than does transcription from P_{RM}? We had anticipated a simple answer: P_{RE} is a more efficient promoter than P_{RM}, and hence more *c*I transcripts are read from P_{RE}. This statement may be true, but recent evidence indicates that a more important factor is differential translation of the messages initiated at the two promoters. Figures 2 and 3 show that the codon corresponding to the amino terminus of repressor is found immediately adjacent to the 5′-terminal AUG of the *c*I message. This is remarkable in that all messages that have been analyzed contain leaders of variable length, preceding the AUG or GUG translational start signals. These leaders have been found to contain short sequences that are complementary to sequences at the 3' end of 16 S ribosomal RNA. It has been argued that pairing of these complementary sequences promotes binding of messages to ribosomes and, hence, efficient translation. Our finding that the cI message transcribed from P_{RM} bears no leader suggests that it may be translated at low efficiency. In contrast, we note that, beginning twelve bases to the right of the translational start point, there is a six-base sequence complementary to a sequence at the 3' end of 16 S ribosomal RNA (see bases marked with an asterisk in Fig. 2). This sequence should be present in the cI message transcribed from P_{RE} and should function as a strong ribosome binding site. Smith et al. (1976) have found that the cI message transcribed from P_{RE} is processed in vivo, but that the cleavage site is to the right of the proposed ribosome binding site. Therefore, we conclude that the message transcribed from P_{RE} bears a strong ribosome binding site and is translated more efficiently than is the message transcribed from P_{RM}.

VII. SUMMARY

Let us consider the action of repressor at one operator, O_R. This operator contains three repressor binding sites designated O_R1, O_R2, and O_R3. Because O_R1 has the highest repressor affinity, at low concentrations repressor will be bound preferentially to O_R1. This has the dual effect of decreasing rightward transcription of gene cro and of enhancing leftward transcription of the repressor gene, cI. At higher repressor concentrations O_R2 is filled, further repressing transcription of cro; at very high repressor concentrations O_R3 is filled and transcription of cI ceases. This sequential interaction of repressor with sites within a single controlling sequence mediates negative control of a function required for lytic growth of the phage (cro) and autoregulates, both positively and negatively, production of repressor. Part of this same operator (O_R3) codes for a sequence that apparently ensures efficient translation of cI, but that sequence is contained in the cI message only if transcription begins at one of two of the possible cI promoters (P_{RE}). The left operator (O_L) is similar in structure to O_R; repressor bound to O_L1 and O_L2 turns off transcription of another gene (N) required for lytic growth. The function of O_L3 is not known.

The functions ascribed to O_L and O_R are not exhaustive. There is strong reason to believe that another repressor, the product of the cro

gene, binds to O_L and O_R during lytic growth to turn down synthesis of N, cro, and cI, but how it does so remains to be seen. Moreover, the N protein is known to recognize some sequence in or near these operators and to render RNA polymerase immune to the blockade found at the end of certain genes. All the functions mediated by these remarkable regulatory sequences have not yet been elucidated.

REFERENCES

Ptashne, M., Backman, K., Humayun, M. Z., Jeffrey, A., Maurer, R., Meyer, B., and Sauer, R. T. (1976). *Science* **194,** 156.
Roberts, J., and Roberts, E. (1975). *Proc. Natl. Acad. Sci. U.S.A.* **72,** 147.
Smith, G. R., Eisen, H., Reichardt, L., and Hedgpeth, J. (1976). *Proc. Natl. Acad. Sci. U.S.A.* **73,** 712.
Walz, A., Pirrotta, V., and Ineichen, K. (1976). *Nature (London)* **262,** 665.

8

Nuclear Magnetic Resonance Studies of Protein–Nucleotide Interactions

MILDRED COHN AND B. D. NAGESWARA RAO

Department of Biochemistry and Biophysics, University of Pennsylvania Medical School,
Philadelphia, Pennsylvania

One of the central problems in understanding the expression of genetic information on the molecular level involves the elucidation of the recognition and interaction processes between proteins and nucleic acids. What is often overlooked is that we do not yet understand the simpler process of the protein–mononucleotide interaction, a phenomenon that occurs in many individual steps in the overall scheme of genetic expression beginning with the synthesis of DNA and proceeding to the termination of the synthesis of specific proteins. Not only are the nucleoside triphosphates involved directly as donors of nucleotidyl groups in nucleic acid biosynthesis catalyzed by DNA and RNA polymerases and in the activation of amino acids in the first step of protein synthesis, but the hydrolysis of ATP or of GTP with cleavage of the terminal phosphate is coupled to a number of steps in synthesis, transcription, and translation.

A number of questions arise concerning the interaction of mononucleotides with proteins, which we will address in this chapter. In what sense, if any, is there a common structural site on proteins for binding nucleotides? Is there a relationship between the structure of the binding site and the type of reaction catalyzed by the enzyme? For a given enzyme, is the nucleotide binding site a static structure or does it

FRONTIERS IN PHYSICOCHEMICAL BIOLOGY

change from its resting state to its active state? If so, what is the nature of the change? What are the amino acid residues involved in binding the different moieties of nucleotides? Finally, what is the relationship between the nucleotide binding site structure and the specificity of the interaction, in both binding constant and catalytic rate of the enzyme?

The answers to these questions have been sought by many investigators using various techniques. For a number of years, magnetic resonance techniques have been used in our laboratory to explore these problems (Cohn and Reuben, 1971; Cohn, 1973, 1975). The enzymes that have been studied in most detail are the kinases, which catalyze the transfer of the γ-phosphoryl group of ATP to various acceptors; a number of such enzymes are listed in Table I. In this chapter, we will illustrate the behavior of the kinases chiefly with arginine kinase and adenylate kinase. Adenylate kinase is of particular interest since it has two nucleotide substrate binding sites and is functionally essential in the cell for the formation of nucleoside di- and triphosphates (Cousin, 1967) and consequently for the biosynthesis of all macromolecules as well as for recycling the nucleoside monophosphates formed in the latter reactions. One protein involved in translation has also been investigated, the elongation factor Tu from *Escherichia coli*, which hydrolyzes GTP to GDP and orthophosphate when bound to the ribosome.

In early experiments, advantage was taken of the requirement of a divalent metal ion for activity of all of these enzymes, and two properties of the paramagnetic manganous ion were used to monitor the environment of the metal ion at the active site: (1) the paramagnetic contribution to the proton relaxation rate of water due to the interaction of the electron spin of manganous ion with the proton nuclear spin and (2) the electron spin resonance spectrum of Mn(II) (Mildvan and Cohn, 1970; Reed and Cohn, 1972, 1973; Price *et al.*, 1973; Buttlaire and Cohn, 1974). In addition, high resolution proton nuclear magnetic resonance

TABLE I

Kinase Reactions

$$\text{AdPP}\textcircled{P} + X \overset{M^{2+}}{\rightleftharpoons} X\text{—}\textcircled{P} + \text{AdPP}$$

X	Bond formed
Creatine, arginine	Guanidinophosphate (N—$\textcircled{P}$)
AMP	Phosphoanhydride (P—O—$\textcircled{P}$)
Pyruvate	Enolphosphate (C=C—O—$\textcircled{P}$)
3-Phosphoglycerate	Carboxyphosphate (R—$\overset{\overset{\displaystyle O}{\|}}{C}$—O—$\textcircled{P}$)

(nmr) spectra of substrates (McLaughlin *et al.*, 1976) and of the enzyme (McDonald *et al.*, 1975) as well as ^{31}P nmr of enzyme-bound nucleotides (Nageswara Rao *et al.*, 1976; Nageswara Rao and Cohn, 1977a) have been used to characterize the nucleotide binding sites and map the active site in various enzyme complexes.

I. TERNARY ENZYME–METAL NUCLEOTIDE COMPLEXES

It has been suggested by X-ray crystallographic studies (Rossmann *et al.*, 1974) that the three-dimensional structure of nucleotide binding sites is very similar in dehydrogenases, kinases, and flavodoxins. Whether this holds in solution for kinases and nucleoside triphosphatases has been examined by measuring a number of parameters.

A. Proton Relaxation Rate Enhancement

The enhancement of the contribution of Mn(II) to the proton longitudinal relaxation rate $(1/T_1)$ of water after adding enzyme and nucleotide is of the same order for most kinases and some synthetases. In addition to the enhancement factors for the E · MnADP complexes shown in Table II (creatine kinase, 20.5; arginine kinase, 18.9), similar values were found for 3-phosphoglycerate kinase (21.0, Chapman *et al.*, 1977), formyltetrahydrofolate synthetase (21.0, Buttlaire *et al.*, 1975), and succinyl-CoA synthetase (17.0, Buttlaire *et al.*, 1977). Enhancement factors for the corresponding ATP complexes are about 25% lower (cf. Table II). Thus, the active site structures of the ternary enzyme–metal nucleotide complexes of all kinases would appear to be very similar as monitored by proton relaxation rates of water.

On the other hand, as shown in Table III, the enhancement factors for the ternary enzyme–Mn nucleotide complexes for the nucleoside triphosphatases, subfragment I of myosin and the elongation factor Tu of *E. coli*, both have relatively low proton relaxation rate (prr) enhancements, a factor of 5–10 lower than those of the kinases. Furthermore, the number *(n)* of water molecules or equivalent protons in the first coordination sphere of Mn(II) in the ternary complexes that are exchangeable with solvent water (estimated from the prr values) is reduced from 2 for arginine kinase (Buttlaire and Cohn, 1974) to 1 for EF-Tu (Wilson and Cohn, 1977) and even less for the myosin subfragment complex (Bagshaw and Reed, 1976), as shown by the data in Tables II and III. A decrease in the value of *n* indicates either (1) that additional protein

TABLE II

Enhancement of the Proton Relaxation Rate of Water of
Enzyme–Metal–Substrate Complexes

Kinase	Complex	Enhance-ment	n (H$_2$O)	Reference
Arginine	E·MnAdPP	18.9	2	Buttlaire and Cohn
	E·MnAdPPP	16.7	2	(1974)
	E·MnAdPP·L-Arg	9.4	1	
	E·MnAdPP·NO$_3^-$·L-Arg	4.6	0.3–0.4	
Creatine	E·MnAdPP	20.5		Reed and Cohn
	E·MnAdPPP	10–15		(1972)
	E·MnAdPP·creatine	12.5		
	E·MnAdPP·NO$_3^-$·creatine	5.1	0.3–0.4	
Adenylate	E·MnAdPPP	14.8		Price et al.
	E·MnAdPPPPPAd	5.0		(1973)

ligands have replaced water ligands in the first coordination sphere of the metal ion or (2) that the number of water molecules in the first coordination sphere remains the same but the molecules exchange more slowly with solvent water, thus lowering their effect on the observed prr. Either phenomenon is associated with an active site structure that is less open to solvent. A comparison of the prr data on kinases (Table II) with those on the two hydrolytic enzymes (Table III) leads to the conclusion that the active sites of the ternary complexes of the latter are less accessible to solvent. It should be pointed out that the two hydrolytic enzymes are in low activity states, that is, myosin in the absence of actin and the elongation factor Tu in the absence of ribosomes.

TABLE III

Proton Relaxation Rate Enhancements (ϵ) and Number of Water Ligands (n) Coordinated
to Mn(II) in Complexes of Nucleoside Triphosphatases

Enzyme	Complex	ϵ	n	Reference
Myosin subfragment I	E·MnADP	1.7	0.2	Bagshaw and Reed (1976)
EF·Tu	E·MnGDP	4.3	1.0	Wilson and Cohn (1977)
	E·MnGTP	4.1	1.0	Wilson and Cohn (1977)

B. Electron Paramagnetic Resonance Spectra

The electron paramagnetic resonance (epr) spectra of the ternary complexes complement the prr data and confirm the conclusions reached from prr measurements. The positions and shape of the lines are determined by the symmetry of the ligand environment of the manganous ion, the motion of its complexes, and the electron spin relaxation processes (Reed and Cohn, 1970; Reed and Ray, 1971). The positions of the spectral lines in ternary enzyme–Mn nucleotide complexes remain the same as those for the manganous ion in aqueous solution for both kinases (Reed and Cohn, 1972, 1973) [with the exception of phosphoglycerate kinase (Chapman *et al.*, 1977)] and the nucleoside triphosphatases (Bagshaw and Reed, 1976; Wilson and Cohn, 1977). However, for the latter, the spectral lines are dramatically sharpened, as expected if the manganous ion is less accessible to solvent water. The electron spin relaxation mechanism involves distortions of the coordination sphere of Mn(II) by collisions with solvent and, consequently, a longer electron spin relaxation time leading to narrower spectral lines would accompany a decrease in the accessibility of solvent water to the Mn(II) coordination sphere.

C. High-Resolution nmr

From the high-resolution ^{31}P nmr spectra of enzyme-bound nucleotides, it is possible to gain information concerning the electronic and geometric structures in the environment of individual phosphorus atoms in close proximity to the site of catalytic action. The theory of ^{31}P chemical shifts and coupling constants does not permit quantitative interpretation. However, empirical correlations have been made. ^{31}P resonance positions in phosphates are particularly sensitive to the O—P—O bond angle (Gorenstein, 1975) and the state of ionization. The ^{31}P spectral lines of nucleoside di- and triphosphates are easily assigned by the coupling-constant patterns and the dependence of the chemical shift on the state of ionization of the terminal phosphate groups (Cohn and Hughes, 1960). Unlike the prr and epr methods, a metal ion is not needed for observation and, thus, it is possible to investigate the effect of the metal ion on the interaction of nucleotide and protein. In Fig. 1, the binary complexes and ternary metal complexes of adenosine di- and triphosphates with arginine kinase are compared with the free nucleotides and Mg nucleotides. It will be noted that, although all resonances of substrate are broadened due to binding to enzyme, as might be expected from the slow rotation caused by the large molecular size of

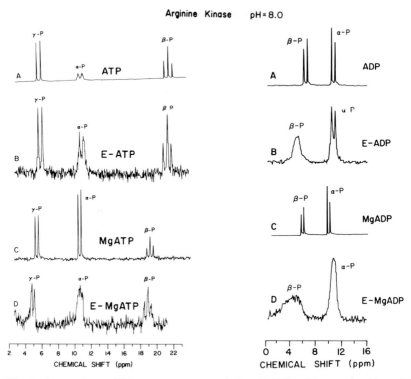

Fig. 1 Comparison of ^{31}P nmr spectra recorded at 40 MHz of free nucleotides with their complexes bound to arginine kinase. Proton decoupling was used for all spectra except the top spectrum of the ATP series, that is, free ATP, which accounts for the difference in the α-P resonance for free ATP (Nageswara Rao and Cohn, 1977a).

the enzyme complex, only the α- and β-P's of ADP are significantly shifted, and the β-P is drastically broadened; the chemical shifts of all three ^{31}P resonances of MgATP are unaffected by binding to enzyme. Although the pattern of shifts for free and enzyme-bound nucleotides is the same for creatine and arginine kinases as shown in Table IV, it is not universal for all kinases. For adenylate kinase, for example, there is a significant shift, ~1.5 ppm downfield, for the β-P of MgATP and a smaller upfield shift of 0.5 ppm for γ-P. The β-P of ADP is found in two positions: one corresponding to the β-P of E–ADP and the other corresponding to E–MgADP. The latter is shifted ~2.5 ppm downfield from the position of free MgADP, which is similar to the analogous complexes of arginine and creatine kinases. The pH dependence of the β-P shift in enzyme-bound MgADP has been investigated for both arginine kinase (Nageswara Rao and Cohn, 1977a) and creatine kinase. In the

TABLE IV

^{31}P Chemical Shifts of Free and Enzyme-Bound Nucleotides at pH 7.0 [a]

Enzyme	MgADP		MgATP		
	α-P	β-P	α-P	β-P	γ-P
None (−Mg)	10.8	7.5	11.0	21.8	7.3
(+Mg)	9.9	5.9	10.8	19.2	5.6
Arginine kinase	11.0	3.3	11.0	19.4	5.6
Creatine kinase	11.0	3.8	10.9	19.0	5.4
Adenylate kinase	10.2 (11.0[b])	3.5 (6.7[b])	10.7	17.8	6.1
Phosphoglycerate kinase	11.0	7.5	11.0	19.4	6.0
Pyruvate kinase	10.0	5.7	10.9	19.2	5.5
EF-Tu	10.1	2.4	10.1	14.7	2.3

[a] Values are parts per million from 85% H_3PO_4.

[b] Chemical shifts of donor ADP, not chelated by Mg, were deduced from spectra obtained under slow turnover condition (Nageswara Rao and Cohn, 1978).

case of the MgADP · arginine kinase complex, the β-P of ADP titrates with a pK_a of 7.5 rather than 6.0 (free MgADP), but the changes in chemical shift between low and high pH are in opposite directions for free and bound Mg nucleotide. The result in the enzyme complex has been tentatively ascribed to the effect of an interacting amino acid residue with a pK_a of 7.5. For creatine kinase, no change in chemical shift of MgADP was observed in the region from pH 6.0 to 9.0. Thus, even in these very closely related enzymes that share many properties of their nucleotide complexes in common, the respective environments of the β-P's of ADP differ.

Pyruvate kinase is unique in that binding to enzyme does not perturb the positions of the five ^{31}P nucleotide resonances. The pattern for 3-phosphoglycerate kinase is different from that of all others in exhibiting an *upfield* shift of ~1.6 ppm for the β-P of MgADP. In contrast to the similarity of those features of the active site structure of all kinases that are reflected by prr and epr parameters, essentially the topography and ligand symmetry in the environment of Mn(II), the electronic structures of the phosphates as monitored by ^{31}P nmr differ among the various kinases. It is of interest to note that in adenylate kinase, where there are two ADP sites, the environments of the two sites on the enzyme are different, as evidenced by a large downfield shift of the β-P of MgADP but a much smaller downfield shift of the β-P of ADP. The difference in the two sites is even more obvious in the proton nmr spectrum of enzyme-bound diadenosine pentaphosphate, Ap$_5$A, a strong inhibitor of adenylate kinase. The ^{31}P nmr spectrum of enzyme-bound MgAp$_5$A has been recorded, and it reveals five well-resolved groups of resonances, one for each phosphate group (Nageswara Rao and Cohn, 1977b). In Fig. 2A, both H$_2$'s, H$_8$'s, and C$_1'$H's, respectively, of Ap$_5$A are observed as single resonances since the molecule is completely symmetrical. As shown in Figs. 2C and 2D, asymmetry is introduced by the enzyme either with or without Mg^{2+}, and each resonance of the adenosine moiety is now split into two and one pair of adenine protons is also broadened much more than the other. The crystal structure of porcine adenylate kinase has been completely determined (Schulz et al., 1974), but the structures of enzyme–substrate complexes in the crystalline state had to be determined indirectly (Sachsenheimer and Schultz, 1977; Pai et al., 1977).

As in the case of the prr and epr parameters, the elongation factor Tu, which is a latent GTPase, shows marked differences in the ^{31}P spectra

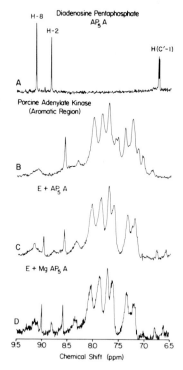

Fig. 2 Comparison of ^{1}H nmr spectra of free Ap$_5$A with the adenylate kinase-bound form. (A) Free Ap$_5$A; (B) aromatic region of porcine adenylate kinase, 1.3 mM; (C) same as B plus 1 eq of Ap$_5$A added; (D) same as C with 1 eq of MgCl$_2$ added. The spectra were recorded at 360 MHz at the Stanford Magnetic Resonance Center.

from all the kinases. The chemical shifts of the α-P resonances appear to be invariant within 1 ppm in all metal nucleotide–enzyme complexes (cf. Table IV). However, the β-P resonance of the MgGDP · EF-Tu complex is at a lower field than any kinases, that is, 2.4 compared to 5.9 ppm for free MgGDP. The pH dependence of this resonance has not yet been determined. The shifts of the β- and γ-P's of bound MgGTP are even more unusual: The γ-P is shifted downfield by 3.3 ppm and the β-P by 4.5 ppm, which are larger than any of the other shifts listed in Table IV. It is probable that the bonds of the substrate are highly distorted in the enzyme-bound form. The binding constants of the nucleotides to EF-Tu are several orders of magnitude higher than those for kinases, and the two phenomena may be related.

II. HIGHER ENZYME–SUBSTRATE COMPLEXES

A. prr Enhancement

The effect of occupying subsites of the active site in addition to that of MnADP on the prr enhancement is exemplified in Table II by the arginine, creatine, and adenylate kinase complexes. The enhancement factor progressively decreases from the high value of the ternary complex as each subsequent substrate is progressively bound. For example, when arginine is added to form an abortive quaternary complex, the enhancement is reduced by a factor of 2 and the number of water molecules in the first coordinate sphere of Mn(II) exchangeable with solvent water is also reduced by a factor of 2. Arginine is not directly liganded to Mn(II) and the reduction represents an induced change in the microenvironment of MnADP due to binding of arginine to the enzyme. A similar pattern is seen for creatine kinase. Addition of nitrate to the abortive quaternary complexes further lowers the enhancement factor and reduces the number (n) of water molecules to less than 1. Nitrate is presumably an analogue of the transferable phosphoryl group and thus forms a transition state analogue. The cogent evidence for this role of nitrate has been discussed in detail by Watts (1973). As mentioned earlier, values of n of less than 1 can be accounted for either by slow exchange of liganded water with solvent water or by the replacement of all water ligands by the protein ligands, with the residual small effects observed arising from outer sphere effects and/or from protein protons in proximity to the Mn(II) and exchangeable with solvent water. At present, there is no way of distinguishing among the possibilities, but all would lead to an active site structure of the transition state analogue complex that is less flexible owing to either additional protein ligands to the metal ion or another structural change that narrows the opening of the active site cleft making it less accessible to the solvent.

The occupancy of all subsites on adenylate kinase by the tightly bound inhibitor Ap_5A (Table II) reduces the prr enhancement by a factor of 3. A similar pattern holds for other systems investigated, such as formyltetrahydrofolate synthetase (Buttlaire *et al.*, 1975) and glutamine synthetase (Villafranca *et al.*, 1976).

B. epr Spectra

The changes observed in the spectrum of Mn(II) as one proceeds from the ternary Mn nucleotide complexes to the abortive quaternary complex to the transition state analogue complex are illustrated for the creatine kinase system in Fig. 3. Increasing asymmetry and immobiliza-

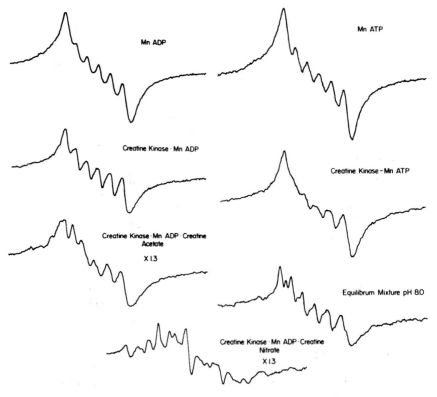

Fig. 3 The epr spectra for solutions of MnADP, MnATP, and their complexes in the creatine kinase system. From Cohn (1973).

tion are observed as the various subsites are progressively occupied. The spectrum of the equilibrium mixture, which is probably a super-position of spectra arising from the $E \cdot MnADP \cdot P$-creatine and $E \cdot MnATP \cdot$ creatine complexes, shows definitely more asymmetry than the abortive quaternary $E \cdot MnADP \cdot$ creatine complex, and the spectrum of the transition state analogue complex has the greatest asymmetry of all. The asymmetry undoubtedly arises from substitution of water ligands by protein ligands in the coordination sphere of Mn(II), which agrees with the conclusion of fewer water ligands in higher com-plexes deduced from prr enhancement data.

C. ^{31}P nmr

The ^{31}P nmr of enzyme-bound MgADP is shown in Fig. 1 for arginine kinase. The addition of arginine shown in spectrum A of Fig. 4 causes no significant shift in either the α- or β-P of MgADP. However, the further addition of nitrate shifts the β-P of MgADP upfield by approxi-mately 3 ppm. At pH 8.0, the β-P resonance of free MgADP has a shift of 5.9 ppm; in the ternary complex, the β-P shifts downfield to 4.4 ppm, but in the transition state it reverses and moves to 7.4 ppm. This upfield shift of the β-P resonance of MgADP in the transition state analogue to 7.4 ppm is consistent with an intermediate structure between enzyme-

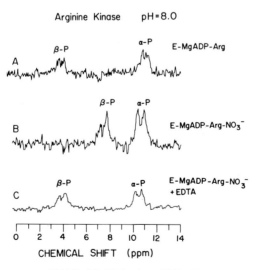

Fig. 4 ^{31}P nmr spectra of (A) $E \cdot MgADP \cdot Arg$, (B) $E \cdot MgADP \cdot Arg \cdot NO_3^-$ (transition state analogue), and (C) $E \cdot ADP \cdot Arg \cdot NO_3^-$ complexes of arginine kinase. From Nageswara Rao and Cohn (1977a).

bound MgADP (4.4 ppm) and enzyme-bound MgATP (19 ppm) that is anticipated in a transition state. The results on the higher ADP complexes of creatine kinase are very similar to those on arginine kinase.

III. WORKING ENZYMES

In the preceding sections, we have considered various complexes of nucleotides with kinases and have demonstrated that the local environment of the nucleotide binding site is greatly affected by occupation of other subsites of the active site of the enzyme. To understand the rates of the catalyzed reactions, one would, of course, like to know the structure of the true transition state and the intermediate complexes of the reacting system. Attempts to obtain such information from the ^{31}P nmr spectra of equilibrium mixtures of kinases are described below.

Figure 5B is a ^{31}P nmr spectrum of an equilibrium mixture of arginine kinase in which all the substrates are bound to the enzyme. This spectrum is considerably different from the catalytic equilibrium mixture in Fig. 5A. The catalytic equilibrium mixture (Fig. 5A) represents the equilibrium of free substrates and products in the overall reaction, that is, MgATP + L-arginine $\rightleftharpoons$ MgADP + L-P-arginine. On the other hand, the spectrum in Fig. 5B represents the equilibrium E · MgATP · L-arginine $\rightleftharpoons$ E · MgADP · P-arginine. Both equilibrium constants may be determined by integration of the respective spectra. It can be readily shown that the line-broadening in Fig. 5B is caused by the interconversion of the substrates as the reaction proceeds by removing the metal (by addition of EDTA) from the equilibrium mixture, thus stopping the reaction (cf. Fig. 5C), although all substrates and products remain bound to the enzyme. Any attempt to determine structural information on the substrates when the enzyme is actually working is thus frustrated by the fact that during the catalysis on the surface of the enzyme the substrates are inherently in a dynamical structure that keeps altering itself as the interconversion takes place, for example, ATP to ADP and arginine to phosphoarginine. On the other hand, a careful analysis of the nmr spectrum in Fig. 5B is capable of yielding worthwhile information on these dynamical processes.

The effect of chemical exchange on the nmr spectrum is illustrated in Fig. 6. ω_A and ω_B are the resonance positions in the absence of exchange between the sites A and B. The effect of exchange on the spectrum is determined by the relative magnitudes of $(\omega_A - \omega_B)$ and the reciprocal lifetimes τ_A^{-1} and τ_B^{-1} in the two states. When $|\omega_A - \omega_B| \gg \tau_A^{-1}, \tau_B^{-1}$ (slow exchange), there is simply a line-broadening without appreciable

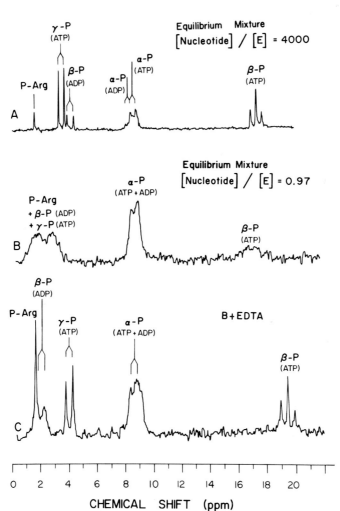

Fig. 5 Comparison of ^{31}P nmr spectra of arginine kinase. (A) Catalytic equilibrium mixture; (B) equilibrium mixture with fully bound substrates; (C) equilibrium mixture of bound substrates with the reaction stopped by removing Mg^{2+} with EDTA.

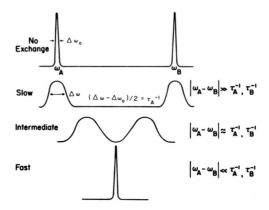

Fig. 6 Effect of chemical exchange on line-shapes and resonance position in high-resolution nmr.

change in the resonance positions; when $|\omega_A - \omega_B| \ll \tau_A^{-1}, \tau_B^{-1}$ (fast exchange), a line is obtained in an intermediate position without appreciable line-broadening. In the intermediate exchange case, both broadening and change in resonance position result, as indicated in Fig. 6.

As the substrates interconvert on the surface of arginine kinase, the following exchanges occur: α-P (ATP) $\rightleftarrows$ α-P (ADP), β-P (ATP) $\rightleftarrows$ β-P (ADP), γ-P (ATP) $\rightleftarrows$ P of P-Arg. An examination of the chemical shifts of these phosphate groups and their respective line-shapes in Figs. 5B and 5C reveals that the exchange of the α-P's ($\omega_A - \omega_B = 75$ sec^{-1}) falls in the fast, that of β-P's ($\omega_A - \omega_B = 3800$ sec^{-1}) in the slow, and that of γ-P (ATP) and P-arginine ($\omega_A - \omega_B = 500$ sec^{-1}) in the intermediate exchange regions. It should be noted that a single rate, namely, the interconversion rate of the substrates, produces all three conditions illustrated in Fig. 6 because of the magnitudes of the relevant chemical shift differences for the three exchanges involved in the reaction. The interconversion rate may be readily calculated from the linewidth difference between the β-P resonance of ATP in Fig. 5B and that in 5C by the appropriate relation for the slow exchange condition:

$$\tau^{-1} (E \cdot MgATP \cdot Arg) = \pi \text{ (linewidth change in hertz of the } \beta\text{-P of ATP)}$$

An important point about the determination of the above exchange rate is that it is exclusively the rate of interconversion of substrates on the surface of the enzyme. Working with an equilibrium mixture in which all substrates are bound to the enzyme allows the nmr spectra to isolate and monitor solely the rate of the phosphoryl transfer step in an

overall kinetic scheme that would include several steps and can therefore provide useful information on the rate of the interconversion step in
the kinetics of the overall reaction. Furthermore, the rate is obtained
from the nmr experiment in a simple and straightforward manner compared to relatively tedious isotope-exchange experiments. In the case of
arginine kinase, the interconversion rate (~ 200 sec^{-1}) is determined to
be about 10 times faster than the overall rate of the reaction (~ 20 sec^{-1})
under the same conditions (pH 7.25, 12°C) and therefore leads to the
conclusion that the former is not the rate-limiting step in the reaction in
either direction. Qualitatively similar results on the relationship between the interconversion rate and the overall rate were obtained for
creatine kinase and adenylate kinase. For pyruvate kinase and
3-phosphoglycerate kinase, it can be shown that the interconversion
rate may be the rate-limiting step in the direction of forming ATP and
not in the opposite direction. It is also possible to determine from the
nmr line-shapes of equilibrium mixtures of fully bound substrates
whether compounds that alter the rate of the overall reaction (e.g., inhibitors) do so by modifying the rate of the catalytic step or one of the
other steps in the kinetic scheme. For example, experiments on
3-phosphoglycerate kinase show that the addition of sulfate ion to the
equilibrium mixture slows down the interconversion rate on the enzyme and alters the equilibrium constant. Experiments on porcine
adenylate kinase indicate that Mg^{2+} concentrations in excess of half the
nucleotide concentrations do not appreciably alter the interconversion
rate and, therefore, the inhibition at such Mg^{2+} concentrations (Noda,
1973) is effected by some other step(s) in the reaction.

The adenylate kinase reaction should be particularly amenable to
study by the ^{31}P nmr method since all the substrates are nucleotides.
Figure 7 shows the ^{31}P nmr spectrum of an equilibrium mixture (obtained with a catalytic amount of porcine adenylate kinase) that displays
the resonances of the different phosphate groups in the nucleotides
AMP, ADP, and ATP. The spectrum of an equilibrium mixture of
enzyme-bound substrates is shown in Fig. 8A, and neither the β-P of
ATP nor the P of AMP is observable. A detailed analysis of the spectrum
with the help of other ^{31}P nmr experiments on various enzyme-bound
complexes shows that the resonances of AMP and of the β-P of ATP are
too broad to be seen in Fig. 8A. The equilibrium is shifted toward ADP
so that the concentration of E · AMP · MgATP is low and, moreover, the
interconversion rate of substrates to products is much faster than for
arginine kinase. The presence of a P_i resonance in the spectrum indicates an ATPase activity. As the concentration of ATP decreases with
time, forming P_i and ADP, the concentration of AMP must increase to

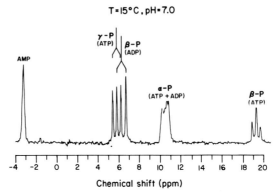

Fig. 7 ^{31}P nmr spectrum of equilibrium mixture of the adenylate kinase reaction with catalytic amounts of enzyme.

maintain the equilibrium. The increase in AMP concentration over that of ATP produces appreciable levels of complexes of the type E · AMP · MgADP, which do not participate in the reaction. Since the overall linewidth in the nmr spectrum is a weighted average of the linewidths of the participating and nonparticipating complexes, the linewidths of the signals of AMP, α-P of ADP, and β-P of MgADP that were too broad to be observed in Fig. 8A are now effectively reduced and can be observed in Fig. 8B. The β-P of ATP, which also should be narrowed, is too low in concentration to be observed in Fig. 8B. The additional resonance seen in the region of β-P of ADP in the spectrum of Fig. 8B, obtained about 8 hours after the spectrum in Fig. 8A, may be assigned to the β-P of MgADP. This assignment can be verified by adding a small amount of ATP to the sample of Fig. 8B. This broadens the resonance of β-P of MgADP (Fig. 8C), which now exchanges at a sufficiently fast rate with β-P of MgATP because of the increase in the concentration of ATP. That the sample in Fig. 8C contains all the nucleotides is demonstrated by the spectrum obtained after stopping the reaction by the addition of EDTA (Fig. 8D).

A detailed description of the results and analysis of the experiments on adenylate kinase is beyond the scope of this chapter and will appear elsewhere (Nageswara Rao and Cohn, 1978). The brief description of the results given above indicates how the equilibrium and some features of the mechanism and kinetics of these reactions may be probed by the ^{31}P nmr method.

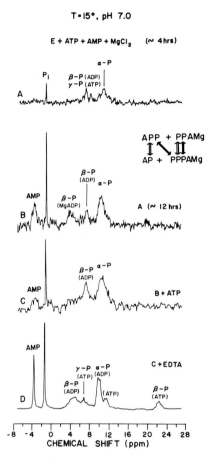

Fig. 8 ^{31}P nmr spectrum of an equilibrium mixture of substrates fully bound to porcine adenylate kinase. (A) After 4 hours of accumulation. (B) Same as A after 12 hours; separate resonances are seen for β-P of E–ADP and E–MgADP. (C) Spectrum of B after further addition of ATP; E–MgADP resonance is broadened. (D) Spectrum of C after the reaction was stopped by adding EDTA (see the text for an explanation of the spectral features).

IV. CONCLUSIONS

The main conclusions of the magnetic resonance investigations of nucleotide binding sites on enzymes are (1) that, for any given enzyme, the nature of the site varies with the occupation of other ligand sites on the enzyme, (2) that, for a given class of enzymes, the topographical features of the site are very similar, (3) on the other hand, the electronic and/or geometric structures of individual phosphates of nucleoside di- and triphosphates differ considerably even within a class of enzymes, (4) in transition state analogue complexes, the ^{31}P chemical shift of the β-P of enzyme-bound MgADP is consistent with an electronic and geometric structure intermediate between the β-P of the ternary E · MgADP complex and the β-P of the ternary E · MgATP complex, and (5) the structure of the enzyme-bound nucleotide during catalysis usually cannot be determined by ^{31}P nmr because the dynamical processes perturb the spectrum; however, the dynamical information obtained leads to evaluation of the catalytic rate constant of the interconversion step of substrates to products on the surface of the enzyme.

The ternary E · Mn nucleotide complexes of most kinases and several synthetases have very similar topographical features as monitored by prr and epr. Two nucleoside triphosphatases that have been investigated have a very different topography of the binding site for the ternary complex than do the kinases by the same criteria. Whether the hydrolytic enzymes as a class will exhibit the same structural features for the nucleotide binding site as the two examined remains to be established. ^{31}P nmr spectra, which reflect the environment in the proximity of the bond cleaved in the reaction, reveal considerable differences within the class of kinases in chemical shift and line-shape. Of all the ^{31}P resonances of enzyme-bound ADP and ATP observed, the least variation from free nucleotides has been found among the α-P's and the most variation among the β-P's of ADP.

A pattern that emerged for higher complexes from the prr and epr studies is confirmed by the ^{31}P nmr studies; that is, as each subsite of the enzyme is progressively occupied, there is a progressive change in the local environment of the phosphate moiety of the nucleotide binding site. The change is particularly striking upon formation of the transition state analogue complexes of the arginine and creatine kinase systems. The active site cleft becomes inaccessible to the solvent water, the ligand environment of the metal ion becomes highly asymmetric probably due to additional protein ligands, and, finally, the resonance peak of the β-P of bound ADP is shifted upfield considerably in the direction of the resonance position of the β-P of bound ATP. It should be pointed out that changes in the conformation of the protein at the active site of various complexes have been inferred although the magnetic resonance parameters observed in these complexes are monitoring changes in the water ligands of the Mn nucleotide complexes (prr), in the structure of the Mn(II) coordination sphere (epr), and in the electronic and geometric structure of the phosphorus atoms. That complementary changes in protein structure occur has been demonstrated for arginine kinase by

Buttlaire and Wong (1976) using ultraviolet spectral changes and for adenylate kinase by McDonald *et al.* (1975) using ^{1}H nmr changes in the protein spectra.

Direct interaction of specific amino acids at the active site with various atoms in the nucleotide can, in principle, be demonstrated by an intermolecular nuclear Overhauser effect, an effect that depends on the strength of the dipole–dipole interaction (and therefore the distance) between a given proton with a well-defined resonance position in the protein and a particular nucleus in the substrate. This method has been applied successfully for nucleotides bound only to creatine kinase, where it was found by James (1976) that an arginine residue interacted with the adenine moiety of ADP and that a lysine residue (James and Cohn, 1974) interacted with the formate ion, an anion that forms a transition state analogue complex in the creatine kinase system. The observation of which residues in the protein are not perturbed in the proton nmr spectrum by addition of nucleotide eliminates many possible residues, but, of the residues that are perturbed, it is not possible to distinguish between direct interaction due to binding and a conformational change remote from the site of interaction (cf. McDonald *et al.*, 1975). Chemical shift variation of the nucleotide resonances with pH, such as that found with the β-P of ADP bound to arginine kinase, also limits the identity of the amino acid residue involved at the binding site.

Thus far, we have discussed the structural changes at the binding site of nucleotides as we approach the active site by forming intermediate complexes. The enzyme-induced structural changes in the nucleotides that are needed to understand the catalytic process must be observed in the working enzyme. The prr parameter of an equilibrium mixture is an average of all species and yields no interpretable information. The Mn(II) epr spectrum of an equilibrium mixture is a superposition of all species and can only show (cf. Fig. 3) that it is different from the intermediate complexes.

The ^{31}P spectra of the equilibrium mixture may provide little explicit information on the structural changes in the substrates during catalysis. These changes are likely to be most pronounced in the transition state of the reaction. In an equilibrium mixture in which the substrates and products are interconverting, the transition state is expected to be the shortest lived and end-point states are the longest lived. This factor and the dynamical features introduced in the spectra discussed above make the equilibrium mixture suitable for a detailed study of structural changes, but only if these changes are large. For example, in the case of pyruvate kinase, the ^{31}P resonance of phosphoenolpyruvate shows the largest chemical shift ($\sim$4.5 ppm) only in the equilibrium mixture. The interference in the spectrum from the reaction is minimal when the interconversion rate between exchanging atoms is slow in comparison with the difference in their chemical shifts. This condition may sometimes be attained by working at higher nmr operating frequencies and at lower temperatures.

An alternative method of observing the substrate in a structure approaching that in the transition state is to use, in cases where it is possible, transition state analogue complexes. The effects produced by the addition of NO_3^- to $E \cdot MgADP \cdot Arg$ and $E \cdot MgADP \cdot creatine$ complexes of arginine kinase and creatine kinase, respectively (see Section III), illustrate this point quite clearly.

REFERENCES

Bagshaw, C. R., and Reed, G. H. (1976). *J. Biol. Chem.* **251,** 1975–1983.
Buttlaire, D. H., and Cohn, M. (1974). *J. Biol. Chem.* **249,** 5741–5748.
Buttlaire, D. H., Cohn, M., and Bridger, W. A. (1977). *J. Biol. Chem.* **252,** 1957–1964.
Buttlaire, D. H., Reed, G. H., and Himes, R. H. (1975). *J. Biol. Chem.* **250,** 254–260.
Buttlaire, D. H., and Wong, P. W. P. (1976). *Fed. Proc., Fed. Am. Soc. Exp. Biol.* **35,** 1433.
Chapman, B. E., O'Sullivan, W. J., Scopes, R. K., and Reed, G. H. (1977). *Biochemistry* **16,** 1005–1010.
Cohn, M. (1973). *Enzymes: Struct. Funct., Fed. Eur. Biochem. Soc., Meet.,* 29th, *1973* pp. 59–70.
Cohn, M. (1975). *Energy Transform. Biol. Syst., Ciba Found. Symp.* **31** (New Ser.), 87–104.
Cohn, M., and Hughes, T. R. (1960). *J. Biol. Chem.* **235,** 3250–3253.
Cohn, M., and Reuben, J. (1971). *Acc. Chem. Res.* **4,** 214–222.
Cousin, D. (1967). *Ann. Inst. Pasteur, Paris* **113,** 309–325.
Gorenstein, D. G. (1975). *J. Am. Chem. Soc.* **97,** 898–900.
James, T. L. (1976). *Biochemistry* **15,** 4724–4730.
James, T. L., and Cohn, M. (1974). *J. Biol. Chem.* **249,** 2599–2603.
McDonald, G. G., Cohn, M., and Noda, L. (1975). *J. Biol. Chem.* **250,** 6947–6954.
McLaughlin, A. C., Leigh, J. S., Jr., and Cohn, M. (1976). *J. Biol. Chem.* **251,** 2777–2787.
Mildvan, A. S., and Cohn, M. (1970). *Adv. Enzymol.* **33,** 1–70.
Nageswara Rao, B. D., Buttlaire, D. H., and Cohn, M. (1976). *J. Biol. Chem.* **251,** 6981–6986.
Nageswara Rao, B. D., and Cohn, M. (1977a). *J. Biol. Chem.* **252,** 3344–3350.
Nageswara Rao, B. D., and Cohn, M. (1977b). *Proc. Nat. Acad. Sci. U.S.A.* **74,** 5355–5357.
Nageswara Rao, B. D., and Cohn, M. (1978). *J. Biol. Chem.* **253,** 1149–1158.
Noda, L. (1973). *In* "The Enzymes" (P. D. Boyer, ed.), 3rd ed., Vol. 8, pp. 279–305. Academic Press, New York.
Pai, E. F., Sachsenheimer, W., Schirmer, R. H., and Schulz, G. E. (1977). *J. Mol. Biol.* **114,** 37–45.
Price, N. C., Reed, G. H., and Cohn, M. (1973). *Biochemistry* **12,** 3322–3327.
Reed, G. H., and Cohn, M. (1970). *J. Biol. Chem.* **245,** 662–664.
Reed, G. H., and Cohn, M. (1972). *J. Biol. Chem.* **247,** 3073–3081.

Reed, G. H., and Cohn, M. (1973). *J. Biol. Chem.* **248,** 6436–6442.
Reed, G. H., and Ray, W. J., Jr. (1971). *Biochemistry* **10,** 3190–3197.
Rossmann, M. G., Moras, D., and Olsen, K. W. (1974). *Nature (London)* **250,** 194–199.
Sachsenheimer, W., and Schulz, G. E. (1977). *J. Mol. Biol.* **114,** 23–36.
Schulz, G. E., Elzinga, M., Marx, F., and Schirmer, R. H. (1974). *Nature (London)* **250,** 120–123.
Villafranca, J. J., Ash, D. E., and Wedler, F. C. (1976). *Biochemistry* **15,** 544–553.
Watts, D. C. (1973). *In* "The Enzymes" (P. D. Boyer, ed.), 3rd ed., Vol. 8, pp. 384–431. Academic Press, New York.
Wilson, G. E., and Cohn, M. (1977). *J. Biol. Chem.* **252,** 2004–2009.

9

The Role of RNA's in the Structure and Function of Ribosomes

J. P. EBEL

Laboratoire de Biochimie, Institut de Biologie Moléculaire et Cellulaire du CNRS, Strasbourg, France

RNA's play an essential role in the structure and function of the ribosome. For a long time, only the structural role was taken into account, but it appears that the ribosomal RNA's might also be directly involved in the functions of ribosomes during protein synthesis. Ribosomal RNA's show this structural–functional role through RNA–protein interactions, but also through RNA–RNA interactions, either between the different ribosomal RNA's or between ribosomal and other RNA species (tRNA's or mRNA's).

This chapter describes the progress that has been made in this field either in our own laboratory or in others. This study is limited to prokaryotic ribosomes, especially those of *Escherichia coli*, where our knowledge is much more advanced than in the field of eukaryotic ribosomes.

I. A RAPID SURVEY OF PRESENT KNOWLEDGE ON THE STRUCTURE OF *E. coli* RIBOSOMAL RNA's

A. 30 S Subunit

16 S RNA

The 16 S RNA contains about 1600 nucleotides, 90% of which have been sequenced in our laboratory (C. Ehresmann *et al.*, 1975). The se-

213

FRONTIERS IN PHYSICOCHEMICAL BIOLOGY
Copyright © 1978 by Academic Press, Inc.
All rights of reproduction in any form reserved.
ISBN 0-12-566960-7

quence data have shown (1) that the 16 S RNA contains few sequence heterogeneities which are often clustered in certain sections, and (2) that most of the 10 methylated nucleotides are concentrated in the 3′-terminal third of the molecule, especially in the last 150 nucleotides (for a general review, see Fellner, 1974).

A model of secondary structure has been proposed (C. Ehresmann *et al.*, 1975), in which the RNA appears to be folded up as a series of hairpins of various sizes, linked by short single-stranded regions. This model is a tentative one and must not be considered as final. However, results have been obtained that confirm it indirectly, at least in one region—region C interacting with protein S8. A comparative study of the binding sites of protein S8 from *E. coli* and from *Bacillus stearothermophilus* showed that an identical secondary structure could be proposed in both cases: When one nucleotide is modified within the base-paired region, the appropriate nucleotide substitution is found on the opposite side of the helix, allowing the overall structure to be maintained (Stanley and Ebel, 1977) (see Fig. 6).

In other sections, the model must be changed because interactions between widely separated regions have been shown to take place. For instance, this is the case within the recognition site of protein S4, where region H″H′ interacts with I″C″ (Ungewickell *et al.*, 1975a). There are also some indications of an additional long-distance interaction between the S4–RNA region (L–C″) and section D′O near the center of the molecule (Ungewickell *et al.*, 1977). Interactions between O′ and A have also been suggested (C. Ehresmann *et al.*, 1975). (For the numbering of the sections of 16 S RNA, see Fig. 1.)

By examining the primary sequence of 16 S RNA, several complementary sequences can be detected within widely separated regions. These could be responsible for the long-distance interactions; for instance, complementary octanucleotide sequences are present in sections O and A (see Fig. 3b). However, additional experimental evidence must be obtained in order to determine the precise complementary regions involved in the long-distance interactions within 16 S RNA.

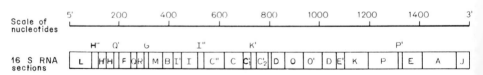

Fig. 1 Scale of nucleotides and numbering of sections in *E. coli* 16 S RNA. The 16 S RNA is divided into sections as described by C. Ehresmann *et al.* (1975).

B. 50 S Subunit

1. 5 S RNA

Since the determination of the nucleotide squence of *E. coli* 5 S RNA (Brownlee *et al.*, 1967), the primary structures of numerous 5 S RNA's from other sources have been established (for a review, see Erdmann, 1976). A comparison of the sequences of bacterial 5 S RNA's has shown that they all have a similar size (about 120 nucleotides) and that they contain no modified nucleotides. In some regions they present conserved sequences, especially the CGAA sequence, which is complementary to the TΨCG sequence common to all prokaryotic tRNA's (Erdmann, 1976).

Several models of secondary structure have been proposed (Erdmann, 1976): All of them present base complementarity between the 5'- and 3'-terminal sequences.

2. 23 S RNA

The 23 S RNA contains about 3000 nucleotides, among which 15 are methylated (Fellner, 1969) and 3 or 4 contain pseudouridine residues. We are studying the nucleotide sequence of 23 S RNA and have obtained the following results: The structure of all T1 ribonuclease digestion products has been determined (Branlant and Ebel, 1977) and the characteristic oligonucleotides have been localized within 24 aligned sections spanning nearly the entire molecule (see Fig. 9) (Branlant *et al.*, 1977c). The sequences of some of these sections, encompassing about 20% of the molecule, have been determined—in particular, those regions interacting with proteins L1 (Branlant *et al.*, 1976a) and L24 (Branlant *et al.*, 1977b). Models of secondary structure have been proposed for these two sections (see Figs. 10 and 11), and there is now evidence for long-range interactions within the binding site of L24 (Sloof *et al.*, 1978). No significant sequence repetition could be detected within the 23 S RNA molecule. The sequence homologies between 16 and 23 S RNA's are very limited. Progress made in the sequence determination showed a possible base pairing between the 3'- and 5'-end sequences of 23 S RNA (see Fig. 3a) (Branlant *et al.*, 1976c).

II. RNA–RNA INTERACTIONS

Interactions between ribosomal RNA's in the 50 S subunit (5 S RNA–23 S RNA) or in the 30 and 50 S subunits (16 S RNA–23 S RNA) have been postulated that could play both a structural and a functional role. On the

216   J. P. EBEL

other hand, arguments have been advanced in favor of specific base
pairing between ribosomal RNA's and other RNA's, such as mRNA or
tRNA, which could play a fundamental role in the initiation or elongation steps of protein synthesis.

A. Interactions between the Different Ribosomal RNA's

1. 5 S RNA–23 S RNA Interactions

In 23 S RNA, Herr and Noller (1975) characterized a sequence that is
complementary to a 12-nucleotide sequence from 5 S RNA (Fig. 2). This
complementarity could be involved in the binding of the 5 S RNA–
proteins L5, L18, and L25 complex to 23 S RNA shown by Gray *et al.*
(1972). The 23 S RNA sequence complementary to 5 S RNA has been
localized within the 500 nucleotides at the 5' end, which have been
shown to interact with protein L24 (Branlant *et al.*, 1977b). The localization of this sequence in the proposed model of secondary structure is
shown in Fig. 11 (see p. 230). It can be seen that, in this model, the
sequence is already engaged in base pairing.

However, it must be emphasized that the proposed model of secondary structure is only a tentative one. Furthermore, changes in the secondary structure could occur in the course of the 50 S subunit assembly,
allowing new base pairing to take place.

2. 16 S RNA–23 S RNA Interactions

We have already mentioned that the 3' and 5' sequences of 23 S RNA
are self-complemenary (Fig. 3a), but they are also complementary to
sequences in the 16 S RNA present in O and A sections (Figs. 3c and 3d),
which are themselves complementary (Fig. 3b) (Branlant *et al.*, 1976c).
The extent of complementarity between the two extremities of the 23 S
RNA and the two regions of the 16 S RNA is far greater than the expected values on a simple random basis. Therefore, it can be proposed
as a working hypothesis that such an exchange between different base

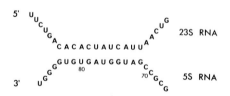

Fig. 2 Possible base pairing between sections of *E. coli* 5 and 23 S RNA's. The 23 S
RNA sequence, which is complementary to residues 72–83 of 5 S RNA, is located within
the L24 binding site, as shown in Fig. 11. From Herr and Noller (1975, *FEBS Lett.* **53**,
248–252).

a)   5' $_p$G̈GUUAAGC   3'   23S RNA
              |||||||
      3' $_{OH}$UUCCAAUUCG   5'   23S RNA

b)   5'   AAGGUUAAA   3'   16S RNA SECTION O
            ||||||||
      3'   UUCCAAUUC   5'   16S RNA SECTION A

c)   5'   AAGGUUAAA   3'   16S RNA SECTION O
            ||||||||
      3' $_{OH}$UUCCGAAUUC   5'   23S RNA

d)   5' $_p$G̈GUUAAGC   3'   23S RNA
              |||||||
      3'   UUCCAAUUCG   5'   16S RNA SECTION A

Fig. 3 Possible base pairing (a) between the 3'- and 5'-terminal sequences of 23 S RNA, (b) between sections O and A of 16 S RNA, (c) between section O of the 16 S RNA and the 3'-terminal sequence of the 23 S RNA, and (d) between section A of the 16 S RNA and the 5'-terminal sequence of the 23 S RNA (Branlant *et al.*, 1976c).

pairings could occur in the course of the association of 30 and 50 S subunits or during different steps of protein synthesis.

Another possibility of base pairing exits between the two 3' extremities of the 16 and 23 S RNA's (Fig. 4) (Van Duin *et al.*, 1976) and could explain the anti-association effect of factor IF-3: As this factor binds to the 16 S RNA near its 3' end, it could prevent the base pairing between the two RNA's.

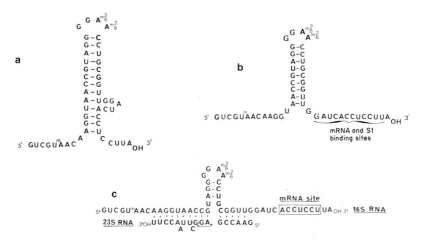

Fig. 4 A tentative model of base pairing between the 3'-terminal sequences of *E. coli* 16 and 23 S RNA's. (a) Model of secondary structure of section J of the 16 S RNA proposed by C. Ehresmann *et al.* (1975). (b) Model of secondary structure of section J of the 16 S RNA in the presence of protein S1: The mRNA site, as proposed by Shine and Dalgarno (1974), is accessible. (c) Possible base-pairing scheme between section J of the 16 S RNA and the 3'-terminal sequence of 23 S RNA (Branlant *et al.*, 1976c; Van Duin *et al.*, 1976).

B. Interactions between Ribosomal RNA's and Other RNA's

1. 5 S RNA–tRNA Interactions

All prokaryotic tRNA's contain the sequence TΨCG, which is complementary to the sequence CGAA found in all prokaryotic 5 S RNA's sequenced so far. It has been postulated that a base pairing between these two regions could take place in the elongation step. Competition experiments between tRNA's and oligonucleotide TΨCG strengthened this concept (Erdmann *et al.*, 1973). Nevertheless, accessibility studies of the 5 S RNA within the 50 S subunit (Noller and Herr, 1974) strongly suggest that a conformational change at the ribosome level is necessary to allow the interaction with the tRNA. A conformational change in the tRNA molecule in the region of the TΨCG loop has also been postulated (Schwarz *et al.*, 1976).

2. 16 S RNA–mRNA Interactions

Shine and Dalgarno (1974) observed a base complementarity between the pyrimidine-rich 3'-terminal sequence of the 16 S RNA ACCUC-CUUA$_{OH}$ (Fig. 4) and a purine-rich sequence frequently found within the initiation site of mRNA's. They also suggested a possible interaction between these two sequences during the initiation process of protein synthesis. This hypothesis has been strengthened by the results of Steitz and Jakes (1975), who isolated a stable complex between the 49-nucleotide fragment located at the 3' end of the 16 S RNA and a fragment of 30 nucleotides present in the initiation site of phage R17.

III. INTERACTIONS BETWEEN RIBOSOMAL RNA's AND PROTEINS

The possible role of these RNA–RNA interactions in ribosome structure and function has only recently been proposed. For a long time the only function attributed to ribosomal RNA's was that they represent the backbone around which the ribosomal proteins were attached during ribosome assembly. This fundamental role, which has been revealed by reconstitution experiments of the ribosome from its constituents (Nomura and Held, 1974; Dohme and Nierhaus, 1976), can only be explained by specific interactions between the ribosomal RNA's and proteins. These experiments have shown that ribosome assembly starts with the binding of a limited number of "initial binding proteins," which are indispensable for the binding of the other proteins. For instance, proteins S4, S7, S8, S15, S20, and possibly S13 and S17 bind directly to the

16 S RNA (Held *et al.*, 1974), L5, L18, and L25 to the 5 S RNA (Gray and Monier, 1972), and L1, L2, L3, L4, L6, L13, L16, L20, L23, and L24 to the 23 S RNA (Garrett *et al.*, 1974). However, other findings suggest that, in the case of 30 S assembly, other proteins could bind individually to the 16 S RNA when the RNA is prepared by acetic acid extraction (Hochkeppel *et al.*, 1976). The method of preparation of proteins could also influence the binding (Littlechild *et al.*, 1977; Dijk *et al.*, 1977).

Numerous studies have been performed to identify the regions that interact with the proteins within the ribosomal RNA's. They are based on four types of techniques. Two of these techniques are based on the study of complexes between the RNA or an RNA fragment and one or several proteins: (1) preparation and characterization of ribonucleoprotein fragments arising from hydrolysis of reconstituted complexes between the RNA and one or several proteins (Branlant *et al.*, 1973; Zimmermann *et al.*, 1975); (2) study of the rebinding of proteins on defined RNA fragments (Zimmermann *et al.*, 1974).

A third type of technique uses an opposite approach and consists of preparing and characterizing ribonucleoprotein fragments arising from hydrolysis of the ribosomal subunit (Yuki and Bromacombe, 1975).

A fourth type of technique can be applied either to reconstituted complexes between isolated RNA and proteins or to the intact ribosomal subunits: the formation of covalent linkages between RNA and proteins by a photochemical reaction, such as uv irradiation (Gorelic, 1975; B. Ehresmann *et al.*, 1975), or by a chemical reaction, such as formaldehyde treatment (Möller *et al.*, 1977).

A. Interactions between 16 S RNA and Proteins from the 30 S Subunit

1. Interactions between the 16 S RNA and Proteins S4 and S20

Using the techniques based on the study of reconstituted complexes between the 16 S RNA or fragments of the 16 S RNA and protein S4, it was possible to characterize within the interaction site a series of RNA fragments lying within a discontinuous region of sequence, located in the 5' half of the 16 S RNA, extending from section L to section C" and encompassing about 500 nucleotides (Fig. 5A) (Zimmermann *et al.*, 1972; Muto *et al.*, 1974; Ungewickell *et al.*, 1975b).

It has been shown (C. Ehresmann *et al.*, 1977) that this RNA region is highly structured, which is in agreement with the long-range interaction described between sections H"H' and I"C" (Ungewickell *et al.*, 1975a), but the most striking result is that this conformation is essentially the same in the presence or in the absence of protein S4. Therefore,

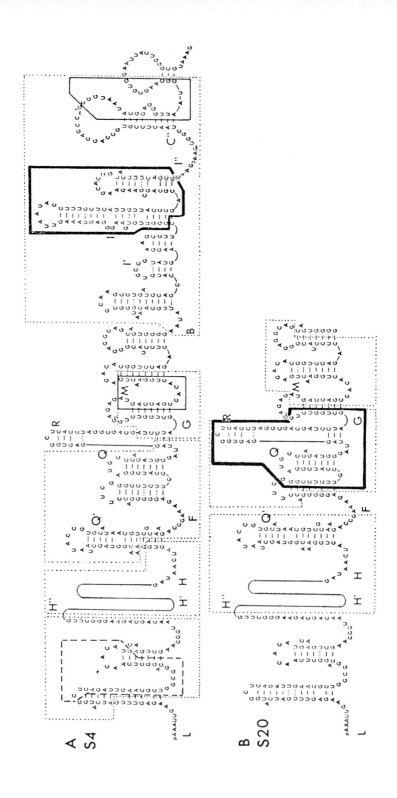

protein S4 binds to this compact structure without affording appreciable additional protection, at least in the presence of carrier-bound pancreatic ribonuclease.

Another new result is that, by submitting the 16 S RNA–protein S4 complex to milder hydrolysis conditions, it is possible to isolate a ribonucleoprotein fragment containing, in addition to the previous RNA fragments (L to C"), three fragments that derive from the center of the 16 S RNA in the D'O region, at a distance of about 200 nucleotides from the L–C" region (Fig. 1). Actually, as already reported, an interaction between the D'O and L–C" regions could be detected in the absence of protein S4, thus giving evidence for a long-distance interaction that facilitates the tertiary folding of the free 16 S RNA (Ungewickell *et al.*, 1977).

The 16 S RNA binding site of protein S20 is shown in Fig. 5B. It contains three groups of subfragments that contain 290 nucleotides lying within the S4 binding site and stretching from section H" to near the end of section M (Ungewickell *et al.*, 1975b). In contrast to protein S4, which does not significantly change the resistance of the compact 5' region against nucleases, protein S20 affords a much more reduced resistance, suggesting destabilization of the conformation of this RNA region in the presence of protein S20.

Mild nuclease digestion of the 30 S subunit yielded two ribonucleoprotein fragments. One of them contains a complex mixture of RNA fragments arising from the 5'-terminal 900 nucleotides of 16 S RNA and proteins S4, S5, S8, S15, S16 (17), and S20 (Rinke *et al.*, 1977). This is an important result, as it shows that, starting from the 30 S subunit, a localization of proteins S4 and S20 is found that is in agreement with the results obtained with reconstituted complexes between RNA and isolated proteins, thus confirming the validity of the latter approach.

A similar result was obtained when 30 S subunits were treated with formaldehyde to induce cross-linking between proteins and RNA. Pro-

---

Fig. 5 The nucleotide sequence and proposed secondary structure of the binding sites of proteins S4 (A) and S20 (B) within *E. coli* 16 S RNA. The sequences enclosed by dotted lines indicate the RNA regions protected by proteins S4 and S20 according to Ungewickell *et al.* (1975b). The sequences enclosed by dark solid lines indicate the RNA fragments strongly covalently cross-linked to protein S4 or S20 after uv irradiation. The sequences enclosed by light solid lines indicate the RNA fragments covalently linked to protein S4 to a weaker and more variable extent. The sequences enclosed by dashed lines indicate the RNA fragment that cannot be shown with certainty to be linked to protein S4 (B. Ehresmann *et al.*, 1977a). The secondary structure model is slightly different from that proposed by C. Ehresmann *et al.* (1975). This model has been corrected according to new information, especially in section C" (C. Ehresmann, P. Stiegler, and J. P. Ebel, unpublished data).

tein S4 was cross-linked, together with proteins S3, S5, S9 (11), and S13, and to some extent proteins S7, S12, and S18, to a region comprising approximately 900 nucleotides at the 5' end (Möller *et al.*, 1977), which fits with the previous localization of S4. Surprisingly, S4 was also found to be covalently linked, together with S9 (11) and S13 and to a lesser extent S7 and S14, to a region comprising about 450 nucleotides near the 3' end of the 16 S RNA, but lacking the 3'-terminal 150 nucleotides, suggesting that S4 has contacts with both the 5'- and 3'-end regions of the RNA.

All these results show that the regions interacting with proteins S4 and S20 are large (500 and 290 nucleotides, respectively). Although it has been shown that S4 and S20 are elongated proteins (Österberg *et al.*, 1977; G'Ulik *et al.*, 1978), they are not likely to interact with such large regions and they probably bind to more restricted sites, which must be maintained in a stereochemical configuration due to RNA–protein and RNA–RNA interactions.

Using the ultraviolet irradiation technique to form covalent bonds between the 16 S RNA and ribosomal proteins, it was possible to characterize much smaller RNA fragments, which are photochemically linked to proteins S4 and S20 (B. Ehresmann *et al.*, 1977). Protein S4 was found to be linked mainly to an RNA fragment corresponding to section I, to a lesser and variable extent to a part of sections C" and M, and perhaps to a segment of section L (Fig. 5A). It is of interest that these attachment sites are all located within the large 500-nucleotide interacting region previously characterized. In the case of protein S20, only one fragment could be linked, corresponding to sections (F)QR(G) arising from the middle part of the S20–RNA-protected region (Fig. 5B). It can be seen that the sites of covalent linkage between the 16 S RNA and proteins S4 and S20 are distinct.

2. Interactions between the 16 S RNA and Proteins S8 and S15

The RNA binding sites of proteins S8 and S15 were determined using mild enzyme digestion of the reconstituted 16 S RNA or RNA fragment–S8 and S15 complexes (Zimmermann *et al.*, 1974; Ungewickell *et al.*, 1975b). As shown in Fig. 6, S8 and S15 bind to very short overlapping and base-paired regions corresponding to the lower halves of the stems of two adjacent hairpin loops.

Protein S8 protects the lower part of hairpin loop C, whereas the upper part of this loop is excised. We have already mentioned (see above) that the base complementarity is conserved in the binding site of protein S8 and the 16 S RNA from *B. stearothermophilus* (Fig. 6), suggesting that the secondary structure proposed for the corresponding region in *E. coli* is correct (Stanley and Ebel, 1977).

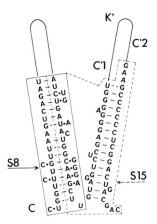

Fig. 6 The nucleotide sequence of the S8–S15 binding site in *E. coli* 16 S RNA (Ungewickell *et al.*, 1975b). The sequences enclosed by solid lines are the S8-protected region. The sequences enclosed by dashed lines are the S15-protected region. The alternative nucleotides shown at the border of the *E. coli* S8-protected region are those that differ in the *B. stearothermophilus* S8-protected region (Stanley and Ebel, 1977).

Protein S15 protects the lower part of the adjacent loop C'_1 and C'_2, whereas the upper part (K') is excised. A part of section C present in the S8-protected RNA fragment is also found to interact with S15. S8 binds more strongly to the 5′ half and S15 to the 3′ half of section C. This could be explained by the opening of these double-stranded regions in the presence of the proteins.

3. Interactions between the 16 S RNA and Protein S7

Protein S7 has been shown to remain associated with an RNA fragment of 8 S arising from the 3′-terminal portion of the 16 S RNA, containing approximately 600 nucleotides and extending from section O′ through section A (Zimmermann *et al.*, 1972). A fragment of about 450 nucleotides near the 3′ end of the 16 S RNA, but lacking the 3′-terminal 150 nucleotides, has been isolated in association with proteins S7, S9, S10, S14, and S19 following mild ribonuclease digestion of unfolded 30 S subunits (Rinke *et al.*, 1977). Another ribonucleoprotein fraction was obtained, which contained only proteins S7 and S19 and fragments from sections O′–E′ and P–A (Rinke *et al.*, 1976). When 30 S subunits were treated with formaldehyde, S7 was found (together with S4 and S13) to be linked not only to the RNA region near the 3′ end (see above), but also to the 900 nucleotides at the 5′ end, suggesting that S7 has contact with both regions (Möller *et al.*, 1977). In uv irradiation experiments, protein S7 was found to be the primary target of cross-linking of protein to the RNA (Möller and Brimacombe, 1975). Taking advantage of this reaction, the RNA fragments containing covalently linked protein S7 were identified and located within the O'–A region and almost certainly in the P–A region (Rinke et al., 1976) (for the numbering of the sections, see Fig. 1).

4. Interactions between the 16 S RNA and Proteins S1 and S21 and Initiation Factor IF-3

Ribosomal proteins S1 and S21 and initiation factor IF-3, which are all involved in the initiation process of protein synthesis, seem to interact with a region near the 3' end of the 16 S RNA, as they could all be covalently cross-linked to the in situ periodate-oxidized 3'-terminal ribose of the RNA (Kenner, 1973; Czernilofsky et al., 1975; Van Duin et al., 1975, 1976). Furthermore, it has been shown (Dahlberg and Dahlberg, 1975) that S1 binds specifically to a region within 12 nucleotides from the 3' end of the 16 S RNA (Fig. 4b): It includes the sequence ACCUCC, which has been reported to interact with a purine-rich sequence on the 5' side of the initiation triplet of messenger RNA's (Shine and Dalgarno, 1974). Several data suggest that S1 may facilitate the formation of or stabilize hydrogen bonding between the 16 S RNA and mRNA. It has been proposed that this is achieved by an unwinding activity of S1 to disrupt internal rRNA secondary structure, thereby exposing the Shine and Dalgarno region (Fig. 4b), or by stabilizing the mRNA–rRNA base pairs once they are formed (Steitz et al., 1977).

As will be seen further, S1 forms a complex, not only with the 16 S RNA, but also with the 23 S RNA, and the interacting 23 S RNA region has been characterized (Krol et al., 1977) (see Fig. 10). It has also been

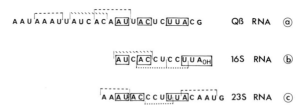

Fig. 7 Comparison of S1 binding sites on (a) Qβ RNA (Senear and Steitz, 1976), (b) E. coli 16 S RNA (Dahlberg and Dahlberg, 1975), and (c) E. coli 23 S RNA (Krol et al., 1977). Dashed lines encompass areas of homology between Qβ RNA and 23 S RNA binding sites. Dotted lines encompass areas of homology between 23 S RNA and 16 S RNA S1 binding sites. Slanted dashed lines delinate Qβ RNA and 16 S RNA homologies in S1 binding sites. The enclosed areas are the patterns that occur in all three S1 binding sites on Qβ, 16 S, and 23 S RNA's.

shown that S1 binds to Qβ RNA (Senear and Steitz, 1976). A comparison of the S1 binding sites in 16S, 23S, and Qβ RNA's shows (Krol *et al.*, 1977) that several restricted oligonucleotide homologies can be found (Fig. 7), especially three sequences (AU, AC, and UUA) that are present in close vicinity in the binding region of S1 in the three types of RNA's and that could represent the interaction site for this protein.

As far as the interaction of protein S21 is concerned, it has been found that this protein can be covalently cross-linked to the 16 S RNA by formaldehyde treatment, but it was not found in the 900 nucleotides at the 5' end or in the 450 nucleotides near the 3' end, which do not include the 150 nucleotides of the 3' terminus (Möller *et al.*, 1977). This result is compatible with an interaction within these 150 nucleotides near the 3' end of 16 S RNA.

B. Interactions between the 5 S RNA and Proteins from the 50 S Subunit

Three proteins of the 50 S subunit interact directly with the 5 S RNA, proteins L5, L18, and L25, and this complex can bind to the 3'-end region of the 23 S RNA (Gray and Monier, 1972)—more precisely, within a region spanning nucleotides 600 to 720 from the 3' end (Branlant *et al.*, 1976b; C. Branlant, unpublished results) (Fig. 9). The regions of the 5 S RNA that are protected against ribonuclease digestion in the presence of each of these three proteins have been identified and localized within the region spanning nucleotide 69 to the 3'-terminal nucleotide 120 (Gray *et al.*, 1973). Weak interactions between the 5 S RNA and proteins L7 (L12), L10, L20, and L30 have also been detected (Erdmann, 1976).

C. Interactions between the 23 S RNA and Proteins of the 50 S Subunit

The study of the interactions between 23 S RNA and the 50 S ribosomal proteins has been facilitated by the fact that 23 S RNA can be easily split within the 50 S subunit by limited digestion with immobilized ribonuclease into two fragments—13 S located at the 5' end and 18 S located at the 3' end (Spierer *et al.*, 1975); the latter can be split further into fragments 8 and 12 S (Spierer *et al.*, 1976). On the other hand, the fact that the 23 S RNA molecule can be divided into 24 aligned sections characterized by their T1 oligonucleotide composition and spanning nearly the entire molecule (see Fig. 9) (Branlant *et al.*, 1977c) allowed a rather accurate localization of the protein binding sites within the 23 S RNA to be achieved.

The following techniques have been used to determine the protein binding sites in 23 S RNA.

1. Hydrolysis of reconstituted 23 S RNA–protein complexes (Branlant *et al.*, 1973, 1975): This method enabled us to localize the binding sites of proteins L1, L20, L23, and L24.
2. Rebinding of proteins on 13 and 18 S fragments (Spierer *et al.*, 1975; Spierer and Zimmermann, 1976).
3. Hydrolysis of the reconstituted 18 S RNA fragment–protein complex (Spierer *et al.*, 1976): The proteins bound to the 18 S RNA fragment were localized within the two subfragments 8 and 12 S.
4. Hydrolysis of 50 S subunits (Newton and Brimacombe, 1974; Branlant *et al.*, 1976b; Chen-Schmeisser and Garrett, 1976): localization of L1–L9 and of the 5 S RNA–L5, L18, and L25 complex within 23 S RNA; characterization of proteins associated with 13, 8, and 12 S fragments.
5. Hydrolysis of LiCl-treated 50 S subunits and rebinding of proteins to the RNA fragments (Branlant *et al.*, 1977a): A more precise localization of the binding site of numerous proteins on 23 S RNA was achieved.
6. Cross-linking studies with formaldehyde on 50 subunits (Möller *et al.*, 1977).
7. Cross-linking studies using uv irradiation (B. Ehresmann *et al.*, 1978): This type of study was applied to the reconstituted complex between 23 S RNA and protein L24.

The results from each of these approaches are shown in Fig. 8. Figure 9 summarizes current knowledge on the localization of the interaction sites of the 50 S proteins on 23 S RNA. It is interesting that, for several proteins, the binding sites are located in different regions, depending on the technique used. This can be interpreted to mean that these proteins have multiple sites on the 23 S RNA. More precise information has been obtained concerning the binding sites of proteins L1 and L24.

1. The Binding Site of Protein L1 on 23 S RNA

Study of the ribonucleoproteins obtained by T1 ribonuclease digestion of reconstituted complexes of protein L1 and 23 S RNA showed that the primary binding site of protein L1 is a continuous RNA fragment of about 110 nucleotides in length, located between the 720th and 830th nucleotides from the 3' end of the 23 S RNA (Fig. 9). The sequence of this region has been determined (Branlant *et al.*, 1976a). Models of secondary structure have been proposed. One of them is given in Fig. 10.

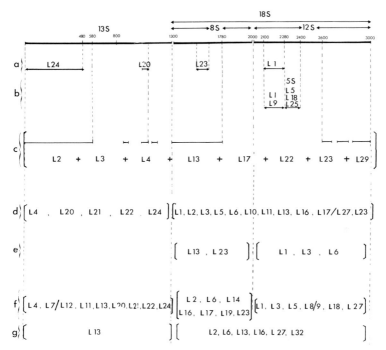

Fig. 8 Localization of protein binding sites on *E. coli* 23 S RNA within the 50 S subunit. The following methods have been used: (a) hydrolysis of reconstituted 23 S RNA–protein cmplexes (Branlant *et al.*, 1973, 1975); (b) hydrolysis of 50 S subunits (Newton and Brimacombe, 1974; Branlant *et al.*, 1976b); (c) hydrolysis of LiCl-treated 50 S subunits and rebinding of proteins to the RNA fragments (Branlant *et al.*, 1977a); (d) rebinding of proteins on RNA fragments (Spierer *et al.*, 1975; Spierer and Zimmermann, 1976); (e) hydrolysis of the reconstituted 18 S RNA fragment–protein complex (Spierer *et al.*, 1976); (f) hydrolysis of 50 S subunits (Chen-Schmeisser and Garrett, 1976); (g) cross-linking with formaldehyde on 50 S subunits (Möller *et al.*, 1977).

We have already mentioned that S1 recognizes a 17-nucleotide fragment of 23 S RNA located within the L1 binding site (see above) (Krol *et al.*, 1977) (Fig. 10). Regarding the possible significance of this interaction *in vivo*, it cannot be stated that it takes places in the intact 50 S subunit. However, it has been reported that the L1 RNA binding site is expected to be at the subunit interface (Stöffler, 1974) and that S1 stimulates the binding of tRNA by the 50 S subunit (Thomas *et al.*, 1975), suggesting that a dynamic interaction is possible.

2. The Binding Site of Protein L24 on 23 S RNA

Upon digestion of the complex formed by 23 S RNA and protein L24, two fragments resistant to ribonuclease were recovered, which both

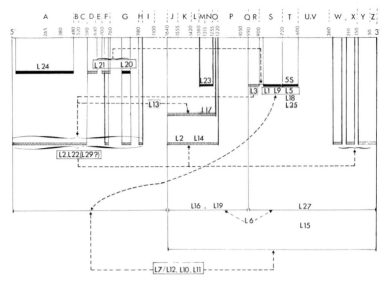

Fig. 9 RNA–protein interactions within E. coli ribosomal 50 S subunits. This scheme summarizes current knowledge on the localization of the interaction sites of 50 S proteins on 23 S RNA. It also shows the 24 aligned sections that have been characterized by their oligonucleotide composition (Branlant et al., 1977c).

contained RNA sections belonging to the 480 nucleotides at the 5′ end of 23 S RNA. The sequence of most of this region has been determined (Branlant et al., 1977b). Evidence has shown that this region has by itself a compact, tertiary structure and that the addition of protein L24 only stabilizes this structure. Long-range RNA–RNA interactions could be demonstrated in this region (Sloof et al., 1978). They led to the model of secondary structure shown in Fig. 11. From this figure, three observations can be made:

1. The nine nucleotides at the 5′ end of the molecule are not involved in the tertiary structure of this section. We have already mentioned (see above) that they are complementary to the 3′-terminal sequence of 23 S RNA (Branlant et al., 1976c) (Fig. 3a). Therefore, base pairing between the 5′- and 3′-terminal sequences could take place.

2. We have already discussed (see above) the possible interaction between 23 and 5 S RNA's, described by Herr and Noller (1975), which involves a sequence located within the L24 recognition site (Figs. 2 and 11).

3. Using the ultraviolet irradiation technique on the reconstituted complex between 23 S RNA and protein L24, one RNA fragment was found to be covalently cross-linked to the protein (B. Ehresmann et al.,

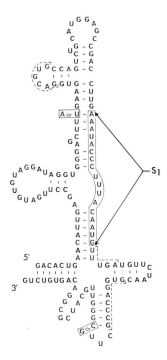

Fig. 10 A possible model of secondary structure for the region of *E. coli* 23 S RNA containing the binding site for protein L1 (Branlant *et al.*, 1976a). The protein S1 binding site in the same region (Krol *et al.*, 1977) is shown.

1978). This fragment could be characterized and is located between the 385th and 402nd nucleotides from the 5' end (Fig. 11). It can be seen that this fragment is very rich in purine nucleotides, especially in the single-stranded regions. This will be discussed later (p. 233).

D. Possible Mechanisms for the Ribosomal RNA–Protein Interaction

The uv irradiation technique has been used to determine the sequences of the tryptic peptides of ribosomal proteins S4 and S7, which are covalently linked to the 16 S RNA by uv irradiation (B. Ehresmann *et al.*, 1975, 1976; Reinbolt *et al.*, 1978). These sequences are shown in Fig. 12.

A comparative study of the cross-linked peptides shown in Fig. 12 shows that these peptides possess some common characteristic features.

1. One common feature is the presence of particularly high amounts of basic residues. In several peptides, clusters of adjacent amino acids

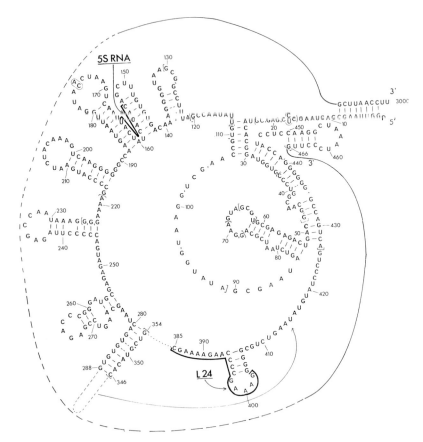

Fig. 11 A possible model of secondary structure for the region of *E. coli* 23 S RNA containing the binding site for protein L24 (Branlant *et al.*, 1977b). The figure also shows (1) the possible base pairing between the 3'- and 5'-end sequences of 23 S RNA (Branlant *et al.*, 1976c) (see also Fig. 3a), (2) the possible 5 S RNA–23 S RNA base pairing within this section (Herr and Noller, 1975; Branlant *et al.*, 1977b) (see also Fig. 2), and (3) the oligonucleotide depicted as L24 which can be cross-linked by ultraviolet irradiation to protein L24 (B. Ehresmann *et al.*, 1978).

are present, such as in protein S4 [Lys-Lys (147–148), Lys-Arg-Lys (180–182)] and in protein S7 [Arg-Lys-Arg (7–9), Lys-Lys-Arg (34–36), His-Arg (40–41)].

2. Another common feature of these peptides is that they contain a significant proportion of the acidic amino acid residues Glu and Asp.

3. The basic and acidic amino acids are often clustered. For instance, in protein S4, two Arg-Glu-Lys sequences are found. The first one (143–145) is close to a basic amino acid (Lys 59) and the second one (143–145)

	55	56	57	58	59
	Arg	-Glu	-Lys	-Gln	-Lys

S4

	143	144	145	146	147	148	149	150	151
	Arg	-Glu	-Lys	-Ala	-Lys	-Lys	-Glu	-Ser	-Arg

	180	181	182	183	184	185
	Lys	-Arg	-Lys	-Pro	-Glu	-Arg

	7	8	9	10	11	12
	Arg	-Lys	-Arg	-Gly	-Asp	-Lys

S7

	30	31	32	33	34	35	36	37	38	39	40	41
	Gly	-Thr	-Ala	-Val	-Lys	-Lys	-Arg	-Glu	-Asp	-Val	-His	-Arg

Fig. 12 Sequences of tryptic peptides in *E. coli* proteins S4 and S7, which are cross-linked to 16 S RNA by uv irradiation (Reinbolt *et al.*, 1977). For protein S4, the numbering of the amino acids corresponds to the protein sequence as determined by Schiltz and Reinbolt (1975). For protein S7, the numbering of the amino acids is that of the C-terminal region, which is the only part to have been sequenced (Tritsch *et al.*, 1977).

is close to another basic and acidic amino acid cluster: Lys-Lys-Glu (147–149). In addition, the sequence Glu-Arg (184–185) is found close to the basic sequence Arg-Lys-Arg (180–182). In protein S7, the sequence Asp-Lys (11–12) is found close to the basic sequence Arg-Lys-Arg (7–9). Another characteristic cluster is Lys-Lys-Arg-Glu-Asp (34–38).

The presence of clusters of basic and acidic amino acid residues in the peptides, which are cross-linked to the RNA and therefore in close vicinity to it, suggests that they could be involved in RNA–protein recognition mechanisms. Therefore, our results provide an experimental basis for an understanding of the mechanisms of the RNA–protein interaction.

The high proportions of basic amino acids could be compatible with the existence of electrostatic interactions between the positively charged amino acid side chains of Lys, Arg, or His and the phosphate group.

The presence of the acidic amino acids Glu and/or Asp in all cross-linked peptides could also be compatible with the existence of electrostatic interactions involving the carboxylic groups of these amino acids and the phosphate group through divalent cations.

But the acidic amino acids could also interact with the bases through hydrogen bonds. Examination of hydrogen-bonding possibilities shows that the formation of pairs of strong hydrogen bonds is possible for the carboxylate groups of Glu and Asp, in either un-ionized or ionized form (Hélène, 1977). In their un-ionized form, they could give rise, like Gln and Asn (Seeman *et al.*, 1976), to hydrogen bonding with A and G in double strands or with all four bases in single strands

(Hélène, 1977). In their ionized form, which is the most likely under physiological conditions, they could form two hydrogen bonds with NH(1) and $NH_2(2)$ of guanine in single-stranded nucleic acids (Hélène, 1977) (Fig. 13).

Another common feature of all cross-linked peptides is the presence of one or several arginine residues. In addition to the possible participation of the arginine residues in electrostatic interactions, these could interact with guanines according to a model (Seeman et al., 1976) (Fig. 14) in which the guanidinium side chain forms hydrogen bonds with guanine, which can itself interact with C in a G-C base pair in double-stranded nucleic acids. In single-stranded nucleic acids, arginine could form hydrogen bonds with cytosine (Hélène, 1977).

The presence in all S4 and S7 cross-linked peptides of Arg-Glu or Glu-Arg close to a basic amino acid or to several adjacent basic amino acids (B. Ehresmann et al., 1975, 1976; Reinbolt et al., 1978) gave an experimental basis for a model (Hélène, 1977) (Fig. 14) involving the couple Arg-Glu, in which ionized Glu interacts through hydrogen bonding with the arginine of the arginine–guanine couple previously described (Seeman et al., 1976). This Arg-Glu couple could interact with guanine in a double-stranded nucleic acid or with both guanine and cytosine in single strands. The presence of a neighboring basic amino acid residue, giving rise to electrostatic interactions with the phosphate group, could favor these interactions.

However, this model cannot be considered to be general, because no Arg-Glu or Glu-Arg couple is present in other sequenced ribosomal proteins that are known to interact directly with ribosomal RNA's, such as the 30 S proteins S8 and S20 or the 50 S protein L25 (Reinbolt et al., 1978). In fact, the most general feature of all of these proteins interacting directly with RNA's is the presence of sequences containing clusters of basic and acidic amino acids that may or may not be adjacent, but there

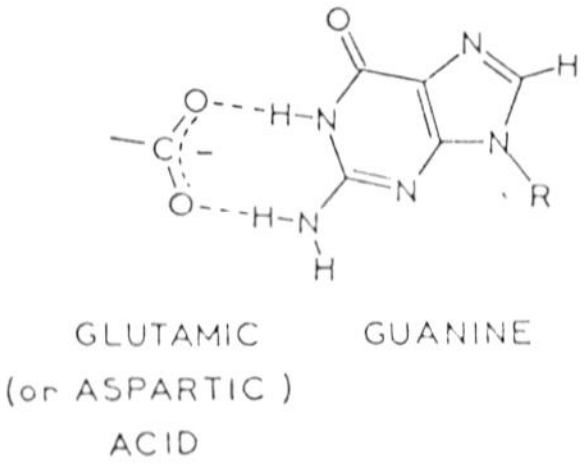

GLUTAMIC   GUANINE
(or ASPARTIC)
ACID

Fig. 13 Possible hydrogen bonding between the carboxylate groups of glutamic or aspartic acid in their ionized form with guanine in single-stranded nucleic acids. From Hélène (1977, FEBS Lett. **74,** 10–13).

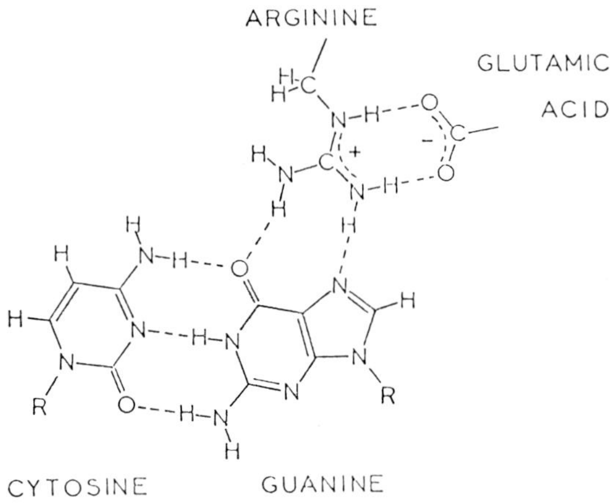

Fig. 14 Model for hydrogen bonding between glutamic acid and the arginine–guanine couple in double-stranded nucleic acids. The arginine–guanine interaction has been proposed by Seeman *et al.* (1976). The interaction between the ionized carboxylate of glutamic acid and arginine has been proposed by Hélène (1977). From Hélène (1977, *FEBS Lett.* **74**, 10–13).

seems to be no specific amino acid couple (e.g., Arg-Glu) present among these sequences. Our results are compatible with the models that show the carboxylate groups of the acidic amino acids Glu and (or) Asp involved in hydrogen bonding either directly with guanine in single strands (Hélène, 1977) or with arginine hydrogen bonded to guanine in single and double strands (Seeman *et al.*, 1976; Hélène, 1977) or to cytosine in single strands (Hélène, 1977). The presence of adjacent or neighboring basic amino acids, often present in clusters, could contribute by ionic interactions with the phosphate group to stabilizing these interactions.

From the sequences of the oligonucleotides interacting with the proteins or covalently bound to them by uv irradiation it has been impossible to draw any conclusion about the validity of this model. The sequence of the 16 S oligonucleotides covalently bound to S4 or S7 (B. Ehresmann *et al.*, 1977) (Fig. 5) did not allow the detection of any special base composition. On the contrary, the sequence of the 23 S RNA oligonucleotide cross-linked to protein L24 (B. Ehresmann *et al.*, 1978) (see Fig. 11) is particularly rich in purine nucleotides, among them several G's, which could fit with the proposed model. But in the case of S1, we have seen that the recognized sequences seem to be rich in pyrimidine bases (Krol *et al.*, 1977) (see Fig. 7). Actually, the rules that govern the nucleic acid–protein interaction may not be the same for all proteins.

REFERENCES

Branlant, C., and Ebel, J. P. (1977). *J. Mol. Biol.* **111,** 215–256.

Branlant, C., Krol, A., Sri Widada, J., Fellner, P., and Crichton, R. (1973). *FEBS Lett.* **35,** 265–272.

Branlant, C., Krol, A., Sri Widada, J., Ebel, J. P., Sloof, P., and Garrett, R. (1975). *FEBS Lett.* **52,** 195–201.

Branlant, C., Korobko, V., and Ebel, J. P. (1976a). *Eur. J. Biochem.* **70,** 471–482.

Branlant, C., Krol, A., Sri Widada, J., and Brimacombe, R. (1976b). *Eur. J. Biochem.* **70,** 483–492.

Branlant, C., Sri Widada, J., Krol, A., and Ebel, J. P. (1976c). *Nucleic Acids Res.* **3,** 1671–1687.

Branlant, C., Krol, A., Sri Widada, J., and Ebel, J. P. (1977a). *J. Mol. Biol.* **116,** 443–467.

Branlant, C., Sri Widada, J., Krol, A., and Ebel, J. P. (1977b). *Eur. J. Biochem.* **74,** 155–170.

Branlant, C., Sri Widada, J., Krol, A., and Ebel, J. P. (1977c). *Nucleic Acids Res.* **4,** 4323–4345.

Brownlee, G., Sanger, F., and Barrell, B. (1967). *Nature (London)* **215,** 735–736.

Chen-Schmeisser, U., and Garrett, R. (1976). *Eur. J. Biochem.* **69,** 401–410.

Czernilofsky, A. P., Kurland, C. G., and Stöffler, G. (1975). *FEBS Lett.* **58,** 281–284.

Dahlberg, A. E., and Dahlberg, J. E. (1975). *Proc. Natl. Acad. Sci. U.S.A.* **72,** 2940–2944.

Dijk, J., Littlechild, J., and Garrett, R. A. (1977). *FEBS Lett.* **77,** 295–300.

Dohme, F., and Nierhaus, K. H. (1976). *J. Mol. Biol.* **107,** 585–599.

Ehresmann, B., Reinbolt, J., and Ebel, J. P. (1975). *FEBS Lett.* **58,** 106–111.

Ehresmann, B., Reinbolt, J., Backendorf, C., Tritsch, D., and Ebel, J. P. (1976). *FEBS Lett.* **67,** 316–319.

Ehresmann, B., Backendorf, C., Ehresmann, C., and Ebel, J. P. (1977). *FEBS Lett.* **78,** 261–266.

Ehresmann, B., Backendorf, C., Branlant, C., and Ebel, J. P. (1978). In preparation.

Ehresmann, C., Stiegler, P., Fellner, P., and Ebel, J. P. (1975). *Biochimie* **57,** 711–748.

Ehresmann, C., Stiegler, P., Carbon, P., Ungewickell, E., and Garrett, R. A. (1977). *FEBS Lett.* **81,** 188–192.

Erdmann, V. A., Sprinzl, M., and Pongs, O. (1973). *Biochem. Biophys. Res. Commun.* **54,** 942–948.

Erdmann, V. A. (1976). *Prog. Nucleic Acid Res. Mol. Biol.* **18,** 45–90.

Fellner, P. (1969). *Eur. J. Biochem.* **11,** 12–27.

Fellner, P. (1974). *In* "Ribosomes" (M. Nomura, A. Tissières, and P. Lengyel, eds.), pp. 169–191. Cold Spring Harbor Lab., Cold Spring Harbor, New York.

Garrett, R., Müller, S., Spierer, P., and Zimmermann, R. (1974). *J. Mol. Biol.* **88,** 553–557.

Gorelic, L. (1975). *Biochemistry* **14,** 4627–4633.

Gray, P. N., and Monier, R. (1972). *Biochimie* **54,** 41–45.

Gray, P. N., Bellemare, G., and Monier, R. (1972). *FEBS Lett.* **24,** 156–160.

Gray, P. N., Bellemare, G., Monier, R., Garrett, R., and Stöffler, G. (1973). *J. Mol. Biol.* **77,** 133–152.

G'Ulik, A., Freund, A. M., and Vachette, D. (1978). *J. Mol. Biol.* **119,** 391–397.

Held, W. A., Ballou, B., Mizushima, S., and Nomura, M. (1974). *J. Biol. Chem.* **249,** 3103–3111.

Hélène, C. (1977). *FEBS Lett.* **74,** 10–13.

Herr, W., and Noller, H. F. (1975). *FEBS Lett.* **53,** 248–252.

Hochkeppel, H., Spicer, E., and Craven, G. (1976). *J. Mol. Biol.* **101,** 155–170.

Kenner, R. A. (1973). *Biochem. Biophys. Res. Commun.* **51,** 932–938.

Krol, A., Branlant, C., Ebel, J. P., and Visentin, L. P. (1977). *FEBS Lett.* **80,** 255–260.

Littlechild, J., Dijk, J., and Garrett, R. A. (1977). *FEBS Lett.* **74,** 292–294.

Möller, K., and Brimacombe, R. (1975). *Mol. Gen. Genet.* **141,** 343–355.

Möller, K., Rinke, J., Ross, A., Buddle, G., and Brimacombe, R. (1977). *Eur. J. Biochem.* **76,** 175–187.

Muto, A., Ehresmann, C., Fellner, P., and Zimmermann, R. A. (1974). *J. Mol. Biol.* **86,** 411–432.

Newton, I., and Brimacombe, R. (1974). *Eur. J. Biochem.* **48,** 513–518.

Noller, H., and Herr, W. (1974). *J. Mol. Biol.* **90,** 181–184.

Nomura, M., and Held, W. A. (1974). *In* "Ribosomes" (N. Nomura, A. Tissières, and P. Lengyel, eds.), pp. 193–223. Cold Spring Harbor Press, Cold Spring Harbor, New York.

Österberg, R., Sjöberg, B., Garrett, R. A., and Littlechild, J. (1977). *FEBS Lett.* **73,** 25–28.

Reinbolt, J., Ehresmann, B., Backendorf, C., Tritsch, D., and Ebel, J. P. (1978). In preparation.

Rinke, J., Yuki, A., and Brimacombe, R. (1976). *Eur. J. Biochem.* **64,** 77–89.

Rinke, J., Ross, A., and Brimacombe, R. (1977). *Eur. J. Biochem.* **76,** 189–196.

Schiltz, E., and Reinbolt, J. (1975). *Eur. J. Biochem.* **56,** 467–481.

Schwarz, U., Menzel, H., and Gassen, H. G. (1976). *Biochemistry* **15,** 2484–2490.

Seeman, N. C., Rosenberg, J. M., and Rich, A. (1976). *Proc. Natl. Acad. Sci. U.S.A.* **73,** 804–808.

Senear, A. W., and Steitz, J. A. (1976). *J. Biol. Chem.* **251,** 1902–1912.

Shine, J., and Dalgarno, L. (1974). *Proc. Natl. Acad. Sci. U.S.A.* **71,** 1342–1346.

Sloof, P., Hunter, J., Garrett, R., and Branlant, C. (1978). Submitted for publication.

Spierer, P., and Zimmermann, R. A. (1976). *J. Mol. Biol.* **103,** 647–653.

Spierer, P., Zimmermann, R. A., and Mackie, G. A. (1975). *Eur. J. Biochem.* **52,** 459–468.

Spierer, P., Zimmermann, R. A., and Branlant, C. (1976). *FEBS Lett.* **68,** 71–75.

Stanley, J., and Ebel, J. P. (1977). *Eur. J. Biochem.* **77,** 357–366.

Steitz, J., and Jakes, K. (1975). *Proc. Natl. Acad. Sci. U.S.A.* **72,** 4734–4738.

Steitz, J. A., Wahba, A. J., Langhrea, M., and Moore, P. B. (1977). *Nucleic Acids Res.* **4,** 1–15.

Stöffler, G. (1974). *In* "Ribosomes" (M. Nomura, A. Tissières, and S. Lengyel, eds.), pp. 615–667. Cold Spring Harbor Press, Cold Spring Harbor, New York.

Thomas, G., Sweeney, R., Chang, C., and Noller, H. F. (1975). *J. Mol. Biol.* **95,** 91–102.

Tritsch, D., Reinbolt, J., and Wittmann-Liebold, B. (1977). *FEBS Lett.* **77,** 89–93.

Ungewickell, E., Ehresmann, C., Stiegler, P., and Garrett, R. A. (1975a). *Nucleic Acids Res.* **2,** 1867–1888.

Ungewickell, E., Garrett, R. A., Ehresmann, C., Stiegler, P., and Fellner, P. (1975b). *Eur. J. Biochem.* **51,** 165–180.

Ungewickell, E., Garrett, R. A., Ehresmann, C., Stiegler, P., and Carbon, P. (1977). *FEBS Lett.* **81,** 193–198.

Van Duin, J., Kurland, C. G., Dondon, J., and Grunberg-Manago, M. (1975). *FEBS Lett.* **59,** 287–290.

Van Duin, J., Kurland, C. G., Dondon, J., Grunberg-Manago, M., Branlant, C., and Ebel, J. P. (1976). *FEBS Lett.* **62,** 111–114.

Yuki, A., and Bromacombe, R. (1975). *Eur. J. Biochem.* **56,** 23–34.

Zimmermann, R. A., Muto, A., Fellner, P., Ehresmann, C., and Branlant, C. (1972). *Proc. Natl. Acad. Sci. U.S.A.* **69,** 1282–1286.

Zimmermann, R. A., Muto, A., and Mackie, G. (1974). *J. Mol. Biol.* **86,** 433–450.

Zimmermann, R. A., Mackie, G., Muto, A., Garrett, R., Ungewickell, E., Ehresmann, C., Stiegler, P., Ebel, J. P., and Fellner, P. (1975). *Nucleic Acids Res.* **2,** 279–302.

10

Mechanism of Initiation of Protein Synthesis

M. GRUNBERG-MANAGO

Institut de Biologie Physico-Chimique, Paris, France

The organelles on which protein synthesis occurs are complex (see Chapter 9), and it is obvious that ribosomes possess the potential of multiple types of RNA–RNA, protein–protein, and RNA–protein interactions, implicating different forces. Because rRNA's are highly negatively charged molecules, electrostatic interactions play a leading part, and it is not surprising that monovalent and divalent ions are strong effectors of the different steps of protein synthesis. Furthermore, intra- or intermolecular base pairing in the different rRNA's allows the selection of different RNA conformations or interactions, but it has only recently been realized that this has implications for the selection of particular regions of the mRNA "initiation signals," which specify the point at which the decoding of the template in RNA begins. This chapter will review our current knowledge on the mechanism involved in selecting these initiation signals and in positioning the initiator tRNA, which provides the N-terminal aminoacyl residue of the nascent polypeptide chain opposite the initiation triplet, AUG or GUG, present in the initiation signal. It is an important step biologically because it sets the readout of mRNA in phase. Because our understanding is primarily derived from work with bacteria and phages, this chapter is limited to prokaryotes. Although we are far from understanding the sequence of events at the microscopic level, the components participating

FRONTIERS IN PHYSICOCHEMICAL BIOLOGY
Copyright © 1978 by Academic Press, Inc.

in the initiation step are now well characterized, and an overall hypothesis
to explain the selection of initiation sites and the role of protein factors
involved in this step has been formulated.

I. INITIATION SEQUENCE: CHARACTERIZATION
OF FACTORS

A brief summary of the initiation sequence is shown in Fig. 1
(Grunberg-Manago and Gros, 1977). The first step is necessary to

$$70 \text{ S} \rightleftharpoons 50 \text{ S} + 30 \text{ S} \tag{1}$$

$$30 \text{ S} + \text{fMet-tRNA}^{\text{fMet}} + \text{mRNA} + \text{GTP} \rightleftharpoons 30 \text{ S} \cdot \text{fMet-tRNA}^{\text{fMet}} \cdot \text{mRNA} \cdot \text{GTP} \tag{2}$$

$$50 \text{ S} + 30 \text{ S} \cdot \text{fMet-tRNA}^{\text{fMet}} \cdot \text{mRNA} \cdot \text{GTP} \rightleftharpoons 70 \text{ S} \cdot \text{fMet-tRNA}^{\text{fMet}}(\text{P}) \cdot \text{mRNA} + \text{GDP} + \text{P}_i \tag{3}$$

initiation complex

Fig. 1 Initiation sequence of protein synthesis.

provide the small 30 S ribosomal subunit on which the complex between the initiator tRNA and mRNA is first formed. The initiator tRNA,
fMet-tRNA^{fMet}, is specific for methionine but its tertiary structure is
different from all the other aminoacyl-tRNA's, including that positioning the internal methionine in the protein chain (Grunberg-Manago
and Gros, 1977). Generally, although there are a few interesting excep-
tions (Delk and Rabinowitz, 1974, 1975; Arnold *et al.*, 1975), the NH$_2$ of
the methionine is formylated after its attachment to the initiator tRNA,
and we believe that the requirement for formylation is related to the
polycistronic structure of many prokaryotic mRNA's (Petersen *et al.*,
1976a,b). Under physiological conditions of temperature and concentration, the ribosomal subunits are in the associated form and step 1 requires two protein initiation factors, designated IF-1 and IF-3
(Grunberg-Manago and Gros, 1977). IF-1, a small basic protein of MW
9500, serves to increase the rate constant for both association and dissociation, but more so for dissociation (Godefroy-Colburn *et al.*, 1975)
(Table I); it could thus be considered to be a dissociation factor. IF-3, a
basic protein of MW 22,000, serves to shift the equilibrium toward
dissociation by strongly binding to the 30 S particle without an effect on
the spontaneous dissociation rate; it is therefore an anti-association factor (Godefroy-Colburn *et al.*, 1975) (Table I). This has been shown by
direct measurements of both association and dissociation rates through
a stopped-flow technique using light scattering to characterize the
couples and the subunits.

Step 2 requires a third slightly acidic factor, that is, IF-2 (MW,
100,000), to direct fMet-tRNA binding. GTP stimulates this binding.

TABLE I

Effect of IF-1 and IF-3 on the Kinetic Constants of Ribosome Dissociation[a]

Concentration (nM)			$k_{1(app)}$ $(\text{sec}^{-1} \times 10^3)$	$k_{2(app)}$ $(\mu M^{-1}\ \text{sec}^{-1})$	$K_{d(app)}$ (nM)
Ribosomes	IF-1	IF-3			
39.6	—	—	10	0.24	42
50.0	—	—	10.9	0.21	52
50.1	—	—	8.0	0.19	42
51.9	30	—	17	0.21	81
51.3	31	—	15	0.21	72
51.0	59	—	20	0.21	96
51.3	65	—	21	0.27	79
51.7	88	—	24	0.28	88
41.8	116	—	21	0.32	65
51.0	144	—	33	0.36	90
41.8	—	117	10.7	0.077	139
38.7	—	117	7.5	0.053	143
63.5	—	146	7.0	0.040	173
51.2	87	58	36	0.19	188
41.0	114	114	19.5	0.060	325

[a] From Godefroy-Colburn et al. (1975). The constants were determined at 2.5 mM Mg, 100 mM NH_4 by a second-order analysis of the recordings. $70\ S \underset{k_2}{\overset{k_1}{\rightleftharpoons}} 30\ S + 50\ S$. $K_{d(app)} = [(50\ S)(30\ S)]/70\ S$.

During this step, IF-3 is released and the 50 S subunits can associated with the 30 S subunits (which is prevented when IF-3 is present on the 30 S subunit). IF-2 assists the reassociation of the two subunits (Godefroy-Colburn et al., 1975).

The last step is accompanied by the release of the other two factors: IF-1 is released when the 50 S joins the 30 S subunit, and IF-2 is released from the 70 S couple. The removal of IF-2 is dependent on GTP hydrolysis. This results in the formation of the initiation complex on the correct site (called P) on the ribosome where fMet-tRNA can form a peptide bond with the next amino acid, positioned on site A (Grunberg-Manago and Gros, 1977).

The initiation factors, isolated independently by three groups of workers (Stanley et al., 1966; Revel and Gros, 1966; Eisenstadt and Brawerman, 1966), exist in a state of loose attachment to the 30 S subunits, from which they can be detached by washing with a high concentration of salt. In a cell growing normally, the factors are not present in stoichiometric amounts with respect to ribosomes (Van Knippenberg et

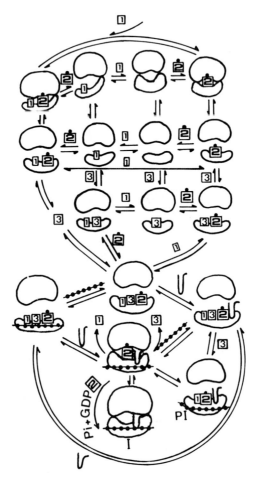

Fig. 2 Ribosomal equilibria in the formation of the initiation complex. Boxed numerals 1, 2, and 3 indicate initiation factors IF-1, IF-2, and IF-3, respectively. ● indicates GTP; –●–●–●–●– indicates mRNA. The hairpin symbol indicates fMet-tRNA^{fMet}. I = initiation complex; PI = preinitiation complex. For equilibria in which ribosomes with two initiation factors are involved, tRNA is optional when one of the factors is IF-2, and mRNA is optional when one of the factors is IF-3. From Grunberg-Manago and Gros (1977).

al., 1973). One 30 S ribosomal subunit out of twenty is equipped with a molecule of IF-3, one out of thirty with a molecule of IF-1, and the amount of IF-2 is comparable to that of IF-3 (although recent data indicate that there might be somewhat more IF-2, but less than ribosomal subunits). These factors must be recycled during protein synthesis.

Figure 2 gives a detailed picture of the equilibria resulting in the formation of the initiation complex. As previously mentioned, IF-1 and IF-3 are involved in the availability of the 30 S subunit. Beyond that stage it appears that, depending on the mRNA structure, two routes are possible (Gualerzi et al., 1977): (1) interaction of the 30 S subunit with the messenger RNA signals before attachment of initiator tRNA and (2) initiator tRNA binding first to the 30 S subunit under the influence of IF-2 to allow correct binding of mRNA. IF-1, in addition to its role as dissociation factor, stabilizes IF-2 on the 30 S subunit and helps to recycle it on the 70 S subunit.

The general requirement of these factors can be shown operationally by the use of an *in vitro* system including the RNA from an RNA phage as messenger, ribosomes washed with a high concentration of salt, protein factors needed for elongation and termination, an energy-yielding system, amino acid tRNA's, and appropriate ions. In this reconstituted system, no protein synthesis occurs unless the three initiation factors are present. Table II shows that lysozyme synthesis in the presence of mRNA from phage T4 requires all three factors. It can also be seen that the positioning of fMet-tRNA on the ribosome requires the initiation factors.

The requirement for IF-3 is more striking when a phage mRNA is used (rather than a synthetic mRNA containing the initiation codon AUG or GUG). IF-3, in addition to its role as an anti-association factor, is believed to be essential for the specific recognition of mRNA initia-

TABLE II

Role of Prokaryote Initiation Factors IF-1, IF-2, and IF-3 in Protein Chain Initiation[a]

| | T4 mRNA-dependent | | AUG-dependent | |
| | | | fMet- | |
Factors added	Lysozyme synthesis (units)	fMet-puromycin formation (pmoles)	puromycin formation (30 S + 50 S) (pmoles)	fMet-tRNA binding (30 S + 50 S) (pmoles)
IF-1	1.2	—	0.28	0.61
IF-3	1.3	—	0.25	—
IF-2	1.4	—	1.20	0.65
IF-1 + IF-2	3.4	0.38	4.60	3.00
IF-1 + IF-3	6.5	—	0.40	—
IF-3 + IF-2	3.5	—	0.95	—
IF-1 + IF-3 + IF-2	12.8	1.06	4.90	—
None	1.5	0.24	0.39	0.49

[a] From Revel et al. (1969).

tion regions (Grunberg-Manago and Gros, 1977), and several regulatory mechanisms have been postulated to be associated with IF-3 action (Lee-Huang and Ochoa, 1971; Revel, 1972).

II. SELECTION OF INITIATION SIGNALS

A. Nature of Initiation Signals

Since the work of Marker and Clark (see the review by Grunberg-Manago and Gros, 1977), it has been known that AUG and GUG direct the binding of N-formyl-Met-tRNA^{fMet}, but these triplets also direct the binding of internal Met-tRNA and Val-tRNA on the ribosome. It was evident that some other feature should distinguish the initiator codon from the many AUG and GUG triplets such as the out-of-phase or in-phase coding for amino acids found internally in the polypeptide chain. It was first believed that ribosomes start protein synthesis directly at the 5' end of an mRNA molecule with no special signal other than the presence of an initiator triplet (Steitz, 1978). However, when the nucleotide sequences of phage RNA and plant virus RNA were determined (Steitz, 1975), it was seen that these naturally occurring mRNA molecules do not begin with AUG or GUG. Moreover, phage mRNA's are polycistronic. Thus, R17, f2, and MS2 RNA's code for three proteins, maturation protein A, coat protein, and replicase (Steitz, 1969; Lodish and Robertson, 1969a,b), and ribosomes can select the three initiation sites. Finally, ribosomes can initiate a polypeptide chain very efficiently on a circular messenger DNA molecule (Bretscher, 1969). Thus, after selection, *Escherichia coli* ribosomes must be capable of binding internally to true messenger RNA initiation signals.

Several groups (particularly Steitz, 1975, 1978) isolated and sequence-analyzed the ribosome binding sites to the single-stranded genomes of RNA bacteriophages. The procedure was based on the results of Takanami *et al.* (1965), which showed that regions of phage RNA are protected from ribonuclease digestion by association with ribosomes. When α-^{32}P-labeled RNA is mixed with washed *E. coli* ribosomes plus initiation factors and initiator tRNA, the ribosomes bind at the initiation sequence and are unable to leave this site because of the lack of internal aa-tRNA. The fragment protected against nucleases can be liberated by a denaturing agent such as sodium dodecyl sulfate (SDS), and its sequence is determined by Sangers's procedure (Barrell, 1971).

While no common features were immediately revealed in the initiation sequences, it was reassuring to find that sites protected by ribosomes under initiation conditions are located exclusively at the beginning of genes (Fig. 3). The protected fragment is about 30–40 nucleotides long; all the sequences include the AUG triplet preceding the first translated codons. The initiator triplet appears approximately in the middle of the protected sequence; knowing the N-terminal peptides of the three proteins, it was easy to verify that previous assignments of the genetic code were correct and that the translated codons are situated at the right of the AUG initiation triplet. In addition, intercistronic spaces exist in the RNA phage genome. Ribosomes can bind to internal initiator regions on a polycistronic mRNA, even in the absence of translation.

It was generally believed that ribosome recognition of initiator regions was a specific protein–nucleic acid interaction (as in the case of tRNA synthetase or methylase recognition); one of the several ribosomal proteins, as well as initiator factor IF-3 or some other protein factor

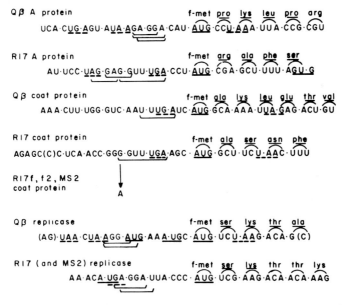

Fig. 3 RNA phage ribosome binding sites. The three ribosome-protected initiator regions from Qβ and R17 RNA are shown. A base change (G to A) in the coat protein initiation site from another variant of R17 (denoted R17f) and from f2 and MS2 RNA is indicated by a solid underline; termination triplets are indicated by dashed underlines. Base-sequence homologies of four or more nucleotides occurring at the 5' extremity of the initiator AUG's are indicated by bridge links. Experimentally determined amino acid sequences are underlined. From Steitz (1975).

called interference factor (Revel, 1972), was presumed to bind selectively to some unidentified important sequence in mRNA.

B. Species Specificity

Lodish and Robertson (1969a, 1970) were the first to demonstrate clearly that the specificity is not due to initiation factors, but lies in the small subunit. *E. coli* ribosomes recognize the beginning of the three cistrons of phage f2 RNA: protein A, coat, and replicase, the coat being the most efficient initiation site. In contrast, *Bacillus stearothermophilus*, a gram-positive bacterium, translates only cistron A, either at 65°C, its physiological temperature, or at 47°C, the temperature at which *E. coli* translates the three cistrons. In general, it appears that gram-positive and -negative bacteria have different specificities.

By crossing the various elements from the translation machinery at 47°C, where both organisms are active, Lodish and Robertson showed that the specificity of action of *E. coli* and *B. stearothermophilus* ribosomes is not altered with heterologous initiation factors, where it is altered by crossing the 30 S particle. Other workers have investigated the components of the small particle responsible for the specific recognition of initiation signals on mRNA (Held *et al.*, 1974; Golberg and Steitz, 1974) using the heterologous 30 S particle reconstitution techniques of Nomura (Higo *et al.*, 1973) and found that the ability to discriminate among the initiation signals present on phage mRNA is based on protein S12 and that, at least according to Nomura's group (Held *et al.*, 1974), the 16 S RNA is also involved. Finally, protein S1 also appears to play a role (Isono and Isono, 1975).

In experiments designed to show that ribosome binding and protection are also limited to the A site with R17 and Qβ RNA instead of phage f2 RNA, Steitz (1973) obtained a different result using *B. stearothermophilus* ribosomes. While with R17 RNA the expected A protein initiator region was obtained, with Qβ RNA the thermophillic ribosomes at temperatures above 65°C ignore all three normal sites and bind instead to two other regions from the RNA molecule. Both binding sites lack the initiator triplet and therefore do not direct the binding of fMet-tRNA^{fMet}; both of these sites are very rich in polypurine tracks (Table III).

C. Shine and Dalgarno's Hypothesis

In 1974, Shine and Dalgarno proposed a revolutionary hypothesis to explain the specificity of selecting the initiation regions that was imme-

TABLE III

Binding and Protection Data Using B. stearothermophilus Ribosomes[a]

Binding temperature[b] (°C)	Site	Sequence[c]	Theoretical T_m of mRNA·rRNA complex[d] (°C)
65	R17 A	CCUAGGAGGUUUGACCU **AUG**	38
	Qβ noninitiator a	CUGAAAGGG̲GAGAUUACUCG	29
	Qβ noninitiator b	GG̲AAGGAGC	29
49	R17 A	CCUAGGAGGUUUGACCU **AUG**	38
	R17 replicase	AUG̲AGGAUUACCC **AUG**	−7
	Qβ replicase	AACUA̲AGGAUGAA **AUG**	−7
	Qβ coat	AAACUUUG̲GGUCAAUUUGAUC **AUG**	−25
Not bound	R17 coat	CAACCGG̲GUUUGAAGC **AUG**	−46

B. stearothermophilus 16 S 3' end   HoAUCUUUCCUCCACUAG₃'

[a] From Steitz (1978, *In* "Biological Regulation and Development," R. Goldberger, ed. Copyright 1978 by Plenum Press).

[b] Note that only the R17 A site is bound strongly by *B. stearothermophilus* ribosomes at 49°C; the other regions are recognized much more weakly by thermophilic than by *E. coli* ribosomes at 49°C.

[c] Underlining indicates phage RNA regions that are complementary to the 3′-end sequence of *B. stearothermophilus* 16 S rRNA aligned below. Dots represent G·U base pairs. Initiator AUG's are in bold type. Italics indicate the portion of the 16 S terminal sequence that differs in *E. coli* and *B. stearothermophilus*.

[d] T_m's calculated according to Gralla and Crothers (1973). Note the correlation between the T_m of the mRNA·rRNA complex and the temperature at which each site binds stably to *B. stearothermophilus* ribosomes.

TABLE IV

R17 Ribosome Binding Site Sequences and Proposed Pairing with the 3′ Terminus of 16 S RNA[a]

Bacterium	R17 cistron	Possible pairing of 3′ terminus with appropriate region of ribosome binding site sequence[b]		Number of base pairs possible	Ribosome binding to unfolded R17, MS2, or f2 RNA
		OH AUUCCUCCACPy	16 S RNA		
E. coli B					
	A protein	(5′) CUAGGAGGUUU(3′)		7	+
	Replicase	ACAUGAGGAUU		4	+
	Coat protein	ACCGGGUUUG → A[c]		3 (4)[c]	+
		OH AUUCCUCUCPy	16 S RNA		
P. aeruginosa					
	A protein	(5′) CUAGGAGGUU(3′)		5	+
	Replicase	UGAGGAUUAC		4	+
	Coat protein	ACCGGGUUU → A[c]		3 (4)	+

246

B. stearothermophilus			
	$_{OH}$ A U C U U U C C U Py	16 S RNA	
A protein	(5′) A U C C U ⌐A G G A G¬ G (3′)		>4 +
Replicase	A C A U G ⌐A G G A U¬		4 +
Coat protein	A A C C G G ⌐G G¬ U U		2 (1) −
	→ A^c		

C. crescentus			
	$_{OH}$ U C U U U C C U Py	16 S RNA	
A protein	(5′) U C C U ⌐A G G A G¬ G (3′)		>4
Replicase	C A U G ⌐A G G A U¬		4
Coat protein	A C C G G ⌐G G¬ U U		2 (1)
	→ A^c		

[a] From Shine and Dalgarno (1974).

[b] The sequences shown represent all or part of the conserved sequence |5′-AGGAGGU-3′| from the A protein, replicase, and coat protein initiator region of R17 RNA. Py = pyrimidine nucleotide.

[c] Two sequences have been reported for this region of the coat protein binding site in the two different R17 stocks used. The G → A transaction represents a spontaneous mutation that occurred after the two stocks were separated. The corresponding sequence from MS_2 and f2 RNA contains the A substitution.

247

diately accepted. Their hypothesis focused attention on and renewed interest in ribosomal RNA's which, after the discovery of mRNA, were assumed to have only a role in the morphogenesis of ribosomes, despite the suggestion by Kurland (1974) to the contrary. Shine and Dalgarno proposed that the 3'-OH-terminal dodecanucleotide of *E. coli* 16 S ribosomal RNA directly participated in the selection of initiation sites by forming several Watson–Crick base pairs with a polypurine-rich sequence located in the vicinity of the 5' end of the initiator triplet in mRNA (Shine and Dalgarno, 1974) (Table IV). It is significant that all coliphage RNA ribosome binding sites contain part of or the whole purine-rich sequence, AGGAGGU, at a relatively similar position on the 5' side of the initiator triplet AUG (Steitz *et al.*, 1977; Steitz, 1978). Seven base pairs could possibly be involved in the interaction between 16 S RNA and the initiation site of protein A, whereas four or five pairs could be used for binding to other initiation sequences. The length of the complementarity can vary between three and nine nucleotides, the average being four or five nucleotides. The number of nucleotides separating the complementary region from the initiator triplet varies, with an average of ten nucleotides to the middle of the potentially base-paired stretch (Steitz, 1978). Finally, it should be pointed out that, in every case where two or more initiator triplets appear in a single initiator region, the triplet preceded by the strongest and most appropriate complementarity is the one that functions in polypeptide chain initiation (Steitz, 1978) [*lacI* mRNA has three GUG sequences in phase and only one is an initiator (Steege, 1977; see Table V)].

In addition to the appearance of a sequence complementarity to 16 S RNA in known ribosome binding sites, other lines of evidence showed that the suggestion of Shine and Dalgarno was quite attractive. Evidence has accumulated that an intact 3' terminus of 16 S RNA is necessary for initiation, as shown by the inhibitory action of colicin E-3, which results in the removal of about 50 nucleotides from the 3' terminus of the 16 S RNA by a single endonucleolytic cleavage (Konisky and Nomura, 1967). Other inhibitors of initiation (streptomycin and kasagamycin) have their sites of action in the vicinity of the 3'-OH terminus of 16 S RNA (Crépin *et al.*, 1973; Melser *et al.*, 1971). Cross-linking experiments also suggest that terminus 16 S RNA is the binding site for initiation factors and ribosomal proteins involved in initiation (Van Duin *et al.*, 1976; S. Langberg and J. W. B. Hershey, personal communication). Random copolymers rich in A and G are the best competitive inhibitors for initiation on phage mRNA (Revel and Greenshpan, 1970). Moreover, the fragment of Qβ RNA bound by *B. stearothermophilus* ribosomes at 65°C, which does not possess an initia-

TABLE V

Initiation Sequences Recognized by *E. coli* Ribosomes[a]

mRNA	Ribosome binding site[b]
R17 A	GAU UCC U<u>AG GAG GU</u>U UGA CCU **AUG** CGA GCU UUU AGU G
MS2 A	GAU UCC U<u>AG GAG GU</u>U UGA CCU **GUG** CGA GCU UUU AGU G
Qβ A	UCA CUG AGU AUA A<u>GA GG</u>A CAU **AUG** CCU AAA UUA CCG CGU
R17 coat	CC UCA ACC <u>G</u>Ġ<u>G GU</u>U UGA AGC **AUG** GCU UCU AAC UUU
Qβ coat	AAA CUU UĠ<u>G GUC</u> AAU UUG AUC **AUG** GCA AAA UUA GAG ACU
f2, MS2 coat	CC UCA ACC <u>GAG GU</u>U UGA AGC **AUG** GCU UCC AAC UUU ACU
R17, MS2 replicase	AA ACA U<u>GA GGA</u> UUA CCC **AUG** UCG AAG ACA ACA AAG
Qβ replicase	AG UAA C<u>UA AGG A</u>UG AAA UGC **AUG** UCU AAG ACA G
f1 coat	UUU AAU <u>GGA</u> AAC UUC CUC **AUG** AAA AAG UCU UU
f1 gene 5	A <u>AGG U</u>AA UUC ACA **AUG** AUU AAA GUU GAA AU
f1 gene (?)	A AAA <u>AAG G</u>UA AUU CAA **AUG** AAA UU
T3[c] *in vitro*	AAC AUG <u>AGG U</u>AA CAC CAA **AUG** AUU UUC ACU AAA GAG
T7 gene 0.3 site a	AAC UGC ACG <u>AGG U</u>AA CAC AAG **AUG** GCU AUG UCU AAC AUG
T7 gene 0.3 site b	GUA CĠ<u>A GGA GG</u>A UGA AGA GUA **AUG** U
λP_R	pppAŬG UAC U<u>AA GGA GG</u>U UGU **AUG** GAA CAA CGC
λC_I	TTG CGG TGA TAG ATT TAA CGT **ATG** AGC ACA AAA AAG
φX174 G	TTT CTG CTT <u>AGG AG</u>T TTA ATC **ATG** TTT CAG ACT TTT ATT
φX174 F	CCT ACT TĠ<u>A GGA</u> UAA AUU **AUG** UCU AAU AUU CAA ACU
φX174 D	ACC ACT AAT <u>AGG T</u>AA GAA ATC **ATG** AGT CAA CTT ACT
φX174 H	ACT TAA GTG <u>AGG TGA</u> TṪT **ATG** TTT GGT GCT ATT
φX174 B	UAA AGG UCC <u>AGG AGC</u> UAA AGA **AUG** GAA CAA CUC ACU
φX174 J	ACG TGC GGA <u>AGG AGT</u> GAT GTA **ATG** TCT AAA GGT AAA
araB	UUU UUU GGA UGĠ AGU GAA ACG **AUG** GCG AUU GCA AUU
trp leader	CAC GUA AAA A<u>GG GUA</u> UCG ACA **AUG** AAA GCA AUU UUC GUG
trpE	GAA CAA AAU UAG AGA AUA ACA **AUG** CAA ACA CAA AAA CCG
trpA	GAA AGC ACĠ <u>AGG</u> GGA AAU CUG **AUG** GAA CGC UAC GAA UCU
lacZ	AAU UUC ACA C<u>AG GAA</u> ACA GCU **AUG** ACC AUG AUU ACG GAU
lacI	AGU CAA UUC A<u>GG GUG GŬG</u> AAU **GUG** AAA CCA GUA ACG
galE	AUA AGC CUA AUG <u>GAG</u> CGA AUU **AUG** AGA GUU CUG GUU ACC
galT	TAT CCC GAT <u>TAA GGA</u> ACG ACC **ATG** ACG CAA TTT AAT CCC

16 S RNA 3' end _HO_A U U C C U C C A C U A G_5'

[a] From Steitz (1978, *In* "Biological Regulation and Development," R. Goldberger, ed. Copyright 1978 by Plenum Press).

[b] Underline indicates contiguous bases complementary to the 3' oligonucleotide of *E. coli* 16 S rRNA. Dots indicates G·U base pairs. Initiator triplets are indicated by bold type.

[c] What was originally thought to be bacteriophage T7 has turned out to be T3.

tor triplet, contains a long polypurine stretch whose sequence is complementary to the 3'-OH end of 16 S RNA (Steitz *et al.*, 1977; see Table
III). Finally, the 3' 16 S terminus is placed on the interface between the
two ribosomal subunits where the mRNA is presumed to bind (Chapman and Noller, 1977; Santer and Shane, 1977).

   More direct evidence of the validity of Shine and Dalgarno's hypothesis comes from the work of Steitz and Jakes (1975), who treated initiation complexes formed by *E. coli* 70 S ribosomes with colicin E-3 in the
presence of the messenger fragment of the initiation cistron for maturation protein A (this fragment was labeled with ^{32}P). After removing the
proteins by adding 1% sodium dodecyl sulfate, they fractionated the
components on a polyacrylamide gel. An mRNA–rRNA hybrid, containing approximately equimolar amounts of the 30-nucleotide mRNA
fragment and of the 49- to 50-nucleotide colicin fragment, was detected
(Fig. 4). This hybrid, which exhibits an electrophoretic mobility different from those of each of the above two fragments, appears only in the
presence of all the components necessary for complex formation, in particular, mRNA. Furthermore, it does not appear if the colicin treatment
is omitted, since the mRNA then cosediments in the sucrose gradient
with intact 16 S RNA. Moreover, aurintricarboxylic acid, an inhibitor of
mRNA binding to ribosomes, lowers the amount of complex formation.

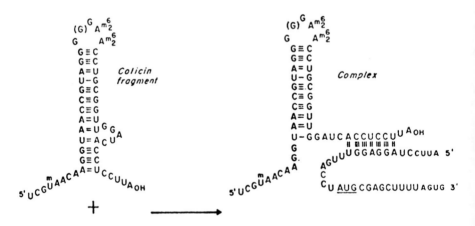

Fig. 4 Postulated hydrogen bonding between the colicin fragment of 16 S RNA and
the R17 A protein initiator region. The secondary structure shown, involving internal base
pairing according to Shine and Dalgarno, is predicted to be stable under physiological
conditions. From Steitz and Jakes (1975).

The hybrid can be dissociated by heat (55°C). With ^{32}P-labeled ribosomes, in addition to the ^{32}P-labeled initiator region, chromatographic analysis revealed that the portion of the gel containing the complex consisted of oligonucleotides assigned to the colicin E-3 fragment and to the R17 protein initiator fragment. These observations strongly support the hypothesis of Shine and Dalgarno that base pairing between mRNA and the 3' terminus of 16 S RNA does occur during the formation of a functional initiation complex.

Figure 4 suggests that a specific secondary structure is assumed by the 3'-terminal region of the 16 S RNA and that after RNA binding some of the intramolecular base pairs may be exchanged for intermolecular hydrogen bonds. Temperature relaxation methods and high-resolution proton magnetic resonance spectra (Baan *et al.*, 1977) are consistent with the structural conformation of the isolated E-3 fragment (there is a stable helical region at $T_m = 80°C$) containing the nine base pairs. Temperature jump studies also reveal a second transition at 21°C that could correspond to the melting of additional base pairs below the loop (Steitz *et al.*, 1977).

Thermal melting studies also support the existence of the intermolecular base pairs in the complex containing the colicin fragment and the R17 "A" protein initiator region (the T_m's are 32°–37°C) (Steitz and Steege, 1977). However, since protein synthesis can begin both *in vivo* and *in vitro* at temperatures far exceeding that of the most stable isolated hybrids, the mRNA–RNA interaction must be supplemented by other mRNA–ribosome contacts, particularly for initiation sites that exhibit much less mRNA–rRNA base pairing. We know that this is the case in the codon–anticodon interaction on the ribosome, where the T_m of the complex far exceeds that of the most stable codon–anticodon interactions in solution (McLaughlin *et al.*, 1968).

Does the Shine and Dalgarno Theory Explain Species Specificity?

After analyzing the twelve residues at the 3' end of *E. coli* 16 S RNA, Shine and Dalgarno (1975) examined comparable regions from other bacterial species (Table VI). At its 3'-OH terminus, *Pseudomonas aeruginosa* has a sequence of six nucleotides identical to those of *E. coli* and has the same specificity for cistron recognition. The gram-positive bacteria these authors examined have pyrimidine-rich 3'-OH termini but with different nucleotide sequences. A complementarity of about four base pairs nevertheless exists between this sequence and the ribosome binding site sequence of both the A protein and the replicase

TABLE VI

3'-Terminal Sequences of 16 S rRNA[a]

Bacterium	Sequence[b]
Escherichia coli B	GAUCACCUCCUUA$_{OH}$
Pseudomonas aeruginosa	G(X)$_2$YCUCUCCUU(A)$_{OH}$[c]
Bacillus stearothermophilus	G(X)$_{-5}$YUCCUUUCU(A)$_{OH}$[c]
Bacillus stearothermophilus	GAUCACCUCCUUUCUA$_{OH}$[c,d]
Bacillus subtilis	G(X)$_{-7}$YCUUUCU$_{OH}$
Caulobacter crescentis (ATCC 15252)	G(X)$_3$YUCCUUUCU$_{OH}$

[a] From Shine and Dalgarno (1975).
[b] X represents any nucleoside other than guanosine; Y is pyrimidine nucleoside.
[c] The variable presence of the 3'-terminal adenosine in 16 S RNA is found in a variety of bacteria and depends on the culture conditions.
[d] Sequence as determined by Sprague *et al.* (1977).

cistrons of R17 RNA. The degree of complementarity with the coat protein binding site sequence is considerably less (one or two base pairs) (Table IV). *B. stearothermophilus,* as discussed above, binds significantly only the A protein site on native f2 or R17 RNA; while appreciable translation of the replicase in addition to the A protein cistrons does occur when the RNA is unfolded by formaldehyde, no coat protein is synthesized under these conditions (Lodish, 1971). The 3' sequence of *Caulobacter crescentis* 16 S RNA is similar to that of *B. stearothermophilus* (Table VI), and the ribosomes of *C. crescentis* do not bind to the coat protein initiation site. It was therefore concluded that the extent of possible interaction between the 3' terminus of 16 S RNA from several different bacteria and the ribosome binding sites of coliphage RNA is consistent with the available data on translation specificity in these bacteria.

However, to directly correlate the mRNA–rRNA complementarity with the translation specificity of prokaryotes, Sprague *et al.* (1977) determined the complete nucleotide sequences of the 3'-terminal oligonucleotide of 16 S rRNA from *B. stearothermophilus* obtained after hydrolysis with T1 nuclease (Table VI) and of the two polypurine stretches from Qβ bacteriophage RNA bound by ribosomes from this species at high temperature (Table III). Their results confirmed that specific recognition of the two Qβ regions can indeed be explained by an extensive mRNA–rRNA pairing. Masking of the two noninitiator Qβ regions by RNA secondary structure probably explains why these sites do not appear among the fragments protected by *B. stearothermophilus* ribosomes at lower temperatures. However, the translational specificity of *B. stearothermophilus* was more difficult to explain. When sequences longer than the extreme 3' termini of *B. stearothermophilus* rRNA were determined, a sequence homology was revealed between *E. coli* and *B. stearothermophilus* 16 S RNA's. The 3' termini of both RNA's are identical, except that the 3'-terminal adenosine of *E. coli* rRNA is replaced by $UCUA_{OH}$ in the thermophile. Thus, *B. stearothermophilus* ribosomes have the same potential of binding R17 or f2 coat protein initiator as do *E. coli* ribosomes, unless the four extranucleotides at the 3'-OH end preclude the availability of that sequence for mRNA recognition. As previously discussed, proteins S12 and S1 are important selectivity determinants, and they might be involved in facilitating the mRNA–rRNA interaction by stabilizing the bonds of the RNA's or by opening the intramolecular base pairs of the colicin E-3 fragment from the 16 S RNA ends. Nevertheless, the possibility exists that the proteins can create other interaction sites between mRNA and ribosomes.

III. SECONDARY STRUCTURE OF mRNA AS NEGATIVE CONTROL

In the preceding sections, only fragments of mRNA adjacent to authentic initiator codons have been considered. However, recognition by ribosomes is influenced by intramolecular interactions within the mRNA molecule. For instance, if the initiator triplet, or the polypurine tract, is made unavailable, a potential initiation signal could be rendered inactive.

The complete nucleotide sequence of the MS2 phage genome has been determined by Fiers *et al.* (1976). They suggest that this mRNA folds into a series of well-defined hairpin loops (Fig. 5). When this sequence is examined, it can be seen that many regions in the viral RNA fulfill the criteria of Shine and Dalgarno base pairing and yet are not bound by ribosomes. For example, in replicase gene segment 2381–2396 (Fig. 5) the sequence UAAGGAGCCUGAUAUG is found; according to the criteria, it should interact with both *E. coli* and *B. stearothermophilus* 16 S rRNA's. That this is not observed may mean that masking accessibility by secondary and tertiary structures may be equally important, and it can be seen that the sequences surrounding this additional triplet are sequestered by the RNA secondary structure. In accordance with this observation, when phage mRNA is treated with formaldehyde additional initiations sites are revealed (Lodish, 1971).

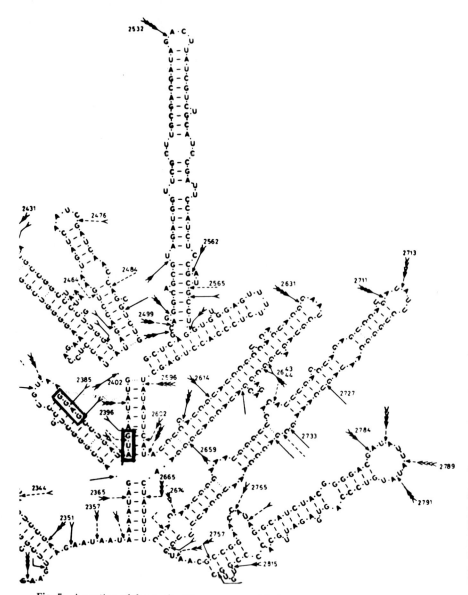

Fig. 5 A portion of the nucleotide sequence with the secondary structure of the repli-case gene. Primarily the sequence is read from the 5' to the 3' end. The numbering starts at the first 5' nucleotide of MS2 RNA and ends at 3'-terminal residue 3569. Arrows point to sites easily split by nucleases during partial digestion of MS2 RNA: solid arrows, T1 ribonuclease; dashed arrows, carboxymethylated ribonuclease. The feathers on the arrows indicate the sensibility of the bond under the conditions used: 0, split very seldom; 1, seldom; 2, split often; 3, split very often; 4, always split. The AUG sequence and Shine and Dalgarno sequence GGAG enclosed in rectangles denote initiation-type sequences not recognized as such by ribosomes. From Fiers *et al.* (1976).

It is obvious that the potential for base pairing and the presence of an initiator triplet are not sufficient to describe a true mRNA initiator region. The secondary structure is an active negative control in limiting initiation to the proper regions. This explains why protein A and replicase are less efficiently translated than coat protein in intact phage RNA, whereas treatment that disrupts the native conformation stimulates both replicase and A protein synthesis. The secondary structure can sometimes play a positive role. The R17 coat protein site as a fragment decreases in initiation potential. It is possible that the coat protein site as an isolated fragment folds into a stable secondary structure not assumed in the intact molecule or that the overall tertiary structure may be designed so that this region is exposed in such a way that it cannot be ignored by the ribosomes (Steitz, 1978).

In vivo "reinitiation" after polypeptide chain termination at the site of a nonsense mutation is another interesting phenomenon that could test the theory of Shine and Dalgarno. In the lactose repressor gene (lacI gene), Weber, Miller, and their collegues (Platt et al., 1972; Files et al., 1974, 1975) have identified three discrete restarting sites that give rise to C-terminal repressor fragments. The codons of valine-23 GUG, methionine-42 AUG, and leucine-62 UUG of the repressor amino acid sequence can each be used to reinitiate translation beginning with N-formylmethionine. Restarted polypeptides are not effective in wild-type E. coli but accumulate in suitable lacI nonsense mutant strains at approximately 10% of the repressor level found in wild-type cells. A single nonsense mutation can activate more than one reinitiation site. To examine the structure of mRNA in the vicinity of the restarting points, Steege (1977) sequenced 214 nucleotides at the 5' terminus of lacI mRNA (Fig. 6).

The wild-type repressor protein is initiated at a GUG codon preceded by a sequence exhibiting substantial complementarity to 16 S RNA; of the three in-phase GUG triplets clustered in this region, the real initiator is that codon most appropriately situated relative to the polypurine stretch.

Relative to the true lacI initiator region, mRNA complementarity to rRNA is weak at the restart sites and is nonexistent at valine-23. Several additional in-phase GUG codons in the sequence that do not serve as reinitiation signals in the nonsense mutant can also be seen. It is clear that, following termination, ribosomes do not restart simply by moving to the next available codon. GUG restart at valine-23 occurs concomitantly with the restart of methionine-42. Binding at the accessible methionine-42 destabilizes the site at valine-23 and makes it available for restart.

----

pppG CAA GAG AGU CAA UUC AGG GUG (GUG AAU) GUG AAA CCA GUA ACG UUA UAC GAU GUC GCA GAG UAU
                                  .   _   Met - Lys - Pro - Val - Thr - Leu - Tyr - Asp - Val - Ala - Glu - Tyr
                                                                                                                      10
                                                          1

GCC GGU GUC UCU UAU CAG ACC GUU UCC CGC GUG GUG AAC CAG CCC AGC CAC GUU UCU
Ala - Gly - Val - Ser - Tyr - Gln - Thr - Val - Ser - Arg - Val - Val - Asn - Gly - Ala - Ser - His - Val - Ser
                                       20   23 restart   30

GCG AAA ACG CGG GAA AAA GUG GAA GCG GCG AUG GCG GAG CUG AAU UAC AUU CCC AAC CGC
Ala - Lys - Thr - Arg - Glu - Lys - Val - Glu - Ala - Ala - Met - Ala - Glu - Leu - Asn - Tyr - Ile - Pro - Asn - Arg
                                       40   42 restart   50

GUC GC(A CAA C)AA CUG GCG GGC AAA CAG UCG (UUG)
Val - Ala - Gln - Gln - Leu - Ala - Gly - Lys - Gln - Ser - Leu
                          60   62 restart

Fig. 6 Sequences of *lacI* mRNA (Steege, 1977) and the repressor protein (Beyreuther *et al.*, 1973), arranged so that the initiator codons for the three restart proteins are aligned under that for the wild-type repressor. All in-phase initiator triplets are shown in large type. Sequences that are complementary to the 3' end of 16 S rRNA and precede the utilized initiators are indicated by underlines (with G · U pairs denoted by dots). The ribosome-protected region (Steege, 1977) is indicated by a dashed line. From Steitz (1978, *In* "Biological Regulation and Development," R. Goldberger, ed. Copyright 1978 by Plenum Press).

IV. PART PLAYED BY RIBOSOMAL PROTEINS AND INITIATION FACTORS

It has been found that the binding sites for all three initiation factors, IF-1, IF-2, and IF-3 (Van Duin *et al.*, 1976; S. Langberg and J. W. B. Hershey, personal communication), and for S1 (Czernilofsky *et al.*, 1975; Dahlberg and Dahlberg, 1975) are close to the 3' end of 16 S RNA; these proteins can be covalently attached to 16 S RNA after oxidation by periodate and reduction of the 30 S ribosomal subunit. This treatment oxidizes the 3' terminus of RNA, and the reduction of the product formed with the amino group of lysine creates a stable bond between the RNA and an appropriately positioned protein. Other chemical cross-linking data also located these proteins in the environment of the 3'-OH of 16 S RNA (Van Duin *et al.*, 1975). Moreover, after treating the ribosomes with colicin E-3 and centrifuging in a sucrose gradient in the absence of Mg, a hybrid containing S1 and the colicin nucleotide fragment was isolated (Dahlberg and Dahlberg, 1975). The work of Dahlberg and Dahlberg further suggests that this protein may interact specifically with the pyrimidine-rich dodecanucleotide terminal of the RNA molecule. Gassen *et al.* (1977) found that S1 interacts with all single-stranded polynucleotides that have a flexible structure whatever the nature of the bases, whereas it has less affinity for stacked bases, such as those in poly(A). S1 will thus react with the polypyrimidine-rich 3'-OH end of 16 S RNA and may correctly position that sequence for subsequent base pairing with the complementary region on mRNA. Other possibilities for S1 action should not be disregarded (see Szer *et al.*, 1976).

Van Duin *et al.* (1976) found that IF-3 can be cross-linked not only to the 3' end of 16 S RNA but also to the 3'-OH end of 23 S RNA, which suggests that the 3' termini of these two molecules are near to each other and to IF-3. This led to the postulate that, in the 70 S couple, 16 S RNA forms complementary base pairs with 23 S RNA, possibly at the 3' termini. This proposal is supported by the complementarity between the two sequences and by the observed differences in the nuclease susceptibility of the 16 S molecule in the 30 S versus the 70 S ribosomal subunit (Chapman and Noller, 1977; Santer and Shane, 1977). Moreover, thermodynamic data of the association of the two subunits are more consistent with the existence of hydrogen bonds and that no hydrophobic forces are involved in the two-subunit association (Hui Bon Hoa *et al.*, 1977).

Upon dissociation, the base pairs between 16 S and 23 S RNA's are replaced by self-complementary base pairs, existing in the 16 S and 23 S

RNA's. The resulting 30 S conformation is appropriate for the binding of mRNA, and the interaction is strengthened in the presence of IF-3. The two reactions, that resulting in 30 S mRNA and the 30 S–50 S association, are thus mutually exclusive (Fig. 7).

From the point of view of this model, the different *in vitro* effects of IF-3 on the stability of 70 S couples and on the binding of mRNA can be seen as two different aspects of a single function. In effect, IF-3 could be responsible either indirectly or directly (by binding to the sequence involved in base pairing between 16 and 23 S RNA's) for a series of transitions of complementary nucleotide interactions, from those between 16 and 23 S RNA's to those between the self-complementary interactions of 16 S RNA and, finally, to those between 16 S RNA and mRNA.

It is expected that S1 and IF-3 would be essential for ribosome binding to initiator regions with a weak hydrogen-bonding potential to 16 S RNA. By contrast, if the mRNA–rRNA interaction is strong (thermodynamically stable under physiological conditions), it may well be able to form without the assistance of this helper protein.

Base pairing offers new possibilities as a selection mechanism for the regulation of protein synthesis. The extent of complementarity between ribosomes and their binding sites may influence the frequency of initiation at certain sites. For instance, for the three R17 initiation regions, ribosome recognition of the coat or replicase site is severalfold more dependent on the presence of initiation factors (Steitz *et al.*, 1977) and of fMet-tRNA than on binding to the A protein initiator (Table VII). Here

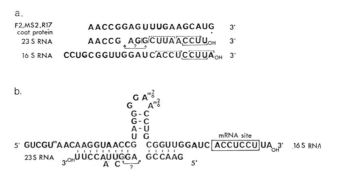

Fig. 7 Possible base pairing between the 3′-OH termini of 16 S RNA and 23 S RNA or mRNA. (a) The nucleotide sequences of the 3′ end of the 16 and 23 S RNA's are compared with those of the bacteriophage RNA ribosomal binding site associated with the coat protein cistron. The hexanucleotide of the 23 S RNA distal to the 3′ end was sequenced by Branlant and Ebel (1977). (b) Hydrogen-bonding schemes are depicted for the interaction of 16 and 23 S RNA's. From Van Duin *et al.* (1976).

TABLE VII

Effect of S1 and Factors on Recognition of R17 Initiator Regions[a]

Experiment	Ribosomes	Factors	A : coat : replicase ratio
I	30 S	IF-2, IF-3	1 : 2.0 : 0.5
	30 S	—	1 : 0.2 : 0.05
	30 S (−S1)	IF-2, IF-3	1 : 0.3 : 0.04
	30 S (−S1)	—	1 : 0.4 : 0.03
II	30 S	Crude	1 : 7.1 : 2.7
	30 S	IF-2, IF-3	1 : 3.6 : 0.8
	30 S	—	1 : 0.2 : 0.2
	30 S (−S1)	IF-2, IF-3	1 : 0.8 : 0.2
	30 S (−S1)	—	1 : 0.1 : 0.06

[a] From Steitz et al. (1977).

it seems likely that the comparatively higher degree of complementarity between the A site and the 16 S RNA 3′ end substitutes to some extent for the usual requirements for the other components of the initiation reaction. The fact that site A can couple with a colicin fragment that has been removed from the ribosome supports this idea.

The weaker dependence on IF-3 observed with poly(A,U,G) versus natural mRNA may be due to the presence of long oligopurine sequences in the synthetic polymers compared to the weak Shine and Dalgarno complementarity of the exposed coat cistron in the phage RNA.

V. SPECIFIC MUTANTS AFFECTING INITIATION

It is obvious that information regarding the theory of Shine and Dalgarno and the question of the essentiality of S1 and initiation factors *in vivo* and their possible regulatory roles in translation could best be obtained with the aid of specific mutants. While genetic loci for many ribosomal proteins and for all elongation factors are known, until recently this information was lacking for mutants affecting the translational initiation. One class of ribosome binding site mutants should be those affecting the triplet interaction between the initiator codon and the fMet-tRNA anticodon sequence.

Weissmann et al. (1977) obtained mutants of the Qβ coat protein cistron in which the third position of the initiator AUG and the next residue (the first position of the succeeding alanine codon) are altered.

They compared the potential base pairing between the initiator tRNA anticodon region and the relevant Qβ RNA sequences with the relative efficiency of *in vitro* ribosome attachment to the same four initiator regions. Since the mutant RNA that lacks the AUG triplet retains considerable ribosome binding activity, it can be concluded that other ribosome–mRNA interactions suffice for the formation of a correctly located initiation complex.

Another very interesting mutant has been obtained and studied (F. W. Studier and J. J. Dunn, personal communication; Steitz, 1977). It affects the T7 gene "0.3 ribosome binding site a" and transforms its five base pairs with 16 S RNA (GAGGU) into a shorter complementary sequence (GAAGU). Concomitantly, a drop of about tenfold is observed in the synthesis efficiency of the gene 0.3 protein in T7-infected cells. This important mutant will provide (when all controls are complete, i.e., mutations in regions other than that of Shine and Dalgarno) direct evidence that mRNA–rRNA base pairing contributes significantly to the binding energy required for ribosome recognition of true initiation signals.

Springer *et al*. (1977a) isolated a thermosensitive mutant that exhibits a thermolabile initiator factor IF-3. When preincubated at 43°C, the 1.5 M NH$_4$ chloride ribosomal wash of the mutant showed inactivation kinetics different from those of the parent strain (Fig. 8). An addition of pure IF-3 from a wild-type strain prior to preincubation of the mixture eliminated the thermolability of the crude IF.

The dissociation activity of the crude IF fraction is also much more thermolabile than that of the parent strain. Finally, the third IF-3 activity, the only one unique to IF-3—the stimulation of polyphenylalanine synthesis in the presence of poly(U) at high concentrations of Mg (Schiff *et al.*, 1974)—is also heat sensitive in this assay, whereas the parent strain is stable under the same conditions.

Genetic data show that the mutation is located near 38' on the new *E. coli* map and is 68% cotransducible with the *aroD* marker. Thermosensitive transductants isolated with phage P$_1$ grown on the thermosensitive strain showed a defective IF-3 activity when compared to the thermoresistant transductants.

A heterodiploid derivative with a *recA* chromosome and an episome carrying the 38' *E. coli* region was found to be resistant *in vivo* and its IF-3 activity was no longer thermolabile. This shows that the genetic defect is recessive. Moreover, *in vitro* studies provide evidence of a correlation between IF-3 activity and the strain phenotype.

A hybrid λ transductant phage carrying the 38' *E. coli* region was prepared *in vitro* and was used to infect the thermosensitive strain

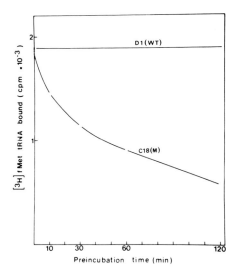

Fig. 8 Inactivation kinetics of crude IF fractions from mutant (M) and parent strains (WT). Preincubation was carried out at 43'C. The preincubation mixture contained the following: Tris–HCl, pH 7.5, 50 mM; NH$_4$Cl, 50 mM; Mg-acetate, 5 mM (TMN buffer); crude IF fraction from mutant C18 (0.49 mg/ml) or crude IF fraction from parent strain D1 (0.32 mg/ml). From Springer *et al.* (1977a).

(Springer *et al.*, 1977b). Thermoresistant transductants were isolated and shown to have a normal IF-3 activity. Final proof that this λ phage carries the structural gene for IF-3 was obtained by SDS polyacrylamide gel electrophoretic analysis of labeled proteins synthesized in UV-irradiated cells after infection with the transducing phage (Figs. 9a and 9b).

It has been further shown that this λ phage carries the information for at least five proteins that are under the control of *E. coli* promoters. The molecular weights of these five proteins are 94,000, 78,000, 38,000, 22,000, and 12,000. The first and third proteins have been identified as α and β subunits of phenylalanyl-tRNA synthetase (Hennecke *et al.*, 1977), and the 22,000 MW protein has been identified as IF-3 by the following criteria: specific cross-reaction with anti-IF-3 antibodies, comigration with pure IF-3 on SDS polyacrylamide gels, and comigration on a two-dimensional gel system separating proteins by charge in the first dimension and by molecular weight in the second dimension.

The availability of both an IF-3 bacterial mutant and a phage carrying the IF-3 gene opens the field of the initiation of translation to genetic and physiological studies.

a

b

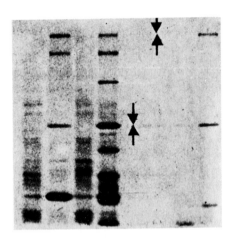

4   3   2   1   1   2   3   4   5   6   7   8   9   10   11   1

Fig. 9 (a) Analysis of IF-3 antiserum cross-reacting material. The extracts sonicated in
1.5 M NH$_4$Cl were centrifuged for 30 minutes at 20,000 g and the supernatants were
centrifuged again for 150 minutes at 50,000 rpm. This high-speed supernatant was concentrated by dialysis against solid sucrose, dialyzed against Tris–HCl, pH 7.5, 50 mM, and
added to 200 μg of rabbit anti-IF-3 γ-globulins. After 12 hours at 4°C, goat γ-globulins
were added to the rabbit serum and this mixture was kept at room temperature for 4 hours
and at 4°C for 18 hours. This two-step precipitation was performed in 50 mM Tris–HCl,
pH 7.5, and 200 μg/ml of sodium azide. The precipitate was centrifuged for 10 minutes at
20,000 g, washed twice with 0.5 ml of 0.15 M NaCl, and resuspended in the SDS sample
buffer. Samples were loaded on a 15% acrylamide gel after 2 minutes at 100°C. Lane 1,
sonicated extract of 159 infected with λp2; lane 2, high-speed supernatant of the extract of
159 infected with λp2; lane 3, proteins precipitated with anti-IF-3 γ-globulins from the
high-speed supernatant of the 159 extract infected with λY199, a control phage carrying
several deletions in the b2 regions (λp2 containing *E. coli* DNA in that region); lane 4,
proteins precipitated with anti-IF-3 γ-globulins from the high-speed supernatant of the
159 extract infected with λp2. From Springer *et al.* (1977b). (b) Analysis of Phe-tRNA
synthetase antiserum cross-reacting material. Scintillation autograph (fluorograph) of the
SDS gel electrophoretic analysis. Samples 1, 2, 11, and 12 contain the nonlabeled marker
proteins, which were used for molecular weight determinations (not shown). Samples 7
and 8 contain nonlabeled purified α and β subunits of PRL, respectively. Their positions
on the Coomassie blue-stained gel (not shown) are given by the arrows. Samples 3–6
contain the following crude extracts: 3, noninfected strain 159 λind^-; 4, 159 λind^- infected
with λp2; 5, 159 λind^- infected with Y199; 6, strain 159 infected with λp2. Sample 9
contains the anti-PRL cross-reacting material from the high-speed supernatant of the
extract of strain 159 infected with Y199, and sample 10 shows the anti-PRL cross-reacting
material from S100 of the extract of 159 infected with λp2. All the samples were loaded on a
10% acrylamide gel. From Hennecke *et al.* (1977).

VI. Conclusion

The brillant discovery of messenger RNA overshadowed the part played by ribosomal RNA. The whole specificity of protein synthesis was thought to proceed from codon–anticodon interactions, and ribosomes were considered to be inert components, somewhat like Sepharose; it was believed that the readout started at the 5′ portion of the messenger RNA and that ribosomes merely stabilized base pairing.

In 1964, thanks to the work of Davies et al. (1964), when reading ambiguities were discovered (errors in base pairing between messenger and transfer RNA's) and when it was found that these errors depended on the nature of the ribosomes, it was considered that ribosome specificity was due to their proteins and, in particular, that two proteins from subunit 30 S, proteins S12 and S4, played a very special part in this respect.

Finally, the attractive hypothesis of Shine and Dalgarno stated that ribosomal RNA played not only a morphological role but also a direct role in the selection of initiation signals. This hypothesis supports the idea that the primitive machinery may have consisted solely of RNA molecules (Crick, 1969). In general, however, the intimate interaction of both RNA and protein contributes to ribosome function. The development of the genetic approach should be primordial for a detailed understanding of the mechanism of selection of initiation signals.

ACKNOWLEDGMENTS

I wish to thank Dr. J. Steitz for a preprint of her review, which was most helpful to me. I also wish to thank Hélène Costinesco and Gisèle Anguérin for their skillful technical assistance in preparing the manuscript.

This work was supported by grants from the Centre National de la Recherche Scientifique, Délégation Génerale de la Recherche Scientifique, and Ligue Nationale Française contre le Cancer.

REFERENCES

Arnold, H. H., Schmidt, W., and Kersten, H. (1975). *Proc. Meet. Fed. Eur. Biochem. Soc.,* 10th, 1975 Abstract No. 388.
Baan, R. A., Hilbers, C. W., Van Chardorp, R., Van Leerdam, E., Van Knippenberg, P. H., and Bosch, L. (1977). *Proc. Natl. Acad. Sci. U.S.A.* **74,** 1028–1031.
Barrell, G. B. (1971). *Procedures Nucleic Acid Res.* **2,** 751–779.
Beyreuther, K., Adler, K., Geisler, N., and Klemm, A. (1973). *Proc. Natl. Acad. Sci. U.S.A.* **70,** 3576–3580.
Branlant, C., and Ebel, J. P. (1977). *J. Mol. Biol.* **111,** 215–256.

Bretscher, M. S. (1969). *J. Mol. Biol.* **42**, 595–598.

Chapman, N. M., and Noller, H. F. (1977). *J. Mol. Biol.* **109**, 131–149.

Crépin, M., Lelong, J. C., and Gros, F. (1973). *Acta Endocrinol. (Copenhagen), Suppl.* **180**, 33–53.

Crick, F. H. C. (1969). *J. Mol. Biol.* **38**, 367–379.

Czernilofsky, A. P., Kurland, C. G., and Stöffler, G. (1975). *FEBS Lett.* **58**, 281–284.

Dahlberg, A. E., and Dahlberg, J. E. (1975). *Proc. Natl. Acad. Sci. U.S.A.* **72**, 2940–2944.

Davies, J., Gilbert, W., and Gorini, L. (1964). *Proc. Natl. Acad. Sci. U.S.A.* **51**, 883–890.

Delk, A. S., and Rabinowitz, J. C. (1974). *Nature (London)* **252**, 106–109.

Delk, A. S., and Rabinowitz, J. C. (1975). *Proc. Natl. Acad. Sci. U.S.A.* **72**, 528–530.

Eisenstadt, J. M., and Brawerman, G. (1966). *Biochemistry* **5**, 2777–2783.

Fiers, W., Contrevas, R., Duerinck, F., Haegeman, G., Iserentant, D., Merregaert, F., Min Jou, W., Molemans, F., Raeymaekers, A., Van der Berghe, A., Volckaert, G., and Ysebaert, M. (1976). *Nature (London)* **260**, 500–507.

Files, J. G., Weber, K., and Miller, J. H. (1974). *Proc. Natl. Acad. Sci. U.S.A.* **71**, 667–670.

Files, J. G., Weber, K., Coulondre, C., and Miller, J. H. (1975). *J. Mol. Biol.* **95**, 327–330.

Gassen, H. G., Linde, R., Nguyen quoc, K., Lipecky, R., and Kolschein, J. (1977). *In* "Translation of Natural and Synthetic Polynucleotides" (A. B. Legocki, ed.), pp. 79–84. University of Agriculture in Poznan Publ.

Godefroy-Colburn, T., Wolfe, A. D., Dondon, J., and Grunberg-Manago, M. (1975). *J. Mol. Biol.* **94**, 461–490.

Goldberg, M. L., and Steitz, J. A. (1974). *Biochemistry* **13**, 2123–2128.

Gralla, J., and Crothers, D. M. (1973). *J. Mol. Biol.* **78**, 301–319.

Grunberg-Manago, M., and Gros, F. (1977). *Prog. Nucleic Acid Res. Mol. Biol.* **20**, 209–284.

Gualerzi, C., Risuleo, G., and Pon, C. L. (1977). *Biochemistry* **16**, 1684–1689.

Held, W. A., Gette, W. R., and Nomura, M. (1974). *Biochemistry* **13**, 2115–2122.

Hennecke, H., Springer, M., and Bock, A. (1977). *Mol. Gen. Genet.* **152**, 205–210.

Higo, K., Held, W., Kahan, L., and Nomura, M. (1973). *Proc. Natl. Acad. Sci. U.S.A.* **70**, 944–948.

Hui Bon Hoa, G., Graffe, M., and Grunberg-Manago, M. (1977). *Biochemistry* **16**, 2800–2805.

Isono, S., and Isono, K. (1975). *Eur. J. Biochem.* **56**, 15–22.

Konisky, J., and Nomura, M. (1967). *J. Mol. Biol.* **26**, 181–195.

Kurland, C. G. (1974). *In* "Ribosomes" (M. Nomura, A. Tissières, and P. Lengyel, eds.), pp. 309–331. Cold Spring Harbor Press, Cold Spring Harbor, New York.

Lee-Huang, S., and Ochoa, S. (1971). *Nature (London), New Biol.* **234**, 236–239.

Lodish, H. F. (1971). *J. Mol. Biol.* **56**, 627–632.

Lodish, H. F., and Robertson, H. D. (1969a). *Cold Spring Harbor Symp. Quant. Biol.* **34**, 655–673.

Lodish, H. F., and Robertson, H. D. (1969b). *J. Mol. Biol.* **45**, 9–21.

Lodish, H. F., and Robertson, H. D. (1970). *Nature (London)* **226**, 705–707.

McLaughlin, C. S., Dondon, J., Grunberg-Manago, M., Michelson, A. M., and Saunders, G. (1968). *J. Mol. Biol.* **32**, 521–542.

Melser, T. L., Davies, J. E., and Dahlberg, J. E. (1971). *Nature (London), New Biol.* **233**, 12–14.

Petersen, H. U., Danchin, A., and Grunberg-Manago, M. (1976a). *Biochemistry* **15**, 1357–1361.

Petersen, H. U., Danchin, A., and Grunberg-Manago, M. (1976b). *Biochemistry* **15**, 1362–1369.

Platt, T., Weber, K., Ganem, D., and Miller, J. H. (1972). *Proc. Natl. Acad. Sci. U.S.A.* **69**, 897–901.

Revel, M. (1972). *In* "The Mechanism of Protein Synthesis and Its Regulation" (L. Bosch, ed.), pp. 87–131. North-Holland Publ., Amsterdam.

Revel, M., and Greenshpan, H. (1970). *Eur. J. Biochem.* **16,** 117–122.

Revel, M., and Gros, F. (1966). *Biochem. Biophys. Res. Commun.* **25,** 124–132.

Revel, M., Herzberg, M., and Greenshpan, H. (1969). *Cold Spring Harbor Symp. Quant. Biol.* **34,** 261–275.

Santer, M., and Shane, S. (1977). *J. Bacteriol.* **130,** 900–910.

Schiff, N., Miller, M. J., and Wahba, A. J. (1974). *J. Biol. Chem.* **249,** 3797–3802.

Shine, J., and Dalgarno, L. (1974). *Proc. Natl. Acad. Sci. U.S.A.* **71,** 1342–1346.

Shine, J., and Dalgarno, L. (1975). *Nature (London)* **254,** 34–38.

Sprague, K. U., Steitz, J. A., Grenley, R. M., and Stocking, C. E. (1977). *Nature (London)* **267,** 462–464.

Springer, M., Graffe, M., and Grunberg-Manago, M. (1977a). *Mol. Gen. Genet.* **151,** 17–26.

Springer, M., Graffe, M., and Hennecke, H. (1977b). *Proc. Natl. Acad. Sci. U.S.A.* **74,** 3970–3974.

Stanley, W. M., Jr., Salas, M., Wahba, A. J., and Ochoa, S. (1966). *Proc. Natl. Acad. Sci. U.S.A.* **56,** 290–295.

Steege, D. A. (1977). *Proc. Natl. Acad. Sci. U.S.A.* **74,** 4163–4167.

Steitz, J. A. (1969). *Nature (London)* **224,** 957–964.

Steitz, J. A. (1973). *J. Mol. Biol.* **73,** 1–16.

Steitz, J. A. (1975). *In* "RNA Phages" (Zinder, ed.), pp. 319–352. Cold Spring Harbor Press, Cold Spring Harbor, New York.

Steitz, J. A. (1978). *In* "Biological Regulation and Development" (R. Goldberger, ed.). Plenum, New York (in press).

Steitz, J. A., and Jakes, K. (1975). *Proc. Natl. Acad. Sci. U.S.A.* **72,** 4734–4738.

Steitz, J. A., and Steege, D. A. (1977). *J. Mol. Biol.* **114,** 545–558.

Steitz, J. A., Sprague, K. U., Steege, D. A., Yuan, R. C., Laughrea, M., Moore, P. B., and Wahba, A. J. (1977). *In* "Nucleic Acid Protein Recognition" (H. J. Vogel, ed.), pp. 491, 508. Academic Press, New York.

Szer, W., Hermoso, J. M., and Boublek, M. (1976). *Biochem. Biophys. Res. Commun.* **70,** 957–964.

Takanami, M., Yan, Y., and Jukes, T. H. (1965). *J. Mol. Biol.* **12,** 761–773.

Van Duin, J., Kurland, C. G., Dondon, J., and Grunberg-Manago, M. (1975). *FEBS Lett.* **59,** 287–290.

Van Duin, J., Kurland, C. G., Dondon, J., Grunberg-Manago, M., Branlant, C., and Ebel, J. P. (1976). *FEBS Lett.* **62,** 111–114.

Van Knippenberg, P. H., Van Duin, J., and Lentz, H. (1973). *FEBS Lett.* **34,** 95–98.

Weissmann, C., Taniguchi, T., Domingo, E., Sabo, D., and Flavell, R. A. (1977). *In* "Genetic Manipulation as It Affects the Cancer Problem," Proceedings of the 9th Miami Winter Symposium (J. Schultz and Z. Brada, eds.), pp. 11–36. Academic Press, New York.

Part III

BIOCHEMISTRY OF OXYGEN AND HEMOGLOBIN

CHAIRMAN: I. C. GUNSALUS
Chapters 11, 12, and 13

CHAIRMAN: K. YAGI
Chapters 14, 15, and 16

11

Molecular Oxygen and Superoxide Dismutase: Environmental Threat and Biological Defense

IRWIN FRIDOVICH

Department of Biochemistry, Duke University Medical Center, Durham, North Carolina

Molecular oxygen is both a blessing and a curse. Abundant and effective as an electron acceptor, it is essential for both energy-yielding and biosynthetic reactions. Respiring organisms cannot survive for more than a few minutes in its absence. Yet oxygen is toxic to all cells and we thrive in its presence by virtue of an elegant system of defenses that have evolved to meet the various components of that toxicity.

Oxygen, in the ground state, is a biradical with two unpaired electronic spins and, as such, cannot accept pairs of electrons without facing a spin restriction. On the one hand, this is fortunate because it markedly limits the reactivity of oxygen. Indeed, were it not for the spin restriction aerobic life would probably not be possible. On the other hand, the spin restriction is unfortunate because it tends to direct the reduction of oxygen into univalent pathways.

The complete reduction of molecular oxygen to water requires four electrons and the univalent pathway presupposes intermediates. The intermediates of oxygen reduction are the superoxide radical (O_2^-), hydrogen peroxide (H_2O_2), and the hydroxyl radical (HO $\cdot$). These intermediates are not well tolerated by living systems because they are very reactive and, by their interactions, can give rise to yet other dangerously

269

FRONTIERS IN PHYSICOCHEMICAL BIOLOGY

reactive entities. It is nevertheless the case that such intermediates of oxygen reduction are commonly generated within respiring cells. Clearly defenses must be at hand, the net effect of which is to prevent the accumulation and blunt the reactivities of these intermediates.

Catalases and peroxidases have long been known to consume H_2O_2. Catalases use H_2O_2 as both a reductant and an oxidant, whereas peroxidases use H_2O_2 only as an oxidant. The overall catalatic and peroxidatic reactions can be written as shown in reactions (1) and (2).

$$H_2O_2 + H_2O_2 \rightarrow 2H_2O + O_2 \tag{1}$$

$$H_2O_2 + RH_2 \rightarrow 2H_2O + R \tag{2}$$

These enzymes are closely related in that both contain hematin as the prosthetic group and they catalyze essentially similar reactions. Indeed, catalase can act as a peroxidase toward such electron donors as ethanol and formate when the concentration of H_2O_2 is low. Catalases are widely but unevenly distributed among respiring cells, perhaps because H_2O_2 is reasonably stable and can thus be excreted from the cell, to be eliminated, in a mixed population, by other cells that do contain catalase or by the action of nonenzymatic catalysts of its decomposition.

In mammals there is a large amount of catalase in erythrocytes, liver, and kidney, but very little in brain, muscle, thyroid, and other tissues (Hartz et al., 1973). The absence of catalase may actually be important for the normal functioning of certain specialized tissues. Thus, the organification of iodide in thyroid depends on the action of a thyroid iodide peroxidase (Stanbury, 1972), and a goitrous cretinism has been associated with a lack of this enzyme (Valenta et al., 1973). The erythrocytes of birds contain less than 1% as much catalase as is found in the corresponding cells in humans and, among respiring microorganisms, the disparities in the distribution of catalase can be even more striking. It is also probable that catalases are often dispensable because redundant defenses exist. Thus, among mammals, glutathione peroxidase is widely distributed and, when coupled with glutathione reductase and a source of NADPH, is a very effective scavenger of H_2O_2 (Arias and Jakoby, 1976).

Most peroxidases are small (MW $\simeq$ 40,000) hemoproteins that are quite specific with respect to the oxidant H_2O_2, but less specific with respect to the reductant. The plant peroxidases, which have been most studied, will oxidize a wide range of arylamines, phenols, enediols, thiols, and other reductants. There are proxidases that are fastidious with respect to the electron donor, and these include the ferrocytochrome c peroxidase of yeast (Coulson et al., 1971) and the glutathione peroxidase already mentioned. The latter is of special interest because it is a selenoprotein rather than a hemoprotein and will utilize organic hydroperoxides as well as H_2O_2 (Arias and Jakoby, 1976).

We shall be primarily concerned with O_2^-, the first intermediate encountered during the univalent reduction of oxygen. This free radical is generated within respiring cells, and a class of enzymes that catalytically scavenge it by promoting reaction (3) has evolved.

$$O_2^- + O_2^- + 2H^+ \rightarrow H_2O_2 + O_2 \tag{3}$$

These enzymes are called superoxide dismutases. They are found in all respiring cells and are enormously efficient catalysts. Indeed, they appear to operate at a rate that approaches the theoretical diffusion limit. Superoxide dismutases containing copper and zinc or manganese or iron have been isolated and studied. These enzymes are essential defenses against the toxicity of oxygen. They are also extremely useful as tools for exposing the participation of O_2^- in oxidation reactions. There has been a widespread interest in these enzymes and an extensive literature has developed. Since this young field has been repeatedly reviewed (Fridovich, 1972, 1974, 1975, 1976; Bors *et al.*, 1974; Halliwell, 1974; Henri, 1975; Asada, 1976; Turkov, 1976; Gutteridge, 1976; Michalski, 1975), this chapter will not attempt to survey it again; rather, it will present, in three sections, some new work.

I. INDUCTION OF SUPEROXIDE DISMUTASE IN *Escherichia coli* K12

E. coli contains three electrophoretically distinct superoxide dismutases (Hassan and Fridovich, 1977a). One of these enzymes contains manganese (Keele *et al.*, 1970) and appears to be a homodimer. There is also an iron-containing enzyme (Yost and Fridovich, 1973) and a third that appears to be composed of one subunit from each of the other two superoxide dismutases (SOD) (Dougherty, 1978). When this organism is grown under rigorously anaerobic conditions it contains only the iron enzyme. Subsequent exposure to oxygen causes induction of the manganese and of the hybrid enzymes (Hassan and Fridovich, 1977a). This response to oxygen is so sensitive that a low concentration of O_2, which is present in deep, quiescent liquid cultures, is enough to cause essentially complete induction. Figure 1 shows the results of linear scanning densitometry of activity-stained electropherograms of cell-free extracts taken at intervals after transfer of *E. coli* K12 from anaerobic to aerobic

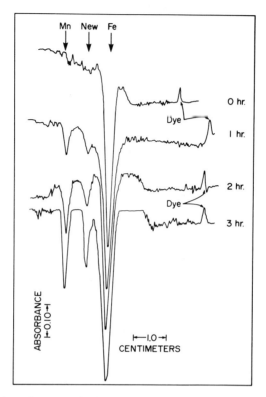

Fig. 1 Induction of superoxide dismutases in *E. coli* K12. *E. coli* K12 bacteria were grown under rigorously anaerobic conditions in a trypticase–soy–yeast extract medium (Hassan and Fridovich, 1977a). At zero time the cells were abruptly transferred to aerobic conditions in this medium, and at intervals cell-free extracts were prepared, subjected to electrophoresis on polyacrylamide gels, stained for superoxide dismutase activity, and examined by linear scanning densitometry. The progressive induction of the manganese and the hybrid (New) enzymes is apparent, as is the constancy of the iron enzyme.

conditions. The rapid induction of the manganese and the hybrid (New) enzymes is apparent, as is the constancy of the iron enzyme.

If this induction is essential in providing a defense against oxygen toxicity, then preventing it should have grave consequences for the cell. Puromycin, which inhibits protein biosynthesis, prevents the induction shown in Fig. 1. Puromycin, at 0.5 mg/ml, was not lethal to anaerobically grown *E. coli* maintained under anaerobic conditions. It was likewise well tolerated by aerobically grown cells kept under aerobic conditions. In contrast, this level of puromycin caused the death of anaerobically grown cells subsequently exposed to air. The prevention of induction of superoxide dismutase by puromycin thus made *E. coli* bacteria susceptible to oxygen toxicity when they contained the low anaerobic level of this protective enzyme, but not when they contained the high aerobic level (Hassan and Fridovich, 1977a). Thus, it was again demonstrated that superoxide dismutase provides an essential defense against oxygen toxicity. It is not known whether the iron- and manganese-containing enzymes play distinct roles in *E. coli*. Certainly the previous view that these enzymes exhibit distinct subcellular localization with the manganese enzyme in the cell matrix and the iron enzyme in the periplasm (Gregory *et al.*, 1973) is now known to be incorrect. Both are present in the matrix space (Britton and Fridovich, 1977). A facultative organism, such as *E. coli*, would do well to have one superoxide dismutase that is responsive to the presence of oxygen, so that the activity can be regulated to suit the need for that activity, as well as a superoxide dismutase that is retained even in the absence of oxygen, to provide some protection in the event of sudden transfer from anaerobic to aerobic conditions. The manganese enzyme seems to meet the former need and the iron enzyme the latter need.

II. INDUCTION OF SUPEROXIDE DISMUTASE IN *E. coli*, AT CONSTANT OXYGENATION, IN RESPONSE TO CHANGES IN METABOLISM

In the previous section the induction of SOD by oxygen was noted and the increased level of this enzyme was correlated with enhanced tolerance for oxygen. This correlation has been made previously under a variety of conditions and is believed to be essentially correct. Nevertheless, it could be argued that oxygen certainly induces the synthesis of proteins other than SOD and that the correlation between SOD and resistance toward oxygen might be fortuitous. Studies of *E. coli* growing in glucose-limited chemostat cultures provide evidence that circumvents this objection (Hassan and Fridovich, 1977b). Thus, under constant and abundant aeration, the growth rate can be manipulated, within wide limits, by changing the rate of pumping of fresh medium into the chemostat. Under these conditions, it was observed that increased specific growth rate correlated with both increased respiration and increased intracellular concentration of SOD. In contrast, catalase and peroxidase were invariant with growth rate or varied inversely with it (Hassan and Fridovich, 1977b). These changes are shown in Fig. 2. Specific growth rate and therefore SOD content correlated with enhanced resistance toward the lethality of hyperbaric oxygen, which is shown in Fig. 3. In the chemostat the level of SOD was manipulated by

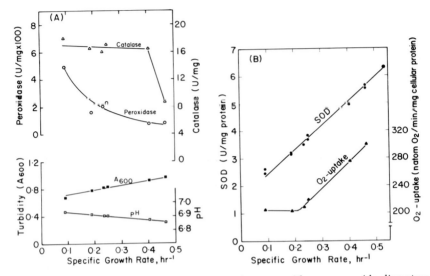

Fig. 2 Effects of specific growth rate on catalase, peroxidase, superoxide dismutase (SOD), and respiration in *E. coli* K12. Cells were grown in a chemostat on a glucose–salts medium and the growth rate was manipulated by varying the rate at which fresh medium was pumped into the constant volume growth chamber. Cells were harvested at different growth rates and their rates of respiration were measured. Cell-free extracts of these samples were also prepared for assays of catalase, peroxidase, and superoxide dismutase.

changing the specific growth rate *under constant oxygenation*. Furthermore, increases in SOD were not accompanied by increases in catalase or peroxidase. The positive correlation between SOD and oxygen tolerance therefore indicates that O_2^- is a major factor in oxygen toxicity and that SOD is an essential defense.

When a surplus of fresh glucose–salts medium was suddenly made available to cells whose growth had been limited in the chemostat, the cells did not abruptly increase their growth rate. The growth rate was unchanged for 60 minutes following this nutritional step-up and during this time the level of intracellular SOD rose. When the SOD level had reached a new plateau the growth rate increased abruptly. These changes, which are shown in Fig. 4, illustrate the importance of SOD. Low growth rate and low respiration rate must be associated with a slow intracellular production of O_2^- and a correspondingly low level of SOD, which suffices to scavenge this O_2^-. Increased growth rate would be accompanied by increased respiration and a greater rate of production of O_2^-. This would necessitate a higher level of SOD. The controls appear to be such that the growth rate cannot increase, even when

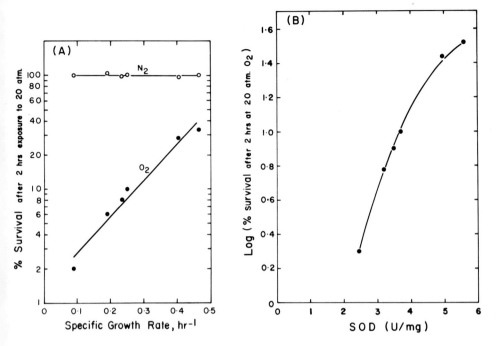

Fig. 3 Effects of specific growth rates and of content of superoxide dismutase (SOD) on resistance toward the lethality of oxygen. Cells were taken from the chemostat at different specific growth rates and were diluted to 10^7/ml in fresh glucose–salts medium containing 0.5 mg/ml of chloramphenicol to prevent subsequent inductions of SOD. The cell suspension was then placed in a shallow petri dish and exposed to 20 atm of O_2 or N_2 for 2 hours at 37°C. Percentage survival was determined by subsequent dilution and plating on agar so that colony-forming units could be counted after 24 hours at 37°C. (A) presents survival under 20 atm of N_2 or O_2 as a function of specific growth rate, whereas (B) presents this survival as a function of the superoxide dismutase content of the cells.

nutritionally feasible, until the SOD level has been raised to that appropriate to the more rapid growth rate.

III. A CONVENIENT NEW ASSAY FOR SUPEROXIDE DISMUTASE

The first assay devised to measure SOD (McCord and Fridovich, 1969), which was also used to guide its isolation, was indirect. This method, which is still widely used, depends on the production of O_2^- by xanthine oxidase and on the reaction of O_2^- with ferricytochrome c. Since the reduction of cytochrome c in this reaction mixture depends

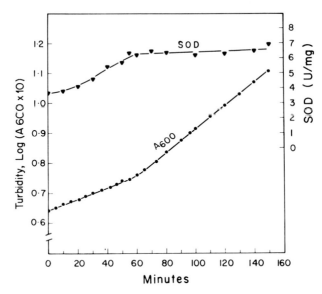

Fig. 4 Changes in culture density ($A_{600\,nm}$) and in content of superoxide dismutase (SOD) (specific activity) during transition from restricted to unrestricted growth conditions. A culture of *E. coli* K12 growing in the chemostat at a specific growth rate of 0.25 cell doubling hr^{-1} was suddenly diluted with an equal volume of prewarmed glucose–salts medium containing 8 mM glucose. At intervals thereafter aliquots were removed for measurement of absorbance at 600 nm and of superoxide dismutase activity.

entirely on O_2^- and since SOD can compete with cytochrome *c* for the available O_2^-, it acts to inhibit this reduction. Under specified conditions, one can define a unit of SOD activity as that amount that causes 50% inhibition of the reduction of cytochrome *c*, followed at 550 nm. As originally described (McCord and Fridovich, 1969), 1 unit of bovine erythrocyte SOD was 0.1 μg/ml. The sensitivity of this assay can be increased by raising the pH, lowering the concentration of cytochrome *c*, and following the reduction in the Soret region of the spectrum (Salin and McCord, 1974).

Many other assays have been described that are mainly inhibition assays, like the original, in that the SOD signals its presence by decreasing the rate of the reaction being directly observed. Thus, indicating scavengers of O_2^-, such as nitroblue tetrazolium, sulfite, luminol, tetranitromethane, or epinephrine, have been used in place of cytochrome *c*. Alternatively, sources of O_2^- other than the xanthine oxidase reaction have been employed. These include photochemical and electrochemical sources or even KO_2 dissolved in dry dimethyl sulfoxide, with the aid of a crown ether. Furthermore, there are numerous autoxidations in which

O_2^- can serve as initiator and/or chain propagator. Such autoxidations, which include those of sulfite, pyrogallol, epinephrine, and 6-hydroxydopamine, can be inhibited by SOD and thus used as the basis for measurement of this enzymatic activity.

All of the foregoing are inhibition assays and have proved to be convenient and sensitive. Nevertheless, positive assays, in which SOD augments the process being observed, would be preferable. Several positive assays have been described, but all leave something to be desired in terms of sensitivity or convenience. In this category are the assays based on detecting O_2^- by electron paramagnetic resonance after freeze quenching of the reaction mixtures, by following the decay of O_2^- in the ultraviolet after pulse radiolysis of oxygenated aqueous solutions, or by following the increase in the polarographic oxygen current, caused by SOD, at a triphenylphosphine oxide-coated mercury cathode.

A positive assay for SOD that combines simplicity with sensitivity has been developed (Misra and Fridovich, 1977a,b) and will be described here. Its advantages are such that it may well become widely used in preference to other assays. It avoids the use of expensive components, such as xanthine oxidase, cytochrome c, or nitroblue tetrazolium, and can be applied to crude soluble extracts. In essence, the photooxidation of dianisidine, sensitized by riboflavin and followed at 460 nm, is augmented by SOD. This effect, shown in Fig. 5, can be used

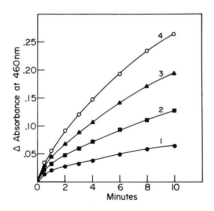

Fig. 5   The effect of superoxide dismutase on the oxidation of dianisidine photosensitized by riboflavin. Reaction mixtures contained 0.2 mM dianisidine, 13 μM riboflavin, and 10 mM potassium phosphate at pH 7.5 and 25°C. Illumination was contrived by exposure to parallel fluorescent lamps in a reflective enclosure and absorbance at 460 nm was measured at intervals. The points on line 1 were observed in the absence of SOD, whereas those on lines 2, 3, and 4 were seen in the presence of 0.2, 0.5, and 1.0 μg/ml of bovine erythrocyte SOD, respectively.

to establish calibration curves for SOD, as shown in Fig. 6. Furthermore, this reaction can be performed on polyacrylamide gel electropherograms, as shown in Fig. 7.

Proteins other than SOD do not affect the riboflavin-photosensitized
oxidation of the dianisidine, and this method can be applied to crude
extracts. Indeed, internal standards of pure bovine erythrocyte SOD
added to crude extracts of liver or of F. coli gave the anticipated increase
in activity. H_2O_2 is, however, slowly generated during the photooxidation of dianisidine, and peroxidase will peroxidize dianisidine. This
leads to a gradual staining of the peroxidase bands when the new
method is applied to acrylamide gels, but peroxidase bands stain much
more slowly than SOD bands. Furthermore, peroxidase bands can be
stained independently by treating the gels with H_2O_2 plus dianisidine.
There is thus no problem in distinguishing peroxidase bands from SOD
bands.

Although the mechanism by which SOD augments the rate of photosensitized oxidation of dianisidine is not known, a very reasonable set

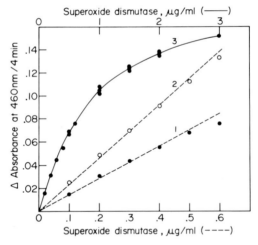

Fig. 6 Assay of superoxide dismutase in terms of augmentation of the photosensitized
oxidation of dianisidine. Reaction mixtures containing 13 μM riboflavin, 0.2 mM
dianisidine, 10 mM potassium phosphate, and the indicated concentrations of bovine
erythrocyte SOD were illuminated for 4 or 8 minutes. The data shown have been corrected
for changes that occurred in the absence of SOD. Lines 1 and 3 present increases in
absorbance at 460 nm during 4 minutes of illumination, whereas line 2 shows the corresponding changes after 8 minutes. Lines 1 and 2 refer to the lower range of SOD concentrations shown on the lower abscissa, whereas line 3 refers to the higher range of concentrations shown on the upper abscissa.

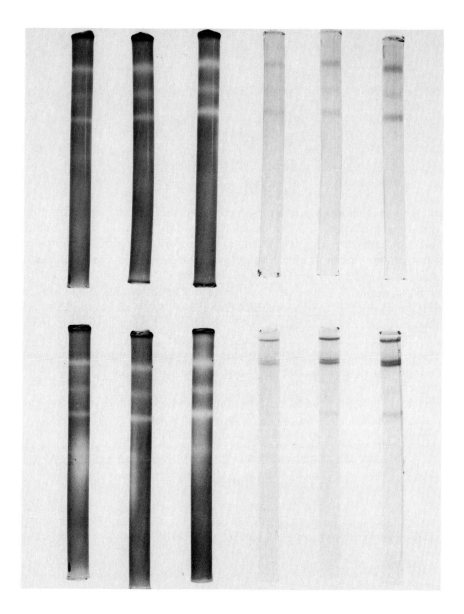

Fig. 7 Staining for SOD activity on electropherograms of cell-free extracts of *E. coli:* a comparison of negative and positive methods. Crude soluble extracts of ultrasonically disrupted *E. coli* K12 were applied to triplicate sets of polyacrylamide gels in graded amounts, that is, 10 µl of extract on the first gel, 20 µl on the second, and 40 µl on the third. The SOD content of these extracts was 1 unit (McCord and Fridovich, 1969) per 10 µl and the peroxidase content was 1.8 units (Decker, 1977) per 10 µl. After electrophoresis the gels were stained for activity by the classic nitroblue tetrazolium method (Beauchamp and Fridovich, 1971) or by the new dianisidine procedure (Misra and Fridovich, 1977b). The upper gels were stained in the presence of 1.0 m*M* cyanide and the lower gels were stained without cyanide.

of reactions can be proposed:

$$Rb + hv \qquad \rightarrow Rb^* \qquad\qquad (4)$$

$$Rb^* + DH_2 \qquad \rightarrow RbH\cdot + DH\cdot \qquad (5)$$

$$RbH\cdot + O_2 \qquad \rightarrow Rb + H^+ + O_2^- \qquad (6)$$

$$DH\cdot + O_2^- + H^+ \rightarrow DH_2 + O_2 \qquad (7)$$

$$DH\cdot + DH\cdot \qquad \rightarrow DH_2 + D \qquad\qquad (8)$$

$$O_2^- + O_2^- + 2H^+ \rightarrow H_2O_2 + O_2 \qquad (9)$$

In reaction (4) riboflavin is electronically excited by absorbing a photon. The excited riboflavin is a more potent oxidant than the ground state and in reaction (5) it oxidizes dianisidine, yielding a flavin semiquinone and a dianisidine radical. The flavin semiquinone reacts with O_2 to give O_2^- in reaction (6) and the O_2^-, in turn, reduces the dianisidine radical in reaction (7). The net effect of these reactions is thus no change. The only reaction that significantly yields the 460 nm-absorbing product is the dismutation of dianisidine radicals as in reaction (8), but it is largely prevented by O_2^- acting as in reaction (7). SOD eliminates this effect of O_2^- by catalyzing reaction (9) and thus increases the amount of DH· available to participate in reaction (8). This is an interesting mechanism because it postulates an internal masking of the true extent of photosensitized oxidation, which is uncovered by the presence of SOD. The utility of the assay described is in no way dependent on the validity of this proposed mechanism.

REFERENCES

Arias, I. M., and Jakoby, W. B., eds. (1976). "Glutathione Peroxidase: Metabolism and Function." Raven, New York.

Asada, K. (1976). *J. Jpn. Biochem. Soc.* **48**, 226.

Beauchamp, C., and Fridovich, I. (1971). *Anal. Biochem.* **44**, 276–287.

Bors, W., Saran, M., Lengfelder, E. Stöttl, R., and Michel, C. (1974). *Curr. Top. Radiat. Res.* **9**, 247.

Britton, L., and Fridovich, I. (1977). *J. Bacteriol.* (in press).

Coulson, A. F. W., Erman, J. E., and Yonetani, T. (1971). *J. Biol. Chem.* **246**, 917.

Decker, L. A., ed. (1977). "Worthington Enzyme Manual," pp. 66–70. Worthington Biochemical Corp., Freehold, New Jersey.

Dougherty, H. (1978). *J. Biol. Chem.* (in press).

Fridovich, I. (1972). *Acc. Chem. Res.* **5**, 321.

Fridovich, I. (1974). *Adv. Enzymol.* **41**, 35.

Fridovich, I. (1975). *Annu. Rev. Biochem.* **44**, 147.

Fridovich, I. (1976). *Adv. Exp. Med. Biol.* **74**, 530.

Gregory, E. M., Yost, F. J., Jr., and Fridovich, I. (1973). *J. Bacteriol.* **115,** 987–991.

Gutteridge, J. M. C. (1976). *Ann. Clin. Biochem.* **13,** 393.

Halliwell, B. (1974). *New Phytol.* **73,** 1075.

Hartz, J. W., Funakosi, S., and Deutsch, H. F. (1973). *Clin. Chim. Acta* **46,** 125.

Hassan, H. M., and Fridovich, I. (1977a). *J. Bacteriol.* **129,** 1574–1583.

Hassan, H. M., and Fridovich, I. (1977b). *J. Bacteriol.* **130,** 805–811.

Henri, J. P. (1975). *Recherche* **6,** 370.

Keele, B. B., Jr., McCord, J. M., and Fridovich, I. (1970). *J. Biol. Chem.* **245,** 6176–6181.

McCord, J. M., and Fridovich, I. (1969). *J. Biol. Chem.* **244,** 6049–6055.

Michalski, W. (1975). *Postepy. Biochem.* **21,** 295.

Misra, H. P., and Fridovich, I. (1977a). *Arch. Biochem. Biophys.* **181,** 308–312.

Misra, H. P., and Fridovich, I. (1977b). *Arch. Biochem. Biophys.* **183,** 511–515.

Salin, M. L., and McCord, J. M. (1974). *J. Clin. Invest.* **54,** 1005–1009.

Stanbury, J. B. (1972). *In* "The Metabolic Basis of Inherited Disease" (J. B. Stanbury, J. B. Wyngaarden, and D. S. Fredrickson, eds.), 3rd ed., p. 226. McGraw-Hill, New York.

Turkov, M. I. (1976). *Usp. Sovrem. Biol.* **81,** 341.

Valenta, L. J., Bode, H., Vickery, A. L., Caulfield, J. B., and Maloof, F. (1973). *J. Clin. Endocrinol. Metab.* **36,** 830.

Yost, F. J., Jr., and Fridovich, I. (1973). *J. Biol. Chem.* **248,** 4905–4908.

12

Indoleamine 2,3-Dioxygenase: A Superoxide-Utilizing Enzyme

OSAMU HAYAISHI

Department of Medical Chemistry, Kyoto University Faculty of Medicine, Kyoto, Japan

I. INTRODUCTION

The superoxide anion is generated by both enzymatic and nonenzymatic univalent reduction of molecular oxygen. It is widely distributed in nature and is generally considered to be a hazardous form of oxygen. Therefore, it is assumed that the biological role of superoxide dismutase is to protect cells from the toxic effect of the superoxide anion. This chapter will deal with indoleamine 2,3-dioxygenase, which utilizes the superoxide anion for the metabolism of tryptophan and various other indoleamine derivatives. As background, the biological function and the molecular properties of this enzyme will be discussed, but most of the chapter will be devoted to the role of the superoxide anion in its catalytic mechanism.

II. BIOLOGICAL FUNCTION

In the early 1900s, tryptophan was discovered and shown to be an indispensable amino acid by Sir Frederick Gowland Hopkins of the University of Cambridge. In mammals, tryptophan is degraded primarily by two major metabolic pathways (Fig. 1). These two pathways are different chemically as well as functionally. The first one involves

FRONTIERS IN PHYSICOCHEMICAL BIOLOGY
Copyright © 1978 by Academic Press, Inc.

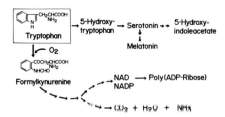

Fig. 1 Major metabolic pathways of tryptophan in mammals.

monooxygenation of tryptophan to 5-hydroxytryptophan followed by decarboxylation to serotonin, one of the most potent biogenic amines. In the pineal gland, serotonin is converted to melatonin, a neurohormone, but in most other tissues it is oxidized by monoamine oxidase and excreted in the urine as 5-hydroxyindoelacetate or its derivatives. It should be stressed that, during the course of biosynthesis and degradation of these indoleamines, the indole nucleus is always preserved.

On the other hand, the second pathway is initiated by the conversion of tryptophan to formylkynurenine, which involves the oxygenative cleavage of the indole ring of tryptophan. This reaction is the first and probably one of the rate-limiting steps in the pathway leading to the biosynthesis of NAD and NADP and ultimately to polyadenosine diphosphate ribose. Alternatively, it may be completely oxidized to CO_2, water, and NH_3. Therefore, it had been thought that the cleavage of the indole ring occurs only with tryptophan and that this reaction is catalyzed by a specific enzyme, tryptophan 2,3-dioxygenase (Fig. 2).

The enzyme that catalyzes this reaction was originally described in 1936 by Kotake and Masayama, who termed it "tryptophan pyrrolase" because the reaction involved a rupture of the pyrrole ring of tryptophan. The primary product is formylkynurenine, as illustrated in Fig. 2, and a second enzyme, "formamidase," hydrolyzes the formyl group of formylkynurenine to yield kynurenine and formic acid. Tryptophan pyrrolase is a hemoprotein containing protoporphyrin IX as its prosthetic group. In 1957, we were able to demonstrate with the use of a heavy oxygen isotope, oxygen-18, that molecular oxygen was directly incorpo-

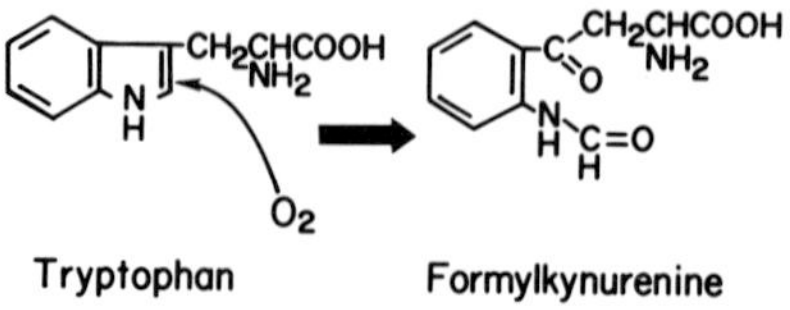

Tryptophan   Formylkynurenine

Fig. 2 Action of tryptophan 2,3-dioxygenase.

TABLE I

Tryptophan 2,3-Dioxygenase (TPO) and Indoleamine 2,3-Dioxygenase (IDO)

	Source	Substrate	Form of oxygen
TPO	Liver	L-Tryptophan	O_2
IDO	Brain,	L- and D-Tryptophan,	O_2^-
	lung,	L- and D-5-hydroxytryptophan,	
	stomach,	tryptamine,	
	intestine,	serotonin,	
	etc.	melatonin, etc.	

rated into formalkynurenine (Hayaishi *et al.*, 1957), and the enzyme was retermed "tryptophan 2,3-dioxygenase," as it is referred to today.

Indoleamine 2,3-dioxygenase is a new hemoprotein and catalyzes a reaction essentially identical to that of the classical tryptophan dioxygenase. However, it is distinctly different from this enzyme in several respects, as shown in Table I. In mammals, the classical tryptophan dioxygenase (TPO) has been found only in the liver. It shows a strict substrate specificity and acts only upon L-tryptophan. As reported by Feigelson, 5-hydroxytryptophan is a potent inhibitor of this enzyme. Molecular oxygen is utilized and incorporated into the product, formylkynurenine, as evidenced by the results of experiments with a heavy oxygen isotope, oxygen-18.

On the other hand, indoleamine 2,3-dioxygenase (IDO) is ubiquitously distributed. It has been isolated and characterized from the brain, lung, stomach, intestine, and almost all other mammalian tissues we have examined so far. Second, it has a much broader substrate specificity and catalyzes the ring cleavage of the indole ring of not only the L and D isomers of tryptophan and 5-hydroxytryptophan but also of tryptamine, serotonin, melatonin, and other indoleamine derivatives. Finally, there is a certain novelty about this enzyme in that the superoxide anion, a univalently reduced oxygen molecule, is involved in the catalytic process.

III. PURIFICATION AND MOLECULAR PROPERTIES

Indoleamine dioxygenase was purified from rabbit small intestine because this organ was found to be the best source of the enzyme. The purification procedure for this enzyme is outlined in Table II. Starting from crude extracts of rabbit intestine, the enzyme was purified by the conventional procedures shown in this table. The specific activity was

TABLE II

Purification of Indoleamine 2,3-Dioxygenase from Rabbit
Small Intestine

Step	Specific activity (mU/mg)	Yield (%)	Purification factor (n-fold)
Crude extracts	7	100	1
Streptomycin	11	109	1.5
Ammonium sulfate	32	87	4.4
P-Cellulose	48	82	6.7
Hydroxylapatite	360	52	50
Sephadex G-100	1200	30	166
Isoelectric focusing	2400	11	332
Sephadex G-100	2700	8	380

enriched about 380-fold with an overall yield of about 8%. The second Sephadex G-100 fraction was essentially homogeneous upon acrylamide gel electrophoresis and ultracentrifugation and was employed in most of the experiments.

The molecular weight was estimated to be 42,000 by ultracentrifugation and by sodium dodecyl sulfate disc gel electrophoresis. It probably consists of one polypeptide chain and contains about 4.8% carbohydrate by weight. There is approximately 1 mole of protoheme IX per mole of enzyme. The molar activity is approximately 2 katals per mole of enzyme; in other words, the turnover number of this enzyme is about 120 moles per minute per mole of enzyme at 24°C under the standard assay conditions. This value is comparable to that of the most purified rat liver tryptophan dioxygenase.

IV. ROLE OF SUPEROXIDE ANION

One of the most interesting features of this enzyme is the fact that the purified enzyme is almost completely inactive unless ascorbic acid and methylene blue are present in the reaction mixture (Hirata and Hayaishi, 1975). When the purified indoleamine dioxygenase was incubated under aerobic conditions, there was no activity unless either ascorbic acid, xanthine oxidase, or glutathione reductase and their respective substrates were added to the incubation mixture (Table III). What is the common denominator among these three systems? Ascorbate, the xanthine oxidase system, and the glutathione reductase sys-

TABLE III

Replacement of Ascorbate by O_2^-- or
H_2O_2-Generating Systems

Addition	Product formation (nmoles)
None	0
Ascorbate	28.0
Xanthine oxidase system	30.0
Glutathione reductase system	25.4
H_2O_2	0
Glucose oxidase system	0
L-Amino acid oxidase system	0

tem are all known to generate the superoxide anion as well as hydrogen peroxide. However, neither hydrogen peroxide nor the glucose oxidase or amino acid oxidase systems, which were known to generate hydrogen peroxide but not superoxide anion, were able to support the enzyme activity. These results provided the first clue indicating that superoxide anion but not oxygen or hydrogen peroxide is involved in this reaction.

If indoleamine 2,3-dioxygenase utilizes the superoxide anion for the oxygenation of substrate, then superoxide dismutase, which catalyzes the disproportionation of superoxide anion, should intercept the superoxide anion and inhibit the reaction catalyzed by indoleamine 2,3-dioxygenase. The results of such experiments are presented in Fig. 3. The open circles indicate the time course of the formation of formylkynurenine from tryptophan. To intercept O_2^-, a superoxide dismutase preparation highly purified from bovine erythrocytes was added to the reaction mixture, and its effect was examined. Superoxide dismutase inhibited the reaction either at the beginning of the reaction or during

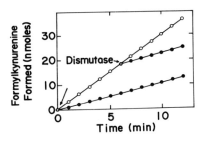

Fig. 3   Effect of dismutase. See Hirata and Hayaishi (1975) for the experimental details.

the steady-state phase of the reaction. In both cases, about 50% inhibition was observed with superoxide dismutase at a concentration of 2 μM. The inhibition was dose dependent and could be completely abolished by inhibitors of the dismutase. The result added further support to our hypothesis that O_2^- is an essential component of the IDO-catalyzed reaction rather than an activator of the enzyme during the initial phase of the reaction.

To verify that the observed inhibition by dismutase was due to its specific enzyme function, the effects of various preparations of bovine erythrocytes dismutase and dismutases of different origins were examined. Under conditions where the native dismutase inhibited about 50%, boiled enzyme, apodismutase, or copper, which is a cofactor of erythrocyte dismutase, failed to inhibit the reaction, whereas the reconstituted dismutase was inhibitory to the same degree as the native dismutase (Fig. 4).

Dismutase preparations from bovine erythrocytes, green peas, and spinach leaves, all of which are copper zinc enzymes, and that of *Escherichia coli*, a Mn enzyme, were all inhibitory to the same extent, although their specific activities and cofactors are different. On the other hand, the scavengers for singlet oxygen hardly inhibited the reaction even though the concentrations used were ten times higher than those employed by previous investigators.

These results strongly indicate that the inhibition by these dismutase preparations was due to their specific ability to disproportionate O_2^- rather than to nonspecific protein–protein interactions and that O_2^- but not singlet oxygen is involved in this system.

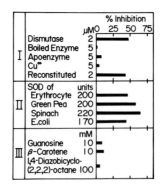

Fig. 4   Effect of dismutases and scavengers for $^1O_2$. See Hirata and Hayaishi (1975) for the experimental details. SOD, superoxide dismutase.

V. EXPERIMENTS WITH KO$_2$

To further establish the participation of the superoxide anion in the catalytic process and to clarify its role in this reaction, a new assay procedure was developed for this enzyme, in which superoxide anion per se was directly infused into the reaction mixture (Ohnishi *et al.*, 1977). For this purpose, potassium superoxide was dissolved in an arpotic solvent such as dimethyl sulfoxide and the solution was directly infused into the reaction mixture. Even under these conditions, the superoxide anion is extremely unstable and, therefore, a sensitive and rapid assay procedure was developed to monitor the enzymatic cleavage of the indole ring of substrate by the use of *ring*-2-^{14}C-labeled substrate. The principle of this assay was based upon the work described by Peterkofsky (1968) for the tryptophan dioxygenase assay.

As can be seen in Fig. 5, when the pyrrole ring of the substrate, for example, tryptophan, is cleaved by the action of indoleamine dioxygenase, formylkynurenine is produced. When formamidase, an enzyme that catalyzes the hydrolysis of the formyl group, is included in the reaction mixture, the radioactive carbon is released as formate, which can be isolated either by distillation or by column chromatography and then quantitatively determined.

The details of the new assay procedure are shown in Fig. 6. The reaction mixture contains DL-[2-^{14}C]tryptophan as substrate, formate as a carrier, catalase to decompose hydrogen peroxide (which is formed from the superoxide anion by spontaneous decomposition), a highly purified formamidase preparation from rat liver, indoleamine dioxygenase (which was purified from rabbit intestine to apparent homogeneity), and buffer in a total volume of 0.2 ml. Potassium superoxide was dissolved in an arpotic solvent, such as dimethyl sulfoxide, and was directly but slowly infused into the reaction mixture, which was being vigorously mixed by means of a vortex mixer. The reaction was terminated by 5% trichloroacetic acid and the acidified reaction mixture was passed through a short column of Dowex 50. The

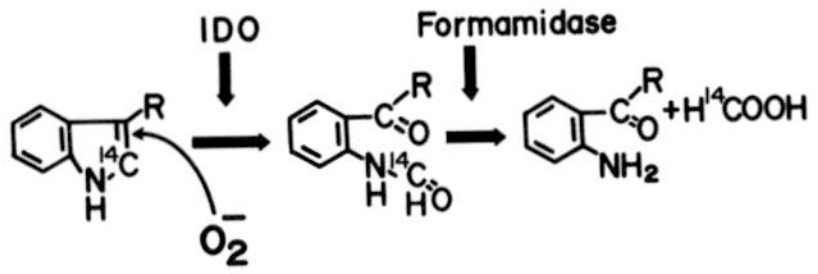

Fig. 5   Principle of the enzyme assay. IDO, indoleamine 2,3-dioxygenase.

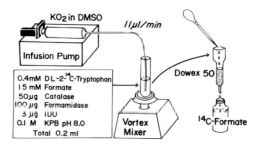

Fig. 6 New assay method for indoleamine 2,3-dioxygenase (IDO). DMSO, dimethyl sulfoxide; KPB, potassium phosphate buffer.

radioactive formate, which was eluted with water, was quantitated with a scintillation counter.

Utilizing this new assay procedure, the requirement for the superoxide anion and the inhibition by superoxide dismutase were re-examined. The time course of the reaction is illustrated in Fig. 7. When the highly purified indoleamine dioxygenase preparation was incubated with the substrate under these conditions, it was completely inactive unless potassium superoxide was infused into the reaction mixture. The reaction proceeded linearly as long as the infusion was continued (as shown by the solid circles in Fig. 7). The reaction stopped almost immediately when the infusion of O_2^- was discontinued or when superoxide dismutase was added to the reaction mixture, indicating that superoxide is absolutely essential for the reaction. These results, taken together, further substantiate our previous conclusion that the superoxide anion is essential for the catalytic activity.

Now that the participation of O_2^- in the catalytic process is well established, the next question concerns the exact role of O_2^- in this reaction. Since O_2^- is a good oxidant and a potent reductant, the question arises of whether O_2^- serves as an oxidizing agent and is directly incorporated

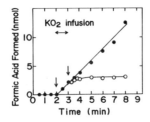

Fig. 7 Time course of the reaction. The experimental details are described in Ohnishi *et al.* (1977).

into the product or reduces the heme coenzyme so that the reduced heme can react with molecular oxygen, which is then incorporated into the substrate. To answer this question, the interaction of the heme coenzyme with the superoxide anion was investigated.

VI. INTERACTION OF ENZYME WITH O_2^-

The native, ferric form of the enzyme exhibits a spectrum typical of a high-spin hemoprotein, as shown by the solid thin line in Fig. 8. It can be reduced chemically or enzymatically to a ferrous form, which is represented by the broken line. The ferrous form is unstable and readily undergoes autoxidation to regenerate mostly the ferric form in the presence of oxygen. However, when the ferric form of the enzyme was treated with the superoxide anion, it was not reduced to the ferrous form of the enzyme. Instead, a new absorption spectrum appeared immediately, with new absorption maxima at 346, 415, 542, 576, and 670 nm, as shown by the solid broad line in Fig. 8 (Hirata *et al.*, 1977). The formation of this new spectral species was completely abolished in the presence of superoxide dismutase, indicating that this process involved the binding of superoxide anion to the ferric form of the enzyme. The positions of the absorption maxima in this new spectrum are analogous to those in the spectra of well-known oxygenated protohemoproteins, as shown in Table IV.

As can be seen from this table, the positions of the absorption maxima of oxyhemoglobin, oxymyoglobin, peroxidase compound III, and the enzyme · substrate · oxygen complex of tryptophan dioxygenase are all similar to those exhibited by the enzyme · superoxide complex, which is therefore interpreted to be the enzyme · O_2^- complex, namely, an oxygenated form of indoleamine dioxygenase.

To recapitulate, the native ferric form of the enzyme binds to the superoxide anion to form the oxygenated form of the enzyme, in which

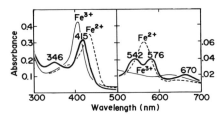

Fig. 8 Spectrum of the enzyme · O_2 complex. The experimental details are described in Hirata *et al.* (1977).

TABLE IV

Absorption Maxima of Oxygenated Heme Proteins

	Absorption maximum (nm)		
	γ	β	α
Enzyme · O₂ complex of IDO	415	542	576
Oxyhemoglobin	415	542	576
Oxymyoglobin	418	544	582
Peroxidase compound III	418	546	581
Enzyme · substrate · O₂ complex of TPO	418	545	576

the ferric heme superoxide complex is in resonance equilibrium with the ferrous oxygen complex (Fig. 9). This interpretation is further supported by the following two observations. First, this apparent $E \cdot O_2^-$ complex can be converted to the ferrous form of enzyme and oxygen by simple degassing. Second, when an excess amount of substrate, such as tryptophan, was introduced into a reaction mixture containing the enzyme · oxygen complex, free enzyme and the product were immediately released. The ferrous form of the enzyme thus formed is autoxidizable and is readily converted to the ferric form, except that a small portion appears to be oxygenated to regenerate the $E \cdot O_2$ complex. If this is the case, molecular oxygen should be incorporated into the products as well as the superoxide anion. To distinguish these two possibilities clearly, $^{18}O_2^-$ and $^{18}O_2$ were used and the incorporation of a heavy oxygen isotope into the product of the reaction was determined.

VII. EXPERIMENTS WITH $^{18}O_2^-$

When the reaction is carried out in the presence of a heavy oxygen, it should be incorporated into the carbonyl and formyl groups of formyl-kynurenine (Fig. 10). Since the oxygen atoms in the carbonyl group are

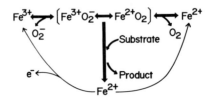

Fig. 9 The interaction of enzyme with O_2 and O_2^-.

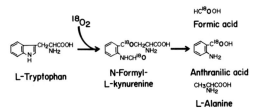

Fig. 10 Incorporation of $^{18}O_2$.

known to exchange readily with those of the water molecule, formyl-kynurenine was converted *in situ* to formate, anthranilate, and L-alanine by the action of kynureninase and formamidase. Anthranilic acid and formate were then isolated and the ^{18}O content was determined using a mass spectrometer.

The results of typical experiments are given in Table V (Hayaishi *et al.*, 1977). Unfortunately, the interpretation is somewhat obscured by the exchange reaction of isotopic oxygen during the incubation and isolation of the products and by the fact that the superoxide anion generates molecular oxygen by spontaneous decomposition. Nevertheless, without imposing too many details, the following can be concluded.

When the reaction was carried out in the presence of $^{18}O_2^-$ and $^{16}O_2$, the ^{18}O enrichment of the incorporated oxygen corresponded to 49.2 and 10.5% of the theoretical values expected in anthranilic acid and formate, respectively. These results indicate that the oxygen atoms derived from $K^{18}O_2$ were incorporated into the products of the reaction. In experiment II (Table V), air was replaced by argon, and glucose oxidase and glucose were included in the reaction mixture to remove any $^{18}O_2$ generated from $K^{18}O_2$. As can be seen here, the percentage enrichment in anthranilic acid and formate increased to 76.6 and 16.6% of the theoretical values, indicating that molecular oxygen must also have been incorporated into the product in experiment I.

To explore this possibility, the reaction was carried out in the presence

TABLE V

Incorporation of $^{18}O_2^-$ and $^{18}O_2$

Experiment	Source of ^{18}O	Percentage enrichment in	
		Anthranilic acid	Formic acid
I	$^{18}O_2^-$, $^{16}O_2$	49.2	10.6
II	$^{18}O_2^-$	76.8	16.6
III	$^{16}O_2^-$, $^{18}O_2$	29.6	8.9

of $^{16}O_2^-$ and $^{18}O_2$. As shown in experiment III (Table V), the percentage enrichment of ^{18}O in anthranilic acid and formate decreased to 29.6 and 8.85%, respectively. These results strongly suggest that about two-thirds of the oxygen atoms incorporated into the products were derived directly from O_2^- and the remaining one-third from molecular oxygen under the conditions employed, although the superoxide anion is absolutely required for the enzyme activity.

VIII. INTERACTION OF ENZYME WITH SUBSTRATE

In the preceding sections, the interaction of the enzyme with the superoxide anion or oxygen was discussed, avoiding a discussion of the interaction of the organic substrate with the enzyme. However, this does not necessarily mean that the sequence of reactions during steady state also involves the $E \cdot O_2$ complex as the primary intermediate. Again, without going into detail, the Cleland-type analysis of steady-state kinetics using potassium cyanide as a dead-end inhibitor indicates that the enzyme binds with the organic substrate first, followed by the binding of the superoxide anion to the $E \cdot S$ binary complex. The ternary complex of enzyme · substrate · oxygen then decomposes to form the enzyme and the final product.

Our current working hypothesis, based on the available evidence, is presented in Fig. 11. Let us start with the native ferric enzyme. It binds to the substrate and then to superoxide to form a hypothetical ternary complex that is in resonance equilibrium with the ferrous oxygen · substrate complex. The latter complex will yield the product regenerating the ferrous form of the enzyme. It is autoxidizable and, in the presence of air, is readily oxidized to the ferric enzyme, which starts the cycle all over again. However, a small fraction of the ferrous enzyme may react with the substrate, and then molecular oxygen and the cycle on the right side may also operate as an auxiliary mechanism.

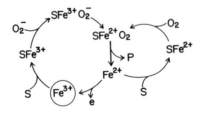

Fig. 11   Hypothetical reaction sequence of enzyme with substrate.

IX. IS O_2^- A SUBSTRATE *IN VIVO*?

Thus far, the molecular and catalytic properties of indoleamine 2,3-dioxygenase have been discussed, with particular emphasis on the evidence indicating that the superoxide anion is utilized by this enzyme. Now, even if this enzyme utilizes the superoxide anion *in vitro*, the possibility still remains that the purified enzyme is an artifact and that the enzyme is maintained in a reduced form by some other mechanism and utilizes molecular oxygen *in vivo*. To carry the point further, one may ask: Does this enzyme really exist and operate *in vivo*? In an attempt to elucidate this question, mucosal cells from rabbit small intestine were isolated and the intracellular dioxygenase activity of dispersed cell suspensions was determined (Taniguchi *et al.*, 1977).

Because superoxide dismutase is quite active in the enterocytes, it may be reasonable to assume that the supply of superoxide anion is one of the rate-limiting factors of intracellular IDO activity. A number of enzyme systems may generate O_2^- in the enterocytes but since xanthine oxidase generates O_2^- *in vitro* and is abundant in the intestine, the effect of inosine, a substrate of xanthine oxidase, on the intracellular dioxygenase activity was examined (Fig. 12). Alternatively, diethyldithiocarbamate (DDC), a copper chelator and a potent *in vivo* inhibitor of superoxide dismutase, was used to inhibit the intracellular dismutase activity, which should increase the level of the superoxide anion inside the enterocytes. The results of such experiments are presented in Fig. 13.

As can be seen here, the addition of inosine alone increased the dioxygenase activity about fivefold, and this effect was completely abolished by the addition of allopurinol, a specific inhibitor of xanthine oxidase. However, neither inosine nor allopurinol had an effect on purified IDO. These results indicate that the superoxide anion generated by the xanthine oxidase system is utilized by the indoleamine dioxygenase and that it is one of the rate-determining factors *in vivo*.

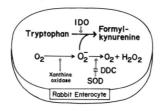

Fig. 12 Intracellular indoleamine 2,3-dioxygenase (IDO) activity and O_2^-. SOD, superoxide dismutase; DDC, diethyldithiocarbamate.

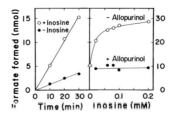

Fig. 13 Effects of inosine and allopurinol on indoleamine 2,3-dioxygenase activity.

In the same manner, the intracellular indoleamine dioxygenase activity in the dispersed enterocytes was markedly enhanced by the addition of 5 mM DDC (Fig. 14). Diethyldithiocarbamate, however, did not show any activating effect on the purified indoleamine dioxygenase activity *in vitro*. The results indicate that DDC probably inhibits superoxide dismutase, thereby increasing the intracellular concentration of the superoxide anion, and stimulates indoleamine dioxygenase activity.

X. CONCLUSION

Indoleamine 2,3-dioxygenase was purified to apparent homogeneity from rabbit intestine and its molecular properties were determined (Fig. 15). The superoxide anion has now been unequivocally demonstrated to be the true substrate of this enzyme, although molecular oxygen is also utilized by this enzyme to a lesser extent. Indoleamine 2,3-dioxygenase may, therefore, be the first example of a new class of enzymes that utilize the superoxide anion as an oxidizing agent just as peroxidases utilize hydrogen peroxide.

Our most recent experiments with dispersed cells of intestinal mucosa indicate that superoxide is also utilized by this enzyme *in vivo* and that

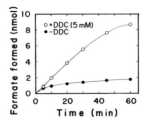

Fig. 14 Effect of diethyldithiocarbamate (DDC) on indoleamine 2,3-dioxygenase activity. See Taniguchi *et al.* (1977) for the experimental details.

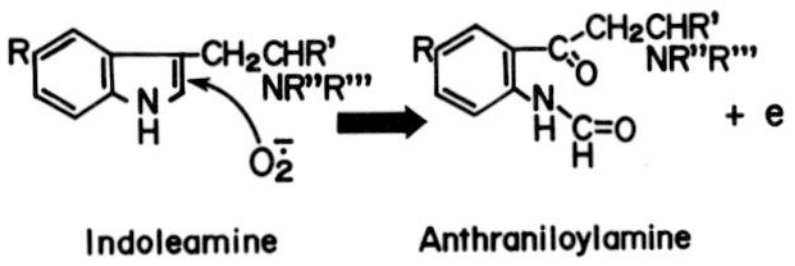

Indoleamine   Anthraniloylamine

Fig. 15   Reaction catalyzed by indoleamine 2,3-dioxygenase.

the intracellular concentration of the superoxide anion is one of the rate-determining factors in the cell.

ACKNOWLEDGMENTS

Part of the experimental work presented is the result of a welcome collaborative effort with Drs. A. M. Michelson and Jean/Pierre Henry of the Institut de Biologie Physico-Chimique and Drs. F. Hirata, T. Ohnishi, T. Shimizu, and T. Taniguchi of the Kyoto laboratory. It is also a pleasure to acknowledge the many useful discussions with Dr. I. Fridovich throughout this work.

REFERENCES

Hayaishi, O. Rothberg, S. Mehler, A. H., and Saito, Y. (1957). *J. Biol. Chem.* **229,** 889–896.
Hayaishi, O., Hirata, F., Ohnishi, T., Henry, J.-P., Rosenthal, I., and Katoh, A. (1977). *J. Biol. Chem.* **252,** 3548–3550.
Hirata, F., and Hayaishi, O. (1975). *J. Biol. Chem.* **250,** 5960–5966.
Hirata, F., Ohnishi, T., and Hayaishi, O. (1977). *J. Biol. Chem.* **252,** 4637–4642.
Kotake, Y., and Masayama, T. (1936). *Z. Physiol. Chem.* **243,** 237–244.
Ohnishi, T., Hirata, F., and Hayaishi, O. (1977). *J. Biol. Chem.* **252,** 4643–4647.
Peterkofsky, B. (1968). *Arch. Biochem. Biophys.* **128,** 637–645.
Taniguchi, T., Hirata, F., and Hayaishi, O. (1977). *J. Biol. Chem.* **252,** 2774–2776.

13

Reaction Intermediates of
D-Amino Acid Oxidase

KUNIO YAGI

Institute of Biochemistry, Faculty of Medicine, University of Nagoya, Nagoya, Japan

I. INTRODUCTION

D-Amino acid oxidase [D-amino acid:oxygen oxidoreductase (deaminating), EC 1.4.3.3] reacts with its substrate, D-amino acid, to form the corresponding keto acid, ammonia, and H_2O_2. The overall reaction is the deaminative oxidation of the substrate with the reduction of oxygen. However, the enzyme catalyzes only the formation of the primary product, imino acid, from D-amino acid, and the hydrolysis of the imino acid is due to a nonenzymatic process (Yagi *et al.*, 1970b). The action of this enzyme is the mediation of the electron transfer from the substrate to oxygen: The enzyme should be reduced by the substrate and then oxidized by oxygen. In this respect, the observation on the oxidoreduction intermediate is important in understanding the reaction mechanism.

Since D-amino acid oxidase is composed of flavin adenine dinucleotide (FAD) and the apoenzyme and contains no other components, this enzyme is one of the simplest flavoproteins in structure. This enzyme can easily be resolved into its coenzyme and apoenzyme and can be easily reconstructed from them. This is advantageous in studying the interaction between apoenzyme and coenzyme in flavoenzymes. Another advantage in studying the reaction mechanism of flavoenzyme with this enzyme is its wide substrate specificity. This enzyme can

299

FRONTIERS IN PHYSICOCHEMICAL BIOLOGY
Copyright © 1978 by Academic Press, Inc.

oxidize a series of D-amino acids, D-hydroxylic acids, glycine, L-proline, sarcosine, etc. The modes of action of this enzyme with different substrates are similar but delicately different from one another. Comparison of the modes of action with different substrates is considered to afford valuable information on the reaction mechanism of flavoproteins.

This chapter deals with the reaction intermediates of this enzyme for the elucidation of the process of flavoprotein-catalyzed oxidation.

II. OCCURRENCE OF TWO TYPES OF OXIDOREDUCTION INTERMEDIATES

To investigate the oxidoreduction intermediates of this enzyme, we wished to obtain the intermediate in crystalline form. We thought that the intermediate should be accumulated in the presence of the appropriate amounts of the reaction products of the enzymatic reaction under anaerobic conditions and that the accumulated intermediate should be crystallized by the addition of ammonium sulfate (Yagi and Ozawa, 1962b). When the enzyme was mixed with D-alanine under anaerobic conditions, the spectrum of the oxidized enzyme (Fig. 1, curve I) was readily converted to that of the fully reduced form (Fig. 1, curve II), with the transient appearance of purple color. In the presence of the products, pyruvate and ammonia, the spectrum shown by curve III in Fig. 1 was obtained. This shows the accumulation of a purple-colored intermediate. From this solution, the purple-colored crystals (Fig. 1, curve IV) were obtained by adding ammonium sulfate to 0.05% saturation. This crystal was found to be composed of equimolar amounts of the substrate and the enzyme (Yagi and Ozawa, 1964) and was spectroscopically identified with the rapidly appearing purple intermediate of this enzyme (Yagi et al., 1972).

Kinetic analysis indicated the occurrence of a complex of the substrate and the oxidized enzyme prior to the formation of the purple intermediate (Yagi et al., 1970a). A study on the kinetic isotope effect using α-deuterated amino acid indicated that the splitting of α-hydrogen occurs prior to or in concert with the formation of the purple intermediate (Yagi et al., 1970c, 1973). The α-hydrogen should be cleaved as proton, hydride ion, or hydrogen atom. The results obtained from our laboratory as well as those of others showed that the hydrogen is transferred as a proton (Yagi et al., 1970c, 1973, 1974a; Walsh et al., 1971; Porter et al., 1973). Therefore, the purple intermediate complex is thought to be formed through a complex between a carbanion of a substrate and the oxidized flavin moiety of the enzyme. Since a model experiment indi-

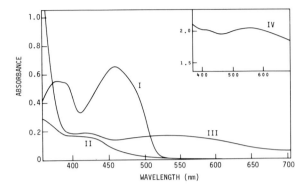

Fig. 1 Absorption spectra of D-amino acid oxidase and its reaction intermediates. (I) The oxidized form of D-amino acid oxidase (58.6 μM with respect to bound FAD). (II) The fully reduced enzyme obtained by mixing the oxidized enzyme with 33 mM D-alanine under anaerobic conditions. (III) The purple complex obtained by mixing the oxidized enzyme with 55 mM D-alanine in the presence of 50 mM ammonium sulfate and 0.1 M lithium pyruvate under anaerobic conditions. (IV) Reflectance spectrum of the crystals of the purple complex.

cated that the amino nitrogen of the substrate interacts with the flavin moiety of the enzyme (Yagi, 1971), electrons of the carbanion are considered to be transferred via the amino nitrogen. Song et al. (1976) demonstrated that the frontier orbital density of N-5 is the highest among the atoms in the isoalloxazine ring, as shown in Fig. 2. Therefore, N-5 seems to be the entrance of the electrons. An investigation of the purple intermediate crystals revealed that this complex is diamagnetic. However, the purple complex gradually changed into the semiquinoid enzyme and the substrate radical (Yagi et al., 1967). By denaturation of the purple complex with trichloroacetic acid, the flavin and substrate moieties readily changed into radical entities (Yagi et al., 1965). Accordingly, the entity of the purple complex was assigned to a

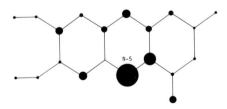

Fig. 2 The frontier orbital density distribution in ribovlavin. The values were calculated by Song et al. (1976) using the frontier electron method (see Fukui et al., 1954). (The area of the closed circle shows the value of the frontier orbital density.)

strong charge-transfer complex, and the electron (or electrons) seems to be shared by the molecules of substrate and flavin.

The rate of dissociation of this intermediate into radical species is so slow that the dissociation process cannot be involved in the enzymatic reaction. The dissociation seems to be prevented by a hydrophobic environment surrounding the site of the charge-transfer interaction.

Although the occurrence of the purple intermediate is commonly observed in the reaction of neutral D-amino acids, we found that this is not the case for basic D-amino acids. Figure 3 shows the change of the absorption spectrum of the enzyme observed by stopped-flow spectrophotometry. When the enzyme was mixed with D-arginine under anaerobic conditions, the purple intermediate was not observed. Kinetic analysis revealed that the concentration of any complex is marginal during the time course of the reaction of the enzyme with D-arginine (Yagi et al., 1974b).

The spectrum obtained by mixing the enzyme with D-arginine is that of the fully reduced form of the enzyme and is identical to that observed with neutral D-amino acid after formation of the purple intermediate. In the fully reduced enzyme, two electrons of the substrate are completely transferred to the coenzyme.

It is obvious that the difference between neutral amino acid and

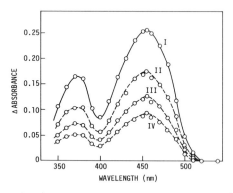

Fig. 3 Changes in the absorption spectra of D-amino acid oxidase in its anaerobic reaction with D-arginine. The final concentrations of the enzyme and D-arginine were 24 μM and 5 mM, respectively. The change in absorbance due to the reaction at pH 8.3 and 20°C was followed at various wavelengths with a stopped-flow spectrophotometer. The circles indicate the difference absorbance against the fully reduced enzyme. The solid line indicates the difference spectrum of the oxidized enzyme against the fully reduced enzyme. I, immediately after mixing; II, 100 milliseconds after mixing; III, 200 milliseconds after mixing; IV, 300 milliseconds after mixing.

basic amino acid is the existence of a positively charged group in the latter compound. In complex formation of the substrate with the enzyme, this charged group would affect the hydrophobicity around the interacting site between the substrate and the enzyme. This would be the reason that the purple intermediate is not observed with basic amino acids: The lifetime of the purple intermediate is considered to be extremely shortened.

III. LIFETIME OF THE PURPLE INTERMEDIATE

The above concept led us to consider that the hydrophobicity around the interacting site between the substrate and the enzyme regulates the lifetime of the purple intermediate. This view was examined by measuring the lifetime of the purple intermediate with different neutral amino acids.

When the enzyme was reacted with D-norvaline under anaerobic conditions, the spectrum of the oxidized enzyme rapidly changed to the purple intermediate and then gradually converted to that of the fully reduced enzyme, as shown in Fig. 4 (Yagi and Ohishi, 1978). From the absorbance change at 550 nm, the rate of conversion of the purple intermediate to the fully reduced enzyme was calculated to be 4.60×10^{-3} sec^{-1}. In the case of glycine, the appearance of the long-wavelength absorption band was minute, showing that the lifetime of the purple intermediate is short. The different rate constants were obtained with different amino acids, as shown in Table I. The rate becomes slower with increasing carbon number of the amino acid. It is slowest with a carbon number of 4–5 and then becomes rapid. In our previous study, we found that the affinity between the enzyme and straight-chain fatty acid increased with increasing number of carbon in the chain, up to 5,

TABLE I

Rate Constants of the Conversion of the Purple Intermediate to the Fully Reduced Form[a]

Amino acid	Rate constant (sec^{-1})
D-Alanine	1.89×10^{-2}
D-α-Aminobutyric acid	3.84×10^{-3}
D-Norvaline	4.60×10^{-3}
D-Norleucine	3.53×10^{-2}

[a] Reaction conditions were the same as those of Fig. 4.

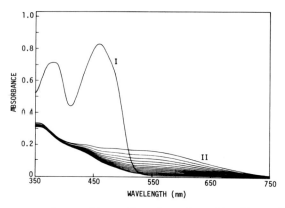

Fig. 4 Time course of anaerobic reaction of D-amino acid oxidase with D-norvaline. D-Amino acid oxidase (73 μM, curve I) was mixed with 50 mM D-norvaline at pH 8.3 and 25°C under anaerobic conditions. Measurements were made at 45-second intervals from 15 seconds (curve II) after mixing.

which was interpreted to mean that the hydrophobicity of the alkyl group determines the affinity of the fatty acid for the enzyme (Yagi *et al.*, 1970d). Taking this result into consideration, the lifetime of the purple intermediate is considered to be regulated by the hydrophobic properties of the alkyl group of neutral amino acids. The decrease in the lifetime with further increases in carbon number might be ascribed to a steric effect, as was considered for straight-chain fatty acids (Yagi *et al.*, 1970d).

IV. EFFECT OF HYDROGEN BONDING ON THE FORMATION OF THE INTERMEDIATE

In 1956, we found that the first absorption band of riboflavin shifted toward the red on complex formation with phenol (Yagi and Matsuoka, 1956). The interaction energy of the phenol–riboflavin complex was calculated to be 6.1 kcal/mole, which indicated that the interaction is due to hydrogen bonding. From this result, the shift of the first absorption band of flavin toward the red after reconstruction of flavoproteins such as old yellow enzyme (Theorell and Åkeson, 1956) and D-amino acid oxidase (Yagi and Ozawa, 1962a) is considered to be ascribed to the hydrogen bonding between coenzyme and apoenzyme.

To reveal the feature of the hydrogen bonds involved in D-amino acid oxidase, we made a theoretical approach to the effect of the hydrogen bonds at possible groups of flavin on its absorption spectrum. First, the

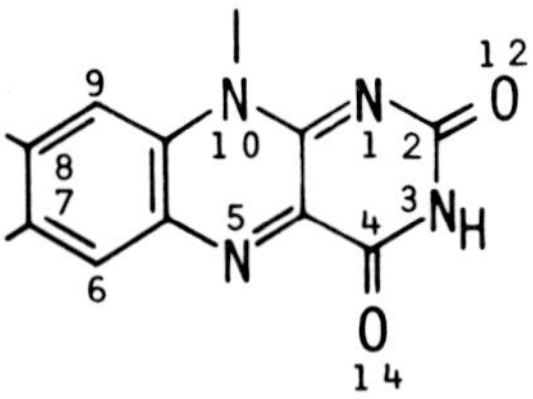

Fig. 5 Numbering of the isoalloxazine ring.

shift of the absorption peak after the formation of hydrogen bonds at definite position(s) of the isoalloxazine ring (see Fig. 5) was calculated as shown in Table II (Nishimoto *et al.*, 1977). Second, the characteristic change in the absorption spectrum of FAD on its complex formation with D-amino acid oxidase apoenzyme was examined. The spectral changes are as follows. The first absorption band shifts toward the red with a slight decrease in absorbance and the second absorption band does not shift but its absorbance decreases slightly (Yagi and Ozawa, 1962a). These spectral changes accord with those of Table II (VIII → VII), obtained theoretically. This implies that the hydrogen bond at N-1 of FAD in aqueous solution disappears after its complex formation with

TABLE II

Calculated Values for the Transition Energy and the Oscillator Strength of Various Types of Hydrogen-Bonded Isoalloxazine Rings

Type of hydrogen bonding	First absorption band		Second absorption band		LUMO[a] energy (eV)
	Transition energy (eV)	Oscillator strength	Transition energy (eV)	Oscillator strength	
I: None	2.99	0.508	3.74	0.083	−3.75
II: N-1—	3.13	0.545	3.74	0.115	−3.83
III: N-5—	2.86	0.509	3.58	0.094	−3.94
IV: N-1,N-5—	3.00	0.555	3.58	0.109	−4.02
V: N-3,O-12,O-14—	2.98	0.514	3.67	0.132	−3.93
VI: N-1,N-3,O-12, O-14—	3.13	0.537	3.67	0.159	−3.99
VII: N-3,N-5,O-12, O-14—	2.85	0.515	3.50	0.133	−4.13
VIII: N-1,N-3,N-5, O-12,O-14—	3.00	0.547	3.50	0.149	−4.19

[a] LUMO, lowest unoccupied molecular orbital.

KUNIO YAGI

TABLE III

Effect of Hydrogen Bonds on the Frontier Electron Density of N-5[a]

Type of hydrogen bonding	Frontier electron density
I: None	0.236
III: N-5—	0.223
VIII: N-3,N-5,O-12,O-14—	0.227
V: N-3,O-12,O-14—	0.242

[a] The frontier electron density was calculated from the atomic orbital coefficient of N-5 in LUMO as reported by Fukui *et al.* (1954).

the apoenzyme, whereas N-5 still interacts with the apoenzyme or remains to interact with the water molecule via hydrogen bonding.

The effect of hydrogen bonds on the electron affinity of the isoalloxazine ring is shown in Table II. The hydrogen bonds involved in D-amino acid oxidase markedly decrease the LUMO (lowest unoccupied molecular orbital) energy; in other words, electron acceptability of flavin is increased by these hydrogen bonds.

As mentioned above, the entrance of electrons into the coenzyme seems to occur at N-5 of the isoalloxazine ring (Song *et al.*, 1976). Therefore, the effect of hydrogen bonds on the magnitude of the frontier electron density of N-5 should be examined. As shown in Table III, the hydrogen bond at N-5 slightly decreases the magnitude of the frontier electron density of N-5. However, when additional hydrogen bonds occur at N-3 H, O-12, and O-14, the electron affinity of the isoalloxazine ring increases and the decrease in the frontier electron density of N-5 is partly recovered. If the hydrogen bond at N-5 is eliminated after complex formation with the substrate, the frontier electron density of N-5 is remarkably increased prior to the initiation of the electron flow.

V. ROLE OF PROTEIN IN THE FORMATION OF THE INTERMEDIATES

From the results mentioned above, the role of the protein moiety of this enzyme can be clearly defined. The protein fixes the flavin and substrate in a sterically favorable way, in which interaction sites for the electron flow can get close to each other. The protein structure affords a hydrophobic environment surrounding the interaction site and it also enables hydrogen bonds to form at definite atoms of the coenzyme flavin. The oxidoreduction seems to be initiated by the abstraction of an α-proton of a substrate with a basic residue of the protein, forming a carbanion. As mentioned previously, the amino nitrogen of a substrate seems to interact with the flavin moiety of the enzyme, and electron flow could occur from the amino nitrogen to flavin N-5, which has the highest electron acceptability. The purple intermediate could be the complex of the substrate and flavin, in which an electron (or electrons) is shared between the two molecules. This strong charge-transfer complex is stabilized by a hydrophobic environment surrounding the interaction site. In addition, this environment seems to cooperate with the hydrophobicity of the alkyl group of the substrate in regulating the lifetime of the purple intermediate. In other words, a hydrophobic environment constructed by the protein moiety is important for the purple intermediate to secure enough time to react with oxygen.

VI. REACTIVITY OF THE INTERMEDIATES WITH MOLECULAR OXYGEN

As mentioned in the previous section, two types of intermediates occur in the reaction of this enzyme. In the case of neutral D-amino acids, the purple intermediate is formed first and is then changed to the fully reduced enzyme under anaerobic conditions. Under aerobic conditions, the purple intermediate is the species that reacts with oxygen in the catalytic process. This was demonstrated in a stopped-flow study (Massey and Gibson, 1964). In addition, the rate of the formation of the fully reduced form cannot explain the molecular activity of this enzyme. The reaction of the purple intermediate with molecular oxygen is of second order, and the rate constant was reported to be 1.2×10^5 M^{-1} sec^{-1} (Nakamura et al., 1963).

On the other hand, in the case of basic D-amino acids, the intermediate, which reacts with oxygen, is the fully reduced enzyme. In this case also, the reaction of the intermediate with molecular oxygen is of second order (Yagi et al., 1974b), and the rate constant was reported to be 1.9×10^4 M^{-1} sec^{-1}. Accordingly, in the reaction of these intermediates with molecular oxygen, the concentration of any complex, if formed, is only marginal.

It was elucidated that, in the process of the reaction of the intermediates with molecular oxygen, the superoxide anion is not involved (Massey et al., 1969; Nishikimi et al., 1972). The reaction between these intermediates and molecular oxygen is a process of two-electron transfer.

It should be noted that the rate of the reaction of the purple intermediate with molecular oxygen is faster than that of the fully reduced enzyme. This seems to indicate that the strong charge-transfer complex has a higher reactivity with molecular oxygen than with the fully reduced enzyme. Accordingly, the above-mentioned hydrophobic environment surrounding the charge-transfer interaction and the hydrophobic interaction between the enzyme and the substrate are considered to provide a favorable situation for electron transfer from the enzyme to the acceptor, oxygen.

REFERENCES

Fukui, K. Yonezawa, T., Nagata, C., and Shingu, H. (1954). *J. Chem. Phys.* **22,** 1433–1442.

Massey, V., and Gibson, Q. H. (1964). *Fed. Proc., Fed. Am. Soc. Exp. Biol.* **23,** 18–29.

Massey, V., Strickland, S., Mayhew, S. G., Howell, L. G., Engel, P. C., Mattews, R. G., Schuman, M., and Sullivan, P. A. (1969). *Biochem. Biophys. Res. Commun.* **36,** 891–897.

Nakamura, T., Nakamura, S., and Ogura, Y. (1963). *J. Biochem. (Tokyo)* **54,** 512–518.

Nishikimi, M., Rao, N. A., and Yagi, K. (1972). *Biochem. Biophys. Res. Commun.* **46,** 849–854.

Nishimoto, K., Watanabe, Y., and Yagi, K. (1977). *Biochim. Biophys. Acta* (in press).

Porter, D. J. T., Voet, J. G., and Bright, H. J. (1973). *J. Biol. Chem.* **248,** 4400–4416.

Song, P.-S., Choi, J. D., Fugate, R. D., and Yagi, K. (1976). *Flavins Flavoproteins, Proc. Symp., 5th, 1975* pp. 381–390.

Theorell, H., and Åkeson, Å. (1956). *Arch. Biochem. Biophys.* **65,** 439–448.

Walsh, C. T., Schonbrunn, A., and Abeles, R. H. (1971). *J. Biol. Chem.* **246,** 6855–6866.

Yagi, K. (1971). *Adv. Enzymol. Relat. Areas Mol. Biol.* **34,** 41–78.

Yagi, K., and Matsuoka, Y. (1956). *Biochem. Z.* **328,** 138–145.

Yagi, K., and Ohishi, N. (1978). *J. Biochem. (Tokyo)* (in press).

Yagi, K., and Ozawa, T. (1962a). *Biochim. Biophys. Acta* **56,** 413–419.

Yagi, K., and Ozawa, T. (1962b). *Biochim. Biophys. Acta* **60,** 200–201.

Yagi, K., and Ozawa, T. (1964). *Biochim. Biophys. Acta* **81,** 29–38.

Yagi, K., Okuda, J., and Okamura, K. (1965). *J. Biochem. (Tokyo)* **58,** 300–301.

Yagi, K., Okamura, K., Naoi, M., Sugiura, N., and Kotaki, A. (1967). *Biochim. Biophys. Acta* **146,** 77–90.

Yagi, K., Nishikimi, M., Ohishi, N., and Takai, A. (1970a). *J. Biochem. (Tokyo)* **67,** 153–155.

Yagi, K., Nishikimi, M., Ohishi, N., and Takai, A. (1970b). *Biochim. Biophys. Acta* **212,** 243–247.

Yagi, K., Nishikimi, M., Ohishi, N., and Takai, A. (1970c). *FEBS Lett.* **6,** 22–24.

Yagi, K., Naoi, M., Nishikimi, M., and Kotaki, A. (1970d). *J. Biochem. (Tokyo)* **68,** 293–301.

Yagi, K., Nishikimi, M., and Ohishi, N. (1972). *J. Biochem. (Tokyo)* **72,** 1369–1377.

Yagi, K., Nishikimi, M., Takai, A., and Ohishi, N. (1973). *Biochim. Biophys. Acta* **321,** 64–71.

Yagi, K., Nishikimi, M., Takai, A., and Ohishi, N. (1974a). *J. Biochem. (Tokyo)* **76,** 451–454.

Yagi, K., Nishikimi, M., Takai, A., and Ohishi, N. (1974b). *Biochim. Biophys. Acta* **341,** 256–264.

14

Biological Aspects of Superoxide Dismutase

A. M. MICHELSON

Institut de Biologie Physico-Chimique, Paris, France

I. INTRODUCTION

It can be confidently stated that studies of superoxide radical anions and superoxide dismutases indeed constitute a frontier in biology, biochemistry, and biophysics. In 1968, this extremely important area in the metabolism of molecular oxygen, and in particular the various activated forms of O_2, was essentially undeveloped. The biological significance of a major free radical of oxygen, the superoxide anion, formed by the reduction by one electron of O_2 was opened to investigation by McCord and Fridovich (1969). They identified an enzymatic activity, dismutation of O_2^- to O_2 and H_2O_2, associated with a copper-containing protein, erythrocuprein (known for many years). Since then, the diverse aspects and roles of superoxide radical anions and the various superoxide dismutases have been developed in an exponential fashion, and the subject is now extremely vast and has penetrated into many areas, ranging from the physiology of hyperoxia and hypoxia, protection against high-energy irradiation and the mechanism of phagocytic killing of bacteria by polymorphonuclear neutrophils, through more or less classical studies of enzyme structure and mechanism, to an examination of the biological role of the Haber–Weiss (1934) reaction:

$$H_2O_2 + O_2^- \rightarrow HO \cdot + HO^- + O_2$$

309

FRONTIERS IN PHYSICOCHEMICAL BIOLOGY

Clinical studies, with respect to both the pathological significance of aberrant levels of superoxide dismutase and the use of this enzyme to diminish inflammation, are now well advanced. Apart from direct studies on O_2^-, the attacks on this particular frontier in physico chemical biology have engendered an interest in the biological significance of other oxygen-containing free radicals.

The history of the initial discovery in 1969 has been fully presented by McCord and Fridovich (1977) in the first book on the subject, but by no means the last, and in this chapter we will not attempt to review either the works of Fridovich and his collaborators or those of the many other groups now extremely active throughout the world, but rather to discuss various aspects developed in Paris.

As is often the case, we became involved in O_2^- and superoxide dismutase (SOD) via a very different domain, that of bioluminescence. In the system *Pholas dactylus,* the luciferase is a copper-containing glycoprotein, MW of 310,000, and the substrate luciferin is also a glycoprotein (MW 35,000). Oxidation of this substrate occurs via formation of O_2^- from the molecular O_2 by the enzyme (Henry *et al.*, 1973) and under suitable conditions light emission from this enzymatic oxidation is inhibited by superoxide dismutase (Isambert and Michelson, 1973). It may be noted that this does not appear to be a general mechanism for bioluminescent systems and may well be unique. Nevertheless, our interests changed direction in 1972, and we became involved in the area initiated in 1969, that of studying O_2^- and SOD's.

II. PRODUCTION OF O_2^-

Superoxide radicals can be produced *in vitro* and *in vivo* by a wide variety of methods. High-energy irradiation, for example, by γ rays, X rays, cosmic rays, and accelerated electrons, of oxygen-containing aqueous solutions gives rise to O_2^-. Chemically, the autoxidation of various molecules and the oxidation of reduced flavins are accompanied by a one-electron reduction of the dissolved molecular oxygen (about $2 \times 10^{-4} M$ at 20°C for aqueous solutions). A number of metal ions in the presence of various ligands can also reduce O_2, and under suitable conditions oxidation of oxyhemoglobin to methemoglobin produces O_2^- (Misra and Fridovich, 1972). Biochemically, isolated xanthine oxidase is a good source of the radical (McCord and Fridovich, 1968), which is produced during the oxidation of the diverse substrates of this enzyme (hypoxanthine, purine, acetaldehyde, etc.). In general, flavoprotein dehydrogenases all produce O_2^- as an intermediate, and microsomal hydroxylations also release this form of active oxygen. Phagocytosis of live bacteria (or other foreign matter) by poly-morphonuclear neutrophils gives rise to a specific production of O_2^- from NADPH and O_2 catalyzed by a membrane enzyme, super-oxide synthetase (Curnutte and Babior, 1974; Salin and McCord, 1974; Bachner et al., 1975; Goldstein et al., 1975). Apart from systems that liberate free O_2^-, many enzymatic oxidations involve enzyme-bound superoxide intermediates.

Thus, all respiring aerobic cells, both prokaryote and eukaryote, con-tain a certain level of O_2^-, which is limited by the rate of production compared with the rate of destruction by spontaneous dismutation or by other processes.

A. Production of Superoxide by Metal Ions

The involvement of superoxide ions in the autoxidation of ferrous to ferric ions was first postulated by Weiss (1935):

$$Fe^{2+} + O_2 \rightleftharpoons Fe^{3+} \cdots O_2^-$$

the reverse process being inhibited by anions complexing with Fe^{3+}, such as phosphate, pyrophosphate, or fluoride. Hydroxylation of vari-ous compounds (Nofre et al., 1961; Goscin and Fridovich, 1972) such as benzoic acid by ferrous ions and oxygen in phosphate buffer has been described, but these reactions are relatively slow and probably arise from secondary reactions derived from an initial pulse of O_2^-.

We have developed some of the properties of the $Fe^{2+}-O_2-$phosphate system (Michelson, 1973a). Injection of Fe^{2+} into phosphate buffer gives rise to a flash of light of low intensity with a half-life of about 1 second. The intensity is a function of various factors, such as the concentration of Fe^{2+} and of phosphate and the pH, and may well be mediated by traces of carbonate anions. If a chemiluminescent product is present, such as luminol, this is oxidized to give a flash of light with similar kinetics but of a much higher intensity. These effects disappear if phos-phate is replaced by acetate. Both light emissions are inhibited by SOD and can be attributed to a pulse of O_2^- formed when the Fe^{2+} ions are injected into the buffer. In contrast, no inhibition occurs when catalase is present. Indeed, the chemiluminescence of luminol oxidized by O_2^- is some 10,000 times greater than that induced by the equivalent quantity of H_2O_2.

Neither Co^{2+} nor Ni^{2+} in the presence of phosphate shows oxidizing properties, and both cyanide and ammonium ions are ineffective as suitable complexants. However, all three metal ions, Fe^{2+}, Co^{2+}, and

Ni^{2+}, when injected into buffers containing dihydroxyfumaric acid (DHF) oxidize luminol with the emission of light (Michelson, 1973b). With Fe^{2+} the reaction is rapid, as with phosphate, but with Co^{2+} and Ni^{2+} a continuous reaction occurs. With Ni^{2+} this is strongly inhibited by SOD and it may be concluded that a continuous production of O_2^- occurs, which is slow due to the low rate of oxidation of the Ni(II)–DHF complex to Ni(III). Since Ni(III) is very unstable, withdrawal of an electron from DHF would then occur to effect a return to Ni(II) and DH· + H^+, perhaps followed by DH· + $O_2 \rightarrow$ D + H^+ + O_2^-. In the case of Co^{2+}, the reaction is inhibited by both catalase and SOD, suggesting production of hydroxyl radicals by a Haber–Weiss reaction.

Tetraglycine and Ni^{2+} form a square planar complex with the terminal amino group and the three peptide nitrogens coordinated. It was therefore of interest to examine this system. At pH 9.5, an immediate rapid reaction occurs after injecting Ni^{2+}, as detected by the chemiluminescence of luminol, followed by an induction period of several minutes when no light emission is observed. There is then an oscillatory pattern of luminol oxidation that continues for at least 20 minutes. SOD, but not catalase, inhibits this reaction. If dihydroxyfumarate is present these oscillatory effects are abolished and a continuous emission of light is observed.

The results indicate that O_2^- is formed by one-electron reduction of O_2 by Fe^{2+}, Co^{2+}, or Ni^{2+} when these ions are in the presence of suitable complexing agents. When the complexant can act as a hydrogen donor, turnover becomes possible and the reaction becomes continuous. This turnover can also be achieved or accelerated when an external hydrogen donor such as DHF is present:

B. Superoxide Production via Riboflavin and Riboflavin 5'-Phosphate

The formation of O_2^- during the oxidation by O_2 of reduced flavins and flavoproteins is well established (Ballou et al., 1969). As with the Fe^{2+}/O_2 system, injecting reduced riboflavin 5'-phosphate (FMN) (or electrolytically prepared aprotic solutions of O_2^-) into aqueous oxygen containing buffer gives rise to a low light emission (Michelson, 1973a). In the presence of luminol or Pholad luciferin a strong chemiluminescence occurs, which is inhibited by SOD, as is light emission in the absence of these compounds. Generally, photoreduction or catalytic reduction of FMN is employed to prepare the reduced flavin. However, under suitable conditions, riboflavin and FMN can be reduced directly (and in obscurity) by NADH (Michelson, 1977). Thus, ferric ions are rapidly reduced to Fe^{2+} by mixtures of riboflavin (or FMN) and NADH but not by either component separately. This reduction is not inhibited by SOD and indeed proceeds equally well under anaerobic conditions. Hence, reduction of Fe^{3+} is direct and is mediated by the production of $FMNH_2$ or FMNH :

$$H^+ + FMN + NADH \longrightarrow FMNH_2 + NAD^+$$
$$Fe^{3+} + FMNH_2 \longrightarrow Fe^{2+} + FMNH^. + H^+$$
$$Fe^{3+} + FMNH^. \longrightarrow Fe^{2+} + FMN + H^+$$

In the same way, both ferricytochrome c and methemoglobin are reduced directly by FMN plus NADH, without intermediate production of O_2^-. However, reduction of nitroblue tetrazolium (another indicator for O_2^-) by this system is strongly inhibited by native SOD, but not by the denatured enzyme. Thus, reduction of this derivative to a formazan occurs via O_2^- resulting from oxidation by O_2 of $FMNH_2$ formed as an intermediate, in contrast to the reduction of cytochrome c and methemoglobin.

Oxidation of luminol by mixtures of flavin and NADH was not detected, probably because of the low steady-state production of O_2^-. However, Pholad luciferin, which is much more sensitive as a detector of O_2^-, was readily oxidized in a continuous fashion, and this was inhibited by SOD. Since ferric ions are rapidly reduced to Fe^{2+} by FMN plus NADH, and Fe^{2+} ions in the presence of phosphate give rise to O_2^-, the oxidation of luminol was reexamined using a mixture of Fe^{3+}, FMN, and NADH. In this case, light emission occurs in a continuous steady-state manner for at least 10 hours, after an initial buildup requiring 30 minutes. This reaction is strongly inhibited by native SOD but not by the denatured enzyme, indicating participation of O_2^- (Michelson, 1977).

It may be noted that, despite the fact that these reactions are nonenzymatic, the nature of the flavin plays an important role. Thus, riboflavin is three times more efficient than FMN and ten times better than FAD (flavin adenine dinucleotide).

Mixtures of flavin and NADH give rise to reduced flavin, which can produce O_2^-. Under certain conditions and with certain acceptors, O_2^- reduces more efficiently than $FMNH_2$, whereas in other cases the inverse is true. For oxidations, the production of O_2^- can be increased by the addition of ferric ions, which are converted to Fe^{2+} as an intermediate. As shown by McCord and Fridovich (1970) using the system ferredoxin–NADP reductase, reduction of cytochrome c mediated aerobically by FMN or FAD is strongly inhibited by SOD. Under anaerobic conditions the rate of reduction is essentially the same, and naturally no inhibition by SOD occurs. Thus, $FMNH_2$ acts in this system directly as a one-electron reductant or via production of O_2^-, depending on the conditions. The nature of the acceptor and whether O_2^- acts as a reducing or an oxidizing agent are shown in studies on reduced methylene blue, which under aerobic conditions reduces cytochrome c directly but oxidizes luminol and epinephrine via formation of O_2^-.

It is clear that a mixture of FMN, NADH, O_2, Fe^{3+}, and phosphate (as complexant) provides a simple chemical model of electron transfer:

$$\text{NADH} \xrightarrow{\ 6\ } \text{FMN} \xrightarrow{\ 6\ } Fe^{3+} \xrightarrow{\ 6\ } O_2 \rightarrow O_2^- \underset{\epsilon}{\overset{6}{\rightleftharpoons}} \begin{array}{l} \text{reduction substrate} \\ \text{oxidation substrate} \end{array}$$

with O_2 above and H_2O_2 below O_2^-.

III. SUPEROXIDE DISMUTASES

Superoxide dismutases are widely distributed metalloproteins that are present in all aerobic cells (Fridovich, 1972, 1974, 1975; McCord et al., 1971), both prokaryote and eukaryote, and even in certain anaerobic bacteria. Three classes of these enzymes are known, all with the same catalytic activity but differing in the nature of the metal involved in the active site. Thus, erythrocuprein (McCord and Fridovich, 1969) is a copper protein (two Cu atoms per molecule, MW of 33,000, consisting of two identical subunits) that also contains zinc as a structural element (2 g-atoms of Zn per mole). This enzyme is generally, but not always, found in the cytoplasm of eukaryotic cells. In these cells, a second enzyme is present, concentrated in the mitochondria but sometimes present in the cytoplasm, which contains manganese. This SOD has a MW of about 80,000 and contains four subunits and 2 g-atoms of Mn per mole (Weisiger and Fridovich, 1973). A similar enzyme is found in bacteria such as *Escherichia coli* (Keele *et al.*, 1970) with a MW of 40,000 (two subunits) containing 1 g-atom of Mn per mole. Finally, certain bacteria and algae contain an SOD with iron as the metallic component.

A. Marine Bacteria

Marine bacteria, particularly bioluminescent species, contain a major iron-containing SOD (MW, 40,000; two identical subunits) containing 1 g-atom of iron per molecule (Puget and Michelson, 1974a). Some preparations contain up to two atoms of Fe, but the second iron is probably an artefact, is readily removed by dialysis at pH 5.5, and plays no role in the catalytic activity of the enzyme. Three distinct classes of this iron SOD have been identified by immunochemical techniques, although with respect to molecular weight, subunits, active iron content, and activity, they are essentially identical but vary with respect to isoelectric points. Class I covers a group of bioluminescent symbiotic bacteria represented by *Photobacterium leiognathi;* "free-living" luminescent bacteria represented by *P. sepia* fall into the second class, whereas Fe-SOD type III is obtained from *P. fischeri*. It is of interest that this classification corresponds entirely with the bacteriological separation based on more classical techniques. In addition, type I Fe-SOD is always associated with a "fast" luciferase (which also characterizes symbiotic bacteria), whereas type II is present in bacteria that possess a different "slow" luciferase. Immunological differences certainly exist between these Fe-SOD's and those isolated from *E. coli* or from algae. There thus exists a range of immunochemically different Fe-SOD's, as is the case for species-distinct erythrocupreins (Cu-SOD) such as bovine, human, and bacteriocuprein. It may be noted that no Mn-SOD could be detected in the species mentioned.

These bacterial Fe-SOD's show the characteristic stability of SOD's in general; for example, *P. leiognathi* Fe-SOD is completely active in 50% ethanol or 50% dimethyl sulfoxide at 22°C. The inactive apoenzyme has been prepared from this Fe-SOD and the holoenzyme has been reconstituted by adding ferrous (but not ferric) ions, with recovery of activity (Puget *et al.*, 1977b). Other metals that can be incorporated into the apoenzyme (but without the resultant catalytic activity) include Cu^{2+}, Mn^{2+}, Zn^{2+}, Ni^{2+}, Co^{2+}, and Cr^{2+}, probably at the active site since they cannot be "chased" by Fe^{2+}. Trivalent metal ions such as Fe^{3+} and Cr^{3+} are not taken up by the apoenzyme, despite the fact that in the resting state the active enzyme contains ferric iron, as shown by electron spin resonance measurements.

Loss of activity and of iron after photodestruction of tryptophan residues (at 254 nm) of *P. leiognathi* Fe-SOD suggests that at least one typtophan residue is involved in the active site as a ligand (Puget and Michelson, 1974a). This is supported by chemical modification of different amino acids in both the apo and holoenzymes (Vanopdenbosh *et al.*, 1977). Alkylation of tryptophan residues in Fe-SOD with hydroxynitrobenzyl bromide caused modifications of 2.6 tryptophans per subunit (each subunit contains 6 of these amino acids) and 3.45 residues per subunit in the apoenzyme. No activity is present in the alkylated enzyme and the modified apoenzyme cannot be reconstituted. Hence, at least one tryptophan residue (difference between 3.45 and 2.6) per subunit is involved as a ligand for the metal and in the active site (loss of activity in the modified Fe-SOD), in accord with fluorescence studies (Puget and Michelson, 1974a).

Amidation of five or six carboxyl groups (out of thirteen or fourteen) per subunit with glycinamide had essentially no effect on activity, and the alkylated apoenzyme could be reconstituted to an active enzyme with Fe^{2+}. The number of groups alkylated was essentially the same in both holo- and apoenzymes. Similarly, extensive acylation of lysine ϵ-amino groups of Fe-SOD with dimethylsuberimidate caused an insignificant loss of activity. Treatment with diethyl pyrocarbonate and with diazonium tetrazole modifies two (of five) histidine residues per subunit in both the holoenzyme and the apoprotein. Enzymatic activity is lost with the Fe-SOD and the apoenzyme cannot be reconstituted. Hence, at least one histidine residue is involved per subunit. We can conclude that in this Fe-SOD (1 g-atom of Fe per molecule of enzyme composed of two identical subunits) the active center contains Fe liganded to two histidine residues and two tryptophan residues (one of each from each subunit):

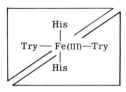

Mechanism. Electron spin resonance studies have shown that the iron is in a high-spin ferric form. Kinetic studies using pulse radiolysis with accelerated electrons to produce a pulse of O_2^- in an extremely short time have shown that the second-order rate constant decreases from 6.1×10^8 M^{-1} sec^{-1} at pH 6.2 to 1.3×10^8 M^{-1} sec^{-1} at pH 10.1, in contrast with erythrocuprein (for which k is constant). The rate constant of Fe-SOD is some ten times less than Cu-SOD ($k = 2.4 \times 10^9$ M^{-1} sec^{-1}); nevertheless, the enzymatic catalysis can be considered extremely rapid compared with the spontaneous dismutation of O_2^-. Turnover experiments (Lavelle *et al.*, 1977) using the absorption band at 350 nm to follow the nature of the iron (ferrous or ferric) and reducing agents [such as ascorbate, H_2O_2, or formate (CO_2^-) radical anions] to convert the native Fe(III)-SOD to a fully reduced Fe(II)-SOD form (which retains the enzymatic activity) have shown that O_2^- alternatively reduces and then oxidizes the iron. Both the oxidation and reduction steps have rate constants identical to that of the enzymatic reaction under turnover conditions measured by the rate of disappearance of O_2^-. The mechanism may thus be written

$$2H^+ + 2O_2^- + \text{Fe-SOD} \xrightarrow{k_1} H_2O_2 + O_2 + \text{Fe-SOD}$$

$$\text{Fe(III)-SOD} + O_2^- \xrightarrow{k_2} \text{Fe(II)-SOD} + O_2$$

$$\text{Fe(II)-SOD} + O_2^- + 2H^+ \xrightarrow{k_3} \text{Fe(III)-SOD} + H_2O_2$$

$$k_1 = k_2 = k_3 = 5.2 \times 10^8 \ M^{-1} \ \text{sec}^{-1} \text{ at pH 8}$$
$$= 4.2 \times 10^{-8} \ M^{-1} \ \text{sec}^{-1} \text{ at pH 9}$$

B. Mushrooms

The first eukaryote organism shown to be devoid of a copper-containing SOD was the bioluminescent fungus *Pleurotus olearius*, which grows on olive trees (Lavelle *et al.*, 1974b). This organism contains two distinct SOD's, but both contain manganese; other Fe- or Cu-SOD's were not detected (Lavelle and Michelson, 1975). Both proteins contain two atoms of manganese per molecule of enzyme and have similar molecular weights (76,000 and 78,000) with four subunits of MW 20,000 per molecule. The specific activities are identical in the two cases, but the minor SOD (20%), probably mitochondrial, contains identical subunits (with respect to electrophoretic properties) and hence is α_4. The major enzyme (80%), probably cytoplasmic, contains two types of subunits (both of the same molecular weight) and hence has an $\alpha_2\beta_2$ structure. This enzyme has been studied further. The band of absorption at 480 nm ($\epsilon_{max} = 850 \ M^{-1} \ cm^{-1}$) characteristic of certain octahedral complexes of high-spin trivalent manganese is not modified by oxidants but disappears when the enzyme is reduced (without loss of activity). The native enzyme shows no electron spin resonance signal [as for other Mn(III)-containing proteins] but the characteristic signal of Mn^{2+} appears after liberating the metal by acid treatment (Lavelle *et al.*, 1974b). The mechanism may be written

$$\text{Mn(III)-SOD} + \text{O}_2^{\overline{}} \longrightarrow \text{Mn(II)-SOD} + \text{O}_2$$
$$\text{Mn(II)-SOD} + \text{O}_2^{\overline{}} + 2\text{H}^+ \rightarrow \text{Mn(III)-SOD} + \text{H}_2\text{O}_2$$

and involves an alternate reduction and oxidation of the metal.

No immunological cross-reactions can be detected between the two enzymes and, given the different types of quaternary structures, this suggests that the two SOD's are isoenzymes rather than two forms derived from the same gene (Lavelle and Michelson, 1975). While *Pleurotus olearius* was the first eukaryote in which a cytoplasmic Mn-SOD was identified, cytoplasmic Mn-SOD's have recently been found in human and baboon (*Papio ursinus*) liver (McCord, 1976).

C. Protozoa

Both *Tetrahymana pyriformis* and paramecia have been examined. No Cu-SOD was detected in either organism in that activity was not inhibited by KCN. In the case of *Tetrahymena*, a major and a very minor band of activity were detected by gel electrophoresis (F. Lavelle and A. M. Michelson, unpublished results, 1975), whereas with paramecia, three distinct almost equal bands of activity are present (K. Puget and A. M. Michelson, unpublished results, 1977). The lack of inhibition by KCN suggests Fe- or Mn-SOD's, which in the case of the bacterial enzymes are destroyed by treatment with chloroform–ethanol. In both *Tetrahymena* and paramecia the bands of activity are resistant to this treatment, as in human Mn-SOD (MW, 80,000) isolated from leukocytes or blood platelets.*

Other SOD's. A superoxide dismutase is present in rat liver nuclei (different from the cytoplasmic enzyme), and two Cu-SOD's are present in human seminal fluid (M. F. Isambert and A. M. Michelson, unpublished results, 1975).

D. Bacteriocuprein

While many eukaryote species contain erythrocuprein-type Cu-SOD's, most prokaryotes do not. The only prokaryote from which a Cu-SOD has been purified and isolated is *P. leiognathi* (Puget and

---

* With respect to the resistance to chloroform–ethanol, it may be noted that the above-mentioned Mn-SOD's have molecular weights of 80,000 to 90,000 (probably four subunits and two Mn atoms) and thus differ from the two-subunit, one-Mn *E. coli* SOD of MW 40,000.

Michelson, 1974b). In addition to the Fe-SOD, this bacterium contains appreciable quantities (28–46% of the total activity) of a copper–zinc SOD, bacteriocuprein. Like erythrocuprein, the enzyme contains Cu_2 and Zn_2; an earlier preparation contained only Cu_1 (and Zn_2) but was only half as active. Bacteriocuprein has about five times the activity per milligram of protein as Fe-SOD; hence, in terms of relative amounts of protein, the cell produces about 7–14% Cu-SOD and 86–93% Fe-SOD (the amounts are a function of the partial pressure of oxygen). A fivefold increase in oxygen pressure (atmospheric to pure O_2) induces a 4.8-fold increase in the Fe-SOD, but only a 2.2-fold increase in Cu-SOD. Bacteriocuprein has been separated into two fractions, one with an isoelectric point (pI) of 8.7 and the other with $pI = 8.15$ (K. Puget, P. Durosay, and A. M. Michelson, unpublished results, 1976). While the fraction of pI 8.7 is stable, that of pI 8.15 is unstable and rapidly degrades (even as a precipitate in concentrated ammonium sulfate solutions) to seven or eight distinct species, all active but with different pI's, thus leading to an apparent microheterogeneity (Puget et al., 1977a,b).

The presence of a typically eukaryotic Cu-SOD in this prokatyote, coupled with the absence of a typically prokaryotic Mn-SOD, raises some interesting problems. It may be that at least in certain branches of aerobic evolution the SOD's of eukaryotes and prokaryotes have a common origin rather than independent lines after passage to aerobic life. However, an alternate explanation is that the symbiotic bacterium P. leiognathi with the pony fish as host may well be a natural example of genetic engineering, in which the eukaryote fish gene for Cu-SOD has been transferred to the bacterium, which subsequently lost a less efficient Mn-SOD. We hope to obtain a pony fish in the near future to study the possible immunochemical similarities as well as other characteristics of the two enzymes to confirm or invalidate this hypothesis.

IV. VARIATION IN SOD LEVELS DURING DEVELOPMENT OF Ceratitis capitata

Insects are an extremely abundant species among terrestial life forms but have been studied little with respect to SOD. We have studied the Mediterranean fruit fly Ceratitis capitata with respect to changes in SOD during development. Respiration by the eggs, larva, and pupa of this species is low in comparison with the high rate of oxygen consumption by the adult insect (about 20 liters of O_2 per gram dry weight per hour). This large increase in oxidative metabolism (within 24 hours after emergence there is a 50-fold increase in the volume of mitochondria) is

probably accompanied by a corresponding increase in production of
O_2^-, which could be toxic for the cell and the total organism. It was
therefore of interest to follow cytoplasmic and mitochondrial superoxide
dismutases at the various stages of development (Fernandez-Sousa and
Michelson, 1976). Three different SOD's are present at all stages. Two
correspond to copper-containing erythrocuprein-type cytoplasmic enzymes with a molecular weight of 35,000, whereas the third (with a MW
of 80,000–90,000), present in smaller quantity, does not contain copper
and corresponds to the eukaryote mitochondrial Mn-SOD. Until the
emergence of the fly, that is, during the development eggs → larva →
pupa, no significant change occurs in the SOD activities, both cytoplasmic and mitochondrial, despite the marked changes in lipid
metabolism that occur at the apolysis stage between larva and pupae.
However, the transition pupa → fly is accompanied by an important
increase (27%) in total SOD activity per gram of material. Half of this
total increase is due to an increase in mitochondrial SOD activity, which
increases 400% (compared with the 13% increase of cytoplasmic Cu-
SOD) and represents 13% of the total activity in the fly, rather than the
4% present in eggs, larva, and pupa. Since the large increase in
mitochondria develops mainly in the thorax, whereas these results concern the entire insect, use of the isolated thorax would undoubtedly
show an even larger increase. It can thus be concluded that the major
defense mechanism exercised by superoxide dismutases is stimulated
by the increase in respiratory processes corresponding to the increased
oxygen consumption of the adult fly.

V. ERYTHROCYTE SOD, CATALASE, AND GLUTATHIONE PEROXIDASE LEVELS IN DIFFERENT ANIMALS

Three major enzymes are present in the erythrocyte to control the
possible toxic effects associated with excessive levels of O_2^-, H_2O_2 (derived from dismutation of O_2^- or by other processes), and, indirectly,
production of hydroxyl radicals. SOD (erythrocuprein) and catalase are
both cytoplasmic, whereas glutathione peroxidase is membrane bound.
In the human, erythrocyte SOD is remarkably constant within a given
population of normal subjects, with a standard deviation of less than
±10% of the mean. It was thus of interest to examine erythrocyte SOD
in different species to determine the interspecies variation of this enzyme (Maral et al., 1977). For comparison, catalase and glutathione
peroxidase were also measured. Ten animals were studied, namely,

TABLE I

Interspecies Variation of Erythrocyte Superoxide Dismutase (SOD), Catalase, Glutathione Peroxidase, and Hemoglobin

	Man ($n=200$)	Mouse ($n=60$)	Sheep ($n=2$)	Pig ($n=2$)	Horse ($n=1$)	Ox ($n=2$)	Pigeon ($n=2$)	Chicken ($n=2$)	Duck ($n=2$)	Rabbit ($n=10$)
SOD										
U/g of Hb	1225	2280	1912	1243	1550	2063	1260	1263	1500	2480
Percentage of human value	100	186	156	101	126	168	102	103	122	200
U/ml of blood	155	329	284	201	330	286	117	71.5	240	316
Percentage of human value	100	212	183	129	213	184	75	46	154	204
Catalase										
μg/g of Hb	4033	1759	834	4590	2760	2207	140	342	31.4	4667
Percentage of human value	100	43	21	114	68	55	3.5	8.4	0.8	115
μg/ml of blood	511	254	124	730	600	317	13.3	20.1	5	565
Percentage of human value	100	49	24	142	117	62	2.6	4	1	110
Gutathione peroxidase										
μmol of NADPH/mm/g of Hb	5.73	154	29	14.4	10.1	11.6	2.69	5.17	75	15.7
Percentage of human value	100	2687	506	251	176	200	47	90	1308	274
μmol of NADPH/mm/ml of blood	0.726	22.4	4.3	2.3	2.19	1.64	0.24	0.3	12.2	1.9
Percentage of human value	100	3080	592	316	302	226	33	41.3	1680	258
Hemoglobin (g/100 ml)	12.67	14.44	14.9	16.3	21.7	14	9.2	5.9	16	12.1

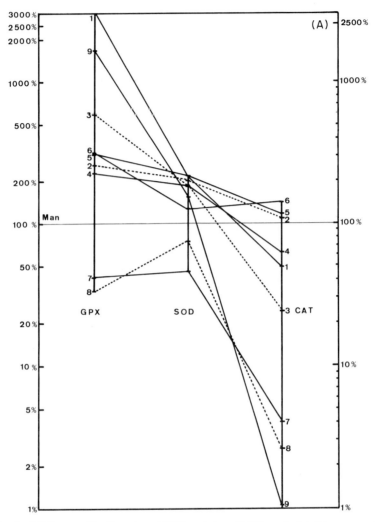

Fig. 1 (A) Superoxide dismutase (SOD), catalase (CAT), and glutathione peroxidase (GPX) activities in erythrocytes from different animals per milliliter of blood compared with man (100%). 1, mouse; 2, rabbit; 3, sheep; 4, ox; 5, horse; 6, pig; 7, chicken; 8, pigeon; 9, duck. Logarithmic scale. (B) Superoxide dismutase, catalase, and glutathione peroxidase activities per gram of hemoglobin in erythrocytes from different animals compared with man (100%). Numerical identifications are the same as in A.

mouse, sheep, pig, horse, ox, pigeon, chicken, duck, rabbit, and man. Hemoglobin (Hb) levels varied from 6 g of Hb/100 ml of blood (chicken) to 22 g of Hb/100 ml (horse). Since variations in size and morphology of the erythrocytes are important, and given the dispersion of hemoglobin

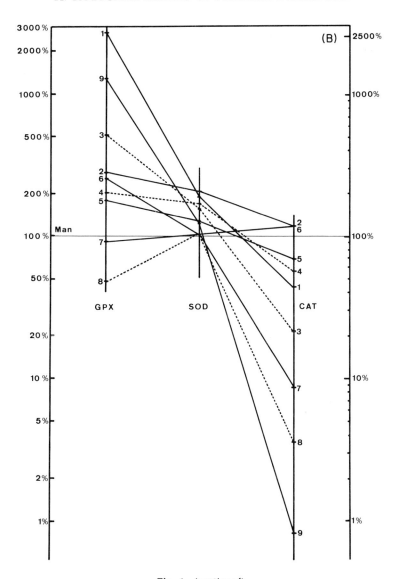

Fig. 1 *(continued)*

values, the results were expressed as activity per gram of hemoglobin as well as per milliliter of blood.

Superoxide dismutase activity per gram of hemoglobin varied from 1200 to 2400 units and thus is remarkably constant among many vertebrate animals (±50% of the median value). In contrast, erythrocyte catalase activities were widely dispersed, the levels ranging from a very high activity (about 4 mg of catalase/g of Hb) shown by rabbit, pig, and man to the virtual acatalasemia of the duck (30 μg/g of Hb), where the catalase activity is difficult to distinguish from the effects of hemoglobin itself. Glutathione peroxidase levels are also widely dispersed, with limits of 154 (mouse) to 2.7 (pigeon) units/g of Hb.

Although a limited number of animals (and of individuals within a given species) are concerned, this survey shows that blood SOD activity levels (factor of 2 between minimal and maximal) are much more constant than catalase and glutathione peroxidase and parallel the limited disparity between erythrocyte SOD levels among different individuals within a given species (man). In contrast, catalase shows a wide variation (factor of 150 between minimal and maximal values) and in birds at least, it is unlikely that catalase plays a major role in protecting erythrocytes against H_2O_2 (see Table I). At low concentrations of H_2O_2, this is probably consumed by glutathione peroxidase-catalyzed oxidation of reduced glutathione. However, there is a large disparity between glutathione peroxidase values (about a 60-fold difference between minimal and maximal values). Within the human species, the variation of individual levels of catalase and glutathione peroxidase is also considerably greater than that of erythrocyte SOD. No immediate correlation is apparent and, while a low level of catalase may be compensated by a high level of glutathione peroxidase in the mouse and duck, this is not true for other species such as chicken and pigeon. The results (Fig. 1) indicate a rather constant level of SOD per gram of Hb in comparison with the large disparity between the other two enzymes, suggesting that the major system of protection against uncontrolled oxidations, which could lead to lysis of the red cell, lies in superoxide dismutase.

VI. BIOLOGICAL UTILITY OF O_2^-

Apart from acting as an essential enzyme-bound metabolic intermediate of molecular oxygen in many enzymatic systems of oxidation or dehydrogenation involving metallo- or flavoproteins, O_2^- in at least one case, indolamine dioxygenase, is a substrate in preference to O_2 (Hirata and Hayaishi, 1975; Hayaishi et al., 1977). This enzyme degrades a variety of indolamines, such as tryptophan, hydroxytryptophan, serotonin, and dimethyltryptamine, and thus perhaps plays an important role in the regulation of various neuromediators. Superoxide radical ions also have a biological role in that phagocytosis by polymorphonuclear leukocytes is accompanied by a burst of production of the radical, due to activation of a membrane enzyme, superoxide synthetase. When SOD-coated latex particles and the live bacterium are ingested by the neutrophil, considerable inhibition of bacterial killing is observed (Johnston and Lehmeyer, 1977). This production of O_2^- is not seen with granulocytes from subjects with chronic granulomatosis, whether or not they are activated (Johnston *et al.*, 1975; Curnutte *et al.*, 1975). Since protection of live bacteria is also afforded by catalase-coated particles, it is likely that, as with a number of biochemical examples, the effective agent is HO·, produced indirectly from O_2^- and H_2O_2. Nevertheless, O_2^- is evidently essential in this particular protection against microbial infection. Indeed, the extracellular release of this radical is probably a major cause of inflammation. As shown by Huber and his collaborators (Huber and Saifer, 1977; Menander-Huber and Huber, 1977), this can be markedly reduced by injections of SOD. It has also been shown that blood platelets (human) produce O_2^-, which is detectable in the surrounding medium (Marcus *et al.*, 1977). This extracellular production of O_2^- is present in both resting and aggregated platelets (5×10^4 molecules of O_2^- per platelet per minute) and is apparently not platelet bound. In comparison with activated granulocytes (10^6 molecules of O_2^- per neutrophil per second), the yield is low, about 1000-fold less. The function or the causes of this constant extracellular production of O_2^- by platelets are as yet unknown. There is a synergistic effect of subthreshold quantities of thrombin in the presence of O_2^- at levels insufficient to promote aggregation, and it may well be that production of the radical by platelets could be implied in hemostasis, coagulation, and occlusive vascular disease, since SOD is not present in the plasma. Thus, as with polymorphonuclear leukocytes, the production of O_2^-, whatever the biological reason, may well result in toxic effects.

VII. TOXICITY OF SUPEROXIDE RADICALS

The role of SOD in diminishing oxygen (and hence in indirectly diminishing O_2^- and HO· species) toxicity and the induction of SOD synthesis by oxygen have been amply demonstrated by Fridovich and his collaborators with respect to bacterial species (Fridovich, 1974; Gregory and Fridovich, 1973a,b; Gregory *et al.*, 1974; Yost and Fridovich, 1976). The adaptation of certain mammalian species to the lethal effects of hyperbaric oxygen (accompanied by an increase in cellular SOD) has been described by Crapo (1977) in a study of the role of SOD in pulmonary oxygen toxicity.

A. Macromolecules

We have studied the effects of systems producing O_2^- on various macromolecules and cells and the possible protection afforded by SOD. The results do not necessarily indicate that O_2^- is directly responsible, but rather that it can also act as a primary radical that, in the presence of H_2O_2, can lead to secondary radicals such as $HO\cdot$, as well as amplifying certain effects by initiating free radical chain reactions. Thus, the loss of activity of ribonuclease in the presence of O_2^- produced by photoreduction of FMN is strongly inhibited when SOD is added (Lavelle *et al.*, 1973). Yeast lysine tRNA ligase is a lipoprotein in which enzymatic activity is a function of the presence and integrity of the lipid moiety. The enzyme is unstable due to autoxidative processes and loses 38% of its activity in 4 days at 4°C. The addition of ascorbate (a typical antioxidant) accelerates this process, and 81% of the activity is lost in the same time interval. However, when 5×10^{-8} M SOD is added to solutions of the enzyme, considerable protection is afforded in that loss of activity is reduced to 4%, demonstrating that O_2^- produced spontaneously (perhaps by traces of metal ions) is responsible.

Protection by SOD is also observed when the nucleoprotein, bacteriophage R17, is subjected to O_2^- produced either by photoreduction of FMN or by the xanthine oxidase–O_2–hypoxanthine system, as measured by the loss of infectivity. At the molecular level, we have observed that puromycin is covalently fixed to ribosomes when a mixture is subjected to the photoreduced flavin system. Such cross-linking is also observed when 30 S ribosome subunits plus IF-3 undergo a similar treatment (Cooperman *et al.*, 1977). In this case, the protein IF-3 is covalently linked to both protein and RNA of the 30 S particle.

B. Cells

Bacterial cells are killed when treated with the FMN photoreduction system, and again exogenous SOD protects against this lethal action (Gregory and Fridovich, 1973a,b; Lavelle *et al.*, 1973). Indeed, irradiation of bacteria (*P. leiognathi*) alone at 365 nm leads to loss of viable cells (probably due to the presence of endogenous flavins) and considerable photoprotection is afforded by external SOD (Lavelle *et al.*, 1973). It should be noted that, in certain of the examples noted above, the protection by SOD was increased when catalase was added, suggesting the role of a Haber–Weiss reaction in the toxic effects studied (Lavelle *et al.*, 1973; Beauchamp and Fridovich, 1970; Cohen, 1977).

The toxicity of O_2^- with respect to eukaryotic cells was also examined.

Calf fetal myoblasts in culture were used because, in addition to lethality, the effects on cell growth, morphology, and differentiation can be readily followed (Michelson and Buckingham, 1974). Calf primary myoblast cultures undergo cell division for the first 42 hours, which is succeeded by a stationary phase (about 10 hours) of alignment of the cells. Cell fusion at about 52 hours then occurs, initially by two or three nuclei fusions followed by long multinucleate cellular fibrils. Production of O_2^- was achieved by photoreduction (365 nm) of FMN present in the medium. After 30 minutes of irradiation (and hence of production of O_2^-) at 24 hours of growth, during the exponential phase, there was considerable cell death, cell division was inhibited, and very few fusions occurred. In addition, a considerable proportion of the surviving cells assume an abnormal hyperplasmic morphology with a very large nucleus and extended cytoplasm. The effects are a function of the quantity of O_2^- generated in terms of either time (accumulative dose) or intensity. The presence of exogenous bovine SOD (1 μg/ml, 3 $\times$ 10^{-8} M) gave almost complete protection against all of the lethal morphological and differentiation effects. When carcinogenic hydrocarbons such as methylcholanthrene or benzpyrene were also present, exactly the same effects were observed at much lower levels of O_2^-. Under the conditions used, neither the hydrocarbon nor the amount of O_2^- produced in the system caused any visible effects within 1 week following treatment. It is thus clear that the biological action of such hydrocarbons is greatly stimulated and amplified by free radicals such as O_2^-. Again, an important protection was observed when SOD was added to the medium.

The moment at which the cells undergo treatment (in the absence or presence of hydrocarbons) is extremely important. Thus, all the effects described above with respect to death, morphology, and differentiation are observed when the O_2^--producing system is applied during cell division, at which time replication of DNA and transcription to unstable messenger RNA (for the synthesis of myosin and other differentiated proteins) occurs. In the stationary period preceding fusion, stable messenger RNA (for the synthesis of contractile proteins) is formed and the cells become much more resistant. Death is considerably reduced, but fusion, although initiated, proceeds abnormally. Once fusion begins, the treatment has no detectable effects at the levels used. These results indicate that the primary toxic effect occurs at the nuclear level during DNA synthesis and RNA transcription and suggest that O_2^- and SOD may play roles in the initiation of certain cancers. Phenotypic modifications can also be envisaged in which DNA polymerases and repair enzymes are rendered defective (for example, by oxidation of tryptophan residues) and, while active, produce base pairing errors during repair or replication, thus indirectly causing genetic changes.

The mutagenic effects of molecular oxygen have been demonstrated (Fenn et al., 1957; Gifford, 1968; Gottlieb, 1971; Yost and Fridovich, 1976). That this is probably related directly to increased levels of O_2^- (or other species such as $HO\cdot$) is shown by the strong mutational effects of O_2^--producing systems on bacteriophage T4 and the protection against such mutations given by SOD (A. M. Michelson and E. Brody, unpublished results, 1974).

VIII. IONIZING IRRADIATIONS

A. Calf Fetal Myoblasts

Since high-energy irradiation of aerobic aqueous solutions is known to produce O_2^- (Behar et al., 1970), the myoblast cultures used above were subjected to γ irradiation (Michelson and Buckingham, 1974). Irradiation during the period of cell division (at 24 hours) with doses of 4000 rad caused cell death, abnormal cells, and reduced fusions, and the changes in cell morphology were similar to those described above using photoreduction of FMN as a source of O_2^-. At 8000 rad, high cell mortality was observed, the remaining cells were mainly abnormal in morphology, and no fusions occurred. The presence of SOD again protected against cell death, reduced the number of abnormal cells, and increased the number of normal fusions. When γ irradiation of the cells was applied at the stationary phase, just prior to fusion, no effects on fusion were observed at 4000 rad. At the higher dose (8000 rad) the morphology of the fusions was abnormal. In both cases, single mononucleate cells were more sensitive and were reduced in number (at 4000 rad) or eliminated entirely (8000 rad). The addition of SOD to the medium before irradiation greatly increased the number of normal fusions and mononucleate cells. The protective action of SOD could be observed even at high doses (10,000 rad) in that cell death was significantly reduced.

It may be concluded that a major (but not unique) cause of the toxicity of γ (or other high energy) irradiation is production of O_2^-, though other radicals are probably involved. Nevertheless, the addition of exogenous SOD confers a considerable protection to the living cell. As with other sources of O_2^-, mammalian cells are most sensitive to γ irradiation at the moment of cellular division, that is, during replication and transcription of DNA.

B. Rhesus Monkey Kidney and Other Cells

We have recently examined (A. M. Michelson and P. Coppey, unpublished results, 1977) the protection afforded by SOD against the lethal action of X rays (400 rad) and γ rays (5.5 MeV; 400 rad; linear accelerator) on cultures of continuous line rhesus monkey kidney cells (CV1). Survival, measured by colony-forming ability, was significantly increased (20%) in the presence of $3 \times 10^{-8} M$ SOD in both cases. An even greater effect was observed if survival was measured by the number of cells in suspension after two or three divisions.

Extracellular SOD also protects *Acholeplasma laidlawii* cells against γ irradiation (Petkau and Chelack, 1974) and against damage to membranes by high-energy irradiation (Petkau and Chelack, 1976). The oxygen enhancement ratios for the survival of *E. coli* after X irradiation in relation to the membrane-associated SOD's of the bacteria have been examined (Oberley *et al.*, 1976). Although no difference was observed when bacteria with high or low internal SOD levels were irradiated (Goscin and Fridovich, 1973), *E. coli* with high levels of periplasmic Fe-SOD showed an oxygen enhancement factor of 1.71, whereas bacteria with low levels of SOD gave a ratio of 2.35. The addition of exogenous enzyme reduced this figure to 1.40, but a further decrease to 1.0 was not possible. An analogy to these results may be found in the purely biological production of O_2^- (corresponding to radiation-induced formation) during phagocytosis by polymorphonuclear neutrophils. The death rate of bacteria with high levels of periplasmic SOD is considerably lower than that for bacteria with low levels of SOD (Yost and Fridovich, 1974).

Catalase and SOD together show a radioprotective effect on chromosome breakage induced in human lymphocyte cultures by γ irradiation (135 rad) when added immediately after irradiation (Nordenson *et al.*, 1976). Chromosomal aberrations decreased significantly (60%) in the presence of both enzymes, but smaller effects were observed in the case of catalase alone, whereas the slight reduction in the presence of SOD was not significant.

C. Erythrocytes

We have examined the lysis of human erythrocytes using several approaches, all based on the production of activated species of oxygen. Three distinct techniques have been used: (1) photosensitization using irradiation at 254 nm (hemoglobin as sensitizer) or at 365 nm using added riboflavin (or FMN) as photosensitizer; (2) treatment with dihydroxyfumaric acid (which, during autoxidation, gives rise to O_2^- and other free radicals) or with sodium perborate; (3) irradiation with γ rays. The first two techniques may be considered photochemical and chemical mimics, or analogues, of high-energy irradiation with respect to the production of active oxygen species.

1. Photosensitized Lysis

Peroxidation of unsaturated lipids in the erythrocyte membrane is generally accepted as a major mechanism in the events leading to red cell lysis. Since, on the average, each normal human erythrocyte contains 260,000 molecules of Cu-SOD (erythrocuprein), 300,000 molecules of catalase, and 290×10^6 molecules of hemoglobin, hypotheses invoking the production of O_2^- by autoxidation of oxyhemoglobin or other sources, and the protective action of the above-mentioned enzymes to deal with this O_2^- and the H_2O_2 formed by dismutation, would seem entirely reasonable. Moreover, it could be supposed that O_2^- itself is the primary species that, by attacking the membrane, causes membrane lysis. We have shown that this concept is highly improbable (Michelson and Durosay, 1977) and that other forms of activated oxygen, such as HO· and excited singlet oxygen ($^1O_2^*$), are much more probable lytic factors.

Irradiation at 254 nm of the hemoglobin in erythrocytes may be expected to give rise by diverse photochemical actions to various free radicals and activated oxygen species, such as O_2^-, HO·, or $^1O_2^*$. An important "oxygen effect" is indeed observed during 254 nm irradiation-induced lysis of erythrocytes, which is about three times slower under anaerobic conditions. Thus, at some stage, activated oxygen species are involved, though this does not exclude alternative slower free radical mechanisms. No protection is observed after adding exogenous SOD or after a 50% increase in the level of internal erythrocyte SOD (use of erythrocytes from trisomy-21 patients) and, conversely, no stimulation of lysis occurs when the SOD contained in the erythrocyte is totally inhibited by KCN. Thus, O_2^- can be eliminated as a participant. That the excited singlet oxygen does not play a role is shown by photohemolysis at 254 nm in D_2O rather than H_2O. The tenfold increase in the lifetime of $^1O_2^*$ in D_2O (Merkel and Kearns, 1973) should lead to accelerated lysis; in fact, there is a measurable protective action.

Since treatment of erythrocytes with 10^{-2} M H_2O_2 does not cause lysis (the H_2O_2 is very rapidly destroyed by the 4 mg of catalase per gram of hemoglobin), it is extremely unlikely that H_2O_2 participates directly in membrane lysis. In confirmation of this, no difference in erythrocyte stability was observed when red cells with a normal level of glutathione peroxidase (a membrane enzyme that is probably the primary protection against H_2O_2-induced damage) and cells from a patient with almost no glutathione peroxidase were compared. However, the addition of exogenous catalase at very low levels (2×10^{-7} M) results in a very significant protection, with a 60% increase in the half-time of lysis. In contrast, lysis is accelerated when sodium carbonate is present in the isotonic medium. Hodgson and Fridovich (1976) have provided evidence that $HO \cdot$ radicals can react with carbonate anions to give carbonate anion radicals:

$$HO \cdot + CO_3^{2-} \rightarrow HO^- + CO_3^-$$

which are strongly oxidant as shown by studies on the chemiluminescence of luminol (Puget and Michelson, 1976a). Since the lifetime of CO_3^- radicals is much longer than that of $HO \cdot$, stimulation of lysis in presence of carbonate would be expected by the above mechanism, as is the case for oxidation of luminol by the xanthine oxidase–O_2–hypoxanthine system.

Thus, it is likely that irradiation of erythrocytes at 254 nm causes hemoglobin-photosensitized production of $HO \cdot$ radicals, which can then attack the membrane directly or via formation of CO_3^- radicals. In support of this, erythrocyte ghosts are much more resistant because hemoglobin is absent (Michelson and Durosay, 1977). Again catalase protects, despite the rapid destruction of H_2O_2 by residual (perhaps membrane bound) catalase in the ghosts. In this case, added SOD somewhat accelerates lysis, perhaps due to inhibition of the loss of $HO \cdot$ via $O_2^- + HO \cdot \rightarrow O_2 + HO^-$.

It may be supposed that external catalase acts as an efficient $HO \cdot$ radical trap (in the absence of other organic material in the medium) and that certain processes involved in membrane lysis occur at the outside surface of the membrane (with respect to erythrocytes). Indeed, this exogenous catalase is destroyed in the system of irradiation at a rate comparable to that of lysis.

An alternate photochemical system, irradiation of erythrocytes in the presence of riboflavin and NADH (as hydrogen donor), which gives rise to large quantities of O_2^-, has also been applied to the study of hemolysis. In this case, once lysis begins (as seen by a small decrease in optical density at 656 nm), the photochemical process can be stopped and the reaction continues in obscurity in an autocatalytic fashion. This suggests that a free radical chain reaction is initiated at various centers in each membrane, which continues until lysis is complete.

No protection is afforded by added SOD, and inhibition of erythrocyte internal SOD by KCN does not accelerate lysis; hence, despite the large production of O_2^- in the system, this radical is not involved. Hemolysis is strongly accelerated in D_2O (more than twice as fast), and hence $^1O_2^*$ produced by riboflavin photosensitization is the effective oxygen species. Carbonate ions also stimulate lysis and this suggests involvement of $HO \cdot$ radicals. Again, external catalase protects, despite the fact that during the time of treatment a maximum of $1.5 \times 10^{-4}\ M$ H_2O_2 could be accumulated by photochemical production, whereas 10^{-2} $M\ H_2O_2$ is without effect. While exogenous SOD is quite stable in this system, catalase is again destroyed (by O_2^-, $HO \cdot$, or $^1O_2^*$) at rates comparable to those of lysis.

Thus, O_2^- does not significantly attack the erythrocyte membrane, but $HO \cdot$ radicals and $^1O_2^*$ (probably produced by direct energy transfer from excited flavin to O_2) are much more efficient and can give rise to an even greater lytic capacity in the presence of carbonate ions by forming CO_3^-, which may well produce oxidative carboxylation of unsaturated membrane lipids.

2. High-Energy Irradiation

Bartosz et al. (1977) have shown that no significant differences exist in the oxygen dependence of radiation-induced (γ irradiation, 10–100 krad) shortening of erythrocyte lysis times between normal human erythrocytes and those from trisomy-21 patients (which contain 50% more SOD), in accord with the results described above with respect to photochemical or photosensitized lysis. Indeed, they also noted a slight, but not statistically significant, acceleration of lysis with the latter cells, as is also seen in photochemical lysis (Michelson and Durosay, 1977). Again, it is concluded that $HO \cdot$ radicals are the main damaging agent rather than O_2^-. The low reactivity of O_2^- with respect to erythrocyte membranes has also been demonstrated in pulse radiolysis studies (Bisby et al., 1975).

These results also confirm the observation by Goscin and Fridovich (1973) that E. coli bacteria containing 20- to 30-fold normal levels of SOD (mainly cytoplasmic) do not show an increased oxygen enhancement ratio, suggesting that intracellular SOD may be of no significance for radioprotection. However, as described earlier (Section VII,B), an increase in the periplasmic enzyme does decrease this ratio.

D. Whole-Animal Irradiation

It has been shown that intravenous administration of SOD (bovine erythrocuprein) protects mice from the lethal effects of ionizing irradiation (Petkau et al., 1975a). The principal defect leading to death in the dose range studied was depression of hematopoiesis. In accord with this, a preirradiation intravenous injection of SOD greatly protected the proliferative capacity of bone marrow stem cells (Petkau et al., 1975b). This protection extends well beyond the lifetime of O_2^- and $HO \cdot$ radicals or other forms of activated oxygen such as $^1O_2^*$, using doses of about 350 rad of 250 kV (15 mA) X rays. Indeed, administration of SOD into mice after irradiation was also very effective (Petkau et al., 1976a). Some penetration of the SOD (^{125}I labeled) into both erythrocytes and bone marrow stem cells occurs (Petkau et al., 1975b, 1976b) and it has been estimated that each erythrocyte receives about 10 molecules of SOD (Petkau et al., 1976b). This is an insignificant increase since each erythrocyte contains about 260,000 molecules of erythrocuprein (Michelson et al., 1977). In addition, although recovery of the erythrocytes was accelerated (twice as many erythrocytes per milliliter at 22 days after irradiation with 675 rad of X rays), the initial phase (decline in erythrocyte count) was not affected (Petkau et al., 1976b), indicating a major effect on the hematopoietic system. It may be noted that excessive doses of SOD diminished the protection afforded rather than reaching a plateau (Petkau et al., 1975b).

We have shown, using low-energy X rays (30 kV) at doses of 1900 rad at the surface, that externally applied SOD protects the skin cells of Swiss mice against such irradiation (P. Jockey and A. M. Michelson, unpublished results, 1977) by a factor of 2 when survival (measured by cloning) of basal layer epidermal cells is followed, compared with controls that receive denatured SOD.

While the postirradiation protective effects of injected SOD may be somewhat surprising at first, a number of possible explanations are possible. As observed by Petkau et al. (1976a), the levels of SOD in lungs, thyroid, thymus, and spleen of irradiated mice are markedly reduced (with a minimum at 3–5 days after irradiation) and hence exogenous SOD could increase the reduced protection against normal cellular metabolically produced O_2^-. In addition, increased production of extracellular O_2^- by activation of the phagocytic capacity of polymorphonuclear neutrophils could be expected to occur as a result of radiation damage, and SOD outside the cells would protect against this, as shown in the clinical studies of inflammation carried out by Menander-Huber and Huber (1977). Finally, it may be expected that radiation would induce relatively slow free radical chain reactions in biological macromolecules and membranes (as well as small molecules) and that certain of these radicals would react with molecular O_2 to produce O_2^-, which then continues the process of destruction.

IX. CHEMILUMINESCENCE OF O_2^-

Luminescence is seen whenever O_2^- is produced under conditions that allow spontaneous dismutation, for example, when Fe^{2+} ions or anaerobic solutions of reduced flavin are injected into phosphate buffer (Michelson, 1973a,b; Hodgson and Fridovich, 1973) or when solutions of O_2^- in aprotic solvents are added to aqueous buffers (Michelson, 1973a). This light emission is strongly reduced if superoxide dismutase is present and is clearly related in some way to the presence of O_2^-. The problem of the exact species giving rise to this emission has been extensively discussed for cases in which oxidation of particular organic components is not involved (Kellog and Fridovich, 1975; Cheson et al., 1976; for reviews, see Henry and Michelson, 1977a,b). In many cases of light emission in biological systems (such as by phagocytosing granulocytes or by human platelets in which O_2^- is postulated), it is likely that secondary excited organic molecules or free radicals are implicated. However, in certain defined cases this is not so, for example, emission by the xanthine oxidase–O_2–substrate system. Earlier studies suggested the formation of excited singlet oxygen as the emitting species, but Hodgson and Fridovich (1976) have shown that the presence of carbonate ions is essential. They proposed that Haber–Weiss-generated HO· radicals oxidize carbonate ions to produce carbonate radicals:

$$HO\cdot + CO_3^{2-} \rightarrow HO^- + CO_3^-$$

the dimerization of which would yield peroxycarbonate:

$$O{=}C{-}O{-}O{-}C{=}O$$
$$\quad\ \ |\qquad\qquad\ |$$
$$\quad\ \ OH\qquad\quad OH$$

The decomposition of peroxycarbonate to CO_2 and H_2O_2 would be the light-emitting step.

We have precisely determined the emission spectra of the xanthine oxidase–O_2–acetaldehyde system in the presence of carbonate ions and shown that λ_{max} occurs at 435 nm (Henry and Michelson, 1977a) and does not correspond to the possible singlet oxygen emissions at 480 or 630 nm. Furthermore, light emission by the xanthine oxidase–O_2–hypoxanthine system in the presence of carbonate is unchanged when H_2O is replaced by D_2O (Puget and Michelson, 1976a), whereas an increase would be expected if $^1O_2^*$ were involved. That O_2^- is directly involved in the production of CO_3^- radicals is shown both by inhibition of the light emission (but not of oxidation of the substrate) by SOD and by the fact that reduction of nitroblue tetrazolium by xanthine oxidase–O_2–hypoxanthine is inhibited 36% in the presence of 0.1 M

carbonate, suggesting a direct reaction of O_2^- (as well as $HO\cdot$) with CO_3^{2-}.

It is of interest that the oxidation of luminol by xanthine oxidase–O_2–hypoxanthine is also strongly stimulated by carbonate ions, with the light emission varying approximately as the square root of carbonate concentration (Puget and Michelson, 1976a). We therefore conclude that luminol can be oxidized (with emission of light) by O_2^-, $HO\cdot$, and CO_3^- with increasing quantum yield in this order of the radicals. The effect of carbonate has also been observed in the stimulation of luminescence of luminol treated with mixtures of perborate–persulfate, H_2O_2–persulfate, Ni^{2+}–tetraglycine–O_2–DHF, or H_2O_2 alone (A. M. Michelson, unpublished results, 1976). We have discussed the role of carbonate in stimulating erythrocyte lysis (Section VII,C,1); it appears that an important carbonate effect can be seen in biological, biochemical, and chemical systems that is probably the general result of creation of carbonate radicals. This increase in efficiency due to formation of CO_3^- radicals (more stable than O_2^- or $HO\cdot$) can be ascribed to the longer lifetime of CO_3^- and, with respect to biological systems, a greater selectivity of attack at specific structures (such as membranes), resulting from the decrease in reactivity of the carbonate radical anion compared with hydroxyl radicals.

Finally, it may be noted that, despite various claims that excited singlet oxygen is a substrate for superoxide dismutase (e.g., Agro *et al.*, 1972), there is no evidence that this is so (Michelson, 1974). We must conclude that the only substrate for SOD is O_2^-.

X. CLINICAL ASPECTS OF SOD LEVELS IN HUMANS

In view of the properties of superoxide radical anions, it is clear that aerobic cells must exercise a strict control over the steady-state level of this species, dangerous not only in itself but also as a precursor of other free radicals, some even more toxic (e.g., $HO\cdot$). Clearly, if too little SOD is present, extensive damage to cellular and subcellular membrane structures could be expected, particularly since amplification by initiation of free radical chain reactions could occur. This could lead to modifications of the "social comportment" of the cell (cf. the effect of O_2^- on calf myoblasts) and ultimately to lysis.

We therefore began a survey of levels of SOD in normal humans and in a wide variety of subjects suffering from various pathological syndromes. In principle, a biochemical explanation for certain types of

degenerative processes in the human exists, but has yet to be associated
with a specific illness or illnesses. Since red cell erythrocuprein is extremely important in view of the production of O_2^- by autoxidation of
hemoglobin, we first considered erythrocyte SOD levels (Lavelle *et al.*,
1974a; Michelson *et al.*, 1977). Other enzymes included in the group of
defense mechanisms against uncontrolled oxidation, such as catalase
and glutathione peroxidase, have also been examined. In the erythrocyte, both catalase and erythrocuprein are cytoplasmic [no SOD activity
is found in erythrocyte ghosts (Michelson *et al.*, 1977)], whereas
glutathione peroxidase is generally considered a loosely membrane-
bound enzyme (Bozzi *et al.*, 1976).

A. Normal Subjects

Levels of SOD and catalase were determined in blood samples from
200 white French rural subjects. The results (Fig. 2) show a very sharp
distribution of SOD values within the population with mean values of
58.1 µg of SOD/ml of blood and 461.4 µg of SOD/g of hemoglobin, with
a standard deviation of less than ±10%. There is thus 1 molecule of

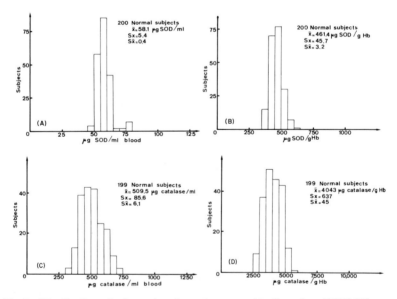

Fig. 2   Distribution of values of erythrocyte superoxide dismutase (SOD) (A) per milliliter of blood and (B) per gram of hemoglobin (Hb); distribution of catalase (C) per
milliliter of blood and (D) per gram of hemoglobin in a normal French rural population
(200 subjects).

erythrocuprein per 1127 molecules of hemoglobin. Similar estimations of catalase show 1 molecule of catalase per 959 molecules of hemoglobin. Since the average normal erythrocyte contains about 290×10^6 molecules of hemoglobin, it can be calculated that about 257,000 molecules of erythrocuprein and 302,000 molecules of catalase are present per erythrocyte.

No significant variations with respect to sex or age were detected for levels of SOD per gram of Hb within the population studied, and values of both catalase and SOD in mothers and newborn babies were strikingly similar. It may be noted that the erythrocuprein content of young and old erythrocytes from a given individual is constant; no variation is observed within the life span of the mature red cell (F. Lavelle, unpublished results, 1977).

Since hemoglobin is perhaps the principal source of O_2^-, the level of SOD was compared with the amount of hemoglobin per 100 ml of blood. An inverse correlation between hemoglobin and SOD was observed. Thus, a group of subjects with an average Hb of 14.4 g/100 ml had 431 μg of SOD/g of Hb, whereas low-Hb subjects (10.8 g/100 ml) had 507 μg of SOD, compared with 461 μg of SOD and 12.7 g of Hb/100 ml for the total normal population. A smaller increase of catalase per gram of Hb with decreasing hemoglobin can also be seen.

B. Hemodialysis

Since hemodialysis reduces the hemoglobin level, it was of interest to study such cases. Blood samples from 24 subjects were examined. The distribution of SOD levels in the population remains remarkably sharp, but there is a significant increase to 540.3 μg/g of Hb, coincident with the low hemoglobin values (mean, 8.01 g of Hb/100 ml).

C. Alcoholic Cirrhosis

Eight cases were examined. The levels of SOD and catalase per gram of Hb were not significantly different from those of a normal population.

D. Vaquez's Disease (Polycythemia Rubra Vera)

Examination of two cases showed a large increase in SOD per milliliter of blood (96.4 and 94.8 μg/ml) corresponding to the large increase in number of erythrocytes. The SOD per gram of Hb also increased (to 634 and 585 μg/g Hb) though, with a normal population, the increase in hemoglobin per milliliter of blood (15.2 and 16.2 g/100 ml in these two cases) is accompanied by a decrease in SOD per gram of Hb. It thus appears that, while the quantity of SOD per erythrocyte is normal, there is an increase in SOD relative to hemoglobin. Catalase levels were essentially normal (4347 and 4581 μg/g of Hb and 661 and 742 μg/ml, corresponding to the increased number of erythrocytes).

Increased fetal hemoglobin (HbF) in polycythemia vera (possibly a clonal disease) has been reported (Hoffman et al., 1977). Although HbF normally decreases to less than 1% of total Hb by the end of the first year of life, the proportion of HbF in an adult case of polycythemia vera is 28–32%. It could well be that HbF produces more O_2^- in the erythrocyte compared with normal Hb, and thus, for survival of the cell, an increase in SOD per gram of Hb is necessary. Since raised levels of fetal Hb have been observed in patients with Fanconi's anemia, aplastic anemia, refractory anemia, and the juvenile form of chronic myelocytic leukemia, it would be of considerable interest to study such cases with respect to the above hypothesis.

E. Trisomy-18

A single case of trisomy-18 was examined. Despite the fact that the gene for erythrocuprein is on chromosome 21 (and that for the mitochondrial Mn-SOD on chromosome 6), and hence a normal erythrocyte SOD level would be expected, the value was excessively high when expressed per gram of Hb (691 μg of SOD); however, due to the low hemoglobin (7.1 g/100 ml) in this case, the SOD per milliliter of blood was quite normal (49.1 μg). Catalase fell within normal limits (323 μg/ml, 4546 μg/g of Hb).

F. Trisomy-21

Since the gene for erythrocuprein is located on chromosome 21 in humans, it was of considerable interest to study trisomy-21 mongoloids. Forty-two subjects were available, but one showed completely normal erythrocyte SOD and catalase levels and was not taken into consideration because it was considered to be a partial trisomy-21 (the karyotype has not as yet been determined), in which the third chromosome lacks that part containing the gene for Cu-SOD. Ages ranged from 6 to 48 years. Hemoglobin levels were normal: $\bar{x}$, 12.55 g of Hb/100 ml; Sx, 1.13; $S\bar{x}$, 0.18.* Erythrocyte catalase levels with respect to both mean values and range and distribution of levels in the population

---

* $\bar{x}$ = mean; Sx = standard deviation; $S\bar{x}$ = standard error of the mean.

were completely normal, thus providing an internal control. (The ratio of mean values of catalase in trisomy-21 to normal was 0.95.) However, the erythrocuprein levels were completely displaced to higher values, with ratios of the mean values trisomy-21/normal = 1.47 for SOD both per milliliter and per gram of hemoglobin, in accord with a gene–dosage relationship (Fig. 3). In view of other cases of high SOD levels, this result cannot be taken as proof of such a relationship and considerable caution must be exercised in interpreting the results due to the control and regulation of what is in fact a very complex system involved in oxygen metabolism.

At least some of the pathological symptoms may be due to the high levels of SOD, since in cases of partial trisomy-21, in which the segment of chromosome containing the gene for erythrocuprein is absent, many of the morphological and mental aberrations are considerably reduced (Sinet *et al.*, 1976).

Similar studies with blood platelets showed that in these cells, the cytoplasmic Cu-SOD also increased about 50% in trisomy-21 (T-21) patients compared with normal subjects (Sinet *et al.*, 1975a). However, the Mn-SOD (at least partially mitochondrial) *diminished* to 67% of the normal value. Regulation of this enzyme may occur as a result of the in-

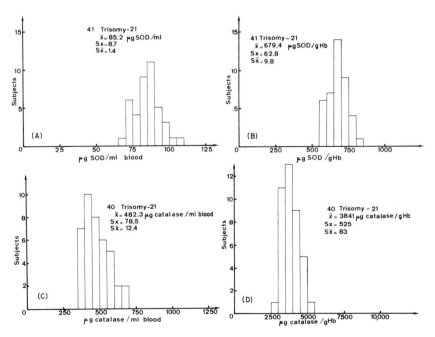

Fig. 3 (A–D) Same as for Fig. 2 for 41 (n = 40 for C and D) cases of trisomy-21.

creased cytoplasmic Cu-SOD, even though such a decrease could lead to increased injury of the mitochondria due to a local higher level of O_2^-. Mitochondria from T-21 platelets would be expected to show increased fragility and perhaps morphological abnormalities.

Glutathione peroxidase activity in erythrocytes from T-21 patients also increased about 50%, compared with normal values (Sinet et al., 1975b). The gene location for this enzyme is considered to be on chromosome 3. The regulation of glutathione peroxidase may be related to SOD levels or H_2O_2 via the hexose monophosphate shunt.

The variation in glutathione peroxidase levels in various cases of hepatic illnesses generally leading to severe hemolysis is currently under investigation in collaboration with Dr. Najman, Hôpital St. Antoine, Paris. No significant relationships between very low levels of glutathione peroxidase and values for SOD or catalase have been noted. Nevertheless, in all cases where the patient showed marked improvement, the glutathione peroxidase level increased and SOD decreased. This suggests that normally an inverse relationship exists between SOD and the peroxidase scavenger of H_2O_2 (in contrast with the parallel increase found in trisomy-21). Similarly, in human diabetics, we found that a low level of glutathione peroxidase is associated with a high level of SOD in erythrocytes, and it appears that some internal feedback system operates between these two enzymes via the metabolism of glucose.

G. Psychiatric Cases

Since in cases of trisomy-21 high SOD levels are associated with mental debility, it was of interest to examine various mental disorders. Again, as in all the previous cases, subjects were rural whites (Normandy), and all suffered severe mental illness. A significant increase in SOD was noted (27% increase of the mean value compared with the mean of a normal population) together with a smaller increase (18%) in catalase.

The highest value of SOD (714 μg/g of Hb) was found in a case of paranoid psychosis, which also revealed a very high catalase level (5872 μg/g of Hb). Schizophrenics (six cases) showed levels of 504–682 μg of SOD/g of Hb, whereas various other psychoses (e.g., manic depressive, chronic hallucinations, Korsakoff syndrome, paranoic delirium, and mental debility) gave values in the range of 541–672 μg of SOD/g of Hb and 4158–5799 μg of catalase/g of Hb. The age range was 23–84 years. No sex-linked differences were detected and hemoglobin levels were normal ($\bar{x}$, 12.54 g of Hb/100 ml; Sx, 0.85; S$\bar{x}$, 0.15).

It thus appears that various mental diseases are related to increased SOD levels, suggesting a possible novel therapeutic approach using inhibitors of SOD. We are presently engaged, in collaboration with Drs. J. Dausset, P. Debray-Ritzen, Q. Debray, and B. Golse, in a detailed study of the superoxide dismutases (both Cu-SOD and Mn-SOD) in platelets from normal subjects, from children with infantile psychoses, and from adult schizophrenics. We have observed that three major Mn-SOD activities (pIs, 7.30, 7.55, and 7.75) are present in human platelets. A fourth band (pI, 7.80) can also be detected. While certain of these SOD's (which together make up about 20% of the total SOD activity) are certainly mitochondrial, others may well be cytoplasmic, in addition to the 80% Cu-SOD activity. The multiplicity of bands of activity after electrofocusing of human platelet Mn-SOD's may be real or may be an artifact due to spontaneous degradation of the original *in vivo* SOD. Instability of the SOD's from protozoa (*Tetrahymena* and paramecia) has also been observed, and formation of multiple protein bands, both active and inactive, occurs after storage of a bacterial Cu-SOD. The possible instability of human platelet Mn-SOD *in vivo* may reflect an *in vivo* situation and could have considerable consequences for the metabolic turnover of superoxide dismutase and the control and regulation of the cellular levels of the differently located SOD's. The *in vivo* rate of turnover of SOD has not been investigated.*

Studies of the distribution of SOD in rat brain (Thomas *et al.*, 1976) have shown that all brain regions contain significant amounts, with highest activities in the brain stem, hypothalamus, and midbrain and lowest activities in the cortex. With respect to subcellular components, synaptosomes show high activity, whereas the myelin fraction has very little activity. This suggests a role for SOD in neuronal function since nerve ends would be particularly susceptible with respect to O_2^- produced by autoxidation of catecholamines.

H. Age and Senility

An estimation of erythrocyte SOD and catalase levels in 23 elderly individuals (70–99 years) in excellent health gave results indistinguishable from those of the normal rural population examined as a control. A somewhat different aspect was seen when 34 elderly individuals (54–94 years, 7 males and 34 females) in very bad health, requiring clinical care and generally in pathological condition, were examined (Fig. 4). The

---

* However, see the report by Yamanaka and Deamer (1974) on the loss of Mn-SOD in SV40-transformed human fibroblasts and in tumor cells.

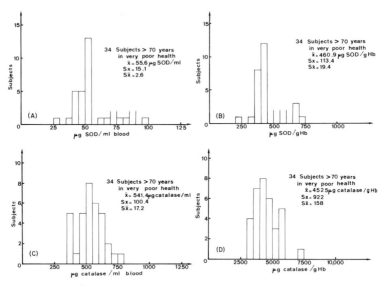

Fig. 4 (A–D) Same as for Fig. 2 for 34 elderly individuals in very poor health.

hemoglobin levels were normal, as were the catalase values, but the distribution of erythrocuprein levels was much more dispersed. Although mean values were normal, standard deviations were more than 2.5 times larger than those of the normal rural Normandy population. Indeed, there are indications of a bimodal distribution of low and high values of SOD about the normal mean. The various pathological disorders included arterial hypertension, coronary insufficiency, myocardial necrosis, arteriosclerosis, arthrosis, respiratory insufficiency, osteoporosis, senile dementia, and coxathrosis. At the highest levels, coronary insufficiency showed values of 615 and 696 μg of SOD/g of Hb, while in a case of hyperlipidic arterial hypertension both SOD and catalase levels were elevated (702 μg of SOD and 5400 μg of catalase/g of Hb). Two cases of hemiplegia showed very low SOD levels—232 and 330 μg/g of Hb—whereas catalase was normal. The first subject, a 75-year-old female, also suffering from hyperlipidemia, was the only individual with an SOD value as low as 50% of the normal mean (461.4 μg/g of Hb), which is also reflected in the value of SOD per milliliter of blood (25.5 μg compared with 58.1 μg). It may well be that while high values of SOD can be supported by the organism to some extent, excessively low values are lethal and lead to accelerated aging and death. No examination of the very rare cases of progeria (with respect to SOD) has been reported; it would also be of interest to study infants who die within the first week after birth when no clear medical explanation is available.

The results described above with respect to erythrocytes may not be applicable to other cells in general. Thus, the activity levels of cytoplasmic SOD in the tissues of aging mice and rats showed only a small reduction in the heart and no reduction in the brain (Reiss and Gershon, 1976a), whereas livers of old rats show a marked reduction of about 60% (Reiss and Gershon, 1976b) compared with young animals. However, using immunochemical methods, it could be demonstrated that a considerable decline in catalytic activity per antigenic unit occurs with increasing age in all three organs in both mice and rats (Reiss and Gershon, 1976a). Thus, brain and heart cells in old organisms synthesize more enzyme but of a defective, lower activity to maintain a fairly constant level of SOD activity.

I. Ethnic Differences

Forty Francophone Parisian Negroes were examined. Both SOD and catalase levels were extremely high, particularly when compared with a white rural population with about the same average hemoglobin per milliliter of blood, since hemoglobin values were generally high. Thus, erythrocyte SOD in these Negroes was about 55% higher than that of the rural white population, whereas the catalase was some 33% higher, indicating a very strong ethnic difference (Table II).

J. Urban–Rural Differences

A series of 20 white, normal Parisian subjects was examined. This group showed much higher hemoglobin levels than the rural population (mean, 15.57 g/100 ml compared with 12.67 g/100 ml), and, in comparison with white rural subjects with the highest hemoglobin levels, erythrocuprein was significantly increased by 38% (per gram of Hb), whereas catalase gave a mean value of 22% (per gram of Hb), greater than that of the rural population (Table II). These results suggest profound differences in oxygen metabolism between rural and city populations and need to be developed further.

K. Toxicity of O_2^- and Benefits of SOD; Toxicity of SOD and Benefits of O_2^-

It is evident that the levels of different SOD activities in a given cell from a particular organism provide a single parameter that in itself is meaningless; this parameter requires integration into a complete, normally balanced system of defenses against activated species of oxygen,

TABLE II

Ethnic and Urban–Rural Differences in Hemoglobin, Superoxide Dismutase (SOD), and Catalase

	Normal French rural subjects ($n = 39$) with low hemoglobin	Tuareg subjects ($n = 7$)	Normal French rural subjects ($n = 46$) with low hemoglobin	Parisian Negroes ($n = 40$)	White Parisians ($n = 20$)
Hemoglobin					
$\bar{x}$[a]	10.83	10.46	14.37	14.26	15.57
Sx	0.58	1.86	0.88	1.85	1.40
$S\bar{x}$	0.09	0.70	0.13	0.29	0.31
SOD ($\mu g/ml$)					
$\bar{x}$	54.9	52.3	62.0	95.5	92.3
Sx	—	6.6	—	13.8	9.8
$S\bar{x}$	—	2.5	—	2.2	2.2
SOD ($\mu g/g$ of Hb)					
$\bar{x}$	506.6	506.7	431.2	672.5	593.6
Sx	45.6	67.8	44.6	81.9	43.1
$S\bar{x}$	7.3	25.6	6.6	13.0	9.6
Catalase ($\mu g/ml$)					
$\bar{x}$	460.7	379.6	554.7	735.2	760.0 ($n = 11$)
Sx	—	70.3	—	130.2	169.9
$S\bar{x}$	—	26.6	—	20.6	51.2
Catalase ($\mu g/g$ of Hb)					
$\bar{x}$	4254	3645	3860	5172	4726
Sx	572	388	654	730	902
$S\bar{x}$	92	147	97	115	272

[a] $\bar{x}$ = mean; Sx = standard deviation; $S\bar{x}$ = standard error of the mean.

which include catalase, glutathione peroxidase, and perhaps other enzymes. Nevertheless, in general terms, if too little SOD were present in the cell, the steady-state level of O_2^- (and associated species such as $HO\cdot$ and $^1O_2$) would rise above normal; in the human this could give rise to increased nonenzymatic oxidation of various neuromediators, leading to mental aberrations. In addition, damage to membranes of both cells and subcellular structures such as mitochondria, particularly in the liver, could cause cell lysis and various hepatic problems.

No significant change in SOD content was found between normal erythrocytes and those from patients with β-thalassemia major and β-thalassemia intermedia (Concetti et al., 1976). These cells contain free α-chains that can give rise to an increased level of O_2^- (Brunori et al., 1975) as well as large quantities of iron (also an intermediate in the production of O_2^- or of $HO\cdot$ radicals by reaction with H_2O_2). Normal SOD levels may thus represent an insufficiency in these cases, leading to increased membrane lipid oxidation and perhaps lysis due to an increased steady-state level of activated oxygen species in this form of congenital hemolytic anaemia. It may be noted that thalassemic erythrocytes often contain considerable quantities of fetal hemoglobin.

We have not found that longevity is associated with low SOD levels, and indeed very low values of SOD may well be associated with senility symptoms and more rapid aging processes due to a diminished protection against accumulated damage by O_2^- either directly or indirectly (via formation of other free radicals). At the molecular level this could be due to modification of DNA repair enzymes and ligases, protein–protein cross-linking and covalent linkage of lipids or nucleic acids to proteins, and direct modification of DNA. Apart from direct oxidative or reductive degradation of biological molecules by O_2^-, this radical can (directly or indirectly) induce mutations. Indeed, the low level of mutation in humans (40 per day, corresponding to 10^6 nucleotide base changes in 70 years, which is 0.34 mm of nucleic acid modified in 10^{10} km of DNA) may well be due to the protective action of SOD. In addition, we have shown (Cooperman et al., 1977) that an O_2^--producing system is extremely efficient in inducing covalent cross-linking of IF-3 with the 30 S particle of the ribosome. The lack of reversibility of this cross-linking affects protein synthesis in general and will lead to diminished cell viability. O_2^- is toxic for the cell at many different levels of components and structures: small molecules, macromolecules, the genetic and translational apparatus, and membranes in general. The action could be amplified by initiation of free radical chain reactions. Levels of SOD must therefore be strictly controlled in order to avoid various pathological lesions due to excessive quantities of O_2^-, and in this sense we may regard SOD as "good" and O_2^- as "bad." However, this control by SOD must be exercised without seriously inhibiting the essential processes involving superoxide radicals as intermediates, such as in various systems of enzymatic oxidation (though, in general, O_2^- as a metabolic intermediate of molecular oxygen is enzyme bound and not susceptible to attack by SOD).

Thus, if there is an excess of SOD in the cell, this could inhibit certain enzymatic oxidations in which O_2^- is a necessary intermediate, perhaps blocking the normal biosynthesis and balance of neuromediators, resulting in mental debility and other aberrations. Furthermore, the substrate for indolamine dioxygenase appears to be O_2^- (Hirata and Hayaishi, 1975) and one of the roles of this enzyme is to degrade serotonin, tryptophan, and dimethyltryptamine (a hallucinogen normally present at low levels as a metabolic intermediate in the brain). For the bad effects of excessively high SOD levels, other processes can also be invoked. It may be noted that in cases of trisomy-21 (50% increase in SOD) there is a frequent occurrence of premature senility (Burger and Vogel, 1973) and leukemia and reduced lifetimes of cultured fibroblasts (Segal and McCoy, 1974; Schneider and Epstein, 1972). This is in agreement with the results presented above concerning elderly individuals in very poor health with high SOD levels. An explanation may lie in the following considerations.

The role of O_2^- as a radical trap for $HO \cdot$ (a much more reactive and dangerous activated form of oxygen) may be important via the reaction $O_2^- + HO \cdot \rightarrow HO^- + O_2$. Excessive levels of SOD could thus increase the concentration of $HO \cdot$ in the cell, and indeed in several systems studied, for example, lysis of erythrocytes, a slight acceleration rather than inhibition was observed when increased SOD was present (Michelson and Durosay, 1977; Bartosz et al., 1977). Oxidation of lysosomal membrane lipids by the xanthine oxidase system (leading to lysis) is stimulated by SOD rather than inhibited (Fong et al., 1973), although an alternate explanation is possible. In addition, an increase in the amount of intravenously injected SOD nullifies the radioprotective effect seen at low concentrations (Petkau et al., 1975b). While the Haber–Weiss reaction (see p. 309) is frequently postulated as the source of biological $HO \cdot$ radicals, other mechanisms, such as the reaction of Fe^{2+} (or oxyhemoglobin, reduced cytochrome c, and other heme proteins) or flavin radicals with H_2O_2, cannot be excluded, for example:

$$Fe^{2+} + H_2O_2 \longrightarrow HO \cdot + HO^- + Fe^{3+}$$
$$FMNH \cdot + H_2O_2 \rightarrow HO \cdot + H_2O + FMN$$

Such reactions can of course be initiated by O_2^- either through dismutation to H_2O_2 or by creation of both H_2O_2 and the intermediate free radical:

$$H^+ + O_2^- + RH_2 \rightarrow RH \cdot + H_2O_2$$
$$RH \cdot + H_2O_2 \longrightarrow R + H_2O + HO \cdot$$
$$H^+ + O_2^- + RH_2 \rightarrow R + H_2O + HO \cdot$$

Another source of $HOO \cdot$ could well be the *reduction* of ferriproteins (e.g., oxidized cytochrome c) or even of the various SOD's by H_2O_2.

In certain cases, only catalase will inhibit, whereas in other cases both SOD and catalase will be effective, as in the postulated Haber–Weiss reaction. The existence of dual inhibition by both enzymes is thus not absolute proof of the Haber–Weiss reaction, but of a rather more generalized concept of the production of $HO \cdot$ radicals. As mentioned above, excessive SOD could, under certain circumstances, increase the amount of $HO \cdot$ available. The same is true of catalase (which also sometimes stimulates instead of inhibiting) by interference with another mechanism of destruction of $HO \cdot$ radicals, in which less reactive O_2^- is produced:

$$H_2O_2 + HO \cdot \rightarrow H_2O + H^+ + O_2^-$$

$$k = 4.5 \times 10^7 \ M^{-1} \ sec^{-1}$$

Finally, the strong stimulation of oxidation of dianisidine by photo-induced riboflavin radicals ($RbH \cdot$) in the presence of SOD (I. Fridovich, personal communication) is a direct example in which O_2^- is useful as a free radical trap for the dianisidine radical, as we have postulated above for hydroxyl radicals. If the generalization is true that SOD can stimulate free radical oxidations that would otherwise be blocked by O_2^-, then excessive amounts of SOD could in fact lead to accelerated aging processes (and cellular damage in general). Thus, we may now consider O_2^- as "good" and SOD as "bad."

It is clear that all assumptions with respect to the direct effects of different levels of such enzymes as SOD, catalase, and glutathione peroxidase in biological systems are necessarily oversimplified at present. The ambiguity of the situation with special reference to SOD is nevertheless seen to be resolved, at least in man, by a rigorous system of control of enzymatic activity leading to the results shown in Fig. 2. Thus, with respect to mental illness, SOD levels in excess of normal levels may be interpreted as a cellular abnormality, as is diminished protection against O_2^-.

L. Clinical Applications of SOD

The pioneering work of Wolfgang Huber and his wife Kerstin Menander-Huber in the use of bovine erythrocuprein as a pharmacological drug is fully described in "Superoxide and Superoxide Dismutases" (Menander-Huber and Huber, 1977). Extensive trials have shown that bovine SOD is completely nontoxic in many animals when administered by any number of different methods. Despite the protein nature of the product, no immunochemical side effects have been observed with Cu-SOD (or with Fe-SOD from *P. leiognathi* when injected into mice, rats, hamsters, or guinea pigs). This is surprising because antibodies have been prepared to a wide range of SOD's with different metal ions (human, bovine, bacterial, fungal, and fish) by injection into rabbits. However, the antigenic properties of SOD in these cases are always increased by using Freund's adjuvant, and it appears that SOD alone is a very poor antigen.

Bovine SOD (and probably all others) appears to be an effective long-lasting anti-inflammatory drug that can be used instead of or in combination with the usual anti-inflammatory drugs, such as aspirin and the corticosteroids (which often have severe side effects), and is unique with respect to the lack of tissue toxicity and other adverse reactions. That exogenous extracellular SOD can be useful (Menander-Huber and Huber, 1977) in reducing the effects of rheumatoid arthritis, degenerative joint diseases such as osteoarthritis, and inflammation as a sequel to radiation therapy of pelvic tumors is readily explained in terms of phagocytotic activity. The release of extracellular O_2^- by activated granulocytes could cause extensive damage to adjacent cells and connective tissue, as well as neutrophil death, since plasma and intracellular liquids are essentially devoid of SOD.

Since very little administered SOD passes to the interior of eukaryotic cells, we are engaged in the preparation and study of liposomes containing ^{14}C-labeled SOD as well as aliphatic long-chain alkylated superoxide dismutases, since such modifications could lead to increased cell penetrability. These and similar approaches will undoubtedly increase the pharmaceutic possibilities of SOD, as will the use of specific antibodies to human Mn- and Cu-SOD's.

M. Techniques of Estimation

In the course of our work, we have incidently developed a number of refined techniques for the estimation of various enzymes and substrates. These all depend on the measurement of light emissions, with all the concurrent advantages, and arise essentially from our earlier studies in bioluminescence.

1. Superoxide Dismutase

Apart from the widely used methods of inhibition of the reduction of cytochrome c or nitroblue tetrazolium by steady-state sources of O_2^-, developed by Fridovich and his collaborators, or inhibition of the autoxidation of pyrogallol (Puget and Michelson, 1974b), the inhibition of steady-state light emission from the oxidation of luminol by xanthine oxidase–O_2–hypoxanthine is an extremely sensitive technique (Hodgson and Fridovich, 1973). Under suitable conditions, reliable estimations can be made of superoxide dismutase activity corresponding to 0.1 to 0.2 ng/ml of erythrocuprein (3 to 6 fmol/ml). Since oxidation of luminol by H_2O_2 produces some 10,000 times less light than does O_2^- under the same conditions, contaminating H_2O_2 does not interfere (Puget and Michelson, 1974b).

2. Catalase

Here oxidation of luminol by H_2O_2 catalyzed by peroxidase is used as the basis. The sample to be estimated is incubated with 2.1×10^{-5} M H_2O_2 and, after 20 minutes at 25°C, residual H_2O_2 is measured by light emission (I_{max} and total light). Calibration curves give the amount of catalase, which can be reliably estimated to about 1 ng/ml of enzyme, that is, about 4 fmol (Michelson et al., 1977). An aliquot of 0.02 μl of blood is sufficient to measure erythrocyte catalase using this approach.

3. Peroxidase

Since oxidation of luminol by H_2O_2 catalyzed by peroxidase is a function of the amount of enzyme, the same system (with a constant excess of H_2O_2) has been used to measure peroxidase activities (Puget et al., 1977a) down to about 10 ng/ml. However, the sensitivity of this technique can be greatly increased by using Pholas luciferin (a glycoprotein) as substrate for the peroxidase (Henry et al., 1973). With this modification it is easy to measure down to about 0.25 fmol of enzyme/ml; under suitable conditions, this lower limit of estimation can be pushed to 1.25×10^{-19} mol/ml, that is, about 75,000 molecules/ml. These techniques have been used in conjunction with peroxidase-coupled antibodies to determine the number of antigenic sites on the surface of mouse lymphocytes (Puget et al., 1977a).

4. Glucose Oxidase

Use of peroxidase-catalyzed oxidation of luminol allows the determination of H_2O_2 down to less than 10^{-10} mol/ml. Since glucose is oxidized by glucose oxidase, producing H_2O_2, a coupled system can be used for the estimation of glucose to about 0.2 nmol. Alternatively, using an excess of glucose, glucose oxidase can be measured by estimating the amount of H_2O_2 produced after 20 minutes of incubation at 25°C. The estimation is linear to about 10 fmol/ml of enzyme (Puget and Michelson, 1976b). We have also used this approach with glucose-labeled antibodies and mouse lymphocytes.

It is clear that essentially the same techniques could be applied to the estimation not only of a variety of substrates that produce H_2O_2 on oxidation, but also of a number of other oxidases that yield H_2O_2 in the presence of a specific substrate, the latter ensuring specificity of the oxidase estimated, even in presence of other oxidases.

N. Conclusions

The above survey of SOD levels, although limited with respect to the numbers of subjects examined, and hence necessarily to be regarded with a certain caution, indicates that relatively high levels of erythrocyte SOD can be associated with various pathological states. Low levels of SOD appear to be even more harmful and excessively low values have been encountered rarely. In normal individuals, the extremely strict control of SOD, with a very narrow distribution of values within a population, and the lack of variation of SOD per gram of hemoglobin among species indicate that the regulation of synthesis of this enzyme in mammalian cells is worth study.

Superoxide dismutases could play an important role in the inhibition of certain aging processes and somatic mutations; we have begun work on the possible effects of SOD on the longevity of eukaryotic cells in culture. It is also possible that diminished protection against O_2^- by a low level of SOD ultimately results in cataract formation (which could be inhibited by external application of the enzyme), since a plausible explanation for cataract formation is cross-linking of proteins of the lens induced by O_2^- or secondary radicals arising from this species. In addition, new types of precise treatments can be envisaged, for example, use of inhibitors of SOD or of specific anti-Mn- or anti-Cu-SOD antibody preparations for trisomy-21 or psychiatric cases; injections of superoxide dismutase when levels are too low; protection of normal cells against high-energy irradiations in the treatment of tumors by externally applied SOD; use of O_2^--producing systems for local application with respect to a differential toxicity for normal and transformed cells, particularly in cases of herpesvirus. Preliminary results indicate that applying O_2^--producing systems or injecting SOD increases survival time when model animal cancer systems (murine sarcoma virus, transplanted melanoma tumors) are studied (A. M. Michelson and S. Cherman, unpublished results; A. M. Michelson, K. Puget, and F. Lacour, unpublished results, 1976). In this respect, it should be noted that catalase activity is absent in Ehrlich ascites tumor cells (Bozzi *et al.*, 1976), although SOD and glutathione peroxidase are present in the cytoplasm. Mitochondria from ascites and Morris hepatoma are devoid of SOD (Dionisi *et al.*, 1975) and this is also true in SV40-transformed cells (Yamanaka and Deamer, 1974) using WI-38 strain fibroblasts derived from human embryonic lung tissue. In this case, despite the loss of Mn-SOD, total cellular SOD activity in SV40-transformed cells was increased compared to WI-38 cells. This suggests that tumor mitochondria, in general, do not contain SOD and that an overcompensation of cytoplasmic Cu-SOD occurs.

It is clear that we are at the beginning of utilizing the medical aspects of O_2^- and of superoxide dismutases, and much work yet remains to be developed both in fundamental exploration and in pragmatic practical applications.

XI. GENERAL CONCLUSION

It is clear from the preceding sections that a new area with vast and far-ranging possibilities has been opened by the initial publication of McCord and Fridovich (1969). These possibilities not only touch the exploration of oxygen metabolism in general, but include medical applications and perhaps industrial utilization in various directions. At the moment, no horizon can be clearly defined; the major work remains to be effected and will undoubtedly provide a number of surprises. Given the intensity of the activity of a large number of groups throughout the world, particularly in the United States and Japan, future developments can be expected at an unusually rapid rate. Rarely has so much stimulation been inspired by the association of an enzymatic activity to a long-known, well-defined, and, until now, uninteresting protein.

ACKNOWLEDGMENTS

I would like to thank my principal collaborators, K. Puget, P. Durosay and F. Lavelle, and others in the department who have so largely contributed to these studies and have

352   A. M. MICHELSON

shown so much patience and perseverence. Various collaborations with medical departments in France, particularly the Centre Regional de Transfusion Sanguine et de Génétique Humaine, Bois-Guillaume (Director, Professor C. Ropartz), are gratefully acknowledged. The work was supported by grants from various French Government agencies and by the Fondation pour la Recherche Médicale Française.

REFERENCES

Agro, A. F., Giovagnoli, C., de Sole, P., Calabrese, L., Rotilio, G., and Mondovi, B. (1972). *FEBS Lett.* **21,** 183–185.
Bachner, R. L., Murrman, S. K., Davis, J., and Johnston, R. B. (1975). *J. Clin. Invest.* **56,** 571–576.
Ballou, D., Palmer, G., and Massey, V. (1969). *Biochem. Biophys. Res. Commun.* **36,** 898–904.
Bartosz, G., Leyko, W., Kedziora, J., and Jeske, J. (1977). *Int. J. Radiat. Biol.* **31,** 197–200.
Beauchamp, C., and Fridovich, I. (1970). *J. Biol. Chem.* **245,** 4641–4646.
Behar, D., Czapski, G., Raboni, J., Dortman, L. M., and Schwarz, H. A. (1970). *J. Phys. Chem.* **74,** 3209–3213.
Bisby, R. H., Cundall, R. B., and Wardman, P. (1975). *Biochim. Biophys. Acta* **389,** 137–144.
Bozzi, A., Mavelli, I., Finazzi-Agro, A., Strom, R., Wolf, A. M., Mondoví, B., and Rotilio, G. (1976). *Mol. Cell. Biochem.* **10,** 11–16.
Brunori, M., Falcioni, G., Fioretti, E., Giardina, B., and Rotilio, G. (1975). *Eur. J. Biochem.* **53,** 99–104.
Burger, P. C., and Vogel, F. S. (1973). *Am. J. Pathol.* **73,** 457–568.
Cheson, B. D., Christensen, R. L., Sperling, R., Kohler, B. E., and Babior, B. M. (1976). *J. Clin. Invest.* **58,** 789–796.
Cohen, G. (1977). *In* "Superoxide and Superoxide Dismutases" (A. M. Michelson, J. M. McCord, and I. Fridovich, eds.), pp. 317–321. Academic Press, New York.
Concetti, A., Massei, P., Rotilio, G., Brunori, M., and Rachmilewitz, E. A. (1976). *J. Lab. Clin. Med.* **87,** 1057–1064.
Cooperman, B. S., Dondon, J., Finelli, J., Grunberg-Manago, M., and Michelson, A. M. (1977). *FEBS Lett.* **76,** 59–63.
Crapo, J. D. (1977). *In* "Superoxide and Superoxide Dismutases" (A. M. Michelson, J. M. McCord, and I. Fridovich, eds.), pp. 231–238. Academic Press, New York.
Curnutte, J. T., and Babor, B. M. (1974). *J. Clin. Invest.* **53,** 1662–1672.
Curnutte, J. T., Kipnes, R. S., and Babior, B. M. (1975). *N. Engl. J. Med.* **293,** 628–632.
Dionisi, O., Galeotti, T., Terranova, T., and Azzi, A. (1975). *Biochim. Biophys. Acta* **403,** 292–300.
Fenn, W. O., Gerschman, R., Gilbert, D. L., Terwiliger, D. E., and Cothran, F. W. (1957). *Proc. Natl. Acad. Sci. U.S.A.* **43,** 1027–1032.
Fernandez-Sousa, J. M., and Michelson, A. M. (1976). *Biochem. Biophys. Res. Commun.* **73,** 217–223.
Fong, K. L., McCoy, P. F., Foyer, J. L., Keele, B. B., and Misra, H. (1973). *J. Biol. Chem.* **248,** 7792–7797.
Fridovich, I. (1972). *Acc. Chem. Res.* **5,** 321–326.
Fridovich, I. (1974). *Adv. Enzymol.* **41,** 35–97.
Fridovich, I. (1975). *Annu. Rev. Biochem.* **44,** 147–159.
Gifford, G. D. (1968). *Biochem. Biophys. Res. Commun.* **33,** 294–298.

Goldstein, I. M., Roos, D., Kaplan, H. B., and Weissmann, G. (1975). *J. Clin. Invest.* **56,** 1155–1163.
Goscin, S. A., and Fridovich, I. (1972). *Arch. Biochem. Biophys.* **153,** 778–783.
Goscin, S. A., and Fridovich, I. (1973). *Radiat. Res.* **56,** 565–570.
Gottlieb, S. F. (1971). *Annu. Rev. Microbiol.* **25,** 111–152.
Gregory, E. M., and Fridovich, I. (1973a). *J. Bacteriol.* **114,** 543–548.
Gregory, E. M., and Fridovich, I. (1973b). *J. Bacteriol.* **114,** 1193–1197.
Gregory, E. M., Goscin, S. A., and Fridovich, I. (1974). *J. Bacteriol.* **117,** 456–460.
Haber, F., and Weiss, J. (1934). *Proc. Roy. Soc. A* **147,** 332–351.
Hayaishi, O., Hirata, F., Ohnishi, T., Henry, J. P., Rosenthal, I., and Katoh, A. (1977). *J. Biol. Chem.* **252,** 3548–3550.
Henry, J. P., and Michelson, A. M. (1977a). In "Speroxide and Superoxide Dismutases" (A. M. Michelson, J. M. McCord, and I. Fridovich, eds.), pp. 283–290. Academic Press, New York.
Henry, J. P., and Michelson, A. M. (1977b). In "Biochemical and Medical Aspects of Active Oxygen" (O. Hayaishi and K. Asada, eds.), pp. 135–151. Univ. of Tokyo Press, Tokyo.
Henry, J. P., Isambert, M. F., and Michelson, A. M. (1973). *Biochimie* **55,** 83–93.
Hirata, F., and Hayaishi, O. (1975). *J. Biol. Chem.* **250,** 5960–5966.
Hodgson, E. K., and Fridovich, I. (1973). *Photochem. Photobiol.* **18,** 451–455.
Hodgson, E. K., and Fridovich, I. (1976). *Arch. Biochem. Biophys.* **172,** 202–205.
Hoffman, R., Donovan, P., and Cuttner, J. (1977). *Lancet* **3,** 866.
Huber, W., and Saifer, M. G. P. (1977). In "Superoxide and Superoxide Dismutases" (A. M. Michelson, J. M. McCord, and I. Fridovich, eds.), pp. 517–536. Academic Press, New York.
Isambert, M. F., and Michelson, A. M. (1973). *Biochimie* **55,** 619–633.
Johnston, R. B., and Lehmeyer, J. E. (1977). In "Superoxide and Superoxide Dismutases" (A. M. Michelson, J. M. McCord, and I. Fridovich, eds.), pp. 291–305. Academic Press, New York.
Johnston, R. B., Keele, B. B., Misra, H. P., Lehmeyer, L. S., Webb, R. L., Baehner, R. L., and Rajagopalan, K. V. (1975). *J. Clin. Invest.* **55,** 1357–1372.
Keele, B. B., McCord, J. M., and Fridovich, I. (1970). *J. Biol. Chem.* **245,** 6176–6181.
Kellog, E. W., and Fridovich, I. (1975). *J. Biol. Chem.* **250,** 8812–8817.
Lavelle, F., and Michelson, A. M. (1975). *Biochimie* **57,** 375–381.
Lavelle, F., Michelson, A. M., and Dimitrijevic, L. (1973). *Biochem. Biophys. Res. Commun.* **55,** 350–357.
Lavelle, F., Puget, K., and Michelson, A. M. (1974a). *C.R. Hebd. Seances Acad. Sci., Ser. D* **278,** 2695–2698.
Lavelle, F., Durosay, P., and Michelson, A. M. (1974b). *Biochimie* **56,** 451–458.
Lavelle, F., McAdam, M. E., Fielden, E. M., Roberts, P. B., Puget, K., and Michelson, A. M. (1977). *Biochem. J.* **161,** 3–11.
McCord, J. M. (1976). *Adv. Exp. Med. Biol.* **74,** 540–564.
McCord, J. M., and Fridovich, I. (1968). *J. Biol. Chem.* **243,** 5753–5760.
McCord, J. M., and Fridovich, I. (1969). *J. Biol. Chem.* **244,** 6049–6055.
McCord, J. M., and Fridovich, I. (1970). *J. Biol. Chem.* **245,** 1374–1377.
McCord, J. M., and Fridovich, I. (1977). In "Superoxide and Superoxide Dismutases (A. M. Michelson, J. M. McCord, and I. Fridovich, eds.), pp. 1–10. Academic Press, New York.
McCord, J. M., Keele, B. B., and Fridovich, I. (1971). *Proc. Natl. Acad. Sci. U.S.A.* **68,** 1024–1027.

Maral, J., Puget, K., and Michelson, A. M. (1977). **77,** 1525–1535.

Marcus, A. J., Silk, S. T., Safier, L. B., and Ullman, H. L. (1977). *J. Clin. Invest.* **59,** 149–158.

Menander-Huber, K. B., and Huber, W. (1977). *In* "Superoxide and Superoxide Dismutases" (A. M. Michelson, J. M. McCord, and I. Fridovich, eds.), pp. 537–549. Academic Press, New York.

Merkel, P. B., and Kearns, D. R. (1973). *J. Am. Chem. Soc.* **95,** 5886–5891.

Michelson, A. M. (1973a). *Biochimie* **55,** 465–479.

Michelson, A. M. (1973b). *Biochimie* 55, 925–942.

Michelson, A. M. (1974). *FEBS Lett.* **44,** 97–100.

Michelson, A. M. (1977). *In* "Superoxide and Superoxide Dismutases" (A. M. Michelson, J. M. McCord, and I. Fridovich, eds.), pp. 87–106. Academic Press, New York.

Michelson, A. M., and Buckingham, M. E. (1974). *Biochem. Biophys. Res. Commun.* **58,** 1079–1086.

Michelson, A. M., and Durosay, P. (1977). *Photochem. Photobiol.* **25,** 55–63.

Michelson, A. M., Puget, K., Durosay, P., and Bonneau, J. C. (1977). *In* "Superoxide and Superoxide Dismutases" (A. M. Michelson, J. M. McCord, and I. Fridovich, eds.), pp. 467–499. Academic Press, New York.

Misra, H. P., and Fridovich, I. (1972). *J. Biol. Chem.* **247,** 6960–6962.

Nofre, C., Cier, A., and Lefier, A. (1961). *Bull. Soc. Chim. Fr.* pp. 530–535.

Nordenson, I., Beckman, G., and Beckman, L. (1976). *Hereditas* **82,** 125–126.

Oberley, L. W., Lindgren, A. L., Baker, S. A., and Stevens, R. H. (1976). *Radiat. Res.* **68,** 320–328.

Petkau, A., and Chelack, W. S. (1974). *Int. J. Radiat. Biol.* **26,** 421–425.

Petkau, A., and Chelack, W. S. (1976). *Biochim. Biophys. Acta* **433,** 445–456.

Petkau, A., Chelack, W. S., Pleskach, S. D., Meeker, B. E., and Brady, C. M. (1975a). *Biochem. Biophys. Res. Commun.* **65,** 886–893.

Petkau, A., Kelly, K., Chelack, W. S., Pleskach, S. D., Barefoot, C., and Meeker, B. E. (1975b). *Biochem. Biophys. Res. Commun.* **67,** 1167–1174.

Petkau, A., Chelack, W. S., and Pleskach, S. D. (1976a). *Int. J. Radiat. Biol.* **29,** 297–299.

Petkau, A., Kelly, K., Chelack, W. S., and Barefoot, C. (1976b). *Biochem. Biophys. Res. Commun.* **70,** 452–458.

Puget, K., and Michelson, A. M. (1974a). *Biochimie* **56,** 1255–1267.

Puget, K., and Michelson, A. M. (1974b). *Biochem. Biophys. Res. Commun.* **58,** 830–838.

Puget, K., and Michelson, A. M. (1976a). *Photochem. Photobiol.* **24,** 499–501.

Puget, K., and Michelson, A. M. (1976b). *Biochimie* **58,** 757–758.

Puget, K., Michelson, A. M., and Avrameas, S. (1977a). *Anal. Biochem.* **79,** 447–456.

Puget, K., Lavelle, F., and Michelson, A. M. (1977b). *In* "Superoxide and Superoxide Dismutases" (A. M. Michelson, J. M. McCord, and I. Fridovich, eds.), pp. 139–150. Academic Press, New York.

Reiss, U., and Gershon, D. (1976a). *Biochem. Biophys. Res. Commun.* **73,** 255–262.

Reiss, U., and Gershon, D. (1976b). *Eur. J. Biochem.* **63,** 617–623.

Salin, M. L., and McCord, J. M. (1974). *J. Clin. Invest.* **54,** 1005–1009.

Schneider, E. L., and Epstein, C. J. (1972). *Proc. Soc. Exp. Biol. Med.* **141,** 1092–1094.

Segal, D. I., and McCoy, E. E. (1974). *J. Cell. Physiol.* **83,** 85–90.

Sinet, P. M., Lavelle, F., Michelson, A. M., and Jérôme, H. (1975a). *Biochem. Biophys. Res. Commun.* **67,** 904–909.

Sinet, P. M., Michelson, A. M., Bazin, A., Lejeune, J., and Jérôme, H. (1975b). *Biochem. Biophys. Res. Commun.* **67,** 910–915.

Sinet, P. M., Couturier, J., Dutrillaux, B., Poissonnier, M., Raoul, O., Rethoré, M. O., Allard, D., Lejeune, J., and Jérôme, H. (1976). *Exp. Cell Res.* **97,** 47–55.

Thomas, T. N., Priest, D. G., and Zemp, J. W. (1976). *J. Neurochem.* **27**, 309–310.

Vanopdenbosch, B., Crichton, R., and Puget, K. (1977). *In* "Superoxide and Superoxide Dismutases" (A. M. Michelson, J. M. McCord, and I. Fridovich, eds.), pp. 199–205. Academic Press, New York.

Weisiger, R. A., and Fridovich, I. (1973). *J. Biol. Chem.* **248**, 3582–3592.

Weiss, J. (1935). *Naturwissenschaften* **23**, 64–69.

Yamanaka, N., and Deamer, D. (1974). *Physiol. Chem. Phys.* **6**, 95–106.

Yost, F. J., and Fridovich, I. (1974). *Arch. Biochem. Biophys.* **161**, 395–401.

Yost, F. J., and Fridovich, I. (1976). *Arch. Biochem. Biophys.* **175**, 514–519.

15

Recent Developments on the Active-Site Structure and Mechanism of Bovine Copper- and Zinc-Containing Superoxide Dismutase

GIUSEPPE ROTILIO

Institute of Biological Chemistry, University of Camerino, Camerino, Italy, and Consiglio Nazionale delle Ricerche Center for Molecular Biology, Rome, Italy

ADELIO RIGO

Institute of Physical Chemistry, University of Venice, Venice, Italy

AND

LILIA CALABRESE

Institute of Biological Chemistry, University of Rome, and Consiglio Nazionale delle Ricerche Center for Molecular Biology, Rome, Italy

I. CHEMICAL STRUCTURE OF THE METAL BINDING SITES AND GENERAL REACTION MECHANISM OF THE ENZYME

The copper- and zinc-containing superoxide dismutase from bovine red blood cells has been the object of very intensive investigation at the molecular level in the last few years. High-resolution X-ray diffraction analysis (Richardson *et al.*, 1975) has shown that one Cu^{2+} and one Zn^{2+} are bound very close to each other on each of the two identical subunits

357

FRONTIERS IN PHYSICOCHEMICAL BIOLOGY
Copyright © 1978 by Academic Press, Inc.
All rights of reproduction in any form reserved.
ISBN 0-12-566960-7

and are bridged by an imidazolato ion; the ligands to the two metals are predominantly histidine and a coordination position open to the solvent is available for the Cu^{2+} ion (Fig. 1). The picture produced by X-ray crystallographers has thus confirmed previous spectroscopic data that had already suggested the close proximity of the two metal centers (Fee and Gaber, 1972; Calabrese *et al.*, 1972; Fee, 1973; Rotilio *et al.*, 1974) and the presence of a fast exchanging water molecule in the first coordination sphere of the Cu^{2+} ion (Rotilio *et al.*, 1971, 1972a; Fee and Gaber, 1972; Gaber *et al.*, 1972; Terenzi *et al.*, 1974). The catalytic reaction of superoxide dismutase has been shown, by pulse radiolysis (Rotilio *et al.*, 1972b; Klug *et al.*, 1972; Klug-Roth *et al.*, 1973; Fielden *et al.*, 1974), to involve alternate copper reduction and reoxidation:

$$E–Cu^{2+} + O_2^- \rightarrow E–Cu^+ + O_2 \tag{1}$$
$$E–Cu^+ + O_2^- + 2H^+ \rightarrow E–Cu^{2+} + H_2O_2 \tag{2}$$

The rate constants of reduction and reoxidation were measured individually (Fielden *et al.*, 1974) and were found to be the same ($\approx 2 \times 10^9$ M^{-1} sec^{-1}) and to agree closely with the turnover rates measured for oxidized and reduced enzyme, as would be expected for a near diffusion-limited reaction with no slower steps than the reduction and reoxidation. Any advance in the understanding of reactions (1) and (2) requires that new experimental facts fit in with the established chemical structure of the metal sites. In this chapter we present further details of the mechanism of action of the enzyme, which can easily be accommodated to the chemistry and geometry of the metal binding region of the protein.

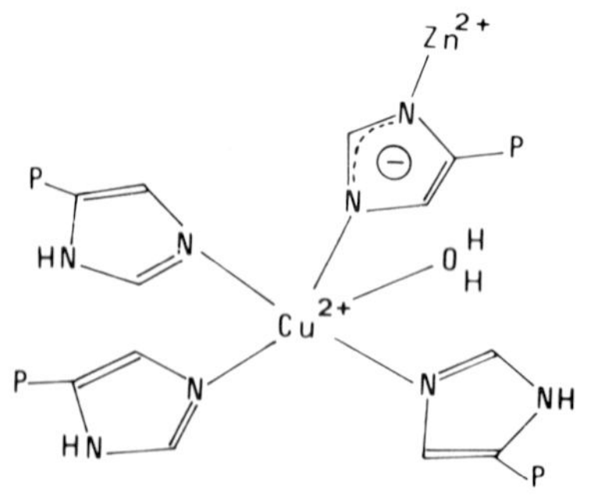

Fig. 1 Schematic picture of the metal binding sites of copper- and zinc-containing superoxide dismutases.

II. ANIONIC INHIBITORS OF THE ENZYME

Anionic inhibitors of the enzyme provide evidence that displacement of a copper-bound water by O_2^- takes place in catalysis. Like any other simple redox reaction involving metal complexes, reaction (1) could be either an outer or an inner sphere electron transfer from O_2^- to Cu^{2+}. Monovalent anions are known from spectroscopic studies (Rotilio et al., 1972a; Fee and Gaber, 1972) to enter the first coordination sphere of the enzyme–Cu^{2+} complex, probably displacing a water molecule. O_2^- could bind to cupric copper of the enzyme because of its anionic character and then donate its electron to the metal ion. Formation of an enzyme–substrate complex in the O_2^-–enzyme reaction is, in fact, supported by kinetic work showing that O_2^- can be produced electrochemically and that the catalytic rate can be measured polarographically (Rigo et al., 1975a). By this method, concentrations of O_2^- as high as 1 mM were obtained and were found to saturate the enzyme, thus allowing the determination of $K_m = 3.6 \times 10^{-4}$ M and $V_{max} = 1 \times 10^6$ sec^{-1}. In previous pulse radiolysis work on the enzyme (Rotilio et al., 1972b), saturation kinetics were not observed due to the lower maximal concentration of superoxide achieved in this type of measurement. Some monovalent anions were tested as inhibitors by the same method. The results are reported in Fig. 2 in the form of classical Lineweaver–Burk plots. It is evident that all the anions tested behave as competitive inhibitors.

The inhibition of the enzyme activity at alkaline pH, which is relieved competitively by increasing the O_2^- concentration, is related to other reversible changes of the enzyme properties occurring between pH 10 and 12. The copper changes its electron paramagnetic resonance (epr) line-shape from rhombic to axial (Rotilio et al., 1971), its effect on the nuclear magnetic relaxation rate of the solvent protons is strongly enhanced (Terenzi et al., 1974), and the enzyme activity decreases (Rigo et al., 1975b). All these phenomena show almost superimposable pH dependence curves and were interpreted in terms of ionization of a group with apparent p$K = 11$–11.5. This group could be the water molecule coordinated to the copper. The data of Fig. 2 and other experiments of cross-inhibition show that OH$^-$ competes with O_2^- and other anions for the copper site, which supports the hypothesis that p$K = 11$–11.5 for the copper-bound water.

The effect of ionic strength on catalytic and inhibition constants was also investigated. Increasing the ionic strength lowered the catalytic constant of the uninhibited enzyme and the inhibition constant of the

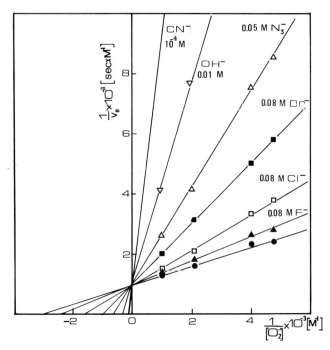

Fig. 2 Double-reciprocal plots of bovine superoxide dismutase in the absence and in the presence of various anions. The initial velocities were determined polarographically in 0.02 M borate buffer, pH 9.8, according to Rigo et al. (1975c). Different O_2^- concentrations were obtained as described by Rigo et al. (1975a). Some experimental points (CN$^-$ and OH$^-$) lie out of this plot. They have been reported on a different scale in Rigo et al. (1975a). Enzyme concentration, 1×10^{-9} M; ionic strength, 0.1 (with NaClO$_4$).

various anions, as expected for complex involving charged species. These results show that reaction of superoxide dismutase with substrate and inhibitory anions may involve the same mechanism, which apparently is the replacement of the water molecule coordinated to the copper with enzyme–substrate and enzyme–inhibitor complexes.

If the various inhibitory anions act by replacing the water molecule coordinated to the Cu^{2+} ion, the same stability constants should be obtained by activity measurements and equilibrium titrations of any spectroscopic parameter of Cu^{2+}. A rough comparison of spectroscopic data (Rotilio et al., 1972a) with activity measurements by pulse radiolysis (Rotilio et al., 1972b) and polarography (Rigo et al., 1975c) seemed to indicate identical constants for CN$^-$ and N$_3^-$. However, the absolute value of the constants was found to vary somewhat in different enzyme batches. The activity and equilibrium measurements were then carried out on the same sample in a parallel set of experiments utilizing polarography as the kinetic method and nuclear magnetic relaxation time of the water protons (T_1) as the parameter for Cu^{2+} equilibrium titration. In fact, this relaxation time increases as the oxidized enzyme is titrated with anions, which apparently replace the copper-bound water molecule (Fee and Gaber, 1972). Table I gives a set of data obtained with five monovalent anions. In all cases, the stability constant as measured by T_1 titration is lower than that calculated from activity data. The components of the polarographic solution have no effect on the T_1 of superoxide dismutase solutions, and therefore this difference has a mechanistic meaning. It can be due to binding of O_2^- and H_2O at different sites in the oxidized protein with different affinities toward the anions. Alternatively, the sites of binding of O_2^- and H_2O could be the same but have different affinities toward anions in the resting and in the working enzyme.

The former hypothesis seems to be ruled out by the results obtained with CN^-. At the pH of the inhibition experiments (pH 9.8), CN^- reacts almost stoichiometrically with the oxidized protein (Rotilio et al., 1972a; see also Fig. 3), which demonstrates that no CN^- is used for reaction with other protein sites. The same should happen with other anions because they behave in the same manner as CN^- with respect to both activity and nuclear relaxation experiments in spite of the much lower values of their stability constants (Table I and Fig. 4). This allows us to postulate that all the tested anions react by the same mechanism. This is particularly important in the case of N_3^-, which has been previously suggested (Fee and Gaber, 1972; Hodgson and Fridovich, 1975) to react

TABLE I

Stability Constants of Superoxide Dismutase Anion
Complexes from Activity and T_1 Measurements[a]

Anion	K (M^{-1})	
	From activity measurements	From water proton relaxation
CN^-	2.5×10^5	1.9×10^4
N_3^-	85	58
Br^-	29	0
Cl^-	12	1.6
F^-	3.2	1
I^-	0	0

[a] Borate buffer, 0.02 M, pH 9.8; ionic strength, 0.1.

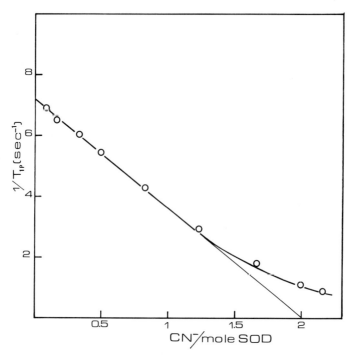

Fig. 3 Titration of the paramagnetic contribution to the nuclear magnetic relaxation time of the water protons (T_{1P}) of a superoxide dismutase (SOD) solution ($9 \times 10^{-4}\ M$) with CN^-. Borate buffer, 0.02 M, $\mu = 0.1$ (with NaClO$_4$), pH 9.8. Temperature, 25°C. The measurements were carried out at 16 MHz according to Terenzi *et al.* (1974).

at a protein site other than that involved in the reaction with CN^-. Our results (Figs. 2 and 4) support an identical mechanism for both anions. As far as the previous conflicting results are concerned, the spectral changes of the copper observed by Fee and Gaber (1972) in the presence of very low concentrations of N_3^-, which did not affect T_1, could never be reproduced. On the other hand, it is likely that Hodgson and Fridovich (1975) saw no inhibition of the enzyme by levels of N_3^- that changed the spectrum of the copper because the assay contained a sufficient amount of O_2^- to displace the N_3^- (Rigo *et al.*, 1975c).

The other hypothesis includes more than one possibility. The increase of the anion binding constants can be due to conformational changes occurring in the working enzyme. The change of valence of the copper occurring in turnover can facilitate anion binding either by increasing the pK of the bound water or because of higher stability constants of the anion complexes of cuprous copper. More than one event may occur at the same time, but all seem to be related, directly or indirectly, to the change of valence of the metal during turnover. Since CN^-, N_3^-, and halide complexes of Cu^+ have higher stability constants than those of Cu^{2+} (Charlot, 1963), it is reasonable to assume that the relative stability of the complexes with the two valence states of the enzyme–copper is directly responsible for the observed difference between the inhibition constants calculated from activity data and those derived from T_1 measurements. If this assumption is true, it is interesting to note that, while for the free Cu^+ the stability constants of halide complexes decrease in the order $I^- \geq Br^- \geq Cl^- \geq F^-$ (Charlot, 1963), in the case of superoxide dismutase Br^-, Cl^-, and F^- have the expected relative inhibitory strength but I^- has no effect. This fact, while supporting the hypothesis of preferential binding of anions to the Cu^+ form of the enzyme, suggests that I^- does not have access to the active site and that the "anion gate" of the copper pocket has dimensions between 2.12 and 1.87 Å, which are the ionic radii of I^- and Br^-, respectively (Cotton and Wilkinson, 1972).

If these conclusions are correct, an anion binding site is present also in

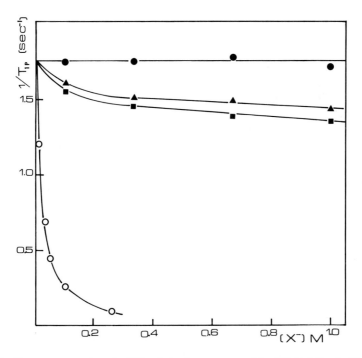

Fig. 4 The same experiment of Fig. 3, in the presence of Br^-(●), F^-(▲), Cl^-(■), and N_3^-(○). Protein concentration, 2×10^{-4} M.

the reduced protein, and this is likely to be a copper-bound water molecule that is displaced by anionic ligands, including O_2^-, during turnover. Results from relaxation of the ^{35}Cl nucleus of Cl^- support this interpretation (Fee and Ward, 1976). It appears that inner sphere electron transfer is compatible with experimental observations in both steps of the enzyme reaction. The estimates of the residence time of the copper-bound water molecule from nuclear magnetic relaxation rate measurements (Gaber et al., 1972; Fee and Ward, 1976) are not contrary to this hypothesis.

III. SPECTRAL STUDIES OF Co²⁺-SUBSTITUTED SUPEROXIDE DISMUTASE

Spectral studies of Co^{2+}-substituted superoxide dismutase suggest that catalysis is accompanied by protonation and deprotonation of the bridging imidazole, which may act as a proton donor in the catalytic formation of peroxide. With respect to reaction (2), once the second O_2^- is bound to Cu^+ according to the inner sphere mechanism suggested above, formation of peroxide requires concomitant electron and proton transfer to O_2^-. The lack of any pH dependence of the enzyme reaction between pH 5 and 10 (Rotilio et al., 1972b) and of a deuterium isotope effect on the catalysis (Hodgson and Fridovich, 1975) may suggest that general acid catalysis by amino acid side chains is not occurring and that protons are donated from water molecules in the region of the active site (Fee and Ward, 1976). In fact, no evidence was available for an activity-linked ionization of an amino acid side chain, which may function as a proton carrier during the catalytic cycle.

We have provided evidence that the bridging imidazolate can be protonated in two different fashions. The Co^{2+}-substituted superoxide dismutase, in which Co^{2+} replaces the zinc ion, is characterized by an optical absorption spectrum in the 500- to 630-nm region, with a maximum at 600 nm, and by lack of an epr spectrum because of magnetic coupling of the Co^{2+} and Cu^{2+} spin systems (Calabrese et al., 1972). If Cu^{2+} is either reduced (Rotilio et al., 1973, 1974) or removed (Calabrese et al., 1976), a Co^{2+} epr signal appears and the maximum of the cobalt optical spectrum moves to shorter wavelengths. The spectral features of Co^{2+} are identical in the Cu^+ and Cu-free protein. This suggests that copper reduction leads to the release of Cu^+ from the binding imidazole, which becomes protonated.

On the other hand, lowering the pH below 5 leads to reversible uncoupling of the Co^2 and Cu^{2+} spin systems of cobalt superoxide dismutase (Calabrese et al., 1975). Under these conditions, the optical spectrum of Co^{2+} is drastically altered, whereas that of Cu^{2+} is practically unchanged. Here, protonation of the cobalt-facing nitrogen of the bridging imidazole is probably occurring. The two types of protonation are shown in Fig. 5.

On the basis of these results, the changes in the optical spectrum of the Co^{2+} chromophore of cobalt–superoxide dismutase were recorded during catalysis in a pulse radiolysis apparatus (McAdam et al., 1977). Under turnover conditions, that is, at high $[O_2^-]/[enzyme]$ ratios, the cobalt spectrum recorded in a time range during which there was neither reduction nor inactivation by H_2O_2 was superimposable on the spectrum obtained by equilibrium reduction of the enzyme Cu^{2+} by H_2O_2 or ferrocyanide to the same extent as it was reduced in turnover [turnover reduction of the enzyme amounts to 25–30% in most preparations (Fielden et al., 1974)]. The next step was to determine if the rates of the spectral changes observed under turnover conditions were the same as the individual rates of reactions (1) and (2). Parallel measurements of absorbance changes were made at 600 nm, where the maximum change of the cobalt spectrum occurs, and at 680 nm, which is the maximum of the Cu^{2+} absorption band, at a low $[O_2^-]/[enzyme]$ ratio, starting from either the oxidized or the reduced enzyme. Under these conditions, initial rates are measured. The results of this experiment demonstrated the following facts.

1. The rates of the optical changes at 600 and 680 nm are the same.
2. The $\Delta A_{601\ nm}/\Delta A_{680\ nm}$ ratio is higher in the Cu,Co–enzyme complex than in the Cu,Zn–enzyme complex; that is, the observed change at 601 nm also involves the Co^{2+} chromophore.
3. The $\Delta A_{601\ nm}/\Delta A_{680\ nm}$ ratio is the same under conditions of individual half reaction rate measurements and turnover reduction.

These results support the conclusion that the cobalt, and by analogy the zinc, environment undergoes chemical changes while copper is reduced and reoxidized catalytically. The chemical structure of the cobalt site, as inferred from spectroscopic evidence, appears to be the same under conditions of equilibrium and catalytic copper reduction. Since the spectrum of the cobalt chromophore in the reduced enzyme is un-

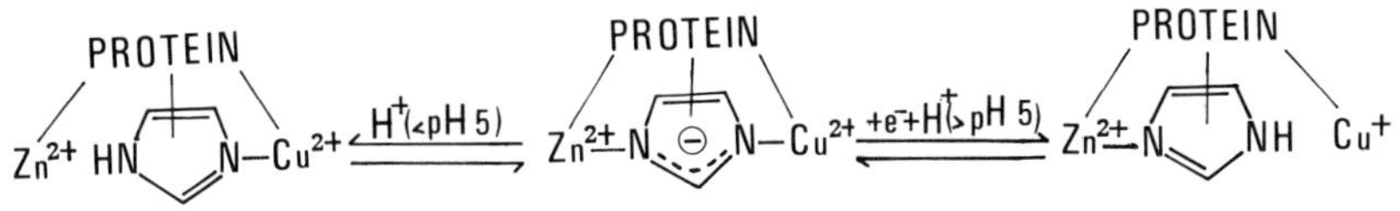

Fig. 5 Protonation of the copper–zinc imidazolate bridge under various conditions.

doubtedly linked to the release of Cu^+ from the intermetal bridge and protonation of Cu-facing nitrogen of the bridging imidazole (see p. 364), protonation and deprotonation of this imidazole group are probably occurring at the same rate of the catalytic copper reduction and reoxidation. Since HO_2^- may be the actual product of the enzymatic dismutation (Hodgson and Fridovich, 1975), the single proton concerned with the catalytic cycle may be donated by the bridging imidazole. In fact, the pK of the second nitrogen of a metal-bound imidazole is high enough (Sundberg and Martin, 1974) to account for the lack of pH dependence of the enzyme in the region pH 5 to 10. Moreover, zinc binding on the other side of the ring could lower this pK to the pH values at which a reversible decrease in superoxide dismutase activity is observed (Rigo et al., 1975b).

IV. CATALYTIC PROTONATION AND DEPROTONATION OF THE BRIDGING IMIDAZOLE

Catalytic protonation and deprotonation of the bridging imidazole facilitate the fast valence changes of the active copper by providing an optimal coordination number to both valence states. Besides its possible

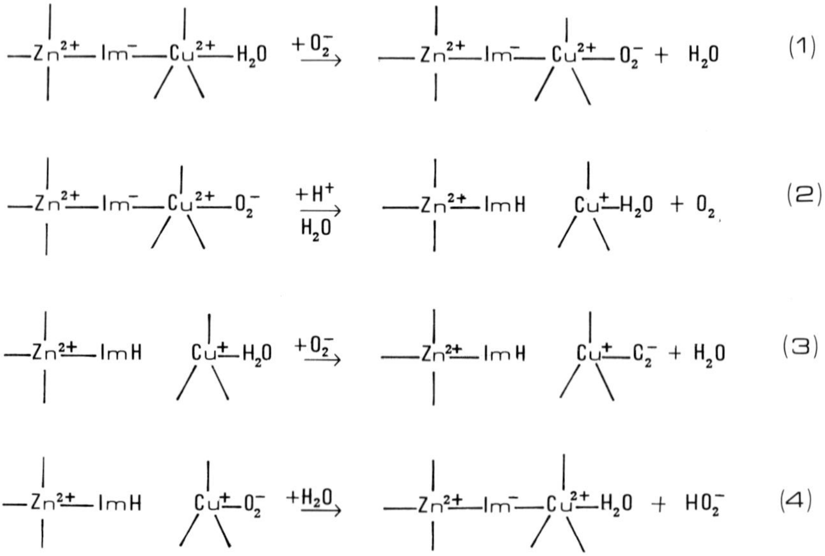

Fig. 6 Proposed mechanism of action of copper- and zinc-containing superoxide dismutases.

role in the proton transfer involved in peroxide formation, it may offer an additional kinetic advantage. Cu^+ likes the tetrahedral coordination, whereas Cu^{2+} is more stable in pentacoordinate complexes. With the mechanism of fast release and rebinding of copper suggested by the above-mentioned pulse radiolysis results, copper–zinc superoxide dismutases can afford a rapid interconversion between pentacoordinate and tetracoordinate structures without any considerable rearrangement of the surrounding nuclei, which would require slow conformational changes. The relevance of the bimetal-bridging structure to this mechanism is further confirmed by the lack of reduction of superoxide dismutase Cu^{2+} at low pH values (Morpurgo *et al.*, 1976), where bridge breaking occurs with protonation of the zinc-facing imidazole nitrogen (Calabrese *et al.*, 1975). On the basis of the results reported above, we propose the mechanism of action for the enzyme shown in Fig. 6.

REFERENCES

Calabrese, L., Rotilio, G., and Mondovì, B. (1972). *Biochim. Biophys. Acta* **263**, 827–829.

Calabrese, L., Cocco, D., Morpurgo, L., Mondovì, B., and Rotilio, G. (1975). *FEBS Lett.* **59**, 29–31.

Calabrese, L., Cocco, D., Morpurgo, L., Mondovì, B. and Rotilio, G. (1976). *Eur. J. Biochem.* **64**, 465–470.

Charlot, G. (1963). "L'analyse qualitative et les réactions en solution," pp. 274–275. Masson, Paris.

Cotton, F. A., and Wilkinson, C. (1972). "Advanced Inorganic Chemistry," p. 458. Wiley (Interscience), New York.

Fee, J. A. (1973). *J. Biol. Chem.* **248**, 4229–4234.

Fee, J. A., and Gaber, B. P. (1972). *J. Biol. Chem.* **247**, 60–65.

Fee, J. A., and Ward, R. L. (1976). *Biochem. Biophys. Res. Commun.* **71**, 427–437.

Fielden, E. M., Roberts, P. B., Bray, R. C., Mautner, G. N., Rotilio, G., and Calabrese, L. (1974). *Biochem. J.* **199**, 49–60.

Gaber, B. P., Brown, R. D., Koenig, S. H., and Fee, J. A. (1972). *Biochim. Biophys. Acta* **271**, 1–5.

Hodgson, E. K., and Fridovich, I. (1975). *Biochemistry* **14**, 5294–5299.

Klug, D., Rabani, J., and Fridovich, I. (1972). *J. Biol. Chem.* **247**, 4839–4842.

Klug-Roth, D., Fridovich, I., and Rabani, J. (1973). *J. Am. Chem. Soc.* **95**, 2786–2790.

McAdam, M. E., Fielden, E. M., Lavelle, F., Calabrese, L., Cocco, D., and Rotilio, G. (1977). *Biochem. J.* **167**, 271–274.

Morpurgo, L., Mavelli, I., Calabrese, L., Finazzi-Agró, A., and Rotilio, G. (1976). *Biochem. Biophys. Res. Commun.* **70**, 607–614.

Richardson, J. S., Thomas, K. A., Rubin, B. H., and Richardson, D. C. (1975). *Proc. Natl. Acad. Sci. U. S. A.* **72**, 1349–1353.

Rigo, A. Viglino, P., and Rotilio, G. (1975a). *Biochem. Biophys. Res. Commun.* **63**, 1013–1018.

368   G. ROTILIO, A. RIGO, AND L. CALABRESE

Rigo, A., Viglino, P., Rotilio, G., and Tomat, R. (1975b). *FEBS Lett.* **50,** 86–88.
Rigo, A., Viglino, P., and Rotilio, G. (1975c). *Anal. Biochem.* **68,** 1–8.
Rotilio, G., Finazzi, Agrò, A., Calabrese, L., Bøssa, F., Guerrieri, P., and Mondovì, B. (1971). *Biochemistry* **10,** 616–620.
Rotilio, G., Morpurgo, L., Giovagnoli, G., Calabrese, L., and Mondovì, B. (1972a). *Biochemistry* **11,** 2187–2192.
Rotilio, G., Bray, R. C., and Fielden, E. M. (1972b). *Biochim. Biophys Acta* **268,** 605–609.
Rotilio, G., Calabrese, L., and Coleman, J. (1973). *J. Biol. Chem.* **248,** 3853–3859.
Rotilio, G., Calabrese, L., Mondovì, B., and Blumberg, W. C. (1974). *J. Biol. Chem.* **249,** 3157–3160.
Sundberg, R. T., and Martin, R. B. (1974). *Chem. Rev.* **74,** 471–517.
Terenzi, M., Rigo, A., Franconi, C., Mondovì, B., Calabrese, L., and Rotilio, G. (1974). *Biochim. Biophys. Acta* **351,** 230–236.

16

Binding of Oxygen to Hemoglobin A in the T-State

QUENTIN H. GIBSON AND CHARLES A. SAWICKI

Department of Biochemistry, Molecular and Cell Biology, Wing Hall, Cornell University, Ithaca, New York

I. MODELS

It is scarcely possible to discuss the hemoglobin–ligand reaction without adopting a model, and, although there is an abundance of these, only one seems worth taking seriously: the two-state model of Monod, Wyman, and Changeux (the MWC model) (Monod *et al.*, 1963). Although Wyman was actively engaged in work with hemoglobin, he scarcely applied the model to it for a curious reason. The model prescribes that the (intrinsic) affinity for successive molecules of ligand shall increase monotonically, and Wyman, who was thoroughly familiar with the hemoglobin literature, knew that what seemed to be the best experiments did not conform to this expectation. Therefore, the model was inappropriate. It was also unfortunate that the MWC model came along at about the same time as the dimer hypothesis (Antonini and Brunori, 1970), according to which the unit of structure and function is an α–β pair of chains, whereas MWC is strictly tetrameric in outlook. It remained for relative newcomers to the hemoglobin field—Hopfield, Shulman, and Ogawa (1971) and Edelstein (1971)—to point out that a large measure of order and sense could be brought to a wide range of kinetic and equilibrium data by applying the MWC model in its simplest form.

FRONTIERS IN PHYSICOCHEMICAL BIOLOGY
Copyright © 1978 by Academic Press, Inc.
All rights of reproduction in any form reserved.
ISBN 0-12-566960-7

In spite of the landmark quality of these papers, the original doubts of Wyman and others had a foundation in experiment that has continued to cause unease, and which surfaces from time to time in reports that some set of observations cannot be reconciled with the MWC model. It is my purpose to examine some of these matters and to report some new experiments that bear upon them.

II. THE VALUE OF L AND PROPERTIES OF R- AND T-STATE HEMOGLOBINS

In the notation of the original authors (Monod et al., 1963), the properties of the system are defined by the equilibrium constant L in the equation $T_0 = R_0 L$, $c =$ the ratio K_T/K_R, and either K_T or K_R, where T_0 and R_0 are the concentrations of the unliganded T- and R-state molecules, and K_T and K_R are the association constants for the binding of ligand to these states. The constant L is, of course, a fundamental quantity, which has proved somewhat elusive, to put it mildly. One might hope to estimate it from the equilibrium curve. As Roughton (1948) pointed out long ago, the ends of a Hill plot at sufficiently high and sufficiently low saturations must approach a slope of unity. Lines drawn with a slope of 1 asymptotic to the ends of the Hill plot will intercept the ordinate zero at the values of log K_R and log K_T, with a separation equal to log c, leaving only L to be determined by curve-fitting. It is implicit that L is of order $>c$ and Lc^4 is of order $<10^{-2}$, which is probably true for mammalian hemoglobins under most conditions. Unhappily, the results of applying this procedure have been inconsistent. For human hemoglobin at pH 7 in phosphate buffer, 20°C, Monod et al. (1963) reported $L = 1 \times 10^4$; the data of Roughton and Lyster (1965) give 1.1×10^6. Hopfield et al. (1971) found 1.5×10^6 in their study of Gibson's (1970) kinetic data, Edelstein (1971) calculated $>3 \times 10^5$ (by another method), and Imai and Yonetani (1975a) have presented data that lead to 7×10^6 for 50 μM hemoglobin. Their estimate would probably exceed 10^7 if the effect of dimers (ca. 20% of the liganded pigment) were taken into account. So wide a range of estimates is disquieting and has contributed to doubts about the validity of the model in some people's minds. In fact, it is a reflection of the interdependence of the constants, which could only be determined uniquely from experimental data more precise than those available at present. It is easy to see that L and c can be traded off against one another to produce any desired $p_{1/2}$ (pO_2 for half-saturation) with only a relatively small change in the shape of the equilibrium curve.

The obvious answer is to seek other methods to supplement the equilibrium determinations. Estimates of K_R are available from kinetic experiments through several routes that agree well with one another, and the values have been stable for 20 years, since Gibson and Roughton (1955) reported a K_R of 8.0 mm^{-1} for sheep hemoglobin type A. This agreed with equilibrium determinations by Roughton et al. (1955) and with indirect comparisons via M (the ratio L_4/K_4) based on carbon monoxide kinetics. For human hemoglobin, the most recent estimates, based on partial flash photolysis of oxyhemoglobin (C. Sawicki and Q. H. Gibson, unpublished) and the O_2 replacement experiments of Olson et al. (1973), give, neglecting any differences between chains, 6.75 mm^{-1} for K_R, which is very close to Imai and Yonetani's (1975a) value at pH 7, 20°C, 0.05 M KP$_i$. This value may be used to stabilize the upper end of the curve.

The situation is quite different for K_T. The past contributions of kinetics have been confusing and often contradictory. We need, of course, data for oxygen binding at low percentage saturation, but because of the time resolution of the stopped-flow method we actually have data rather late in the binding reaction. In fact, the data of Gibson (1970) at 125 μM O_2 begin at about 50% saturation and really establish only that oxygen dissociation from the T-state must be quite rapid as compared with the R-state [500/second (Hopfield et al., 1971) or 2000/second (Gibson, 1970) for T-state, 15/second (Olson et al., 1973) for R-state]. In addition, Gibson (1973), using his oxygen-pulse technique, measured a dissociation rate of 1100/sec at 20°C; he interpreted his results as being due to only one kind of chain. On the other hand, Salhany et al. (1975) could see little difference between the chains in hemoglobin (Hb) Kansas in the presence of inositol hexaphosphate (IHP) in the oxygen-pulse experiment, but also found considerably higher rates than for HbA. The reality of these high rates was confirmed in experiments on the rate of approach to equilibrium in standard mixing experiments, so there is considerable doubt that Hb Kansas with IHP closely resembles HbA in its absence.

If the kinetics of oxygen binding to the T-state was to be used to supplement the equilibrium information, it was clear that some new experiments were needed and that the stopped-flow method was not equal to the task. A series of flash photolysis experiments has now been performed using a dye laser giving 1 J at 540 nm. This was able to produce 97–99% photodissociation in 200 nanoseconds when used with 50 μM oxyhemoglobin, while raising the temperature of the solution by only 0.2°–0.5°C. In these experiments it is essential to work with quite low initial saturations because oxyhemoglobin is in the R-state, and flash photolysis populates the R-deoxy state. Although at pH 7 this relaxes quickly to the T-deoxy state, a disturbing transient would be introduced just after the flash. In addition, with practical concentrations of hemoglobin (50–100 μM heme), up to 20% of liganded dimers would be present, which would remain in the R-state on the time scale of the experiments. For these reasons, observations were restricted to the range 0–15% saturation at equilibrium. Even at these saturations, calculation (Fig. 1) shows that appreciable amounts of liganded dimer would be present, but its effects are much less severe than with fully liganded hemoglobin. Suppose, for example, that at 2% total saturation 0.5% O_2 is bound to dimers; then after flashing 50 μM heme, there will be 0.25 μM as deoxy dimer, but 49.75 μM as deoxy tetramer. It is the reaction of the tetramer that will predominate because there is 200 times more of it, most of which was not bound to oxygen before flashing.

A sample of the results is presented in Fig. 2, which shows three records with different initial saturations. The reaction is strikingly biphasic, with about half showing a time constant of about 1000/ second, and the rest spread out over about 0.5 second, and so not shown on the scale used. It is this slow phase alone that can be observed at 20°C by the stopped-flow method, and which Roughton and I used to call the "drift."

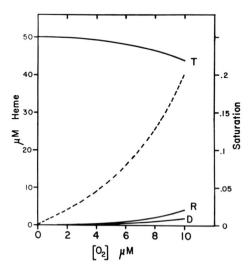

Fig. 1 Calculated distribution of intermediates in the T-state (T), in the R-state (R), as dimers (D), and as a function of saturation (---). Calculations were performed using the two-state model with $L = 1.6 \times 10^7$, $c = 0.0025$, $K_R = 3.6/\mu M$, and $K_{4,2} = 1 \times 10^{-6}$ M.

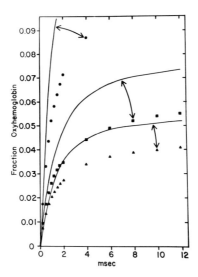

Fig. 2 Time course of binding of oxygen to deoxyhemoglobin at three different oxygen concentrations (▲, ■, ●) after laser photolysis. Hemoglobin concentration, 46 μM; 20°C, pH 7, 0.05 M KP$_i$; 436 nm. The lines were calculated using the parameters of Hopfield *et al.* (1971) ($L = 1.45 \times 10^6$, $c = 0.0033$, $K_R = 2.5/\mu M$, and $k_T = 500$/second). The arrows connect corresponding families of points and lines.

In an attempt to analyze the data, they were first compared with the values generated by the parameters of Hopfield *et al.* (1971), expanding the equations to include an allowance for dimers. The misfit of Fig. 2 is not necessarily surprising because Hopfield *et al.* (1971) did not attempt an optimization procedure, and, as already mentioned, Gibson's (1970) stopped-flow data simply do not include the rapid phase of the reaction.

What was surprising was the information from the next step, in which the data were compared with the MWC model using a standard Davidon–Fletcher–Powell (Bevington, 1969) optimization procedure. The results of Fig. 3 show a typical optimum with systematic misfitting, the data for the highest oxygen lying entirely below the computed curve, whereas the data with lower oxygen lie above the computed curves. Several minima were noted, each corresponding to a good fit to one data set, and convergence was erratic. This behavior is rather characteristic of a situation in which either the data are internally inconsistent or the wrong model is being used. As the experiments are highly reproducible, and the equilibrium curve measured from the amplitude of the excursion agrees well with the lower part of published curves

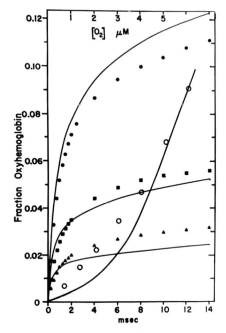

Fig. 3 Data of Fig. 2 fitted by the two-state model of Monod *et al.* (1963) with $L = 5.4 \times 10^6, c = 0.0095, k_R = 3.6/\mu M, k_R' = 59/\mu M \text{ sec}, k_T = 2470/\text{sec}, \text{ and } k_{4,2} = 1 \mu M.$

(Roughton and Lyster, 1965; Imai and Yonetani, 1975a), it seems more likely that the model is incorrect.

Three modifications of the two-state model have been considered. The simplest, and one that has some experimental justification, is that binding to T and binding to R are associated with different absorbance changes. The spectral distribution of these differences would presumably be similar to that seen by Adams and Schuster (1974) on the addition of IHP to oxyhemoglobin and would correspond to a small shift in the position of the bands of the liganded form. It is scarcely possible to make even a guess about the size of effect to be expected, especially in the Soret region: It is probably small and should reverse in sign on passing the 414-nm maximum. At 436 nm it should be quite small, because the gradient of the oxyhemoglobin spectrum is small there also. It would be predicted that the change on binding to T should be less than on binding to R.

This scheme was implemented by collecting the terms for R and T forms separately in the optimization program. The extinction coefficient for binding to R was assigned from the total excursion in absorbance: That for T was allowed to vary. Reasonable, but not striking, fits were obtained: The approach to the optimum was irregular and uncertain. The value of the extinction coefficient for binding to T was 0.75 of that for binding to R, at 436 nm, and the sum of squared residuals was about 30 times that of the best result obtained by another model.

A corollary of this modification of the MWC model is that there should be some wavelength dependence of the results, and indeed there is, though the smallness of the changes involved makes this difficult to establish. The effects observed in direct experiments seemed to correlate closely, however, with what would be expected if differences between R and T *deoxy* were dominant. Thus, the greatest differences were seen between 430 and 442 nm, where the deoxy R–deoxy T difference is opposite in sign absolutely and relative to the ligand binding absorbance excursion. At the same time, results at 414 and 436 nm agreed closely, and these are wavelengths at which the deoxy R to T changes should be small. At low percentage saturations (10% and less), the differences as a function of wavelength are quite minor, but increase rapidly with an increase in liganded R in the equilibrium solutions and with an increase in the proportion of ligand bound to dimers. Experiment, then, does not support the idea that there are considerable differences in absorbance between R-liganded and T-liganded forms.

The next modification tried sought to make it easier to reconcile the equilibria and kinetics by giving up the assumption that the R and T forms are always at equilibrium during oxygen binding. A case for this can be made arising from the nuclear magnetic resonance (nmr) observations of Ogawa and Shulman (1971) of the slow interconversion of the R and T forms of deoxy valence hybrids. The problem here is to decide on values for the new rate constants (five independent constants) that must be assigned. The program used assumed that all tetramers were in the T-state after the flash and that $L > 1/c^2$, which is reasonable at pH 7. There is no need, then, to consider the rate of the T $\rightarrow$ R transition for $Hb_4(O_2)$ or $Hb_4(O_2)_2$, since the R-states are not favored thermodynamically. Further, it is not material what state $Hb_4(O_2)_4$ is in since it is fully liganded in any case: It was assumed to be in the R-state. This leaves $Hb_4(O_2)_3$ as the key compound that might not be at equilibrium. In the program, both $Hb_4(O_2)_2$ and $Hb_4(O_2)_3$ were permitted to bind or release oxygen at the T-state rates or to transfer to the R-state at definite rates. Trials showed that the rate for $Hb_4(O_2)_3$ must be set quite large; indeed, when adjustment was left to the program it tended to float upward and destabilize the integration routine. This program was able to fit the kinetics somewhat better. The running of the program proved to be sensitive to the weight given the equilibrium curve, and it was this curve that contributed most of the residual variance.

The last model tried was reached by considering the errors in Fig. 3, which suggested that a better fit might be obtained if only a part of the hemoglobin participated in the initial rapid reaction. The total hemoglobin was therefore divided into two species given equal weight (and so corresponding formally to the α- and β-chains) and the curve-fitting was repeated, with the results shown in Fig. 4.

This solution has residuals that are almost randomly distributed and, at less than 0.0004 in absorbance, are of the order of experimental error, the fit being more than 100 times better, in terms of variance, than the best result with the scheme used in generating Fig. 3. The most serious difficulty is that admitting chain differences increases to 11 the number of parameters to be determined. Of these 11, 6 were assigned from other experiments as follows. Two R-state dissociation velocities were taken from Olson *et al.* (1973), and the R-state association velocities were set equal because there is at present no reason to do otherwise. The tetramer–dimer dissociation constant was assigned by measuring the proportion of fast and slow reactions observed over a tenfold range of hemoglobin concentrations at constant oxygen concentration following

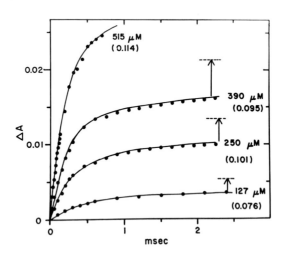

Fig. 4 Reaction of hemoglobin with oxygen after laser photolysis. The values in parentheses give the final saturation: 436 nm; 20°C, pH 7, 0.05 M KP$_i$. Points are experimental; lines are calculated using an extended two-state scheme with $Tk'_\alpha = 2.9/\mu M$ sec, $Tk_\alpha = 183/\text{sec}$, $Tk'_\beta = 11.8/\mu M$ sec, $Tk_\beta = 2480/\text{sec}$, $Rk'_\alpha = 59/\mu M$ sec, $Rk_\alpha = 12/\text{sec}$, $Rk'_\beta = 59/\mu M$ sec, $Rk_\beta = 21/\text{sec}$, $L = 1.43 \times 10^7$, $K_{4,2} = 1 \ \mu M$, $k'_{\text{dimer}} = 30/\mu M$ sec.

full photolysis. The same experiment yielded the combination rate for the dimer: The dissociation velocities for the dimer were assumed to be statistical and equal to the R-state rates, in line with Olson *et al.* (1973).

The remaining five parameters were assigned by the optimization procedure. Repeated runs from different starting points have shown that the solution presented in Fig. 4 is apparently unique: It was the only solution obtained when the program was started from each point on a grid in which all combinations of three times the value found, and one-third the value found, were used for each of the five parameters.

A second difficult problem is to assign standard errors to the parameters. When only some of the constants are varied, the standard computation using the sum of residuals and the diagonal of the inverse error matrix gives the errors on the assumption that the constants to which fixed values have been assigned are precisely known—which is not true. As an approximate expedient, standard errors were assigned to the six fixed constants from the literature, when available, or from experience in repeated experiments. Repeated runs (about 50) were then made in which the five variable parameters were optimized using randomly chosen values for the six fixed parameters, each of which was assumed to be normally distributed about its estimated value. The different values of the five parameters then give an estimate of the variance contributed by the uncertainty in the values of the six fixed parameters. The variance computed in the usual way from the matrix has been added to this to give the final estimates of the rates and standard errors in Table I.

TABLE I

Rates and Standard Errors of the Parameters Varied[a]

Constant		V1[b]	V2[c]	SD[d]
$k'_\alpha T$	2.9	0.04	0.05	±0.30
$k_\alpha T$	183	185	20	±14
$k'_\beta T$	11.8	0.04	0.08	±0.35
$k_\beta T$	2,480	11,700	400	±110
$L (\times 10^{-6})$	14.3	0.04	21	±4.6

[a] Fixed parameters: $k'_\alpha R = 59 \pm 5$; $k'_\beta R = 59 \pm 5$; $k_\alpha R = 12 \pm 1.5$; $k_\beta R = 21 \pm 2$; $k_{4,2} = 1 \pm 0.3$; $k_{dimer} = 30 \pm 3$. All second-order rates are in units $\mu M^{-1}/sec^{-1}$; first-order rates are per second. $\Delta E_\alpha = 0.00475/\mu M$; $\Delta E_\beta = 0.00555/\mu M$.

[b] V1, contribution to variance calculated from error matrix (Berington, 1969).

[c] V2, contribution from fixed parameters.

[d] SD, standard deviation.

It is clear that in this model the two components (α- and β-chains) are quite different from one another kinetically: The dissociation velocity for the fast component is 50 times that for the R-state; for the slow component, 5 times may be an overestimate. Correspondingly, the binding rate for the fast component is $1/5$th of that for the R-state; that for the slow component is $1/25$th. If we accept the model on general principle and identify its components with chains, the question arises, which chain is which? The experiments do not answer this. The difference between the chains in the R-state is not sufficient, and a similarly good fit is obtained whichever way around the chains are distributed. The properties assigned to the two types of chains are scarcely altered in the T-state, but the value of L is about two times larger than in the case shown in Table I. The assignment of rapid behavior to the β-chain is by analogy only: It seems to be generally true that β-chains react faster than α-chains when the relative rates are known.

To continue, the two-state model in any of the variants considered here also explains the extent and rate of the "drift" of Roughton and Gibson in an entirely natural way. The slow phase is made up of a complex series of reactions that are slightly different in the cases of stopped-flow and flash photolysis reactions. In the flash case, preformed dimers recombine with oxygen quite slowly because their concentration and that of oxygen are low. The dimer rate is of the same order as that of binding to the slower chains and contributes to the uncertainty in that rate (Table I). Finally, a slow rearrangement of oxygen within tetramers occurs as cooperativity leads to the population of the high-affinity species $Hb_4(O_2)_3$ and $Hb_4(O_2)_4$. Although the rates of the individual steps in this reaction are high, the overall process is slow because when $Hb_4(O_2)_2$, say, is formed it is much more likely to break down to $Hb_4(O_2)$ than to progress to $Hb_4(O_2)_3$. This offers a close analogy to a replacement reaction such as $HbO_2 + CO \rightleftharpoons HbCO + O_2$ and may be written

$$Hb_4O_2 + O_2 \underset{k_2}{\overset{k_2'}{\rightleftharpoons}} Hb_4(O_2)_2 + O_2 \overset{k_3'}{\rightarrow} Hb_4(O_2)_3 \tag{1}$$

The overall rate has the form

$$k_3'[O_2]k_2'[O_2]/(k_2'[O_2] + k_2) \tag{2}$$

and contains the square of the (small) oxygen concentration divided by a term including the large value k_2.

As it happens, in 1957 I did an experiment at pH 7 by the stopped-flow method that showed that the apparent second-order rate constant for the observed reaction of oxygen with deoxyhemoglobin was pro-

portional to the square of the oxygen concentration. This result, together with the slow rates actually observed, could not be explained,
and Roughton and I turned to the apparently easier case of pH 9 borate
buffer. It is now clear that what is seen by the stopped-flow method is
not directly determined by the reaction of deoxyhemoglobin with oxygen, where equilibrium is reached too quickly to be observed but reflects the rearrangement process. It has taken rather a long time to reach
this conclusion, both because of the very high velocity of dissociation of
oxygen from the T-state and because of continuing uncertainty about
the amounts and properties of hemoglobin dimer present in the solutions. It might have been supposed that in all of the computer simulation of the hemoglobin–oxygen reaction that has gone on someone
might have noticed that a slow drift in saturation was appearing in the
solutions, but most, if not all, of the computation has been related to
experiments—and the experiments have been deliberately truncated so
as to avoid interference from tetramer to dimer dissociation. Whatever
the reason, the increase in fully liganded tetramer continues for a long
time—in the model, at least.

In sum, the work outlined here shows that it is possible to account
rather precisely (to about 0.1% saturation) for the kinetics and equilibria at pH 7, 0.05 M KP$_i$, using rate constants that are appropriate to
describe the replacement of oxygen by carbon monoxide with similar
precision, as well as the binding of oxygen to R-state hemoglobin, so
including all these reactions within the scope of the two-state model as
extended to include chain differences.

Is this then the end of the road? I doubt it. In the kinetic experiments
that can be so satisfactorily reconciled with the equilibrium curve, one
is looking *either* at T-state hemoglobin or at R-state hemoglobin predominantly, but *not* at a situation in which hemoglobin changes state.
It is therefore possible, and reasonably legitimate, to assume that the R-
and T-states are at equilibrium throughout these kinetic experiments.
If, however, there is a change of state, as must happen when
deoxyhemoglobin is fully oxygenated, the assumption of $R–T$ equilibrium is less secure. At present there are only straws in the wind. The
rate $R_0 \rightarrow T_0$ has been measured at 20°C in borate buffer, pH 9, as
6500/second by Sawicki and Gibson (1976), and Ferrone and Hopfield
(1976) have given estimates for $R_3 \rightleftharpoons T_3$ with CO as the ligand of the
order 2000/second. ($Lc^3 = 0.3$ at pH 7 with KP$_i$ buffer.) Under Ferrone
and Hopfield's conditions, Sawicki and Gibson found $R_0 \rightarrow T_0$ was so
rapid as to be difficult to measure and suggested a rate of the order of
5×10^4, in agreement with earlier estimates of Gibson and Parkhurst
(1968). The rate of the $T_0 \rightarrow R_0$ reaction at pH 7 would then be of the

order 5/second, and perhaps 1/second at pH 9. This leaves the rates for $R_1 \rightleftharpoons T_1$ and $R_2 \rightleftharpoons T_2$ entirely unknown except for the clue that $R_1 \rightarrow T_1$ was apparently considerably smaller than $R_0 \rightarrow T_0$ at pH 9 (Sawicki and Gibson, 1976). A very interesting speculation is due to Ogawa and Shulman (1971), who reported slow interconversion of the valence hybrids between the R and T forms. Their observations by nmr were later confirmed by stopped-flow methods that showed that the rate of interconversion of the hybrids could not exceed 0.2/second in Bistris buffers, pH 7, 20°C. Similarly Gibson and Nagel (1974) found that the mutant hemoglobin Chesapeake in Bistris buffers has a rapidly reacting component in CO binding that could not be attributed to dimers, since Chesapeake is only slightly dissociated and since the rapid component was converted to a slowly reacting one within the dead time of the stopped-flow apparatus on the addition of inositol hexaphosphate. It seems, in fact, that when L is near unity, which is of course the condition for seeing two kinetic phases, the rate of interconversion between R and T may be quite slow. If similar relations hold in the case of other buffers, quite striking kinetic effects may be expected, but these have not yet been observed, partly perhaps because the only ligand that can be studied usefully is CO, which unfortunately binds rather slowly. (The behavior of O_2 is complicated by the high T dissociation velocity, and NO binds at the same speed to R and T forms, while the isocyanides combine the disadvantages of CO and O_2.)

III. EFFECTS OF TEMPERATURE ON THE OXYGEN–HEMOGLOBIN REACTION

Preliminary determinations analogous to those summarized in the experiments of Table I have been made at temperatures between 2° and 40°C. The unexpected result indicated at present is that L is strongly influenced by temperature, increasing by more than tenfold as the temperature is lowered from 40° to 2°C. The temperature dependence of the rate constants is similar to that reported for R-state hemoglobin, with large apparent activation energies for dissociation rates and small energies for association reactions (Hopfield et al., 1971).

A whole series of observations may be explained on this basis. It was established by Roughton et al. (1955) that temperature changes the shape of the oxygen equilibrium curve, and an attempt was made to calculate the heat of oxygenation for each step in binding from the Adair constants. It was found that if the curves are made to coincide at 50% saturation, the 19°C curve lies below the 0.2°C curve at low saturation and above it at high saturation. It has never been easy to explain this result, and so it has been systematically ignored by theoretical writers on hemoglobin. There is, however, no reason at all to question the work, and, indeed, Imai and Yonetani (1975b) confirmed the effect. Applying the numerical values from the kinetic experiments, Roughton's result is reproduced qualitatively as a natural consequence of the shift in the occurrence of the T → R transition to higher percentage saturations. The toe of the equilibrium curve is prolonged at the lower temperature so that when the curves are scaled to coincide at 50% saturation the lower parts are above the high-temperature curve.

It follows that calculations of the heats of oxygenation made from the Adair model do not reflect the desired quantity, the heat of reaction of oxygen with the site, but reflect the temperature effect on L: The actual heat of reaction is much the same at each step if the effect of L is removed. The apparent paradox that the effects of temperature on the kinetics—so far as these were known—and on the equilibria did not agree thus appears as a natural consequence of the two-state model.

IV. EFFECT OF pH AND EFFECTORS

There are too few experiments for detailed discussion but, as with temperature, the major effect is upon L, which is increased on lowering pH and adding effector. There is also a considerable effect upon the rates of the two components used in the analysis, and with inositol hexaphosphate, in particular, the rates of dissociation of oxygen are substantially increased, especially from the slower component. This, of course, represents a deviation from the strict construction of the two-state model, but it is a relatively minor departure.

V. CONCLUSIONS AND PROBLEMS

The first conclusion is that the two-state model in its simplest form is not adequate to describe the reactions of oxygen with hemoglobin. This is not a matter of a minor discrepancy, as with the earlier and unsound objections discussed in Section I, but a serious and systematic misfit between experiment and calculation. It does not follow that the two-state model should be discarded—it is more appropriate to try to modify it to take account of the new observations. Of the three modifications attempted, adjustment of the extinction coefficients assigned to the R- and T-states had the least effect on the misfit and has the least justification from independent experiments. To be sure, the extinction coefficients for liganded T and liganded R differ, but not by much, especially at the wavelength most used for data collection, where the liganded form has a small extinction.

The second modification, dropping the assumption of equilibrium between the R- and T-states during oxygen rebinding, is rather more successful, but not entirely so. The difficulty here is that the oxygen equilibrium curve is sufficiently well defined by the experiments and by the assumed properties of the R-state to give quite well-fixed values of L and c and, it may be added, a good fit to the data. The kinetic experiments alone are also quite well represented by the scheme if one assumes slow conversion from $T \rightarrow R$ for $Hb_4(O_2)_1$ and $Hb_4(O_2)_2$; the data do not require that a view be taken about the rate for $Hb_4(O_2)_3$ because the species is simply not sufficiently populated for this to matter. Unfortunately, however, quite a different value for c is required to fit the kinetic data, about half that needed for the equilibrium curve. This would not matter much if there were no data at the bottom of the equilibrium curve—but there are. In view of the importance of the equilibrium results in pointing to an appropriate model, it is very desirable to compare our data with those of others. There are only two sets of suitable data, those of Roughton and Lyster (1965) and of Imai and Yonetani (1975a). If one makes allowance for the likely effect of dimerization, our data agree very closely indeed with Roughton's; they lie to the right (lower affinity) of Imai's. This is quite a reasonable result since Imai and Yonetani took no steps to remove CO from their blood and had up to 4% of methemoglobin in their samples. Both of these factors would act to shift the curve to the left and to shorten the "toe" at its foot, but quantitation is scarcely possible. For the present purpose, if one repeated the kinetic curve-fitting with Roughton and Lyster's equilibrium curve, there would be no difference worth considering: With Imai and Yonetani's curve, the results would be much worse. The conclusion seems to be that the data are not adequately represented by the model—but it is better than the simple MWC model.

The third modification, assuming two T-state species with similar affinities but different kinetics, works excellently and represents the full range of data, yielding unique solutions within experimental error. This does not, however, specify the chemical nature of the two species, nor does it guarantee that the model is correct. Identification with chains is purely speculative, but does have the advantage of prescribing the amounts of the two species and so agrees with chemical data.

Even if, as is almost certain, the results described here prove to be no more than an intermediate step on the road toward the final aim of describing the reaction of hemoglobin with oxygen, it is a matter of no small interest to be able to explain the effects of temperature and of pH on the equilibrium curve, as well as the slow association of oxygen first seen in the stopped-flow apparatus long ago, and this explanation does not depend on the exact form of the model chosen. The same result could be obtained qualitatively from any of the versions of the two-state model discussed here.

Finally, what is the status of the two-state model and of the objections to it? Were Edelstein, Hopfield, and Shulman right in advancing it, or were the objectors right in pointing out that it did not really fit? As often turns out to be the case, both were right on their own terms, and the discussion and controversy that followed have led to a clarification of the problems that might have been long delayed without it. As I see it, the upshot is that the behavior of hemoglobin in its reactions with oxygen can indeed be described by a two-state model, but not that advocated by Hopfield *et al.* (1971). One requires the extension that two distinct entities with different kinetic properties be assumed to be present in equal amounts, and these may be the α- and β-chains.

Such an extension leaves the original form of the MWC model in place as a logical tool and as an aid to thought, but also disqualifies it for use in any but the crudest of qualitative work. In a sense, then, Hopfield and Edelstein may be seen as losing the battle (the MWC model in its simplest form does not work) but winning the war.

Finally, is the model with two types of T-state hemoglobin complicated enough, or will a more complete description be required soon? It is certainly adequate for the present data, which will not bear anything more elaborate. A good many years ago I heard Dr. Gomez of the Rockefeller University give a paper in which he described how he had substituted for the Adair equation a power series of similar form— perhaps a quotient of series would be better—and found that when he included terms in p^3 ($p = pO_2$) he could fit Roughton's best data as well as Roughton could with terms going up to p^4. If Gomez was right, and there is no reason to doubt him, Roughton's attempt to fit four constants was doomed to fail because the information could be fully utilized by three constants. Broadly, this is where we seem to be with oxygen kinetics. There certainly are conformation rate constants, but the data presented so far, which mainly refer to the T-state alone or the R-state alone, can be interpreted without them.

ACKNOWLEDGMENTS

The research presented in this chapter was supported by Grants USPH 14276 and NSF 8309.

REFERENCES

Adams, M. L., and Schuster, T. M. (1974). *Biochem. Biophys. Res. Commun.* **58,** 525–533.
Antonini, E., and Brunori, M. (1970). *Annu. Rev. Biochem.* **39,** 977–1042.
Bevington, C. R. (1969). "Data Reduction and Error Analysis for the Physical Sciences." McGraw-Hill, New York.
Edelstein, S. J. (1971). *Nature (London)* **230,** 224–227.
Ferrone, F. A., and Hopfield, J. J. (1976). *Proc. Natl. Acad. Sci. U.S.A.* **73,** 4497–4501.
Gibson, Q. H. (1970). *J. Biol. Chem.* **245,** 3285–3288.
Gibson, Q. H. (1973). *Proc. Natl. Acad. Sci. U.S.A.* **70,** 1–4.
Gibson, Q. H., and Nagel, R. L. (1974). *J. Biol. Chem.* **249,** 7255–7259.
Gibson, Q. H., and Roughton, F. J. W. (1955). *Proc. R. Soc. London, Ser. B.* **143,** 310–346.
Hopfield, J. J., Shulman, R. G., and Ogawa, S. (1971). *J. Mol. Biol.* **61,** 425–443.
Imai, K., and Yonetani, T. (1975a). *J. Biol. Chem.* **250,** 2227–2231.
Imai, K., and Yonetani, T. (1975b). *J. Biol. Chem.* **250,** 7093–7098.
Monod, J., Wyman, J., and Changeux, J.-P. (1963). *J. Mol. Biol.* **12,** 88–118.
Ogawa, S., and Shulman, R. G. (1971). *Biochem. Biophys. Res. Commun.* **42,** 9–16.
Olson, J. S., Andersen, M. E., and Gibson, Q. H. (1973). *J. Biol. Chem.* **248,** 1623–1630.
Parkhurst, L. J., and Gibson, Q. H. (1968). *J. Biol. Chem.* **243,** 5521–5524.
Roughton, F. J. W. (1948). "Barcroft Memorial Symposium," pp. 67–82, Butterworth, London.
Roughton, F. J. W., and Lyster, R. L. J. (1965). *Hvalradets Skr.* **48,** 185–198.
Roughton, F. J. W., Otis, A. B., and Lyster, R. L. J. (1955). *Proc. R. Soc. London, Ser. B* **144,** 29–54.
Salhany, J. M., Castillo, C. L., McDonald, M. J., and Gibson, Q. H. (1975). *Proc. Natl. Acad. Sci. U.S.A.* **72,** 3998–4002.
Sawicki, C., and Gibson, Q. H. (1976). *J. Biol. Chem.* **251,** 1533–1542.

Part IV

STUDY OF ORGANIZED SYSTEMS

CHAIRMAN: S. MARICIC
Chapters 17, 18, and 19

CHAIRMAN: M. KASHA
Chapters 20 and 21

17

The Structure and Assembly of the Membrane of Semliki Forest Virus

*KAI SIMONS, HENRIK GAROFF, ARI HELENIUS, AND
ANDREW ZIEMIECKI*
European Molecular Biology Laboratory, Heidelberg, Germany

I. SEMLIKI FOREST VIRUS

Biological membranes are supramolecular assemblies of lipids and proteins with a crucial role in most cellular phenomena. The study of the molecular events involved in various membrane functions has been hampered by the complexity of most cellular membranes. Much of the work in the field has therefore been directed to the study of simple membranes that carry out single functions, such as the purple membrane from *Halobacterium* (Henderson and Unwin, 1975), the sarcoplasmic reticulum in muscle cells (Inesi, 1972), the rod outer segment membranes in the retina (Hall *et al.*, 1969), or the postsynaptic membranes in electric organs (Changeux, 1974). These membranes contain only one major protein. The mammalian red blood cell membrane has also been widely studied. Although more complex than the single-function membranes, it has proved to be a useful experimental system for structural studies due to the ease with which it can be isolated and handled (Bretscher, 1973).

There is a third group of biological membranes with useful properties for research at the molecular level. These are the animal, plant, and

FRONTIERS IN PHYSICOCHEMICAL BIOLOGY

bacterial viruses, which contain lipid membranes (Lenard and Compans, 1974; Franklin, 1974). They are simple in structure, easy to purify, and have advantages that will become apparent below for studies of membrane synthesis and assembly.

We have been studying one of the simplest in structure, namely, Semliki Forest virus (SFV) (Simons *et al.*, 1977; Kääriäinen and Renkonen, 1977). SFV belongs to the alphavirus group of the togavirus family (this virus group was formerly known as the arbo A group). Another well-studied member of this virus group is Sindbis virus. In nature, the alphaviruses are spread by blood-sucking arthropods (mainly mosquitoes) and have a wide host range (Pfefferkorn and Shapiro, 1974). They can be grown in avian, mammalian, or insect cell cultures and grow well within a wide range of temperatures (25°–41°C). SFV contains a spherical nucleocapsid (diameter, 40 nm) (von Bonsdorff, 1973) composed of a single-stranded (42 S) RNA molecule (MW about 4500K*) (Strauss and Strauss, 1978) and one lysine-rich protein species, the capsid protein (Simons and Kääriäinen, 1970). The nucleocapsid is covered by a lipid membrane (cross section, 4.5 nm) that is studded by glycoprotein spikes with lengths of 7.5 nm (Harrison *et al.*, 1971).

In infected vertebrate cells the nucleocapsids are assembled in the cytosol. They attach to the cytoplasmic surface of the plasma membrane and bud out of the cell into the extracellular fluid and are thereby enveloped by a modified segment of the host cell plasma membrane (Acheson and Tamm, 1967). In invertebrate cells, virus replication and assembly take place within cytoplasmic vesicles (Gliedman *et al.*, 1975). The vesicles contain ribosomes and complex laminate structures reminiscent of the Golgi apparatus. When the virus is matured, the entire content of the vesicle is emptied into the extracellular fluid.

II. COMPOSITION OF SFV

The SFV membrane contains lipid and protein at a weight ratio of about 2 : 3 (Laine *et al.*, 1973). As expected from the way the viral membrane is formed from the host cell surface, its lipid composition is similar to that found in the host cell plasma membrane (Renkonen *et al.*, 1971). The lipid compositions of viruses grown in different host cells can therefore be quite disimilar (Quigley *et al.*, 1971; Hirschberg and Robbins, 1974). The most dramatic differences have been documented by Renkonen *et al.* (1974), who found that SFV grown in baby hamster

---

* Throughout this chapter K indicates a value of 1000. For example, 4500K = 4,500,000.

kidney (BHK21) cells and in insect (*Aedes albopictus*) cells have only 36% of their phospholipids in common. The virus grown in BHK cells contains about 1.8×10^4 phospholipid and 1.6×10^4 cholesterol molecules (Laine et al., 1973) arranged in a bilayer around the nucleocapsid (Harrison et al., 1971).

In contrast to the lipids, all the proteins in the virus are specified by the viral genome (Pfefferkorn and Clifford, 1964). Four polypeptides have been found (Garoff et al., 1974): one in the nucleocapsid, the capsid protein (MW, 30K), and three in the membrane, E1 (50K), E2 (50K), and E3 (10K). These proteins occur in equimolar amounts in the virus particle. The E proteins are all glycoproteins, E1 containing 7.5% carbohydrate, E2 11.5%, and E3 45.1%. The number of polypeptides per virus particle has not been determined with certainty because the exact molecular weight of SFV is not known. Compositional studies suggest that there are approximately 240 copies of each polypeptide.

The oligosaccharide chain of E1 isolated after Pronase digestion from SFV grown in BHK cells has a molecular weight of 3.4K and consists of N-acetylglucosamine, mannose, galactose, fucose, and sialic acid. E3 also has one carbohydrate side chain of the same type with a molecular weight of about 4K, whereas E2 has one side chain of complex compositional type (MW of about 3.1K) and, in addition, another carbohydrate unit with a molecular weight of about 2K containing only N-acetylglucosamine and mannose (Mattila et al., 1976). Glycosylation of the SFV E proteins is dependent on host cell functions. Differences in size and composition can be induced in the oligosaccharide chain of the alphavirus proteins by growing the virus in different cells (Luukkonen, 1976; Schlesinger et al., 1976; Keegstra et al., 1975).

III. GROWTH CYCLE OF SFV

The growth cycle of SFV in vertebrate cells is summarized in Fig. 1. Virus infection leads to a rapid inhibition of cellular nucleic acid and protein syntheses (Strauss et al., 1969; Mussgay et al., 1970). Within 3 hours after infection the cell is essentially a factory for virus production. Virus release from the cell is detectable about 2 hours after infection, and it increases exponentially for about 3 to 4 hours. After a further period of linear growth for 2 to 3 hours the virus production decreases and the host cell begins to die (Kääriäinen and Renkonen, 1977). In contrast to vertebrate cells, alphavirus infection of invertebrate host cells does not appreciably affect host cell function (Davey et al., 1973).

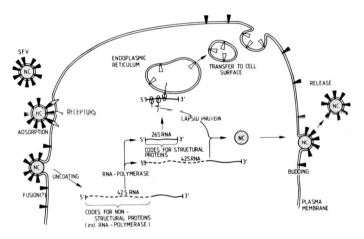

Fig. 1 Growth cycle of Semliki Forest virus in vertebrate cells. NC, nucleocapsid.

Infected insect cells grow and divide normally while producing virus. The cells can thus support a chronic infection.

The molecular events involved in the early interaction of the incoming virus with the host cell are poorly understood (Lonberg-Holm and Philipson, 1974). The spike proteins are needed both for attachment and for penetration. Binding to the host cell surface follows saturation kinetics (Fries, 1978). The chemical nature of the surface receptors to which the spikes attach is not known. If a specific receptor is recognized it is probably widespread in nature because SFV infects most vertebrate and invertebrate cells (in culture) tested. Alternatively, the virus spikes may have more than one specificity and may therefore bind to a range of structures on the cell surface (cf. Cameron and Erlanger, 1977). What follows after virus attachment to the cell is not known, but, by analogy with paramyxo- and myxoviruses, fusion of the virus membrane with the host cell plasma membrane is most likely to occur.

Once inside the cell the nucleocapsid must be uncoated. The 42 S RNA released from the nucleocapsid then serves as messenger for translation of proteins needed for virus RNA synthesis (Lachmi et al., 1975; Clegg et al., 1976; Lachmi and Kääriäinen, 1976; Brzeski and Kennedy, 1977). Two RNA molecules are produced: new 42 S RNA molecules and 26 S RNA molecules (see Strauss and Strauss, 1978). The 26 S RNA represents one-third of the genome and is located inward from the 3' end of the 42 S RNA (Kennedy, 1976; Wengler and Wengler, 1976). The 26 S RNA serves as a messenger RNA for the structural proteins of the virus (Simmons and Strauss, 1974; Cancedda and Schlesinger, 1974; Wengler et al., 1974; Clegg and Kennedy, 1974).

As soon as there is enough capsid protein made, the 42 S RNA molecules are rapidly packed into nucleocapsids by binding to newly synthesized capsid protein (Söderlund, 1973; Scheele and Pfefferkorn, 1969). In this way the 42 S RNA molecules are removed from their messenger function. It is apparently only early in infection that they have the possibility of functioning in protein synthesis. Thus, there appears to be a clear-cut division of labor between the two viral messenger RNA's. The 42 S molecule functions as an (early) messenger for the nonvirus structural proteins necessary for RNA replication, whereas the 26 S RNA acts as messenger for the virus structural proteins throughout the growth cycle. Although the 42 S RNA contains the same information as the 26 S RNA in its 3'-terminal region, initiation of structural protein synthesis on the 42 S RNA is blocked.

By using labeled formylmethionyl-tRNA to initiate protein synthesis *in vitro*, only one formylmethionyl-tRNA–peptide has been found from protease digests of the translation products from 26 S RNA (Clegg and Kennedy, 1975a; Glanville *et al.*, 1976; Cancedda *et al.*, 1975). Thus, there is only one initiation site for protein synthesis on the 26 S RNA. Nevertheless, all four structural polypeptides are read from this RNA. This is accomplished by sequential synthesis of the polypeptides as a polyprotein. A 130K protein that contains all the structural proteins can be found to accumulate in cells infected under nonpermissive conditions with a class of temperature-sensitive mutants of alphaviruses (Schlesinger and Schlesinger, 1973; Lachmi *et al.*, 1975). This protein is not, however, detected in infections with wild-type virus, and it cannot be seen if 26 S RNA is used to make protein *in vitro*. Two major products are then obtained under optimal conditions: the capsid protein (30K) and a 97K protein that contains E1, E2, and E3 (Simmons and Strauss, 1974; Cancedda and Schlesinger, 1974; Clegg and Kennedy, 1975b; Glanville *et al.*, 1976; Garoff *et al.*, 1978).

Other temperature-sensitive mutants of the virus have been found to accumulate two other proteins in the infected cell: one having an apparent molecular weight of about 97K containing E1 and E2, and another with an apparent molecular weight of about 87K containing the nucleocapsid protein and E2 (Schlesinger and Schlesinger, 1973; Lachmi *et al.*, 1975). Whether these proteins contain E3 has not been studied. In wild-type infection and in a number of mutants an additional smaller precursor 62K protein is found containing both E2 and E3 (Schlesinger and Schlesinger, 1973; Simons *et al.*, 1973a). All of the polyprotein or bits of it can thus be found using SFV mutants under nonpermissive conditions.

More information on how the SFV proteins are synthesized has been obtained by treating infected cells with high salt concentrations, which inhibits initiation of protein synthesis but allows elongation to proceed (Saborio et al., 1974). When the cells are restored to isotonicity, synchronous reinitiation of protein synthesis is obtained. Using this method it has been possible to demonstrate that the viral structural proteins are sequentially synthesized (Clegg, 1975; Söderlund, 1976). The synthesis of the SFV proteins starts with the capsid protein followed by E2 and E1. Recent refinements of the method in our laboratory show that E3 is made after the capsid protein but before E2 (Garoff and Söderlund, 1978).

The capsid protein is released from the polysome shortly after its synthesis (Clegg, 1975). The mechanism of release is not known, but most likely the capsid protein is cleaved from the nascent polypeptide chain by a protease. After release, the capsid proteins combine with the 42 S RNA to form the nucleocapsid. The E proteins are inserted into the endoplasmic reticulum to form the virus spikes. How this is done can be studied in vitro (Garoff et al., 1978). If 26 S RNA is translated in vitro, two major proteins are made: the capsid protein and the 97K protein. If vesicles of dog pancreas endoplasmic reticulum are added to the same system, the 97K protein is cleaved to form E1 and p62, the precursor for E2 and E3. If protease is added the capsid protein is degraded, whereas E1 and p62 are not; they are apparently protected against the protease by insertion into the vesicles. The membrane vesicles have to be present during translation for insertion and cleavage to occur. If the vesicles are added after the 97K protein has been finished, neither protection nor processing is seen. From previous work it is known that most of the ribosomes that translate the 26 S RNA are membrane bound in the cell (Wirth et al., 1977; Martire et al., 1977). How the ribosomes become membrane bound is not known, but one possibility is that the release of the capsid protein reveals a signal on the amino terminus of the p97 protein for ribosome binding to the endoplasmic reticulum. The signal on the N terminus of p97 would have the same function as those found on the N termini of secretory proteins necessary for transfer of the nascent polypeptide chains to the cisternal side of the endoplasmic reticulum (Blobel and Dobberstein, 1975). With the development of systems where the insertion of proteins into membranes can be studied in vitro, answers to the mechanisms involved should be forthcoming. Similar studies of the insertion of another virus spike glycoprotein from vesicular stomatitis virus into endoplasmic reticulum have been reported (Katz et al., 1977).

Glycosylation of the E proteins begins during translation (Sefton, 1977). This involves the transfer of 1.8K oligosaccharide units from a lipid intermediate to the growing polypeptide chain. Further addition of sugars takes place as E1 and p62 migrate to the plasma membrane. The route of E1 and p62 to the plasma membrane (Richardson and Vance, 1976) is presumably similar to that taken by the cell surface proteins. How the flow of molecules from the cell interior to the surface is accomplished is still not known.

The virus spike proteins arrive at the plasma membrane within 30 minutes of their synthesis (Scheele and Pfefferkorn, 1969). The proteolytic cleavage of p62 to yield E3 and E2 seems to occur at the cell surface; it can be inhibited by antibodies to the spike proteins added to the infected cells (Bracha and Schlesinger, 1976; Jones *et al.*, 1977; Ziemiecki *et al.*, 1978). After this final cleavage the stage is set for the terminal event in virus morphogenesis, the budding of the nucleocapsid through the modified plasma membrane; however, before we can attempt to explain this process more information on the structure of the virus membrane is needed.

IV. STRUCTURE OF THE VIRUS MEMBRANE GLYCOPROTEINS

A. Subunit Structure

The glycoproteins E1, E2, and E3 form the spikelike projections on the external surface of the SFV membrane. In electron micrographs the surface projections cannot be resolved sufficiently to see single spikes. The morphology is clearly different from that of influenza virus, where two different types of spike structures can easily be distinguished (Laver and Valentine, 1969).

To tackle the problem of the spike subunit structure by biochemical methods, we have used the general approach outlined in Fig. 2 (Ziemiecki and Garoff, 1978). As a first step we solubilized the membrane with Triton X-100, which is a mild nonionic detergent known to be inefficient in breaking protein–protein interactions (Helenius and Simons, 1975). The spike proteins were recovered as 150K complexes containing equimolar amounts of E1, E2, and E3 bound to a micelle of Triton X-100, which accounts for 50K of the complex (Simons *et al.*, 1973b). The properties of these Triton–glycoprotein complexes as determined by centrifugation and gel filtration are consistent with a trimer structure (one copy of E1, E2, and E3 each). However, because of the similarity in molecular weight of E1 and E2, the existence of complexes with similar molecular weights but different polypeptide compositions could not be ruled out. Detergent–protein complexes composed of

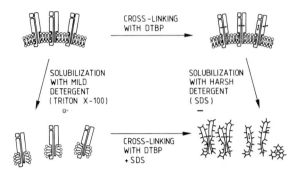

Fig. 2 General scheme for determining the quaternary structure of membrane proteins. DTBP, dithiobispropionimidate; SDS, sodium dodecyl sulfate.

(E1,E3)$_2$ or (E2,E3)$_2$ or other combinations would be indistinguishable from (E1,E2,E3) complexes. Furthermore, it could not be excluded that the solubilization itself caused rearrangements in the spike quaternary structure. This became more apparent when another mild detergent, sodium deoxycholate, was used to solubilize the spikes (Helenius *et al.*, 1976). After solubilization of the SFV membrane with this detergent, E1, E2, and E3 disaggregated from each other, making their isolation possible under nondenaturing conditions.

To circumvent the problem of possible rearrangement and dissociation during solubilization, dithiobispropionimidate, a cleavable cross-linker (see Peters and Richards, 1977), was used to cross-link neighboring polypeptides in the viral membrane. After cross-linking, a harsh dissociating detergent must be used for solubilization to break all noncovalent protein–protein interactions. We solubilized the cross-linked products with sodium dodecyl sulfate (SDS) and isolated them by sucrose density gradient centrifugation in the presence of the detergent.

The major cross-linked species had an apparent molecular weight of 100K, and it was shown to contain equimolar amounts of E1, and of E2 after reversal of the cross-linking. E3 was not cross-linked at all, presumably because it contains only one lysine and a bulky carbohydrate part (Garoff *et al.*, 1974). These results confirmed that in intact virus the spikes occur as 100K complexes in the membrane, but it was still not possible to tell whether the 100K cross-linked complexes were homo- or heterodimers of E1 and E2 due to their similar molecular weights. Immunoprecipitation of the isolated cross-linked 100K complexes with antisera specific against either E1 or E2 precipitated both polypeptides at an equimolar ratio. This can be explained only by assuming that one copy each of E1 and E2 make up the 100K complex.

By a similar approach, the 150K Triton–glycoprotein complexes obtained after Triton X-100 solubilization of the virus were also shown to consist of an E1,E2 heterodimer (Fig. 2). Thus, Triton X-100 solubilization did not lead to any dissociation and rearrangement of the native glycoprotein complex. The fact that E3 could be demonstrated to be associated with the 150K Triton–glycoprotein complex indicates that trimers of (E1,E2,E3) form a single type of spike in the viral membrane (Ziemiecki and Garoff, 1978).

A methodological point worth mentioning is that precipitation of the 150K complexes obtained after solubilization with Triton X-100 with antibodies against E1 or E2 causes dissociation of the trimers (Ziemiecki and Garoff, 1978). Cross-linking is necessary to keep E1 and E2 connected during antibody precipitation.

B. Amphiphilic Properties

1. Protease Treatment

When SFV is treated with the proteolytic enzyme thermolysin all three membrane polypeptides (E1, E2, and E3, including their oligosaccharide side chains) are cleaved off, leaving virus particles without spikes but otherwise intact. Peptide fragments representing about 10% of the membrane proteins are shielded from attack and left in the membrane (Utermann and Simons, 1974). These peptides are hydrophobic. Their polarity index (hydrophilic amino acids/total amino acids) calculated from the amino acid composition is as low as 0.28, compared to 0.46, 0.46, and 0.41 for E1, E2, and E3, respectively. The peptides are soluble in chloroform–methanol. Peptide mapping shows that both E1 and E2 have their own hydrophobic fragments; E3 does not. We have been able to separate the hydrophobic fragments from each other by electrophoresis in a gel gradient in the presence of SDS (Fig. 3). Three major bands were found with apparent molecular weights of 9K, 6K, and 5K (H. Garoff, unpublished results), two of which are derived from E2 (see Section IV,C).

2. Triton X-100 Binding

The capacity of the spike proteins to bind Triton X-100 is another indication of their amphiphilic character (Helenius and Simons, 1975; Tanford and Reynolds, 1976). The SFV spike protein binds 75 molecules of Triton X-100, which corresponds roughly to one micelle of the detergent. The nucleocapsid does not bind Triton X-100 (Helenius and Söderlund, 1973). To study the detergent binding of the individual

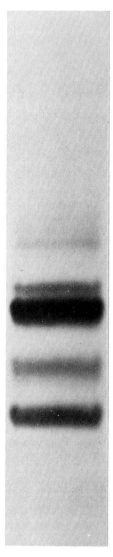

Fig. 3 High-resolution gradient (17 to 22% acrylamide) gel electrophoresis of the hydrophobic fragments of E1 and E2 obtained after thermolysin treatment of SFV. Migration downward. The band at the top is capsid protein.

polypeptides of the spike, we used the charge-shift electrophoresis method developed in our laboratory (Helenius and Simons, 1977). This is a simple and rapid screening method to distinguish between amphiphilic and hydrophilic proteins. The proteins to be studied are subjected to agarose gel electrophoresis in the presence of a nonionic detergent (Triton X-100), a mixture of a nonionic and an anionic detergent (Triton X-100 and sodium deoxycholate), and a mixture of a nonionic and a cationic detergent (Triton X-100 and cetyltrimethylammonium bromide). The electrophoretic mobility of proteins that do not bind detergent is unaffected in the three detergent mixtures. The mobility of an amphiphilic protein shifts anodally in the Triton X-100–deoxycholate system and cathodally in the Triton X-100–cetyltrimethylammonium bromide system compared to the mobility in Triton X-100 alone. E1 and E2 show shifts in mobility and are thus amphiphilic, whereas E3 does not, which suggests that it is hydrophilic (Fig. 4).

Taken together, these results show that the SFV glycoprotein spikes have a hydrophobic moiety containing segments from E1 and E2. This moiety attaches the spikes to the lipid bilayer. E3 is peripheral and is apparently bound to the membrane by binding to E2, to E1, or to both.

3. Protein Micelles

Triton X-100 can be removed from the detergent–spike protein complexes by centrifuging the complexes into detergent-free sucrose

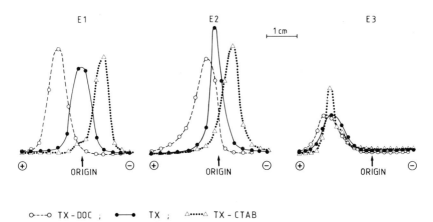

Fig. 4 Charge-shift electrophoresis of the SFV membrane polypeptides isolated as described by Helenius et al. (1976). The proteins had been radioactively labeled by [³⁵S]methionine. For details of the electrophoresis method, see Helenius and Simons (1977). TX-Doc, Triton X-100 and sodium deoxycholate; TX, Triton X-100 only; TX-CTAB, Triton X-100 and cetyltrimethylammonium bromide.

gradients. This results in the formation of monodisperse and water-soluble 900K complexes containing eight copies of E1, E2, and E3 each (eight spikes) practically devoid of Triton X-100 (Helenius and von Bonsdorff, 1976). Apparently the SFV glycoproteins mutually cover up their hydrophobic segments to form micellelike complexes with a sufficiently polar surface to remain soluble. Similar protein micelles have also been obtained from other amphiphilic membrane virus glycoproteins (Laver and Valentine, 1969; Scheid *et al.,* 1972) as well as from cytochrome b_5 (Calabro *et al.,* 1976), thy-1 surface glycoprotein (Küchel *et al.,* 1977), and *Bacillus licheniformis* membrane penicillinase (Simons and Sarvas, 1978). The concentration of protein monomer in equilibrium with the micelle has been determined for cytochrome b_5 (Calabro *et al.,* 1976), but for most other protein micelles the monomer concentration is so low that it is difficult to measure.

If hydrophobic interactions are mainly responsible for the formation of these protein complexes, it should be possible to produce mixed micelles between different amphiphilic proteins. We have been able to make such complexes from SFV spikes and membrane penicillinase. By mixing the protein–Triton X-100 complexes of the two proteins in different ratios and applying the mixtures to sucrose gradients, complexes are obtained containing differing amounts of virus spikes and penicillinase. These mixed micelles vary in their molecular weights between those of the pure SFV spike micelle (900K) and those of the pure penicillinase micelle (660K), depending on the mixture of spike protein and penicillinase used. Laver and Valentine (1969) have shown that mixed micelles can be formed from the two different spike structures present on influenza virus membranes.

Micelle formation appears to be another property typical of amphiphilic membrane proteins having a rather large hydrophilic domain outside the lipid bilayer with a hydrophobic tail sticking into the bilayer (Fig. 5).

C. Orientation of the Spike Glycoprotein in the Membrane

One way to determine whether the amino- or carboxy-terminal end of a surface membrane protein lies on the outside is to attempt to shave off the protein with a protease near the bilayer to release the hydrophilic part fairly intact and determine its end group amino acids (cf. Henning *et al.,* 1976). This has not been possible with SFV because all proteases tested digest the hydrophilic part of E1 and E2 into small fragments. Another way to approach this problem is to label the polypeptide chains with a gradient of radioactivity increasing from one end of the molecule to the other and then locate the hydrophobic fragments within E1 and

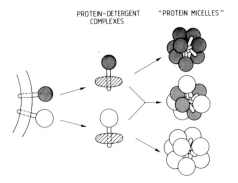

PROTEIN–DETERGENT COMPLEXES "PROTEIN MICELLES"

Fig. 5 Scheme to visualize what may happen when amphiphilic membrane proteins (with a fairly large hydrophilic domain) are solubilized with mild detergents and the detergent is removed gently. The shaded area around the hydrophobic protein domain indicates the detergent micelle.

E2 from their radioactivity. For this purpose, Garoff and Söderlund (1978) have refined the labeling method previously used to determine the order of capsid E1, E2, and E3 protein synthesis. If infected cells are treated with high salt concentrations (0.33 M NaCl) to inhibit initiation of protein synthesis, a population of polypeptide chains can be labeled with [^{35}S]methionine, which contains a gradient of increasing amounts of radioactivity toward the carboxy-terminal end of the protein. The labeled polypeptide chains can be chased with unlabeled methionine into virus particles if the cells are returned to isotonicity to start protein synthesis again. To determine the gradient accurately, another batch of virus is uniformly labeled with [^{3}H]methionine and mixed with the [^{35}S]methionine gradient-labeled virus. The E1 and E2 polypeptides are isolated and split into fragments, of which the ^{35}S/^{3}H ratios are determined. As expected, the ^{35}S/^{3}H ratio increases linearly from the amino-terminal end of E2 to the carboxy-terminal end of E1. When the hydrophobic fragments of E1 and E2 are isolated from the same mixture of [^{35}S]- and [^{3}H]methionine-labeled virus, their location within E1 and E2 can be determined from their ^{35}S/^{3}H ratios (Fig. 6). The 6K peptide has a ^{35}S/^{3}H ratio of 0.79 and must derive from the carboxy-terminal end of E1. Both the 9K and the 5K peptides have almost identical ^{35}S/^{3}H ratios of 0.56 and 0.59, respectively, and are therefore both located in the carboxy-terminal end of E2. The two peptides obtained from E2 are probably overlapping, the 9K peptide presumably resulting from an incomplete proteolytic cleavage.

Thus, both E1 and E2 are attached by their carboxy-terminal ends to the bilayer and have their amino-terminal portion on the outside. This is

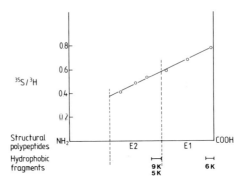

Fig. 6 Location of the hydrophobic fragments 5K, 6K, and 9K within E1 and E2. SFV proteins E1 and E2 were labeled either by [^{35}S]methionine in the infected cell under high salt conditions or by [^{3}H]methionine under uniform labeling conditions. E1 and E2 were isolated and cleaved into fragments by 2-(2-nitrophenylsulfenyl)3-methyl-3-bromo-indolenine (see Fontana, 1972). These fragments were separated by gel electrophoresis as shown in Fig. 3. The ^{35}S/^{3}H ratios of the fragments were determined to construct the graph.

the orientation that has been found for all other surface glycoproteins studied so far (Marchesi *et al.*, 1976; Henning *et al.*, 1976; Skehel and Waterfield, 1975). Only one exception to this rule is known and that is the intestinal aminopeptidase that seems to be oriented the other way around (Maroux and Louvard, 1976)—but this needs confirmation.

D. Membrane Location

The first experiments suggesting that the surface glycoprotein spikes of SFV penetrate the bilayer showed that most of the SFV spikes can be cross-linked to the nucleocapsid when intact virus particles are treated with the cross-linking reagent dimethylsuberimidate (Garoff and Simons, 1974). As dimethylsuberimidate maximally bridges a distance of 11 Å, which is about one-fourth of the lipid bilayer thickness in SFV, such extensive cross-linking is possible only if the viral glycoproteins lie close to the nucleocapsid, that is, span the lipid membrane.

We have also found a nonionic detergent, octylglucoside, that solubilizes the lipid from the virus but leaves most of the spikes attached to the nucleocapsid. Solubilization must be done at low ionic strength; if the ionic strength (50 m*M* NaCl) is increased the spikes fall off (Helenius, 1978).

A third approach is to label the glycoproteins in SFV with the radioactive surface label formyl-[^{35}S]methionyl sulfone methylphosphate

(Garoff and Simons, 1974) from the outside in intact virus or from both sides using membrane "ghost" preparations obtained from the virus by adding enough Triton X-100 to barely lyse the membrane (Helenius and Söderlund, 1973). When membrane ghosts are labeled, the peptide map of the E2 glycoprotein contains one additional basic, heavily labeled peptide that is not seen in the peptide map derived from the E2 glycoproteins labeled only from the outside using intact virus (K. Simons, unpublished results). The E1 peptide maps are identical. This suggests that at least E2 penetrates the lipid bilayer, but does not, however, exclude E1 penetration. A negative result (no additional peptide labeled from the inside) indicates only the absence of a reactive amino group.

A further observation in keeping with spike penetration is the finding that lowering the pH to about 6 causes contraction of the SFV particles with a decrease in diameter of about 100 Å (von Bonsdorff, 1973). This effect is due to a decrease in size of the nucleocapsid (Söderlund et al., 1972). The viral membrane apparently adheres to the nucleocapsid during the contraction and excess membrane is extruded from the contracted virus particle into blebs. Spikes are not seen on the surface of these blebs, suggesting that the blebs are mainly formed by viral lipids and that the viral spike glycoproteins remain bound to the nucleocapsid. This effect is similar to that seen with erythrocyte membranes. Conditions that lead to spectrin aggregation have been shown to cause blebbing of protein-free vesicles from the membrane surface (Elgsäter et al., 1976).

E. Three-Dimensional Structure

The surface organization of the alphavirus spike glycoproteins has been studied using electron micrographs of both negatively stained preparations and freeze-etched preparations (von Bonsdorff and Harrison, 1975). These studies are consistent with a $T = 4$ icosahedral surface lattice. A $T = 4$ structure has not yet been demonstrated directly for the nucleocapsid, but the regularity of its structure argues strongly for an icosahedral structure. Furthermore, chemical analysis yields about 240 capsid proteins per nucleocapsid, the number of subunits required for a $T = 4$ structure, and there is good evidence for an equimolar ratio of capsid proteins to glycoprotein spikes.

Electron micrographs of thin sections of pellets produced by ultracentrifuging SFV show a remarkably regular array of particles (Fig. 7). Analyses by Wiley and Bonsdorff (1978) of such pictures prove that they represent three-dimensional crystals with the space group $F23$, the largest crystals found in the pellets being about 20 μm. The virus appears

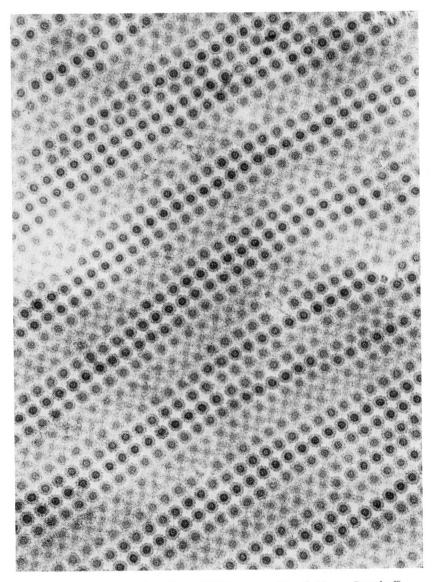

Fig. 7 Thin section of a pellet of SFV (courtesy of Dr. C.-H. von Bonsdorff).

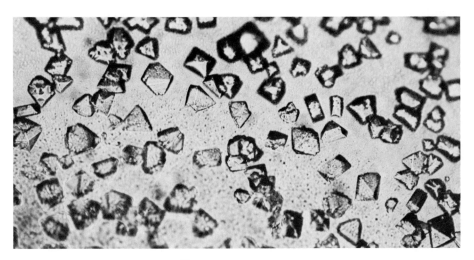

Fig. 8 Crystals of SFV.

to sit on the tetrahedral position in the lattice; that is, it shares the tetrahedral subgroup of its icosahedral symmetry with the crystal lattice. These findings demonstrate the remarkable three-dimensional order possessed by this complex structure. In fact, R. Leberman, in this laboratory, has succeeded in crystallizing SFV by conventional methods, the largest crystals so far being about 0.1 mm (Fig. 8).

V. THE FINAL EVENT IN VIRUS MATURATION: THE BUDDING PROCESS

From the structure of SFV, a plausible mechanism to explain the budding process emerges (Garoff and Simons, 1974).

The most notable feature of the budding is that the nucleocapsid is enveloped by a segment of the plasma membrane that contains only virus spike proteins and from which most proteins are excluded. How this is achieved can easily be visualized from the findings of equimolar stoichiometry of the virus proteins and the transmembrane penetration of the spike proteins (Fig. 9). The spike proteins are inserted into the cell surface before the first nucleocapsids are attached to the plasma membrane (see Strauss and Strauss, 1978). The first glycoproteins arrive at the surface of a chick fibroblast about 2 hours after infection; the concentration of virus proteins at the surface increases during the infection and reaches its maximum about 6 hours after infection. Viruses are then budding out at a maximal rate. The nucleocapsids could attach to

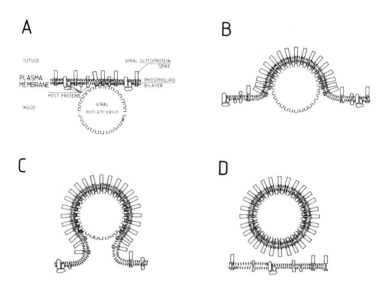

Fig. 9 Four hypothetical stages in the budding of the nucleocapsid through the modified plasma membrane of the infected host cell.

the cytoplasmic ends of the virus spike glycoproteins in the cell surface membrane. Interaction between the nucleocapsid and the virus spikes would thus be mediated by the transmembrane segments of membrane glycoprotein. The spike proteins have been shown to move freely in the host cell membrane (Birdwell and Strauss, 1974), and, every time spike proteins diffuse into the area to which the nucleocapsid is already bound, new spikes could be captured by attaching to vacant binding sites on the capsid proteins of the nucleocapsid (Fig. 9). The association of the viral spikes with the nucleocapsid proceeds until all binding sites have been filled. A mature virus particle buds off perhaps simply due to the strain imposed on the bilayer by the nucleocapsid–viral spike interaction (Fig. 9). Host proteins that have no affinity to the nucleocapsid or the spike glycoproteins may simply be excluded from the budding site for steric reasons.

Local modification of the cell membrane composition resulting from a specific interaction between submembraneous structures and transmembrane proteins may not be restricted to the budding of membrane viruses. Other examples that may be mediated through a similar mechanism are known. In milk fat-globule formation, modified segments of the glandular cell surface membrane envelop the fat globule (Patton and Keenan, 1975). In the vesiculation involved in phagocytotic vesicle formation, the plasma membrane is modified to include only a limited number of cell surface proteins (see Silverstein *et al.*, 1977). Eisen (1978) found that, when the nucleus is expelled from Friend ery-throleukemic cells, spectrin accumulates at the site of the plasma membrane through which the nucleus buds out, and most of the spectrin in the cell is expelled with the nucleus.

The protein–protein interactions involved in such membrane modifications may be due to peripheral proteins binding to integral membrane proteins on either side of the membrane. If antibodies to the surface spikes of Sindbis virus are used to induce cap formation in the infected cells, thin sections reveal a cluster of nucleocapsids underlying the cap in the electron micrographs (Smith and Brown, 1977).

REFERENCES

Acheson, N. H., and Tamm, I. (1967). *Virology* **32**, 128.
Birdwell, C. R., and Strauss, J. H. (1974). *J. Virol.* **14**, 366.
Blobel, G., and Dobberstein, B. (1975). *J. Cell Biol.* **67**, 835–862.
Bracha, M., and Schlesinger, M. J. (1976). *Virology* **74**, 441–449.
Bretscher, M. S. (1973). *Science* **181**, 622–629.
Brzeski, H., and Kennedy, S. I. T. (1977). *J. Virol.* **23**, 420.
Calabro, M. A., Katz, J. T., and Holloway, P. W. (1976). *J. Biol. Chem.* **251**, 2113–2118.
Cameron, D. J., and Erlanger, B. E. (1977). *Nature (London)* **268**, 763–765.
Cancedda, R. L., and Schlesinger, M. J. (1974). *Proc. Natl. Acad. Sci. U.S.A.* **71**, 1843–1947.
Cancedda, R. L., Villa-Komaroff, L., Lodish, H., and Schlesinger, M. J. (1975). *Cell* **6**, 215–222.
Changeux, J. P. (1974). *In* "Cell Surface Development" (A. Moscona, ed.), pp. 207–220. Wiley, New York.
Clegg, J. C. S. (1975). *Nature (London)* **254**, 454.
Clegg, J. C. S., and Kennedy, S. I. T. (1974). *FEBS Lett.* **42**, 327–330.
Clegg, J. C. S., and Kennedy, S. I. T. (1975a). *J. Mol. Biol.* **97**, 400–411.
Clegg, J. C. S., and Kennedy, S. I. T. (1975b). *Eur. J. Biochem.* **53**, 175–183.
Clegg, J. C. S., Brzeski, H., and Kennedy, S. I. T. (1976). *J. Gen. Virol.* **32**, 413–430.
Davey, M. W., Dennett, D. P., and Dalgarno, L. (1973). *J. Gen. Virol.* **20**, 225.
Eisen (1978). Personal communcation.
Elgsäter, A., Shotton, D. M., and Branton, D. (1976). *Biochim. Biophys. Acta* **426**, 101–122.
Fontana, A. (1972). *In* Methods of Enzymology" (C. H. W. Hirs and S. N. TImasheff, eds.), Vol. 25, p. 419. Academic Press, New York.
Franklin, R. M. (1974). *Curr. Top. Microbiol. Immunol.* **68**, 108.
Fries, E. (1978). In preparation.
Garoff, H., and Simons, K. (1974). *Proc. Natl. Acad. Sci. U.S.A.* **71**, 3988.
Garoff, H., and Söderlund, H. (1978). Submitted for publication.
Garoff, H., Simons, K., and Renkonen, O. (1974). *Virology* **61**, 493.
Garoff, H., Simons, K., and Dobberstein, B. (1978). In preparation.
Glanville, N., Ranki, M., Morser, J., Kääriäinen, L., and Smith, A. E. (1976). *Proc. Natl. Acad. Sci. U.S.A.* **73**, 3059–3063.
Gliedman, J. R., Smith, J. F., and Brown, D. T. (1975). *J. Virol.* **16**, 913.
Hall, M. O., Bok, D., and Bacharach, A. C. E. (1969). *J. Mol. Biol.* **45**, 397–406.

406   KAI SIMONS ET AL.

Harrison, S. C., David, A., Jumblatt, J., and Darnell, J. E. (1971). J. Mol. Biol. 60, 523.
Helenius, A. (1978). In preparation.
Helenius, A., and Simons, K. (1975). Biochim. Biophys. Acta 415, 29.
Helenius, A., and Simons, K. (1977). Proc. Natl. Acad. Sci. U.S.A. 74, 529–532.
Helenius, A., and Söderlund, H. (1973). Biochim. Biophys. Acta 307, 287.
Helenius, A., and von Bonsdorff, C.-H. (1976). Biochim. Biophys. Acta 436, 895.
Helenius, A., Fries, E., Garoff, H., and Simons, K. (1976). Biochim. Biophys. Acta 436, 319.
Henderson, R., and Unwin, P. N. T. (1975). Nature (London) 257, 28–32.
Henning, R., Milner, R., Reske, K., Cunningham, B. A., and Edelman, G. M. (1976). Proc. Natl. Acad. Sci. U.S.A. 73, 119.
Hirschberg, C. B., and Robbins, P. W. (1974). Virology 61, 602.
Inesi, G. (1972). Annu. Rev. Biophys. Bioeng. 1, 191–210.
Jones, K. J., Scupham, R. K., Pfeil, J. A., Wan, K., Sagik, B. P., and Bose, H. R. (1977). J. Virol. 21, 778–787.
Kääriäinen, L., and Renkonen, O. (1978). In "Membrane Assembly and Turnover" (G. Poste, ed.), Vol. IV (in press).
Katz, F., Rotham, J. E., Lingappa, V. R., Blobel, G. V., and Lodish, H. F. (1977). Proc. Natl. Acad. Sci. U.S.A. 74, 3278–3282.
Keegstra, K., Sefton, B. M., and Burke, D. (1975). J. Virol. 16, 613.
Kennedy, S. I. T. (1976). J. Mol. Biol. 108, 491.
Küchel, P. W., Campbell, D. G., Barclay, A. N., and Williams, A. F. (1977). Biochem. J. 169, 411–417.
Lachmi, B. E., and Kääriäinen, L. (1976). Proc. Natl. Acad. Sci. U.S.A. 73, 1936–1940.
Lachmi, B. E., Keränen, S., Glanville, N., and Kääriäinen, L. (1975). J. Virol. 16, 1615–1629.
Laine, R., Söderlund, H., and Renkonen, O. (1973). Intervirology 1, 110.
Laver, W. G., and Valentine, R. C. (1969). Virology 38, 105–119.
Lenard, J., and Compans, R. W. (1974). Biochim. Biophys. Acta 344, 51.
Lonberg-Holm, H. K., and Philipson, L. (1974). Monogr. Virol. 9, 1–148.
Luukkonen, A. (1976). Ph.D. Thesis, University of Helsinki.
Marchesi, V. T., Furthmayr, H., and Tornita, M. (1976). Annu. Rev. Biochem. 45, 667–698.
Maroux, S., and Louvard, D. (1976). Biochim. Biophys. Acta 419, 189–195.
Martire, G., Bonatti, S., Aliperti, G., De Gull, C., and Cancedda, C. (1977). J. Virol. 21, 610–618.
Mattila, K., Luukkonen, A., and Renkonen, O. (1976). Biochim. Biophys. Acta 419, 435.
Mussgay, M., Enzmann, P.-J., and Horst, J. (1970). Arch. Gesamte Virusforsch. 31, 81.
Patton, S., and Keenan, T. W. (1975). Biochim. Biophys. Acta 415, 273–309.
Peters, K., and Richards, F. M. (1977). Annu. Rev. Biochem. 46, 523–552.
Pfefferkorn, E. R., and Clifford, R. L. (1964). Virology 23, 217.
Pfefferkorn, E. R., and Shapiro, D. (1974). Compr. Virol. 2, 171–230.
Quigley, J. P., Rifkin, D. B., and Reich, E. (1971). Virology 46, 106.
Renkonen, O., Kääriäinen, L., Simons, K., and Gahmberg, C. G. (1971). Virology 46, 318.
Renkonen, O., Luukkonen, A., Brotheurs, J., and Kääriäinen, L. (1974). In "Control of Proliferation in Animal Cells' (B. Clarkson and R. Baserga, eds.), pp. 495–504. Cold Spring Harbor Lab., Cold Spring Harbor, New York.
Richardson, C. O., and Vance, D. E. (1976). J. Biol. Chem. 251, 5544–5550.
Saborio, J. L., Pond, S.-S., and Koch, E. (1974). J. Mol. Biol. 85, 195–211.
Schelle, C. M., and Pfefferkorn E. R. (1969). J. Virol. 3, 369.
Scheid, A., Caliguri, L. A., Compans, R. W., and Choppin, P. W. (1972). Virology 50, 640–652.
Schlesinger, M. J., and Schlesinger, S. (1973). J. Virol. 11, 1013.

Schlesinger, S., Gottlieb, C., Feil, P., Gelb, N., and Kornfeld, S. (1976). *J. Virol.* **17,** 239.
Sefton, B. M. (1977). *Cell* **10,** 659–668.
Silverstein, S. S., Steinman, R. M., and Cohn, Z. A. (1977). *Annu. Rev. Biochem.* **46,** 669–722.
Simmons, D. T., and Strauss, J. H. (1974). *J. Mol. Biol.* **86,** 397.
Simons, K., and Kääriäinen, L. (1970). *Biochem. Biophys. Res. Commun.* **38,** 981.
Simons, K., and Sarvas, M. (1978). In preparation.
Simons, K., Keränen, S., and Kääriäinen, L. (1973a). *FEBS Lett.* **29,** 87.
Simons, K., Helenius, A., and Garoff, H. (1973b). *J. Mol. Biol.* **80,** 119.
Simons, K., Garoff, H., and Helenius, A. (1977). *In* Membrane Proteins and their Interactions with Lipids" (R. A. Capaldi, ed.), pp. 59–86. Dekker, New York.
Skehel, J. J., and Waterfield, M. D. (1975). *Proc. Natl. Acad. Sci. U.S.A.* **72,** 93–97.
Smith, J. F., and Brown, D. T. (1977). *J. Virol.* **22,** 662–678.
Söderlund, H. (1973). *Intervirology* **1,** 354.
Söderlund, H. (1976). *FEBS Lett.* **63,** 56–58.
Söderlund, H., Kääriäinen, L., von Bonsdorff, C.-H., and Weckström, P. (1972). *Virology* **47,** 753.
Strauss, J. H., and Strauss, E. G. (1978). *In* "Molecular Biology of Animal Viruses" (D. Nayak, ed.). Dekker, New York (in press).
Strauss, J. H., Burge, B. W., and Darnell, J. E. (1969). *Virology* **37,** 367.
Tanford, C., and Reynolds, J. (1976). *Biochim. Biophys. Acta* **457,** 133–170.
Utermann, G., and Simons, K. (1974). *J. Mol. Biol.* **85,** 560.
von Bonsdorff, C.-H. (1973). *Comment. Biol., Soc. Sci. Fenn.,* **74,** 1.
von Bonsdorff, C.-H., and Harrison, S. C. (1975). *J. Virol.* **16,** 141–145.
Wengler, G., and Wengler, G. (1976). *Virology* **73,** 190–199.
Wengler, G., Beato, M., and Hackenack, B. A. (1974). *Virology* **61,** 120–128.
Wiley, D., and von Bonsdorff, C.-H. (1978). *J. Mol. Biol.* (in press).
Wirth, D. F., Katz, F., Small, B., and Lodish, H. F. (1977). *Cell* **10,** 253–266.
Ziemiecki, A., and Garoff, H. (1978). *J. Mol. Biol.* (in press).
Ziemiecki, A., Garoff, H., and Simons, K. (1978). In preparation.

18

Oscillations: A Property
of Organized Systems

BENNO HESS

Max-Planck-Institut für Ernährungsphysiologie, Dortmund, Germany

I. GENERAL MECHANISTICS OF DYNAMIC
ORGANIZATION

The property of self-organization is a fundamental feature of living systems. Macroscopically, it is reflected in the phenomena of evolution, of differentiation, and of other numerous biological and biochemical functions. Indeed, the mechanistics of all these processes are the subject of intensive studies in many laboratories. The quality of organization results from basic thermodynamic and kinetic constraints, to which the occurrence of biological systems is fundamentally bound, and its theoretical frame lies in the concept of dissipative structures (Glansdorff and Prigogine, 1971), as a new science of motion.

It has been clearly recognized that the characteristic biological boundaries favor the existence of oscillations of the density distribution of chemicals in time and space, leading to the establishment of organized states. Oscillations have been widely observed in living systems (Hess and Boiteux, 1971; Chance *et al.*, 1973; Goldbeter and Caplan, 1976). Although their mechanistics may be just as widely different from and dependent on the degree of complexity involved, some common features have been recognized in biochemical oscillations, which will be discussed here.

Biological order results from the initiation of chemical trans-

FRONTIERS IN PHYSICOCHEMICAL BIOLOGY

formations and transport phenomena that are maintained far from equilibrium; these processes are open by a continuous supply of energy and matter and quasi-closed with respect to a large number of chemical structures (such as lipids, proteins, and nucleic acids) in a unicellular or multicellular space. The "far-from-equilibrium condition" implies the irreversibility of the processes and unidirectionality of the evolution of time. Under equilibrium conditions, all living phenomena disappear.

Biological systems consist of complex chemical networks of interactions of a vast number of different chemicals, resulting in highly nonlinear kinetics. The kinetic nonlinearity arises from different sources:

1. Multiple types of positive or negative feedback interactions are organized in the form of enzymatic cycles such as metabolism, biosynthetic pathways, or the predicted evolutionary hypercycles.

2. Many processes are controlled by allosteric enzymes responding to small changes in substrate, products, and controlling ligands in a cooperative manner by changing their conformation states.

3. The organization of transport of chemical particles in living systems implies not only free diffusion but to a large extent processes, which occur upon coupling between vectorial transmembrane events and enzyme functions. The activity of membrane-bound enzymes is controlled not only by the components of the membrane itself but, in addition, by the nature of the membrane protein-directed diffusion processes. Electrical fields have been observed to be an additional large-scale force affecting the transport of charged particles and controlling the microenvironment of catalytic centers.

Thermodynamic conditions and nonlinear chemical transformation and transport processes result in the evolution of a number of dynamic states, some of which are the domains of periodic behavior and spatial organizations. It has been pointed out that, at far-from-equilibrium conditions, the relationship between chemical kinetics and the space–time structure of reacting systems cannot be described in strictly local terms; it involves global features of the systems environment. In addition to the values of the kinetic constants and transport coefficients, the solution of the kinetic equations depends on such features as the size, form, and geometry of the reacting medium (Hess *et al.*, 1977).

II. NONEQUILIBRIUM DYNAMICS

A variety of dynamic states of nonlinear chemical transformations can be conceptualized for the homogeneous case neglecting diffusion with

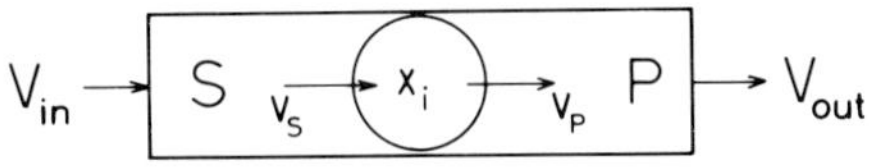

$$V_{in} \rightarrow \boxed{S \xrightarrow[V_S]{} \left(X_i\right) \xrightarrow[V_P]{} P} \rightarrow V_{out}$$

$$V_S (S,P) = V_P (S,P) = V (S,P)$$

$$\frac{dS}{dt} = V_{in} - V (S,P) \qquad\qquad \frac{dP}{dt} = V (S,P) - V_{out}$$

$$\frac{d\alpha}{dt} = \frac{V_{max}}{K_S} \left(V_{in} - v\,(\alpha,\gamma)\right) \qquad \frac{d\gamma}{dt} = \frac{V_{max}}{K_P} \left(v(\alpha,\gamma) - \kappa \cdot \gamma\right)$$

$$\alpha = \frac{S}{K_S} \qquad\qquad \gamma = \frac{P}{K_P}$$

Fig. 1 Model describing the kinetics of yeast phosphofructokinase. From Hess (1975).

the help of a phenomenological model (Fig. 1) demonstrating the input and output relationship of a chemical transformation system connected by the function of an enzyme. Here, a substrate with the velocity v_{in} is fed into the system supplying a substrate pool (S), which reacts with an enzyme (X_i) at a rate of V_S and is transformed at a rate of v_P into a product (P), which leaves the enzyme system and enters a sink at a given rate, v_{out}. X_i designates a variety of conformational states of an allosteric enzyme according to the Monod–Wyman–Changeux mechanism (Monod et al., 1965). Depending on the level of a constant supply of substrate and sink constant as well as the allosteric characteristics of the enzyme, the following dynamic states are observed in a time analysis of the concentrations of S and P as illustrated in Figs. 2, 3, and 4:

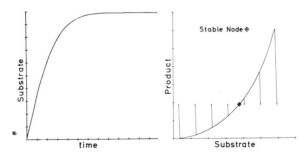

Fig. 2 Computer representation of the solution of the equations given in Fig. 1 with $L_0 = 10^9$ and $c = 10^{-3}$. Left, monotonic transition to a steady state; right, concentration ratios—the system passes to a stable node irrespective of the initial conditions. From Hess (1975).

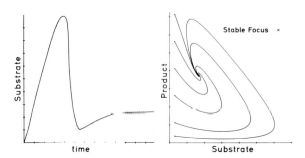

Fig. 3 Computer representation of the solutions of the equations given in Fig. 1 with $L_0 = 10^9$ and $c = 10^{-2}$. Left, variation and substrate concentration with time; right, despite several initial conditions, a stable focus is reached. From Hess (1975).

1. Multiple steady states with monotonous or overshooting transitions from one state to another (stable node or stable focus, respectively)
2. Rotation on a limit cycle around an unstable singular point

In addition, a chaotic behavior has been described that never reaches a stable motion (Rössler, 1976). For the nonhomogeneous case, transport processes by diffusion and active transport must be considered and, indeed, the delicate coupling between the catalytic transformation and transport processes results readily in the occurrence of standing or moving chemical waves, depending on the geometry and size of the systems as well as the critical wavelength of its reacting components (Prigogine and Nicolis, 1971). This simplified model holds for enzymatic as well as pure chemical systems. In the latter case, X_i designates a variety of states of a catalyst. In addition, the model allows the description of a periodic

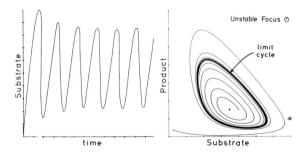

Fig. 4 Computer representation of the solution of the equation given in Fig. 1 with $L_0 = 10^9$ and $c = 10^{-2}$. Left, periodic change in concentration of substrate with time; right, unstable focus: Concentrations approach the limit cycle. From Hess (1975).

coupling between reaction centers of different enzymes or catalysts. All these states have been recorded experimentally in chemical and biochemical as well as biological systems. It is interesting to note that, because of the general feature of their occurrences, the mere observation of such states is of the greatest heuristic value in the search for mechanisms of coupling and control.

III. EXPERIMENTAL OBSERVATION OF OSCILLATIONS

A. Variety of Systems

Immediately after its presentation, the Belousov–Zhabotinsky reaction became a classic model of spatiotemporal pattern formation of a pure chemical system. In this process, the oxidative decarboxylation of malonic acid in the presence of bromate is catalyzed by cerium ions, leading to the production of brominated derivatives, CO_2 and H_2O. In a given domain, the process is periodic and results in the occurrence of chemical waves. The process can be broken down into twelve elementary reactions and includes the autocatalytic formation of bromous acid. The complex elementary reaction system can be reduced to a scheme of five reactions that are adequately described by a system of nonlinear partial differential equations. Here, the autocatalytic step is the source of the nonlinearity (for summaries, see Faraday Symposia of the Chemical Society, 1974; Murray, 1976).

The many components of oscillating chemical catalytic processes are also a feature of enzymatic systems. Oscillations have been described in a number of single- and multienzyme systems. Horseradish peroxidase and lactoperoxidase are cases of open soluble enzyme systems. Here, the source of the nonlinearity is the autocatalytic character of the oxidation of NADH catalyzed by the peroxidase. In the case of horseradish peroxidase oscillations, bistability and the chaotic state have been observed under conditions of quasihomogeneity maintained by constant stirring of the observation sample (Degn, 1969; Olsen and Degn, 1977). The delicate coupling between transport processes and enzyme catalysis leads to oscillations in the case of the artifical papain membrane oscillator. The enzyme is immobilized in a membrane and hydrolyzes benzoyl-L-arginine ethyl ester. The overall reaction leads to an acidification of the system upon hydrolysis. Under proper conditions, the rate of the reaction is intrinsically enhanced by the activating acid influence on the enzyme activity. Because of the binding of the enzyme to the membrane, this enhancement is strongly favored by the critical membrane-dependent diffusion processes, which lead to a local accumulation of protons, which in turn favors the reaction autocatalytically. If the rate of supply of substrate to the enzyme through the membrane is constant, the system shuttles between diffusion- and enzyme-limited states, which can be described by a limit cycle (Goldbeter and Caplan, 1976). These two single-enzyme systems might illustrate an apparent simplicity of the biochemical oscillators; however, it should not be overlooked that the underlying mechanism is just as complex as in the case of the Belousov–Zhabotinsky reaction.

The number of participating components is greatly increased in the case of oscillations of multienzyme systems. It is interesting to remember that oscillations have been observed in all three classical pathways of energy supply, glycolysis, photosynthesis in the dark, and mitochondrial respiration, the latter process being a case of membrane-bound oscillations of energy-dependent vectorial ion transport events coupled to electron flow and ATPase activity (Hess and Boiteux, 1971; Boiteux and Hess, 1974; Goldbeter and Caplan, 1976). An even higher degree of complexity is the basis of the periodic functions of enzymes and transport processes observed during morphogenesis of the slime mold *Dictyostelium discoideum*. This biological system illustrates the intra- and intercellular functions of several oscillating systems as a mechanism of the organization process. The periodic production, ejection, reception, and relay amplification of pulses of cyclic AMP, which are coupled to periodic chemotactic activity, are perfectly synchronized within the life cycle of the cells. Here a large number of functions are controlled by a common chemical intermediate transported extracellularly by long-range diffusion in chemical waves as well as by intracellular processes (Gerisch and Hess, 1974; Gerisch *et al.*, 1975; Hess *et al.*, 1977). In the context of this presentation, the oscillation of glycolysis will be discussed in greater detail to illustrate that the dynamic states of the whole process can be ascribed to an appreciable extent to the function of one single enzyme in a complex enzymatic pathway.

B. Glycolysis

The demonstration of oscillation in glycolysis of yeast cells and in cell-free extracts of yeast was the first indication that periodic activities of biological systems might be caused by the function of a single enzyme in a chemical pathway consisting of a sequence of enzymatic reactions. It was shown that in a given flux domain the concentrations of all glycolytic intermediates oscillate. The analysis of this phenomenon was greatly facilitated by the use of spectrophotometric and potentiometric techniques, allowing a continuous readout of components of the pathway, namely, reduced pyridinnucleotide, pH, and CO_2 as indicators of the dynamic states of the process. Using a continuous substrate injection technique to induce dynamic states like those observed in classic stirred flow reactor systems, we have shown that the frequency and amplitude of a glycolytic NADH oscillation in a given yeast extract are flux dependent, and a critical flux range was established in which oscillations are recorded. The range of this rate was found to extend over 1.5 orders of magnitude, well within the physiological flux range (Hess and Boiteux, 1971; Boiteux and Hess, 1974; Boiteux *et al.*, 1975; Hess *et al.*, 1977).

The influence of random perturbation of the glycolytic system is of interest with respect to general biological conditions. It was found that the stochastic addition of the substrate of glycolysis led to periodic behavior within a narrow range around the period length, which was observed if a continuous rate of substrate supply was used. It was also shown that by periodic addition of substrate the glycolytic system can be readily entrained, with the oscillation period of glycolysis synchronizing with the period of substrate supply. Furthermore, synchronization to a subharmonic of the driving frequency of the periodic substrate supply was recorded. To illustrate this relationship, domains of entrainment of the fundamental frequency and two subharmonics are given in Fig. 5 for the relevant model in accordance with experimental observations. The figure shows that subharmonic synchronization occurs if the driving frequency is approximately an integral multiple of the frequency recorded, if a continuous rate of substrate supply is used (Boiteux *et al.*, 1975). This phenomenon is also known as subharmonic

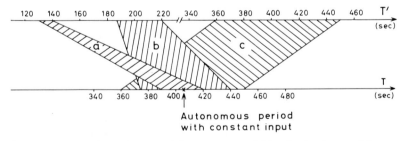

Fig. 5 Domains of entrainment of the enzyme model by the fundamental frequency (c), half harmonic (b), and one-third harmonic (a) of a sinusoidal source of substrate, σ_1, which equals $[0.5 + 0.25 \sin(2\pi t/T')]$. In tne given range no entrainment takes place outside these domains, which extend symmetrically about the autonomous period $T_0 = 406$ seconds. From Boiteux *et al.* (1975).

resonance or frequency demultiplication. Furthermore, it should be mentioned that it is an important feature of circadian rhythms.

The observation of subharmonic synchronization proves the non-linear nature of the glycolytic oscillator, the mechanism of which has been analyzed to a considerable extent by biochemical experimentation (Boiteux and Hess, 1974). From a large number of observations, it followed that the "primary oscillophore" of glycolysis in yeast is the enzyme phosphofructokinase, and it could be shown that its activity under oscillating conditions varies between a maximum and a minimum (Hess and Boiteux, 1971; Boiteux and Hess, 1974).

The molecular weight of the enzyme was found to be 720,000, with at least four protomers, each associating into two subunits with molecular weights of approximately 86,000 and 94,000. The kinetics of enzyme activity are in agreement with an allosteric model having the homotropic effector fructose 6-phosphate and ATP as substrates and a strong heterotropic activator, AMP, indicating a large cooperative response to the enzyme function. Under critical glycolytic flux conditions, the enzyme operates periodically, generating pulses of ADP and fructose 1,6-bisphosphate. The periodic activity change of the enzyme is propagated along the glycolytic pathway through the adenylate system, which also affects the enzymes phosphoglycerate kinase and pyruvate kinase. Both these latter enzymes generate the product ATP, which feeds back to the enzyme phosphofructokinase. Phosphofructokinase operates not only in a feedback cycle via the adenylate system but also through an intrinsic feedback structure, supplied by its oligomeric composition and non-linear allosteric function. Thus, the regulatory mechanism of the enzyme phosphofructokinase itself has autocatalytic features (Boiteux and Hess, 1974).

A theoretical analysis of the dynamic behavior of the phosphofructokinase reaction as an open system was based on an allosteric model of the enzyme as described in the frame of the concerted transition theory of Monod *et al.* (1965). Neglecting diffusion, a set of ordinary differential equations permits the description of the time evolution of the metabolite and enzyme concentrations. The treatment of the concentrations of the various enzyme species as defined by their variable affinities to substrates and products of the reaction yields a total of 13 equations, which can be simplified by assuming a quasi steady state for the enzymatic forms. This simplification is justified because of the observation that the concentration of metabolites exceeds that of the enzyme by several orders of magnitude. With this alteration, the system reduces to two equations:

$$d\alpha/dt = \sigma_1 - \sigma_M \Phi$$

and

$$d\gamma/dt = q\sigma_M \Phi - k_s \gamma$$

where

$$\Phi = \frac{[\alpha e(1 + \alpha e)](1 + \gamma)^2 + L\Theta \alpha c e'(1 + \alpha c e')}{L(1 + \alpha c e')^2 + (1 + \alpha e)^2(1 + \gamma)^2}$$

with the definitions (in detail, see Boiteux et al., 1975) $\gamma = P/K_P$, $\alpha = S/K_{R(S)}$, $c = K_{R(S)}/K_{T(S)}$, $K_P = d_2/a_2$, $K_{R(S)} = d/a$, $K_{T(S)} = d'/a'$, $\Theta = k'/k$, $q = K_{R(S)}/K_P$, $L = k_1/k_2$, $\sigma_1 = v_1/K_{R(S)}$, $\sigma_M = (2kD_0)/K_{R(S)}$, $\epsilon = k/d$, $\epsilon' = k'/d'$, $e = (1 + \epsilon)^{-1}$, and $e' = (1 + \epsilon')^{-1}$.

The function Φ is a rate function (v/V_{max}) for the phosphofructokinase reaction. The dominant source of the nonlinearity in this function is the allosteric constant (L) as defined by Monod et al. (1965). On lowering this constant, the allosteric property of the system vanishes, and the function finally reduces to a simple Michaelis–Menten rate law. The term $(1 + \gamma)^2$ describes the activation of the enzyme by its product. It is interesting to note that this term is decisive for the evolution of the property of nonequilibrium instability (Hess et al., 1977). This mechanism illustrates the autocatalytic character of the function of the enzyme. Furthermore, we would like to add that the existence of an asymptotically stable limit cycle for general enzyme-catalyzed reactions with positive feedback as given here has been proved mathematically by application of the Poincarè–Bendixson theorem (Erle et al., 1978).

The stability analysis of the system demonstrates an excellent agreement between the dynamic behavior of the allosteric model of glycolytic oscillation and biochemical experiments. These results justify the assumption that, to a large extent, the dynamic property of the glycolytic system can be reduced to the properties of phosphofructokinase as a master enzyme within the set of glycolytic enzymes and that it can be described by a global master equation (Boiteux et al., 1975).

Although the analysis of glycolytic oscillations so far has mainly been carried out under homogeneous experimental conditions, where transport processes can be neglected, it is to be expected that spatial structures do occur whenever the evolution of enzymatic processes is allowed to couple with transport via appropriate diffusion gradients. Indeed, patterns of spatiotemporal organization have been computed on the basis of the phosphofructokinase model given above (Goldbeter, 1973; Hess et al., 1977). In addition, in confirmation of the theoretical predictions, the occurrence of a macroscopic periodic structure formation synchronized with glycolytic oscillations has been recorded in cell-free extracts of yeast (A. Boiteux and B. Hess, unpublished experiments).

IV. GENERAL REMARKS

The demonstration of stable periodic reactions in enzymatic systems raises a number of questions with respect to its physiological significance. Although the physiological function of periodic reactions is clearly evident in the case of the slime mold *D. discoideum*, in many other systems their function is not known. We are still searching for direct evidence showing whether simple enzymatic reaction systems, such as those observed in metabolic and epigenetic processes displaying limit cycle behavior, might serve as a molecular clock for processes like the mitotic cycle, pathway separation, differentiation, and morphogenesis. Furthermore, such molecular processes might be the basis of circadian rhythms, which are abundantly observed in living systems of higher order.

The analysis of many of these properties of "large" living systems is experimentally and theoretically limited by currently available techniques. While the lower limits of reductionism in the description of biological order seem clearly defined in the concept of dissipative structures, an understanding of the function of the large complexity of biological systems is lacking. Although it is clearly recognized that biological systems are irreversibly organized (they do not explode or decay under ordinary conditions but seem to grow and evolve toward a singularity or limit cycle if energy and matter is available), the rules of their network composition, their underlying network hierarchy, and their order in time and space remain to be understood. These rules must be recognized clearly to understand the stability versus complexity problem of molecular processes. The minimum number of elementary reactions to exhibit stable dynamic behavior in space and time and the upper limit of stability with respect to complexity should be analyzed in the future as a problem of nonlinear dynamics.

ACKNOWLEDGMENT

I am grateful to Dr. Thomas Reed for reading the manuscript and for his helpful suggestions.

REFERENCES

Boiteux, A., and Hess, B. (1974). *Faraday Symp. Chem. Soc.* **9**, 202–214.
Boiteux, A., Goldbeter, A., and Hess, B. (1975). *Proc. Natl. Acad. Sci. U.S.A.* **72**, 3829–3833.
Chance, B., Pye, E. K., Ghosh, A. K., and Hess, B. (1973). "Biological and Biochemical Oscillators." Academic Press, New York.
Degn, H. (1969). *Biochim. Biophys. Acta* **180**, 271–290.
Erle, D., Mayer, K. H., and Plesser, T. (1978). *J. Math. Biol.* (in press).
Faraday Symposia of the Chemical Society. (1974). "Physical Chemistry of Oscillatory Phenomena," No. 9. Faraday Div. Chem. Soc., London.
Gerisch, G., and Hess, B. (1974). *Proc. Natl. Acad. Sci. U.S.A.* **71**, 2118–2122.
Gerisch, G., Hülser, D., Malchow, D., and Wick, U. (1975). *Philos. Trans. R. Soc. London* **272**, 181–192.
Glansdorff, P., and Prigogine, I. (1971). "Thermodynamic Theory of Structure, Stability and Fluctuations." Wiley, New York.
Goldbeter, A. (1973). *Proc. Natl. Acad. Sci. U.S.A.* **70**, 3255–3259.
Goldbeter, A., and Caplan, S. R. (1976). *Annu. Rev. Phys. Chem.* **5**, 449.
Hess, B. (1975). *In* "Energy Transformation in Biological Systems," pp. 369–392. Elsevier, Amsterdam.
Hess, B., and Boiteux, A. (1971). *Annu. Rev. Biochem.* **40**, 237–258.
Hess, B., Goldbeter, A., and Lefever, R. (1977). *Adv. Chem. Phys.* (in press).
Monod, J., Wyman, J., and Changeux, J. P. (1965). *J. Mol. Biol.* **12**, 88–118.
Murray, J. D. (1976). *J. Theor. Biol.* **56**, 329–353.
Olsen, L. F., and Degn, H. (1977). *Nature (London)* **267**, 177–178.
Prigogine, I., and Nicolis, G. (1971). *Q. Rev. Biophys.* **4**, 107–148.
Rössler, O. E. (1976). *Z. Naturforsch., Teil A* **31**, 259–264.

19

Cation-Induced Regulatory Mechanism of Enzyme Reactions

PATRICK MAUREL AND PIERRE DOUZOU

INSERM, Montpellier, France, and Institut de Biologie Physico-Chimique, Paris, France

I. INTRODUCTION

Enzyme systems are now generally recognized as possessing self-regulatory mechanisms controlling their activity (Monod and Jacob, 1961; Jacob *et al.*, 1963; Monod *et al.*, 1963, 1965). Feedback inhibition, enzyme induction, and repression account for, to a certain extent, the complex behavior of many systems but fail to explain the potential involvement of ions in enzyme activity.

Besides their essential role in promoting and maintaining the active conformation of many biological molecules (von Hippel and Schleich, 1969; Felsenfeld and Miles, 1967; Träuble *et al.*, 1976), it is well established that ions strongly influence the catalytic behavior of enzyme systems (Webb, 1963; Bygrave, 1967). In particular, it appears that for many systems the "response" of the enzyme to ionic strength depends on the experimental conditions under which the reaction proceeds, that is, pH, concentration of other ionic species, concentration of ligands, state of the enzyme, etc. While these observations are in most cases not clearly

421

FRONTIERS IN PHYSICOCHEMICAL BIOLOGY

understood, they indeed stress the existence of a possible ion-induced regulatory mechanism of enzyme reactions. Although some aspects of the ion's role have been discussed from a theoretical point of view, such a mechanism has never been seriously analyzed at the molecular level.

There are presumably many possible approaches to the study of the ionic control of enzyme reaction. We decided to study this problem in light of the polyelectrolyte theory for two reasons. First, enzymes as well as many other biological molecules are polyelectrolytes. Moreover, most of the enzyme systems function *in vivo* or *in vitro* inside polyelectrolytic (generally polyanionic) microenvironments provided either by biological membrane or by certain macromolecules, such as nucleic acids, especially in gene expression systems. Second, the polyelectrolyte theory deals with the interaction of ions with polyelectrolytes, as well as with the consequences of this interaction in terms of changes in certain local physicochemical parameters and structures. It is therefore expected that the analysis of the observations made on enzyme reactions in light of this theory could be helpful in the understanding of the ionic control of these reactions.

In this chapter we will report and summarize some of our investigations (Maurel, 1976; Maurel and Douzou, 1976; Douzou and Maurel, 1976, 1977a,b) on membrane-bound enzymes as well as on enzyme systems involved in gene expression.

It appears that an oversimplified treatment of the catalytic implications of the polyelectrolyte theory (i.e., only at the level of the environmental factors: pH, ionic content) provides a suitable physicochemical approach to the problem of the influence of ions on enzyme reactions. Since this treatment allows the satisfactory interpretation of many observations from this laboratory, as well as from the literature, a general nonspecific mechanism of ionic regulation common to all enzyme systems functioning within polyelectrolytic environments (membrane-bound systems and enzymes involved in DNA, RNA, and protein synthesis) is presented and discussed.

This chapter is divided into four main sections. In Section II we present a simplified treatment of the catalytic implications of the polyelectrolyte theory. Sections III and IV are devoted to experimental studies and to their interpretation in terms of this treatment. Section V is a general critical conclusion that shows that the correct interpretation of the more specific aspects of the effects of ions requires considering further implications of the polyelectrolyte theory, that is, structural implications at the level of membranes, nucleic acids, and proteins and related aspects such as the influence of the electrostatic potential heterogeneity and the influence of the structure of water on the electrostatic interactions.

II. CATALYTIC IMPLICATIONS OF THE POLYELECTROLYTE THEORY: AN OVERSIMPLIFIED TREATMENT

Very few studies have been devoted to the influence of polyelectrolytic environments on enzyme catalysis. McLaren (1957) and McLaren and Esterman (1957) reported that α-chymotrypsin functioning at the water–solid (kaolinite) interphase exhibited a pH profile different from that obtained with the free enzyme in solution. Later Katchalsky and co-workers (see Goldstein et al., 1964) and Goldstein (1972, 1977), studying the kinetic behavior of enzymes attached to a polyelectrolytic matrix (trypsin and α-chymotrypsin covalently bound to copolymers of maleic acid and ethylene), introduced a theoretical treatment of the catalytic implications of the polyelectrolyte theory. Engasser and Horvath (1975) made a theoretical extension of this treatment in which they took into account the influence of substrates, products, and/or inhibitors in the cases where these agents significantly affect the ionic strength of the medium.

We shall consider here the simplified treatment developed by Katchalsky and co-workers (see Goldstein et al., 1964). This treatment is based on the following restrictions.

1. The enzyme is supposed to be maintained within the polyelectrolytic environment through a binding that does not affect its catalytic functions.
2. The electrostatic potential generated within the environment is assumed to be homogeneous and pH independent (i.e., the pK of the ionizable groups responsible for this electrostatic potential is considered as either very low or very high).
3. It is supposed that the electrostatic potential developed by the polyelectrolytic carrier does not influence the structure of the latter or the structure of the enzyme. In other words, the treatment does not take into account any structure–function relationship introduced by the electrostatic potential of the environment.
4. The possible influence of the high density of the electric charges on the structure of water in the environment is not considered.
5. The contribution to the total ionic strength, brought about by substrates or inhibitors when these are charged molecules, is assumed to be negligible.

This treatment actually only concerns the influence of the electrostatic potential prevailing within the environment on the local concentration of protons and substrates.

In the environment of any polyelectrolyte, an electrostatic potential Ψ prevails. This electrostatic potential is a result not only of the high density of permanent charges ρ_P borne by the polyelectrolyte, but also of the density of the mobile charges ρ_m brought about by the ionic species present in the solution, which form an ionic atmosphere around the polyelectrolyte. According to the Boltzmann distribution law, the local concentration of any ionic species contributing to ρ_m is given by

$$X_{local} = X_0[\exp(-Z_X\epsilon\Psi/kT)] \tag{1}$$

where Z_X is the algebraic value of the charge borne by X, ϵ is the unit electric charge, and kT has its usual significance. Charge density and electrostatic potential are related through the fundamental Poisson–Boltzmann equation (see Rice and Nagasawa, 1961):

$$\Delta\Psi = (4\pi/D)(\rho_P + \rho_m) \tag{2}$$

where Δ is the Laplace operator and D is the dielectric constant of the medium.

Let us now consider a polyelectrolyte–enzyme system. The catalytic implications of the polyelectrolyte theory become evident when Eq. (1) is applied to protons, substrates, and inhibitors (as long as these are charged molecules).

A. Local Distribution of Protons

Equation (1) shows that the average concentration of protons in the polyelectrolytic environment is different from that of the bulk medium:

$$[\overline{H^+}]_{local} = [H^+]_0[\exp(-\epsilon\overline{\Psi}/kT)] \tag{3}$$

Theoretically, Ψ is a decreasing function of the distance. However, for the sake of clarity and simplicity, we shall consider here the Ψ within the environment has an average value denoted $\overline{\Psi}$. It follows from this relation that any ionizable group submitted to the influence of $\overline{\Psi}$ has an apparent $pK(\overline{\Psi})$ given by

$$pK(\overline{\Psi}) = pK_0 - 0.43(\epsilon\overline{\Psi}/kT) \tag{4}$$

where pK_0 is the pK of the group in the absence of any electrostatic potential. Thus, the pK of all the ionizable groups of the enzyme (or of only part of them, depending on whether the enzyme is completely or partly exposed to the electrostatic potential of the environment) is modified according to Eq. (4), as is, in particular, the pK of those groups apparently involved in the active site. Therefore, as the primary consequence of the presence of the electrostatic potential, the pH profile of the enzyme located inside the polyelectrolytic environment is different from that observed with the free enzyme in solution: It appears to be shifted by the amount $0.43(\epsilon\overline{\Psi}/kT)$. The direction of the shift depends on the nature of the environment. In the case of a polyanionic environment surrounding the enzyme [$\overline{\Psi} < 0$, and accordingly pK ($\overline{\Psi}$) > pK_0], the pH optimum of the reaction will be displaced toward alkaline pH values. The reverse shift will be observed, toward acidic pH values, in a polycationic environment ($\overline{\Psi} > 0$).

B. Local Distribution of Substrates

Most of the natural substrates of enzyme systems are charged molecules. Their local concentration within any polyelectrolytic environment is thus given by

$$\bar{S}_{local} = S_0[\exp(-Z_S\epsilon\overline{\Psi}/kT)] \tag{5}$$

Therefore, as a secondary consequence of the presence of the electrostatic potential, the apparent Michaelis constant of the reaction within the polyelectrolytic environment is different from that obtained with the enzyme free in solution:

$$K^{\overline{\Psi}}_{m(app)} = K^0_{m(app)}[\exp(Z_S\epsilon\overline{\Psi}/kT)] \tag{6}$$

We can see that, in a polyanionic environment, positively charged substrate molecules will tend to concentrate and the resulting K_m will be lower than K^0_m obtained in the absence of any electrostatic potential. The same development can be seen concerning the inhibitor constant when the inhibitor is a charged molecule.

C. Ionic Control of Enzyme Reactions

With respect to the above treatment, an ionic control of the enzyme reactions at the level of polyelectrolytic microenvironments provided by the biological organization (biological membranes, nucleic acids, ribosomes, etc.) can therefore be expected and understood. This ionic control, based on the following points (Douzou and Maurel, 1977a,b; Maurel, 1976), is depicted in Fig. 1.

1. A solution of any polyelectrolyte–enzyme system might be considered to consist of two phases in equilibrium, although, within limits, the polyelectrolyte is soluble. Due to the local electrostatic potential, the "inner" polyelectrolytic phase, or environment, possesses its own local physicochemical parameters, which differ from those of the bulk phase.

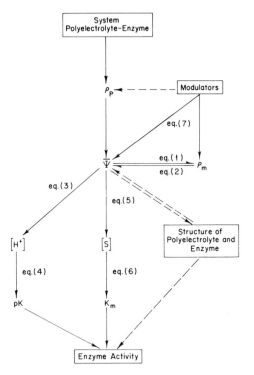

Fig. 1 Representation of the catalytic implications of the polyelectrolyte theory and of the suitable control modulators that might mediate an enzyme reaction. The solid arrows indicate implications dealt with in quantitative terms by the simplified treatment given in Section II. The dashed arrows indicate further implications of the polyelectrolyte theory that cannot be dealt with quantitatively at present. The system consists of a polyelectrolyte carrier on which the enzyme is bound during the catalytic reaction. ρ_P is the density of permanent charges borne by the polyelectrolyte, ρ_m is the density of mobile charges contributing to the ionic atmosphere around the polyelectrolyte, $[H^+]$ is the local concentration of protons, and $\bar{S}$ is the local concentration of substrate (when the substrate is not the polyelectrolyte).

$$X_{\text{local}} = X_0[\exp(-Z_X\epsilon\bar{\Psi}/kT)] \tag{1}$$
$$\Delta\Psi = (4\pi/D)(\rho_P + \rho_m) \tag{2}$$
$$[\overline{H^+}]_{\text{local}} = [H^+]_0[\exp(-\epsilon\bar{\Psi}/kT)] \tag{3}$$
$$pK(\bar{\Psi}) = pK_0 - 0.43(\epsilon\bar{\Psi}/kT) \tag{4}$$
$$\bar{S}_{\text{local}} = S_0[\exp(-Z_S\epsilon\bar{\Psi}/kT)] \tag{5}$$
$$K^{\bar{\Psi}}_{m(\text{app})} = K^0_{m(\text{app})}[\exp(Z_S\epsilon\bar{\Psi}/kT)] \tag{6}$$
$$|\bar{\Psi}(I)| = |\bar{\Psi}(I \to 0)| - A(\log I) \tag{7}$$

2. The enzyme activity in the system is therefore fully controlled by the local electrostatic potential.

3. Any change in this electrostatic potential affects the equilibrium between the inner phase and the bulk phase, and as a result the enzyme activity of the system changes. Those agents able to modulate the electrostatic potential of the system are obviously ionic species. According to their electrostatic affinity for the ionic groups of the polyelectrolyte, there are two types of modulators: modulators of moderate affinity, essentially monovalent ions contributing to ionic strength, and modulators of high affinity, di- and polyvalent ions, polyelectrolytes, etc. The first type of modulator modifies the electrostatic potential of the system by affecting its surrounding ionic atmosphere [$\overline{\Psi}$ affected via ρ_m, Eq. (2)], that is, "externally" with respect to the polyelectrolyte; the second type modifies the electrostatic potential by interacting strongly with the polyelectrolyte, thereby affecting (neutralizing) its permanent charge distribution [$\overline{\Psi}$ affected via ρ_P, Eq. (2)], that is, "internally" with respect to the polyelectrolyte (Fig. 1). We shall see in Section IV,A that, although acting in the same direction, these two kinds of modulators might be strongly differentiated on both qualitative and quantitative levels.

D. Experimental Aspects

We will consider here ionic strength as the modulator since it is the only modulator dealt with in both qualitative and quantitative terms by the polyelectrolyte theory (Rice and Nagawasa, 1961).

The catalytic implications of this theory can be experimentally detected at three levels: the pH profile of the enzyme reaction in the polyelectrolyte–enzyme system at various ionic strengths (or various concentrations in modulators); activity–ionic strength plots at various pH values; the ionic strength effect on the Michaelis constant of the enzyme.

The ionic atmosphere created around the polyelectrolyte in solution, whose distribution is given by Eq. (1), tends to minimize the electrostatic potential through the so-called "electrostatic screening effect." Its action on the electrostatic potential is reasonably accounted for by the following semi-empirical relation (Kotin, 1963).

$$|\overline{\Psi}(I)| = |\overline{\Psi}(I \to 0)| - A(\log I) \tag{7}$$

where absolute magnitude of the potential is considered, A is a positive constant, and I is the ionic strength.

Obviously, this relation holds only over a limited range of ionic strengths, which, nevertheless, roughly coincides with the most frequently used conditions (10^{-4}–10^{-1} M).

Now, combining Eqs. (7) and (4), it can be predicted that, as ionic strength increases in the medium, Ψ decreases, and the pH optimum of the enzyme reaction inside the environment tends, linearly with log I, toward the pH optimum of the same reaction with the enzyme free in solution. This process, illustrating the pH ionic strength interdependence existing within any polyelectrolytic environment, is depicted in Fig. 2 for a polyanionic environment.

An alternate representation of this interdependence is the plot of enzyme activity (as V_{max}, for instance) against ionic strength for various pH values, as shown in Fig. 3 for a polyanionic environment. Such a plot is

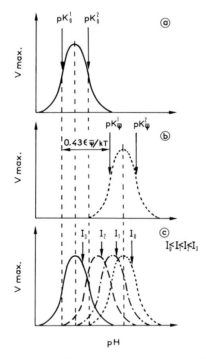

Fig. 2 Catalytic implications of the polyelectrolyte theory: pH profiles. (a) pH profile of an enzyme in the absence of any electrostatic potential. The ionizable groups, apparently involved in the active site, are characterized by pK_0^1 and pK_0^2. (b) pH profile of the same enzyme placed within a polyanionic environment (i.e., under the influence of a strong negative electrostatic potential, $\bar{\Psi}$) at low ionic strength, I_0. (c) Shift of the pH profile accompanying the increase in ionic strength from I_0 to I_3. Within the range of applicability of Eq. (7) we have $pH_{opt}(I) = pH_{opt}(I \to 0) - B(\log I)$. At high ionic strength, $\bar{\Psi}$ vanishes, and the normal pH profile of the enzyme is restored.

frequently reported in the literature. Bell-shaped curves are generated. This representation is actually the direct consequence of the pH profile shift (Fig. 2) as ionic strength increases: the lower the pH, the higher the ionic strength giving the optimum. However, in contrast to what is shown in Fig. 3a, ionic strength in a real system cannot be decreased to zero. A certain amount of salt I_m has to be kept in the solution to avoid denaturing the systems and to maintain their active conformation. Thus, ionic strength increases not from zero but from a minimum value I_m. Accordingly, parts of the curves in Fig. 3a are missing on the experimental activity–ionic strength plots (Fig. 3b) and two types of curves are then generated: type I curves, bell shaped; type II curves, monotonically decreasing, depending on the pH of the experiment. It can be seen that, if this pH is lower than the pH optimum the reaction would have had at minimal ionic strength (I_m), a type I curve is generated. A type II curve is obtained when the pH is higher than the pH optimum the reaction would obtain at I_m.

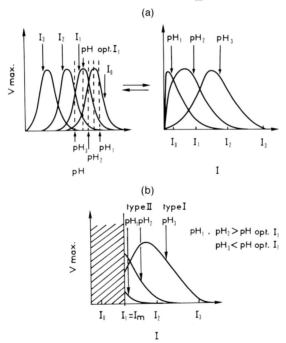

Fig. 3 Catalytic implications of the polyelectrolyte theory: plot of catalytic activity versus ionic strength. It can be seen that such a plot, obtained at pH_2 for instance, is the consequence of the pH profile shift with respect to pH_2 as ionic strength varies from I_0 to I_3, the enzyme being placed within a polyanionic environment. a, theoretical curves; b, real curves.

On the other hand, combinations between Eqs. (6) and (7) show how (via modification of the electrostatic potential) ionic strength affects K_m and K_i (the inhibition constant) and thus the enzyme activity.

In the following sections of this paper, the kinetic behavior of enzyme systems in their natural milieu is analyzed in terms of these catalytic implications of the polyelectrolyte theory. As we shall see, the experimental results are in complete agreement on a qualitative level with the treatment presented above. However, it will appear that some observations are not interpretable in terms of this treatment, an obvious consequence of its oversimplification. To be understood, these unexplained observations require a more realistic analysis of the catalytic implications of the polyelectrolyte theory, which would take into account the various aspects contained in the restrictions on which the simplified treatment developed here is based.

III. MEMBRANE-BOUND ENZYMES

A large number of cellular enzyme systems function in membranes (Rothfield and Romeo, 1971). According to Singer's (1972) model, as well as other studies (Brown, 1974; Hackenbrock, 1975; Hallan and Wrigglesworth, 1976; Rahmann et al., 1976), under normal conditions the membrane surface carries a strong electric charge that is provided by the phospholipid polar head, structural proteins, terminal sialic acid residues of the carbohydrate chains of glycoproteins, etc. It is thus expected that, depending on the local chemical composition of the membrane, bound enzyme systems might be located within highly charged environments and therefore behave according to the catalytic implications of the polyelectrolyte theory. Two examples of such enzymes investigated in our laboratory are presented in this section: the lytic activity of lysozyme on bacterial cell walls (Maurel and Douzou, 1976), and the activity of cytochrome c oxidase from the inner membrane of mitochondria (Douzou and Maurel, 1976, 1977a and Maurel et al., 1978).

A. The Lytic Activity of Lysozyme

Lysozyme (mucopeptide N-acetylmuramoylhydrolase, EC 3.2.1.17) cleaves the $\beta(1 \rightarrow 4)$ glycosidic bonds between N-acetylmuramic acid and N-acetyl-D-glucosamine in the polysaccharide network of bacterial cell walls (Imoto et al., 1972). In addition to these natural substrates, lysozyme is also active on low molecular weight molecules, among which the $\beta(1 \rightarrow 4)$-linked hexasaccharide of N-acetylglucosamine and the corresponding hexasaccharide extracted from cell walls digests have been the most widely investigated. A great deal of work has been realized with these substrates and the mechanism of the reaction is now well established. In particular, the pH profile of the hydrolysis of oligosaccharides presents a pH optimum at 5.2, resulting from the presence in the active site of the enzyme of two ionizable groups, a deprotonated carboxyl of residue Asp-52 and a protonated carboxyl of residue Glu-35, both identified by X-ray diffraction studies (Imoto *et al.*, 1972; Banerjee *et al.*, 1973; Blake *et al.*, 1967). Moreover, the pH profile is independent of ionic strength (Davies *et al.*, 1969).

Because of its physiological importance, many experimental studies were done on the reaction of bacterial cell wall lysis (Imoto *et al.*, 1972; Davies *et al.*, 1969; Chang and Carr, 1971; Jensen and Kleppe, 1972; Saint-Blancard and Jollès, 1972; Neville and Eyring, 1972). Most of these studies show that, in contrast to the reaction with oligosaccharides, the lytic activity of lysozyme presents a number of peculiar characteristics, which we can summarize as follows (Maurel and Douzou, 1976).

1. At low ionic strength (0.01), the pH profile of the lytic reaction has an optimum at about 9.0–10.0, that is, in a range where both carboxyl groups of the active site would be deprotonated and, consequently, the enzyme inactivated (Fig. 4).
2. As ionic strength increases, the pH optimum of the reaction shifts toward acidic pH values; at sufficiently high ionic strength (0.16), it seems to be stabilized near pH 5.5 (Fig. 4). Moreover, it appears that pH optimum is a linear function of the logarithm of ionic strength.
3. As lytic activity is plotted against ionic strength, at constant pH, the reaction is first activated, reaches a maximum, and then is inhibited; the lower the pH at which this plot is obtained, the higher the ionic strength giving the maximal activity (Fig. 5).
4. Finally, the $K_{m(\text{app})}$ of the reaction increases strongly as ionic strength increases.

Indeed, these results are qualitatively similar to those depicted in Figs. 2 and 3, and it is therefore tempting to ascribe them to the presence, on the bacterial cell walls, of polyanionic environments surrounding the sites where lysozyme cleaves the polysaccharide network. However, the point is that, although several authors have concluded from various experimental observations that the cell walls of bacteria like *Escherichia coli* and *Micrococcus luteus* are predominantly negatively charged (Davies *et al.*, 1969; Katchalsky *et al.*, 1953; Salton, 1964), the complexity of the bacterial cell wall architecture means that little is

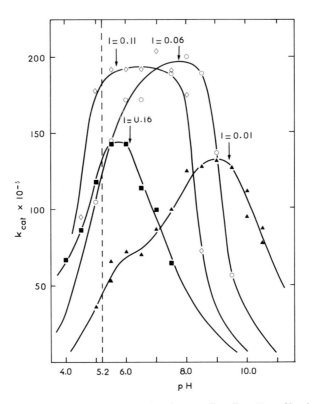

Fig. 4 Lytic activity of lysozyme toward *M. luteus* cell walls. pH profile of the reaction at various ionic strengths (NaCl). The dotted line at pH 5.2 represents the pH optimum of the reaction, using hexasaccharide as substrate. From Maurel and Douzou (1976).

known about the arrangement of its ionizing groups and, consequently, about its electrical configuration. We therefore planned a series of experiments to determine the existence of polyanionic sites on the cell walls of *M. luteus*, a finding that would strongly support the interpretation of the previous results in terms of the polyelectrolyte theory.

These experiments are reported in Figs. 6 and 7 and are based on a well-known property of these cell walls: their aggregation in the presence of polycations such as poly(Lys) in the suspending medium (Katchalsky *et al.*, 1953; Sela and Steiner, 1963). This aggregation, monitored by the increase in turbidity of the suspension, is specifically produced by polycations: Polyanions such as poly(Glu) are inefficient in this process. As the concentration of polycations increases in the medium, the turbidity of the suspension first increases linearly and then

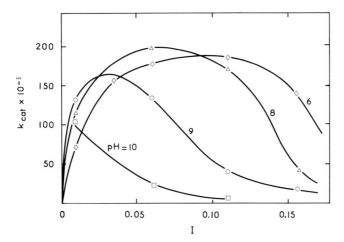

Fig. 5 Lytic activity of lysozyme toward *M. luteus* cell walls. Plots of activity versus ionic strength (NaCl) at various pH values. From Maurel and Douzou (1976).

reaches a plateau, indicating that the aggregation is at its maximum. If lysozyme is simultaneously added to the suspension and the rate of reaction determined, a strong inhibition is observed as a function of the polycation concentration. Figure 6 shows that the plots of change in

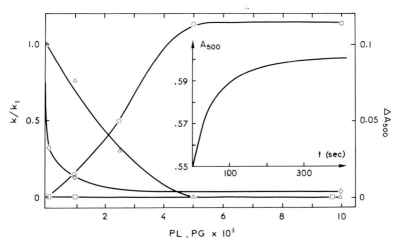

Fig. 6 Inhibition of the lytic activity of lysozyme k/k_0 and of the turbidity increase of *M. luteus* suspensions (ΔA at 500 nm) as a function of increasing concentrations of polyelectrolytes. $\triangle$, k/k_0, poly(Lys) (PL); $\bigcirc$, ΔA of PL; $\diamondsuit$, k/k_0, poly(Glu) (PG) [this strong inhibition results from the binding of lysozyme on poly(Glu) molecules (Seda and Steiner, 1963)]; $\square$, ΔA of PG. Inset: turbidity increase of *M. luteus* suspension after adding poly(Lys). From Maurel and Douzou (1976).

turbidity against poly(Lys) concentration and those of percentage inhibition against poly(Lys) concentration are symmetrical, pointing out that lysozyme and poly(Lys) compete for the same sites on the cell wall. Thus, it can be reasonably concluded that the bonds susceptible to lysozyme are located within (or in the immediate proximity of) polyanionic regions. Further support comes from the experiments described in Fig. 7. In water at normal temperature, lysozyme activity is monitored by the decrease in turbidity that follows the addition of enzyme to the bacterial suspension. Now, considering the facts that (a) lysozyme at neutral pH is a polycationic protein (isoionic point, 11.2) and (b) very low temperatures would stop the lytic reaction according to its high energy of activation, we expected that the addition of lysozyme to a supercooled fluid cell wall suspension would produce an increase in turbidity, reflecting the aggregation induced by the polycationic lysozyme molecule. This is what we observed in a water–methanol mixture (3:2, v/v) at −30°C. Furthermore, we could show that, as with poly(Lys) under normal conditions, the aggregation reached a plateau as the concentration of lysozyme increased.

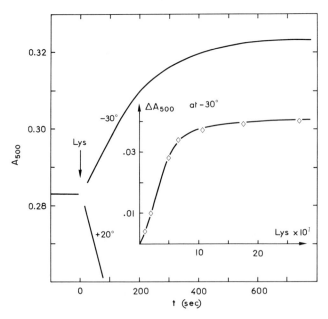

Fig. 7 Changes in optical density of a suspension of *M. luteus* following the addition of lysozyme. At 20°C, lysis occurs (rapid decrease in turbidity). At −30°C, cell wall-induced aggregation (increase in turbidity) without lysis occurs. Inset: turbidity increase of a *M. luteus* suspension as a function of increasing concentrations of lysozyme. From Maurel and Douzou (1976).

It is therefore apparent from these experiments that the $\beta(1 \rightarrow 4)$ linkages in the polysaccharide network of the cell walls of M. *luteus* are located within polyanionic microenvironments. The positively charged molecules of lysozyme are electrostatically attracted by these environments, inside which the lytic reaction proceeds under physicochemical conditions (especially proton concentration) different from those of the bulk medium. The kinetic observations reported in Figs. 4 and 5 and the $K_{m(app)}$ variations with ionic strength can therefore be satisfactorily understood in terms of the catalytic implications of the polyelectrolyte theory.

B. Cytochrome c Oxidase

Cytochrome c oxidase (ferrocytochrome c : oxygen oxidoreductase, EC 1.9.3.1) is the terminal enzyme of the electron transport chain located in the inner membrane of mitochondria. This enzyme is widely distributed among animals, higher plants, and aerobic microorganisms and has been extensively investigated because it is physiologically important in living systems (Nicholls and Chance, 1974; Capaldi and Briggs, 1976).

Hackenbrock (1975), by combining electron microscopy and kinetic measurements, showed that cytochrome c oxidase and other enzyme systems of the inner mitochondrial membrane were located within, or in the proximity of, polyanionic sites of this membrane. We therefore decided to undertake a study of the pH–ionic strength effects on the kinetics of this enzyme to test whether, bound on the membrane, it behaved according to the catalytic implications of the polyelectrolyte theory. A detailed report (Maurel *et al.*, 1978) of this investigation will appear soon and only the essential results will be presented here.

V_{max}–pH profiles obtained at various ionic strengths with a particulate preparation extracted from rat liver (a preparation from beef heart gave the same results) are presented in Fig. 8. For a comparison, the same experiments carried out with the purified enzyme, prepared according to the method of Yonetani (1961), are reported in Fig. 9. It can be seen that the membrane-bound and purified enzymes behave differently with respect to pH and ionic strength. At low ionic strength, their pH optima are, respectively, 7.5 and about 4.5. While the membrane-bound form exhibits a pH profile shift from alkaline to acidic as ionic strength increases, a shift of smaller amplitude in the opposite direction is observed with the purified enzyme. It is very interesting to note that at high ionic strength (0.21 M) both forms have the same pH optimum, that is, 5.5. Indeed, these results agree with the theoretical previsions,

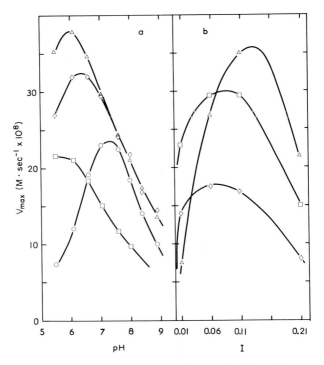

Fig. 8 Membrane-bound cytochrome c oxidase. (a) pH profile at various ionic strengths (in KCl): $\bigcirc$, 10 mM; $\Diamond$, 60 mM; $\triangle$, 110 mM; $\square$, 210 mM. (b) Activity against ionic strength (in KCl) at various pH values: $\Diamond$, 8.4; $\square$, 7.0; $\triangle$, 5.5. The preparation was extracted from rat liver. Enzyme concentration, 30 mg/ml. The reaction rate was spectrophotometrically determined at 550 nm.

which is seen by comparing Fig. 8 with Figs. 2 and 3. They provide strong support for Hackenbrock's observations.

The observations reported here on the membrane-bound form might therefore be attributed to the influence of the polyanionic environment provided by the membrane. On the other hand, the slight pH profile shift obtained with the purified enzyme could be the consequence of the polycationic character it acquires in this rather acidic pH range (4.5–5.5), where a number of anionic (carboxyl) groups become protonated, whereas all the cationic groups (histidine, lysine, arginine) bear a positive charge. The resulting positive electrostatic potential repels the protons so that the apparent pK of the ionizable groups involved in the reaction is decreased [Eq. (4) with $\overline{\Psi} > 0$]. This effect disappears as ionic strength increases. Such behavior has been observed with other proteins (Goldstein *et al.*, 1964; Sluyterman and Degraff, 1972) and shows

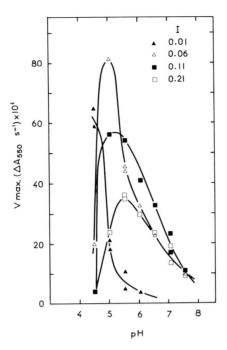

Fig. 9 Purified cytochrome *c* oxidase; pH profile at various ionic strengths (in KCl). The enzyme was extracted from beef heart according to the method of Yonetani (1961) in 1% Tween-20 phosphate buffer. Enzyme concentration, 5 n*M*.

that, even when soluble, an enzyme is able to generate a substantial electrostatic potential in its environment, which in turn influences its catalytic behavior according to the polyelectrolyte theory.

Two questions must be raised at this point: (a) Which ionizable groups of the membrane are responsible for the polyanionic environment? (b) Why is the purified enzyme strongly stimulated, under certain conditions, by the addition of phospholipids? Many authors have documented this stimulation but have not been able to explain it (Nicholls and Chance, 1974; Capaldi and Briggs, 1976; Yonetani, 1961; Chuang and Crane, 1973; Chuang *et al.*, 1973; Roberts and Hess, 1977; Awasthi *et al.*, 1971). Some authors suspect a conformational change of the enzyme upon phospholipid binding (Chuang and Crane, 1973; Chuang *et al.*, 1973; Robinson and Capaldi, 1977). Since conformational changes in proteins might involve concomitant changes in the p*K* of ionizable groups, it should be possible to ascribe the differences in the behavior of purified and membrane-bound forms to this conformational change, rather than to the polyanionic environment.

In an attempt to answer these questions, the experiments shown in Figs. 10 and 11 were carried out. Figure 10 compares the pH profile of the purified enzyme, at low ionic strength (i.e., 0.01 *M*), with the pH profile obtained after the enzyme had been incubated (1 hour at 0°C) in the presence of liposomes of phosphatidylcholine and of cardiolipin, for which it has a high affinity (Chuang and Crane, 1973; Chuang *et al.*, 1973). While liposomes of phosphatidylcholine (zero net charge in the pH range investigated here) involve only a slight activation of the enzyme without modifying its pH profile, liposomes of cardiolipin (negative charge) induce a strong change in the pH optimum of the reaction, which appears to lie between 6.5 and 7.0, that is, as observed with the particulate preparation. While both types of liposomes provide the enzyme with a hydrophobic environment, only those developing a strong negative electrostatic potential modify the pH profile in a manner similar to that of the membrane. These observations and their interpretation in terms of the polyelectrolyte theory allow one to understand why, at neutral pH and moderate ionic strength, anionic phospholipids are more powerful activators of the enzyme than the zwitterionic phos-

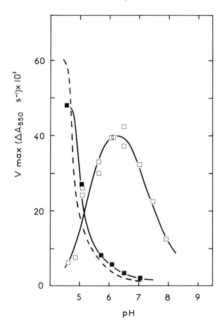

Fig. 10 pH profiles of purified cytochrome *c* oxidase at low ionic strength (10 m*M* in KCl). ---, enzyme incubated in 1% Tween-20 phosphate buffer; ■, enzyme incubated under the same conditions in the presence of phosphatidylcholine (7 mg/ml) liposomes; □, enzyme incubated under the same conditions in the presence of cardiolipin (7 mg/ml) liposomes. Enzyme concentration, 5 n*M*.

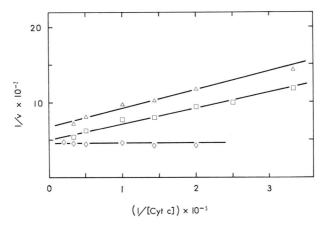

Fig. 11   Lineweaver–Burk plots of purified cytochrome c oxidase activity at pH 7.0 and 10 mM in KCl. $\triangle$, purified enzyme; $\diamond$, purified enzyme incubated in the presence of cardiolipin liposomes (enzyme concentration, 5 nM); $\square$, in the presence of phosphatidylcholine liposomes.

pholipids (Roberts and Hess, 1977). Furthermore, it can be seen in Fig. 11 that, while the $K_{m(app)}$ of the enzyme (relative to cytochrome c) is not significantly affected after incubation in liposomes of phosphatidylcholine, it is strongly reduced by incubation in cardiolipin liposomes. Since cytochrome c is a very basic protein (Roberts and Hess, 1977), it is attracted by the negative electrostatic potential in the cardiolipin liposome environment, where it becomes more concentrated. According to Eqs. (5) and (6) the $K_{m(app)}$ is therefore decreased.

It can reasonably be concluded from this series of experiments that the pH–ionic strength effects observed on the catalytic behavior of cytochrome c oxidase in the inner membrane of mitochondria must be attributed to the influence of the polyanionic environment created around the enzyme by the ionic heads of phospholipids. As further proof in favor of this interpretation, let us emphasize that cytochrome c oxidase is found to be strongly associated with phospholipids, 75% of which are cardiolipins (Chuang and Crane, 1973; Awasthi *et al.*, 1971).

These considerations therefore seem to be of interest for further studies concerning the ionic regulation of the electrons transport chain of mitochondria.

C. Conclusion

The satisfactory interpretation of the results discussed in this section in terms of the catalytic implications of the polyelectrolyte theory suggests that membranes could provide their bound enzyme systems with polyelectrolytic environments, thereby allowing an ionic regulation of their activity.

Such environments are presumably created by the charged polar heads of phospholipids. However, since only some of them bear a non-zero net charge (i.e., cardiolipin, phosphatidylserine, phosphatidylglycerol, and phosphatidylinositol), it must be assumed that polyanionic environments result from the local heterogeneous distribution of phospholipids, which leads to the formation of domains inside of which negatively charged phospholipids prevail.

At a time when one of the main objectives in the study of the molecular biology of membranes is to understand the relationship between structure and function (Rothfield, 1971), such conclusions appear to be of particular interest.

IV. GENE EXPRESSION SYSTEMS

Synthesis of the fundamental biological macromolecules is carried out by highly organized systems such as DNA polymerase, RNA polymerase, and ribosomes, which in their active states involve both nucleic acids and proteins in close interaction. It is well established that these systems are strongly influenced by the ionic environment. Slight modifications in mono-, di-, and even polyvalent ion concentration involve corresponding marked changes in activity, generally following the same pattern regardless of the reaction considered. This apparent uniformity in the "response" to the ionic environment of so widely different systems raises the problem of a common mechanism of regulation. Since all of these systems associate enzymes to nucleic acids, which in the usual range of pH values are highly charged polyanions, it is reasonable to expect that their behavior obeys the catalytic implications of the polyelectrolyte theory.

With respect to the pH–ionic strength and polyamine effects, we have shown (Douzou and Maurel, 1977b) that the RNase–RNA reaction, which can be considered to be a model of the more complex reactions involved in DNA, RNA, and protein synthesis, behaves according to the polyelectrolyte theory. In this section we report further experimental results on this reaction. Our main objective in presenting these data is not to provide additional support to our previous conclusions, but to (a) emphasize the inability of the simplified treatment developed here to explain certain observations frequently made about the systems investigated (a criticism of this treatment based on some of the results reported here will be made in Section V) and (b) provide some "standard observations," made with a simple model reaction behaving according to the polyelectrolyte theory, with which the results reported for more complex reactions involved in gene expression could be compared. From an extensive review of the literature on the subject, such a comparative study allowed us, first, to explain satisfactorily the effect of polyamines on these reactions and, second, to assess the general applicability of our treatment of the catalytic implications of the polyelectrolyte theory.

A. The RNase I–RNA Reaction: Polyamines as "Modulators" of the Electrostatic Potential of the System

The reaction between ribonuclease I (ribonucleate 3′-pyrimidino-oligonucleotidohydrolase, EC 3.1.4.22) and RNA is particularly suited for study in the light of the catalytic implications of the polyelectrolyte theory since, as is lysozyme, in addition to its physiological substrate it is also active on cytidine 2′,3′-phosphate, a low molecular weight molecule (Richards and Wickoff, 1971). The actual influence of the electrostatic potential generated by RNA on the reaction can therefore be directly evaluated by comparing the behavior of the enzyme with these two substrates. Figure 12 shows, for the RNase–RNA reaction, the plots

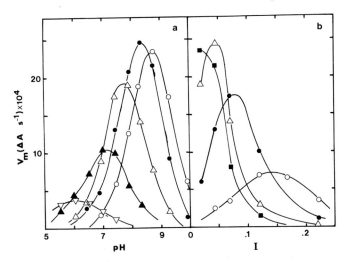

Fig. 12 pH–ionic strength effects on the RNase I–RNA reaction. (a) V_{max} as a function of pH for various ionic strengths (in NaCl): $\bigcirc$, 20 mM; $\bullet$, 45 mM; $\triangle$, 70 mM; $\blacktriangle$, 0.12 M; $\triangledown$, 0.22 M. (b) V_{max} as a function of ionic strength for various pH values: $\blacksquare$, 8.75; $\triangle$, 8.3; $\bullet$, 7.5; $\bigcirc$, 6.5. Enzyme concentration, 1.82 nM.

V_{max}–pH at various ionic strengths and V_{max}–ionic strength at various pH values. Except for the important inhibition observed on V_{max} (pH optimum) as ionic strength increases (a consequence of the concomitant change in the structure of RNA), which will be analyzed in Section V, the system behaves according to the theoretical previsions in Figs. 2 and 3. RNase is electrostatically attracted by RNA and hydrolysis occurs under the influence of the negative electrostatic potential developed by the phosphate groups. As expected, when cytidine 2'3'-phosphate is used as a substrate, the plots of V_{max}–pH are not shifted by ionic strength, and the pH optimum 6.8 is not far from that obtained for RNA in the presence of high ionic strength, that is, in the absence of electrostatic potential.

These results are in good qualitative agreement with other reports in the literature (Kalnitsky et al., 1959; Irie, 1965). However, some disagreements appear. For example, we did not observe the profile shift toward basic pH values as ionic strength increases found by Irie (1965) with cytidine 2',3'-phosphate. Moreover, in contrast with our observations, as well as with those of Kalnitsky et al. (1959), the pH profiles reported by Irie at 0.1 and 0.25 M in NaCl (using RNA as the substrate) present a shoulder in their acidic portion. We have no explanation for these disagreements.

A much more interesting aspect of this work concerns the effect of polyamines on this reaction and, by extension, on the other more specific reactions involved in gene expression.

Polyamines such as putrescine, spermidine, and spermine are polycations (bearing, respectively, two, three, and four positive charges) found in most living systems (Tabor and Tabor, 1976). As a general rule, their synthesis is strongly accelerated in highly proliferating tissues. With respect to the relationship observed between their biosynthesis and that of nucleic acids and proteins, it has been suggested that these polyamines are directly responsible for the increase in macromolecular synthesis occurring during growth and neoplasia (Fujita et al., 1976). The effects of these molecules on the behavior of a variety of gene expression systems have been widely investigated (Stirpe and Novello, 1970; Pegg, 1971; Igarashi et al., 1973, 1974, 1975; Schuber and Pinck, 1974; Konecki et al., 1975; Atkins et al., 1975; Santi and Webster, 1975; Takeda and Ohnishi, 1975; Carr et al., 1975; Moruzzi et al., 1975; Lillehang and Kleppe, 1975; Raae et al., 1975; Yoshida et al., 1976; Rose and Jacobs, 1976; Evans and Deutscher, 1976; Fuchs, 1976; Kumagai et al., 1977; Raymondjean et al., 1977; Levy et al., 1973, 1974). The following is a summary of the most frequently reported observations.

1. There is no absolute requirement for polyamines in most of the gene expression reactions.
2. Polyamines modify, and actually decrease, the pH optimum of the reactions.
3. They modify the ionic requirement for maximal activity. As a general observation from all investigated systems, as the polyamine concentration increases, the ionic concentration giving maximal activity decreases.
4. Depending on the conditions of pH and ionic strength, polyamines appear as either activators or inhibitors.
5. They can replace divalent metallic cations.
6. There is a correlation between their action and their charge number.
7. Their action depends on nucleic acid concentration and structure, but is generally not affected by enzyme concentration.

Although some of these observations are in most cases ascribed to polyamines binding to nucleic acids, they generally are not fully understood. Considering these reactions in the light of the polyelectrolyte theory, let us see how most of the aspects of the effect of polyamines can be understood.

Due to their polycationic nature, polyamines bind tightly to nucleic acids through an electrostatic interaction with their negative charges (Tabor and Tabor, 1976). It is therefore expected that this binding involves a local decrease in the electrostatic potential created by the phosphate groups. Polyamines thus appear to be "modulators" of the electrostatic potential developed by these systems. Their action can then be understood, according to the catalytic implications of the polyelectrolyte theory, in terms of local changes in the electrostatic potential and, therefore, in terms of perturbations of the interdependence of pH–ionic strength on the biological activity of the systems (Douzou and Maurel, 1977b; Maurel, 1976). We obtained the following results with the RNase I–RNA reaction, which show that these previsions are indeed confirmed.

Figures 13 and 14 show the pH profiles of the reaction obtained at low ionic content in the presence of increasing concentrations of spermidine and putrescine, respectively. It appears that, as shown in Fig. 12 for NaCl, polyamines produce a strong pH profile shift toward acidic pH values, which results from decreasing the electrostatic potential generated in the environment of the enzyme. However, here again, the strong inhibition observed on V_{max} (pH optimum) as the polyamine concentration increases cannot be explained by the simplified treatment of the catalytic implications of polyelectrolyte theory. This aspect as well as the

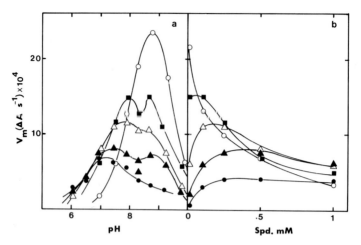

Fig. 13 pH–spermidine (Spd.) effects on the RNase I–RNA reaction. (a) V_{max} as a function of pH for various concentrations of spermidine: ○, 0; ■, 0.1 mM; △, 0.25 mM; ▲, 0.5 mM; ●, 1 mM. (b) V_{max} as a function of spermidine concentration for various pH values: ○, 8.5; ■, 8.0; △, 7.5; ▲, 7.0; ●, 6.5. Enzyme concentration, 1.82 nM.

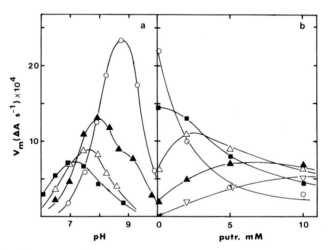

Fig. 14 pH–putrescine (putr.) effects on the RNase I–RNA reaction. (a) V_{max} as a function of pH for various concentrations of putrescine: ○, 0; ▲, 2 mM; △, 5 mM; ■, 10 mM. (b) V_{max} as a function of putrescine concentration for various pH values: ○, 8.5; ■, 8.0; △, 7.5; ▲, 7.0; ▽, 6.5. Enzyme concentration, 1.82 nM.

presence of two optima in some of the pH profiles will be discussed in detail in Section V. A comparison of Figs. 12, 13, and 14 shows that polyamines are much stronger "modulators" than monovalent ions: A concentration of 100 mM NaCl was necessary to decrease the pH optimum by 1.5 units, whereas concentrations of only 1 and 10 mM, respectively, of spermidine and putrescine were necessary to produce the same effect. This efficiency, which must be ascribed to their charge number, shows that polyamines do not behave in the same manner as does ionic strength. While ionic strength modulates the electrostatic potential of the system by affecting the mobile charge distribution surrounding the polyelectrolytic phase, polyamines act internally on the permanent fixed charge distribution of the polyelectrolyte.

Figures 13 and 14 show plots of V_{max} versus polyamine concentration. As in the case of ionic strength, these plots are the consequence of the pH profile shift in the presence of an increasing concentration of polyamines. It can be seen that, depending on the pH, these molecules involve either activation or inhibition of the reaction.

Let us now consider how polyamines perturb the interdependence of the activity, pH, and ionic strength of the system. Figure 15 shows plots of the pH optimum against the logarithm of ionic strength for the RNase–RNA reaction in the absence and in the presence of spermidine

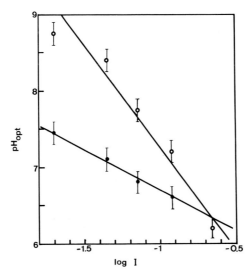

Fig. 15   pH optima of the RNase I–RNA reaction as a function of the logarithm of ionic strength (NaCl) in the absence of spermidine (O) and in the presence of 1 mM spermidine (●).

(1 mM). One first sees that these graphs are linear, as expected from polyelectrolyte theory. Moreover, it should be emphasized that they cross at high ionic strength (0.2–0.3 M) and near the pH optimum obtained with cytidine 2′,3′-phosphate as substrate, that is, when no substantial electrostatic potential influences the enzyme.

We can see that the graph generated in the presence of 1 mM spermidine has a lower slope than that generated in the absence of this reagent, which means that once the electrostatic potential has been partly lowered by the binding of spermidine, ionic strength can still have a screening effect on the "internally modified" system. Polyamine and ionic strength effects on the electrostatic potential of RNA are synergistic.

Figure 16 presents the same aspect of polyamine action from another point of view. In this figure, the response of the activity of the system (as V_{max}) to ionic strength at a constant pH (7.0) is shown for various putrescine concentrations [similar results obtained with spermidine

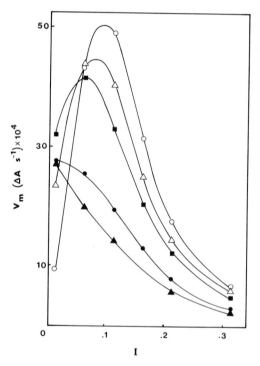

Fig. 16 V_{max} of the RNase I–RNA reaction as a function of ionic strength (NaCl) at pH 7.0 and for various concentrations of putrescine: ○, 0; △, 2 mM; ■, 5 mM; ●, 10 mM; ▲, 20 mM.

have been reported elsewhere (Douzou and Maurel, 1977b)]. In the absence of polyamine, a type I curve is generated, since, according to the patterns presented in Fig. 2, the working pH (here 7.0) is lower than the pH optimum of the reaction (8.8) at minimal ionic strength (0.02). In the presence of a high concentration of putrescine (20 mM), the pH optimum of the reaction is, at minimal ionic strength, about 6.5 (Fig. 14), that is, lower than the working pH. Therefore, increasing ionic strength generates a type II curve. For intermediate increasing polyamine concentrations, a continuous conversion of the response from the type I to the type II curves is observed. An alternative interpretation of this process can be made as follows. As putrescine binds to RNA, it decreases the local electrostatic potential and, according to Eq. (3), decreases the local proton concentration. Increasing concentrations of putrescine should then have the same effect as increasing the pH of the solution, that is, a decrease in the ionic strength giving the optimum. Comparing Figs. 12 and 16, it appears that this is indeed the case.

These results and their interpretation allow us to understand some of the effects polyamines produce on the behavior of RNase–RNA systems and on the other systems involved in DNA, RNA, and protein synthesis. It is obvious, however, that other aspects of the polyamine effects, such as the effect on the structure of nucleic acids or on the enzyme itself, as has been shown by Levy *et al.* (1973, 1974), cannot be excluded and could contribute to the perturbation of the reactions as well.

B. Catalytic Implications of Electrostatic Potentials and the Systems Involved in DNA, RNA, and Protein Synthesis

Like the RNase I–RNA model, the highly organized systems involved in DNA, RNA, and protein synthesis all associate an enzyme to a polyanionic carrier. Does the polyelectrolyte theory apply to these systems? If it does, the typical patterns described for the RNase I–RNA model should also be found for these systems. A review of the relevant literature showed us that these patterns do exist, and a few cases will now be described in more detail.

1. DNA Polymerase

The results obtained by Byrnes *et al.* (1973) on the behavior of a cytoplasmic DNA polymerase from bone marrow as a function of pH and ionic strength are particularly interesting. These authors show that, at a given pH, as ionic strength increases, the activity of the enzyme increases, passes through a maximum, and then decreases (type I curve). The higher the pH, the lower the ionic strength at which maximal activity occurs. It is also apparent that, as ionic strength increases, the pH optimum of the reaction is shifted toward the acidic values. These effects have been related to the reversible dissociation of DNA polymerase from a dimer to monomer, and it has been suggested that this interconversion might be part of a regulatory mechanism of DNA synthesis brought about by physiological variations in pH and ionic strength in the cytoplasm. However, the close similarities observed between these results and those reported for RNase I–RNA, where no enzyme dissociation can be suspected, strongly suggest that the *in vitro* regulation of this system might be explained in terms of the polyelectrolyte theory.

On the other hand, Yoshida *et al.* (1976) showed that the salt requirement for both α and β forms of DNA polymerase from calf thymus was markedly dependent on polyamine concentration. Type I curves were obtained and, as observed in Fig. 16, the maximum appeared to be shifted toward lower salt concentrations as the polyamine concentration increased. This polyamine effect was found to be independent of the form (α or β) of the enzyme and was observed with either natural DNA or a synthetic homopolymer–oligomer duplex. This apparent nonspecificity of the polyamine effect with regard to both enzyme and template favors an interpretation in terms of the polyelectrolyte theory.

2. RNA Polymerase

Stripe and Novello (1970) studied the influence of the concentration of mono- and divalent ions and polyamines on the activity of an RNA polymerase solubilized from rat liver nuclei. They showed that all these additives stimulate RNA synthesis and that their effects are interconnected. Type I curves were found to be modified by divalent ions and polyamines, and in some cases the conversion to type II was observed. These authors demonstrate that no unambiguous interpretation in specific molecular terms can account for all of these observations. In particular, it was shown that the stimulatory effect of the ammonium ion was not due to the detachment of histones from the DNA template or to a specific effect on the enzyme. Since there are striking similarities between these results and those found in the present work, it is reasonable to suppose that they, too, could be explained in terms of the catalytic implications of the polyelectrolyte theory.

3. Ribosomes

It has been shown in this laboratory (Kliber *et al.*, 1976) that the ionic exchanges between the external medium and the 30, 50, and 70 S ribosomal particles from *E. coli* can be understood in terms of the polyelectrolyte theory. The electrostatic potential of each particle is provided mainly by the phosphate groups of the ribosomal RNA and can be modulated by the screening effects of either monovalent or divalent ions (Mg^{2+}), as well as by the cationic groups of the ribosomal proteins. With this picture in mind, it is interesting to consider the results reported by Voigt *et al.* (1974) and Arai and Kasiro (1975) on the effects of NH_4^+ and K^+ ions on the activity of the ribosomal subunits in the EF (elongation factor)-G- and EF-T-dependent GTP hydrolysis. Both groups of workers showed that, when only the 50 S subunit is used, the reaction is inhibited by increasing either the NH_4^+ or K^+ concentration and a type II curve is obtained. It should be noted that magnesium (10 to 14 mM) is also present in the solution. When the same experiment is repeated in the presence of the 30 S particle (reconstituted ribosomes), the reaction is first stimulated by increasing ionic strength, passes through a maximum, and then is inhibited, giving a type I curve. The maximum is around 80 and 20–40 mM for NH_4^+ and K^+, respectively. Furthermore, the higher the EF-G concentration with respect to the 50 S subunits, the higher the ionic concentration giving the maximum. Arai and Kasiro (1975) suggested that these results could be explained by a conformational change of the 50 S particle as it binds to the 30 S particle. This interpretation cannot be excluded, but a satisfactory analysis in terms of the polyelectrolyte theory can also be given.

When 50 S particles alone are used, the local electrostatic potential is strongly reduced because magnesium (which generates a screening effect) is present. Under these conditions, the ascending part of the activity–ionic strength curve is likely to be missing (Fig. 3), so that a type II curve is generated. In other words, according to the theory, the pH optimum of the reaction under these conditions is lower than the working pH of 7.6. After adding 30 S particles, the screening effect of the magnesium ions is lowered because the total number of phosphate groups increases. Under these new conditions, the electrostatic potential is increased and, accordingly, the pH optimum of the reaction is shifted toward alkaline values (higher than 7.6). Increasing the ionic strength then gives rise to a type I curve. Thus, the 30 S particle behaves as a modulator of the 50 S particle by increasing the local proton concentration. The EF-G protein can also be treated as a modulator since, as its concentration rises, the maximum of the type I curves is shifted toward a higher ionic strength. Repeating these experiments at different pH values within the stability range of the ribosomes would provide the data needed to test the applicability of the polyelectrolyte theory to this system more definitely.

4. Other Reactions

Schuber and Pinck (1974) showed that alkaline hydrolysis of the ester bond of amino acid residues linked to tRNA is under the influence of the electrostatic potential promoted by the polynucleotide. Their results were then satisfactorily interpreted on the basis of the polyelectrolyte theory.

C. Conclusion

The results reported and discussed in this section show that the highly organized systems involved in the replication, transcription, and translation of genetic information can be treated in terms of the catalytic implications of the polyelectrolyte theory. This interpretation, formulated in physicochemical terms, accounts for many of the characteristics these systems have in common.

On the other hand, the catalytic implications of the polyelectrolyte theory shed new light on the behavior of nucleic acid–enzyme systems. They also suggest a number of experiments such as the determination of pH profiles (as long as pH may be varied sufficiently without denaturing the system) at various ionic strengths with or without modulators, the use of model polyelectrolytes to perturb the system, and the use of more appropriate molecular probes such as pH-dependent chromophores or label compounds to evaluate the local conditions in specific sites.

V. GENERAL CONCLUSION

The detailed kinetic studies of the systems described in this chapter clearly indicate that one of the essential functions of polyelectrolyte microenvironments is to provide proteins with physicochemical conditions different from those of the bulk medium, and which could be widely varied through modulations of the local electrostatic potentials induced by cations. The data suggest that the control of reactivity by changes in electrostatic potentials may be an important regulatory mechanism brought about by the ionic environment.

In addition to the systems investigated and reported here, an extensive review of the literature on the subject shows that a large number of enzyme systems behave in their natural milieu according to the catalytic implications of the polyelectrolyte theory applied in its simplest form, and a tentative classification has been proposed (Douzou and Maurel, 1977a) based on the examples given in Table I.

TABLE I

Enzyme Reactions Apparently Behaving according to the Catalytic Implications of the Polyelectrolyte Theory[a]

Class	Reaction	Polyelectrolytic carrier	Modulators	Reference
I	Cytochrome c oxidase	Inner membrane of mitochondria	NaCl, KCl	This work
	Lysozyme–$M.$ $luteus$	$M.$ $luteus$ cell walls	NaCl	This work
	ITP hydrolysis in sarcoplasmic reticulum membranes	Sarcoplasmic reticulum membranes	$CaCl_2$	Verjovski-Almeida and de Meis (1977)
	GTP hydrolysis FG dependent on 50 S ribosomal particles	rRNA	NH_4Cl, KCl, 30 S particle, protein factor G	Voigt et $al.$ (1974); Arai and Kasiro (1975)
II	RNase I–RNA	RNA	NaCl, KCl, spermidin, putrescine	This work
	RNase I–RNA	RNA	KCl	Kalnitsky et $al.$ (1959)
	RNase I–RNA	RNA	NaCl, KCl, $CaCl_2$	Irie (1965)
	RNase I–poly(A)	Poly(A)	NaCl	Imura et $al.$ (1965)
	DNase–DNA	DNA	$CaCl_2$	Cuatrecasas et $al.$ (1967)
	DNA polymerase (bone marrow)	DNA	NH_4Cl, $(NH_4)_2SO_4$, NaCl, KCl, $MgCl_2$	Byrnes et $al.$ (1973)
	DNA polymerase (thymus)	DNA	$MgCl_2$, spermine, spermidine, putrescine	Yoshida et $al.$ (1976)
	RNA polymerase (from nucleus of hepatic cells)	DNA	$MgCl_2$, $MnCl_2$, $(NH_4)_2SO_4$, spermine	Stirpe and Novello (1970)
	Hydrolysis of the ester aminoacyl–tRNA bond	tRNA	NaCl, spermidine	Schuber and Pinck (1974)
	Sialidase–sialoglycolipids	Sialoglycolipids, liposomes	NaCl	Barton et $al.$ (1975)
III	Trypsin–BAEE	Trypsin	NaCl	Goldstein et $al.$ (1964)
	Papain–BAEE	Papain	KCl	Sluyterman and Degraaf (1972)
IV	Hemoglobin, Böhr effect	Inositol hexaphosphate		Rollema et $al.$ (1976)
	Glutamate dehydrogenase	Glutamate dehydrogenase	Na_2HPO_4, ATP, GTP	di Prisco (1975)

[a] These reactions were collected according to a previously defined classification (Douzou and Maurel, 1977a). The nature of the carrier providing the polyelectrolytic environment is indicated, as are the reagents (modulators) used by the authors to perturb the response of the system.

The present results and interpretation permit us to consider the potential involvement of metal ions in the regulation of enzymatic and metabolic activities and to gain a detailed knowledge concerning the biology of these ions.

Although the simplified treatment given in Section II and next applied to several systems satisfactorily can account, in qualitative terms, for most of the results, it does not allow a full understanding of some of them. For instance, Fig. 2 shows that, as ionic strength increases, the pH profile is shifted as a whole; the theory does not foresee any modification of the activity of the system at the optimum. All the experimental results, however, show such a modification of activity. In addition, the simplified theoretical treatment adopted does not explain the existence of two optima in some pH profiles, particularly in the RNase–RNA reaction in the presence of polycations, and does not allow an understanding of the specificity that characterizes certain ions. It is beyond doubt that such failures are to be ascribed to the restrictions on which the simplified theoretical treatment is based, as well as to the fact that we only considered the most obvious implications of the electrostatic potentials (local proton and substrate concentrations).

Our oversimplified model did not consider the fact that metal ions are more or less specific in influencing enzyme specificity or the fact that changes in electrostatic potentials will induce conformation changes at the level of the polyelectrolyte carrier, of the enzyme, as well as structural changes of water involved in the solvation shells.

It is well established that proton concentration and ionic content are essential in promoting and maintaining the conformation of the highly charged biopolymers, that is, polyelectrolytes as well as enzymes. It is then obvious that any change in proton concentration will involve further perturbations, particularly in enzyme activity. In this connection, the strong inhibition observed on V_{max} (pH optimum) in Figs. 12, 13, and 14 as ionic strength or polyamine concentration increases could result from a change in the tertiary structure of RNA. As reasonably expected, the random-coil RNA molecule should be more fully extended at low ionic content than at high ionic content. Such a structural change could affect the susceptibility of RNA to RNase action. Unfortunately, it is difficult to relate these perturbations experimentally to recordable conformational changes. The problem of the perturbations possibly induced by changes in water structure is even more inaccessible by present techniques. Under these unfavorable conditions, it will be difficult to explain a number of quantitative observations at the level of the highly organized systems characterized by highly specific reactions involving sequentially several metal ions and charged ligands.

However, progress in understanding certain systems' responses to these agents could be made by simple refinements of the theoretical model. Our initial theoretical treatment implicitly admits that the electrostatic potential is homogeneous within the polyelectrolytic environment and that the monovalent cations form a homogeneous ionic atmosphere. This is, of course, a rough approximation. The following two types of heterogeneities in electrostatic potentials can be expected.

1. A genuine heterogeneity can result from secondary and tertiary conformations of the polyelectrolyte as well as from its permanent interactions with ions and molecules of opposite charges in the resting state of the system. This should be the case for the highly organized systems carrying out the synthesis of biopolymers and for biomembranes characterized by their heterogeneous chemical composition, in which an enzyme operating in different domains would behave differently. In this connection, the shoulder observed on the lysozyme–*M. luteus* pH profile (Fig. 4) at about pH 5.5 and at low ionic strength, as well as its progressive narrowing as ionic strength increases, could result from heterogeneous electrostatic potentials at the level of the peptidoglycan networks.

2. An induced heterogeneity can result from the specific binding of highly charged molecules or ions to the polyelectrolyte, in contrast to the diffuse binding of cations of spherical symmetry forming a homogeneous ionic atmosphere. Indeed, this emphasizes the importance of the distinction we made in Section II between external and internal modulators, the former being characterized by a poor affinity constant for the polyelectrolyte, and the latter forming tight electrostatically stabilized complexes. For the internal modulators, the value of the electrostatic potential will be neutralized at the binding site and, here again, enzyme molecules operating in such a heterogeneous microphase will be split into two populations: one operating under minimum electrostatic potential in the proximity of the binding site (region 1) and the other operating under an electrostatic potential controlled by ionic strength (region 2). Then, depending on pH and ionic strength conditions, the effect of internal modulators would direct the reaction toward either region 1 or region 2, a behavior that would explain the changes in specificity some reagents induce in a reaction, which are otherwise unexplained in a different context.

With respect to the second type of heterogeneity, let us return to Figs. 13 and 14, where the pH profiles of the RNase I–RNA reaction are presented at low ionic strength (0.01 M NaCl) in the presence and absence of polyamines (i.e., internal modulators). In these experiments, the

maximal concentration of RNA (expressed as phosphate) was on the
order of 0.3 mM. When no polyamine is present in the medium, a
narrow single peak is observed at pH 8.8. At low concentrations of a
polyamine, for instance, 0.1 mM spermidine, two optima appear: one at
8.8 and the other at a lower pH. As spermidine concentration increases,
the optimum at 8.8 does not move but progressively disappears,
whereas the other optimum regularly moves toward acidic pH. At a
high concentration of spermidine, only one optimum remains
significantly shifted in the acidic range.

Indeed, these results reflect the heterogeneity induced in the electrostatic potential of RNA through the binding of small amounts of
polyamines. Since their concentration is low, we can suspect that some
regions of the RNA molecules are still statistically free of polyamines
(i.e., region 2) so that the reaction proceeds at a rate the same as that in
the absence of polyamines and the optimum appears at pH 8.8. In those
regions (region 1) adjacent to the binding sites of polyamines, the reaction occurs under a low electrostatic potential, due to the local neu-
tralization of the phosphate groups, and the pH optimum decreases
according to Eq. (4). As the concentration of polyamines increases, all
available sites on the RNA molecule are progressively filled and a
moderate homogeneous electrostatic potential results, as indicated by
the presence of only one peak on the pH profile. Let us recall that such
perturbations are not observed when the same experiment is performed
with increasing concentrations of NaCl (external modulator) instead of
polyamines (Fig. 12).

Regardless of the way in which it is generated, heterogeneity in the
electrostatic potential of a polyelectrolyte–enzyme system means that
not all of the enzyme molecules are active simultaneously. It follows
that, when the environmental factors (e.g., pH and ionic strength) are
changed within reasonable limits, there always remains a fraction of
fully active molecules so that the resulting activity of the system, although moderate, is not greatly affected.

It might be concluded from such observations that heterogeneous
electrostatic potentials of a polyelectrolyte phase involve some kind of
"buffering" effect on the activity of bound enzymes against changes in
the environmental factors.

We have seen that, when ionic strength is increased at constant pH,
the RNase–RNA reaction is activated, reaches an optimum, and is then
progressively inhibited (Fig. 3, type I curve) and that, after polyamine
binding, the response of the reaction becomes of type II, due to the shift
of the curve toward lower ionic strengths. A similar result can be obtained in the absence of polyamines by increasing the pH of the

medium, and we have explained the conversion of type I curves to type II curves by a drastic local decrease in electrostatic potential values. We could add that such curves and the conversion of such curves might be used to investigate in organized systems the main effect of the binding of highly charged ligands such as protein factors. On the other hand, it should be stressed that heterogeneity in the electrostatic potential values would be a rule in the highly organized systems synthesizing biopolymers. We are just beginning to explore the implications of electrostatic potentials both at the level of biomembranes and at the level of systems involved in gene expression, where there are many interrelated regulatory mechanisms waiting to be discovered. The preliminary evidences presented here of these implications should pave the way to further investigation in that emerging domain.

ACKNOWLEDGMENTS

The work presented in this chapter was supported by grants from the Institut National de la Santé et de la Recherche Médicale (Groupe U-128), the Délégation Générale à la Recherche Scientifique et Technique (Contrat 75.7.0745), and the Fondation pour la Recherche Médicale Française.

REFERENCES

Arai, N., and Kasiro, Y. (1975). *J. Biochem. (Tokyo)* **77**, 439–447.

Atkins, J. F., Lewis, J. B., Anderson, C. W., and Gesteland, R. F. (1975). *J. Biol. Chem.* **250**, 5688–5695.

Awasthi, Y. C., Chuang, T. F., Keenan, T. W., and Crane, F. L. (1971). *Biochim. Biophys. Acta* **226**, 42–52.

Banerjee, S. K., Kregar, I., Turk, V., and Rupley, J. A. (1973). *J. Biol. Chem.* **248**, 4786–4792.

Barton, N. W., Lipovak, V., and Rosenberg, A. (1975). *J. Biol. Chem.* **250**, 8462–8466.

Blake, C. C. F., Johnson, L. N., Mair, G. A., North, A. C. T., Phillips, D. C., and Sharma, V. R. (1967). *Proc. R. Soc. London* **167**, 378–388.

Brown, R. H., Jr. (1974). *Prog. Biophys.* **28**, 341–370.

Bygrave, F. L. (1967). *Nature (London)* **214**, 667–671.

Byrnes, J. J., Downey, K. M., and So, A. G. (1973). *Biochemistry* **12**, 4378–4384.

Capaldi, R. A., and Briggs, M. (1976). In "The Enzymes of Biological Membranes" (A. Martonosi, ed.), Vol. 4, pp. 86–102. Wiley, New York.

Carr, A. C., Igloi, G. L., Penzer, G. R., and Plumbridge, J. A. (1975). *Eur. J. Biochem.* **54**, 169–173.

Chang, K. Y., and Carr, C. W. (1971). *Biochim. Biophys. Acta* **229**, 496–503.

Chuang, T. F., and Crane, F. L. (1973). *J. Bioenerg.* **4**, 563–578.

Chuang, T. F., Awasthi, Y. C., and Crane, F. L. (1973). *J. Bioenerg.* **5**, 27–72.

Cuatrecasas, P., Fuchs, S., and Anfinsen, C. (1967). *J. Biol. Chem.* **242**, 1541–1547.

Davies, R. C., Neuberger, A., and Wilson, B. M. (1969). *Biochim. Biophys. Acta* **178**, 294–305.

di Prisco, G. (1975). *Arch. Biochem. Biophys.* **171,** 604–612.

Douzou, P., and Maurel, P. (1976). *C.R. Hebd. Seances Acad. Sci., Ser. D* **282,** 2107–2110.

Douzou, P., and Maurel, P. (1977a). *Trends Biochem. Sci.* **2,** 14–17.

Douzou, P., and Maurel, P. (1977b). *Proc. Natl. Acad. Sci. U. S. A.* **74,** 1013–1015.

Engasser, J. M., and Horvath, C. (1975). *Biochem. J.* **145,** 431–435.

Evans, J. A., and Deutscher, M. P. (1976). *J. Biol. Chem.* **251,** 6646–6652.

Felsenfeld, G., and Miles, H. T. (1967). *Annu. Rev. Biochem.* **36,** 407–448.

Fuchs, F. (1976). *Eur. J. Biochem.* **63,** 15–22.

Fujita, K., Nagatsu, T., Maruta, K., Ito, M., Senba, M., and Miki, K. (1976). *Cancer Res.* **36,** 1320–1324.

Goldstein, L. (1972). *Biochemistry* **11,** 4072–4084.

Goldstein, L. (1977). *In* "Methods in Enzymology" (K. Mosbach, ed.), Vol. 44, pp. 397–443. Academic Press, New York.

Goldstein, L., Levin, Y., and Katchalsky, E. (1964). *Biochemistry* **3,** 1913–1919.

Hackenbrock, C. R. (1975). *Arch. Biochem. Biophys.* **170,** 139–148.

Hallan, C., and Wrigglesworth, J. M. (1976). *Biochem. J.* **156,** 159–165.

Igarashi, K., Hikami, K., Sugawara, K., and Hirose, S. (1973). *Biochim. Biophys. Acta* **299,** 325–330.

Igarashi, K., Sugarawa, K., Izumi, I., Nagayama, C., and Hirose, S. (1974). *Eur. J. Biochem.* **48,** 495–502.

Igarashi, K., Kumagai, H., Watanabe, Y., Toyoda, N., and Hirose, S. (1975). *Biochem. Biophys. Res. Commun.* **67,** 1070–1077.

Imoto, T., Johnson, L. N., North, A. C. T., Phillips, D. C., and Rupley, J. A. (1972). *In* "The Enzymes" (P. D. Boyer, ed.), 3rd ed., Vol. 7, pp. 665–868. Academic Press, New York.

Imura, N., Irie, M., and Ukita, T. (1965). *J. Biochem. (Tokyo)* **58,** 264–272.

Irie, M. (1965). *J. Biochem. (Tokyo)* **57,** 355–362.

Jacob, F., Brenner, S., and Cuzin, F. (1963). *Cold Spring Harbor Symp. Quant. Biol.* **28,** 329–348.

Jensen, H. B., and Kleppe, K. (1972). *Eur. J. Biochem.* **28,** 116–122.

Kalnitsky, G., Hummel, J. P., and Dierks, C. (1959). *J. Biol. Chem.* **234,** 1512–1516.

Katchalsky, E., Bichowsky-Slomnitski, L., and Volcani, B. E. (1953). *Biochem. J.* **55,** 671–680.

Kliber, J. S., Hui Bon Hoa, G., Douzou, P., Graffe, M., and Grunberg-Manago, M. (1976). *Nucleic Acids Res.* **3,** 3423–3438.

Konecki, D., Kramer, G., Pinphanichakarn, P., and Hardesty, B. (1975). *Arch. Biochem. Biophys.* **169,** 192–198.

Kotin, L. (1963). *J. Mol. Biol.* **7,** 309–311.

Kumagai, H., Igarashi, K., Yoshikawa, M., and Hirose, S. (1977). *J. Biochem. (Tokyo)* **81,** 383–388 and 389–394.

Levy, C. C., Mitch, W. E., and Schmukler, M. (1973). *J. Biol. Chem.* **248,** 5712–5719.

Levy, C. C., Hieter, P. A., and Legendre, S. M. (1974). *J. Biol. Chem.* **249,** 6762–6769.

Lillehaug, J. R., and Kleppe, K. (1975). *Biochemistry* **14,** 1225–1229.

McLaren, A. C. (1957). *Science* **125,** 697.

McLaren, A. C., and Esterman, E. F. (1957). *Arch. Biochem. Biophys.* **68,** 157–160.

Maurel, P. (1976). Thèse Doctorat d'État, Université de Paris VII.

Maurel, P., and Douzou, P. (1976). *J. Mol. Biol.* **102,** 253–264.

Maurel, P., Douzou, P., Woldman, J., and Yonetani, T. (1978). *Biochim. Biophys. Acta* (in press).

Monod, J., and Jacob, F. (1961). *J. Mol. Biol.* **3,** 318–356.

Monod, J., Changeux, J. P., and Jacob, F. (1963). *J. Mol. Biol.* **6**, 306–329.

Monod, J., Wyman, J., and Changeux, J. P. (1965). *J. Mol. Biol.* **12**, 88–118.

Moruzzi, G., Barbiroli, B., Moruzzi, M. S., and Tadolini, B. (1975). *Biochem. J.* **146**, 697–703.

Neville, W. M., and Eyring, H. (1972). *Proc. Natl. Acad. Sci. U.S.A.* **69**, 2417–2419.

Nicholls, P., and Chance, B. (1974). *In* "Molecular Mechanisms of Oxygen Activation" (O. Hayaishi, ed.), Chapter 12, pp. 479–534. Academic Press, New York.

Pegg, A. E. (1971). *Biochim. Biophys. Acta* **232**, 630–642.

Raae, A. J., Kleppe, R. K., and Kleppe, K. (1975). *Eur. J. Biochem.* **60**, 437–443.

Rahmann, H., Rosner, H., and Breer, H. (1976). *J. Theor. Biol.* **57**, 231–237.

Raymondjean, M., Bogdanovsky, D., Bachner, L., Kneip, B., and Schapira, G. (1977). *FEBS Lett.* **76**, 311–315.

Rice, S. A., and Nagasawa, M. (1961). "Polyelectrolyte Solutions," 1st ed., Chapter 7, pp. 283–290. Academic Press, New York.

Richards, F. M., and Wickoff, H. W. (1971). *In* "The Enzymes" (P. D. Boyer, ed.), 3rd ed., Chapter 24, pp. 647–806. Academic Press, New York.

Roberts, H., and Hess, B. (1977). *Biochim. Biophys. Acta* **462**, 215–234.

Robinson, N. C., and Capaldi, R. A. (1977). *Biochemistry* **16**, 375–381.

Rollema, H. S., de Bruin, S., and van Os, A. J. (1976). *Biophys. Chem.* **4**, 223–228.

Rose, K. M., and Jacobs, S. T. (1976). *Arch. Biochem. Biophys.* **175**, 748–753.

Rothfield, L. I. (1971). *In* "Structure and Function of Biological Membranes" (L. I. Rothfield, ed.), Chapter 1, pp. 3–9. Academic Press, New York.

Rothfield, L. I., and Romeo, D. (1971). *In* "Structure and Function of Biological Membranes" (L. I. Rothfield, ed.), Chapter 6, pp. 251–285. Academic Press, New York.

Saint-Blancard, J., and Jollès, P. (1972). *Biochimie* **54**, 7–15.

Salton, M. R. J. (1964). *Proc. Int. Symp. Fleming's Lysozyme, 3rd*, p. 5/RT.

Santi, D. V., and Webster, R. W., Jr. (1975). *J. Biol. Chem.* **250**, 3874–3877.

Schuber, F., and Pinck, M. (1974). *Biochimie* **56**, 397–403.

Sela, M., and Steiner, L. A. (1963). *Biochemistry* **2**, 416–421.

Singer, S. J. (1972). *Ann. N.Y. Acad. Sci.* **195**, 16–23.

Sluyterman, L. A. A., and Degraaf, M. J. M. (1972). *Biochim. Biophys. Acta* **258**, 554–561.

Stirpe, F., and Novello, F. (1970). *Eur. J. Biochem.* **15**, 505–512.

Tabor, C. W., and Tabor, H. (1976). *Annu. Rev. Biochem.* **45**, 285–306.

Takeda, Y., and Ohnishi, T. (1975). *J. Biol. Chem.* **250**, 3878–3882.

Träuble, H., Teubner, M., Woolley, P., and Eibl, H. (1976). *Biophys. Chem.* **4**, 319–342.

Verjovski-Almeida, S., and de Meis, L. (1977). *Biochemistry* **16**, 329–334.

Voigt, J., Sander, G., Nagel, K., and Parameggiani, A. (1974). *Biochem. Biophys. Res. Commun.* **57**, 1279–1286.

von Hippel, P. H., and Schleich, T. (1969). *In* "Structure and Stability of Biological Macromolecules" (S. N. Timasheff and G. D. Fasman, eds.), Chapter 6, pp. 417–574. Dekker, New York.

Webb, J. L. (1963). "Enzyme and Metabolic Inhibitors," Vol. 1, pp. 815–839. Academic Press, New York.

Yonetani, T. (1961). *J. Biol. Chem.* **236**, 1680–1688.

Yoshida, S., Masaki, S., and Ando, T. (1976). *J. Biochem. (Tokyo)* **79**, 895–901.

20

The Proton Pumps of
Photosynthesis

A. R. CROFTS AND S. SAPHON

Department of Biochemistry, Medical School, University of Bristol, Bristol, United Kingdom

I. INTRODUCTION

Since the first demonstration of light-induced H^+ uptake by chloroplasts (Jagendorf and Hind, 1963; Neumann and Jagendorf, 1964) and by chromatophores from photosynthetic bacteria (von Stedingk and Baltscheffsky, 1966), much effort has been devoted to characterizing the nature of the process of H^+ uptake, its relation to electron transport and to the mechanism of coupling to phosphorylation, and the molecular mechanisms of the H^+ pumps themselves. We will be concerned here with a brief review of the way in which the arrangement in the membrane of the photosynthetic electron transport chains of chloroplasts and chromatophores determines that these act as proton pumps, and we will describe our work on some of the protolytic reactions involved.

II. PROTOLYTIC REACTIONS IN GREEN PLANT PHOTOSYNTHESIS

It seems fairly well established (see reviews, Walker and Crofts, 1970; Witt, 1971, 1975; Junge, 1975; Trebst, 1974; Bendall, 1977) that the protons released on oxidation of H_2O and of plastodihydroquinone appear in the inner aqueous phase of the thylakoid and that the protons in-

FRONTIERS IN PHYSICOCHEMICAL BIOLOGY
Copyright © 1978 by Academic Press, Inc.
All rights of reproduction in any form reserved.
ISBN 0-12-566960-7

volved in the reduction of plastoquinone and of system 1 acceptors are lost from the external medium. In none of these cases can it be claimed that the mechanism of the reaction involved is known, even at the level of the identity of reaction partners. Nevertheless, the general "architecture" of the noncyclic electron transport pathway is clear, and the role of protons in these reactions can be convincingly explained in terms of classical Mitchellian proton pumps operating through an anisotropic distribution of redox components arranged to act as alternating H-carrying and electrogenic arms of two H^+ pumping loops. The H-carrying arms are the water/oxygen couple and the plastoquinone/plastohydroquinone couple, with a balancing stoichiometry of external H^+ involved in the reactions on the acceptor side photosystem I. The electrogenic arms of the loops are the photochemical reactions with their immediate donors and acceptors.

Figure 1 shows a schematic representation of the noncyclic electron transport chain of chloroplasts arranged as summarized above. In the following sections we will review some of the evidence for this point of view and examine in greater detail the data available on the sites at which protons are thought to interact with the oxygen-evolving apparatus.

A. Arrangement of the Oxygen-Evolving Apparatus

Parts of the electron transport chain on the oxygen side of photosystem II (Girault and Galmiche, 1974; Giaquinta and Dilley, 1975; Lien and Racker, 1971; Cramer and Horton, 1975; Horton and Cramer, 1974; Izawa and Ort, 1974; Selman et al., 1973; Giaquinta et al., 1974; see

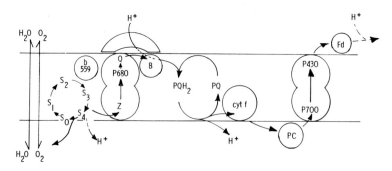

Fig. 1 The electron transport chain of green plant photosynthesis arranged as two proton pumps (no attempt has been made to show correct proton stoichiometries). Fd, ferredoxin.

review by Arntzen and Briantais, 1975) appear to be available to reagents on either side of the thylakoid membrane. However, the protolytic reaction sites are in equilibrium with the aqueous phase inside the thylakoid. Manganese released by relatively mild treatments appears to be discharged into the inner aqueous phase (Babcock and Sauer, 1975). However, in treated chloroplasts, added Mn^{2+} can act as an effective electron donor to photosystem II, as can ferrocyanide (Izawa and Ort, 1974). These are both highly charged ions; in the case of the ferri–ferro cyanide couple, it is known that neither component is able to cross liposome membranes (Deamer et al., 1972), and in the case of Mn^{2+} the ion released after treating the chloroplasts with Tris buffer does not become available to the external phase for several hours (Babcock and Sauer, 1975). Cytochrome b_{559}, which acts as a donor to photosystem II at low temperature, is itself readily and rapidly oxidized by ferricyanide (Horton and Cramer, 1974; Cramer et al., 1974; Ikegami and Katoh, 1973); the water-splitting reactions and cytochrome b_{559} are modified by trypsin and by a variety of other reagents that are unable to cross the membrane. We may conclude that the oxygen-evolving apparatus is bulky and that it spans the membrane, but that the sites of proton release are within the thylakoid. Babcock and Sauer (1975) have proposed a topological model that accounts for many of these observations, the main features of which are included in Fig. 1. Since little more is known of the detailed reaction mechanism, it would be premature to speculate further.

B. Involvement of Protons in the Reactions on the Oxygen Side of Photosystem II

The oxidation of water by electron transfer to photosystem II leads to the release of $4H^+$ per O_2. There has been strong evidence from direct measurements of pH changes (Schwartz, 1968; Fowler and Kok, 1974; Graber and Witt, 1975, 1976; Ausländer and Junge, 1974; Junge and Ausländer, 1974) and from the effects of internal pH on electron transport from water (Siggel, 1975; Bamberger et al., 1973; Wraight et al., 1972) and on delayed fluorescence (Wraight and Crofts, 1971) that the protons released on oxidation of water are in equilibrium with the inner aqueous phase. They appear rapidly outside the thylakoid only in the presence of excess amine or ionophores such as gramicidin, nigericin, or dianemycin, which catalyze the rapid equilibration of H^+ concentration gradients without inhibiting the water-splitting reactions (Renger, 1972a–c; Étienne, 1974). We will discuss the involvement of protons in the reactions of the water-splitting apparatus by reference to the model proposed by Kok *et al.* (1970) and Fowler and Kok (1974), which is summarized in Scheme 1.

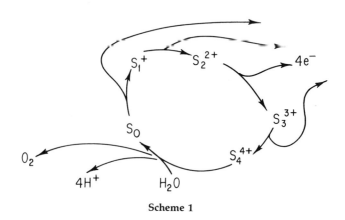

Scheme 1

$$S_4^{4+} + 2H_2O \rightleftharpoons S_0 + 4H^+ + O_2 \qquad (1)$$

C. Reactants in the Transition of S_4 to S_0 *

The mechanism for O_2 evolution is usually shown (see Scheme 1) as involving the four-stage accumulation of charge to give a complex able to react directly with water so that a concerted reaction occurs, leading to the liberation of oxygen and the regeneration of the "uncharged" state. While this is obviously a convenient and simple representation of a reaction generally acknowledged to be considerably more complex, it

* Abbreviations used: S states, the states of the oxygen-evolving apparatus (S_0, S_1, S_2, S_3, S_4) reached through flash illumination from the dark-adapted state; Q, the primary acceptor of photosystem II, so named because in its oxidized form it quenches fluorescence; P, P^+, the special chlorophyll pigment acting as the primary donor of the reaction center, and its oxidized form; P-700, P-680, P-870, the special chlorophylls of photosystem I, photosystem II, and the bacterial reaction center, respectively identified by numbers corresponding to the wavelength maxima of their absorbance change (in nanometers); UQ · Fe (or Fe · UQ), the complex of ubiquinone and iron acting as the primary acceptor in bacterial reaction centers; DBMIB, 2,5-dibromo-3-methyl-6-isopropyl-p-benzo-quinone; DCMU, 3-(3,4-dichlorophenyl)-1,1-dimethylurea; LDAO, lauryl dimethyl amine oxide; FCCP, 4-trifluoromethoxyphenylhydrozone; PMS-S, N-methylphenazonium-3-sulfonate methosulfate.

may also be misleading in an important sense, since the assumed reaction of *water* with the S_4 state of the system has no basis in experimental fact.

Sinclair and Arnason (1974) and Arnason and Sinclair (1976) have attempted to measure directly the involvement of water in these reactions by substituting ^2H_2O for H_2O under conditions in which inhibition would have been expected if reactions involved in the splitting of OH bonds in the system had been rate limiting. They observed no marked isotope effect on the transitions of any of the S states and concluded that the reactions of water involving the splitting of OH (or O^2H) bonds were not rate determining at any step in the cycle.

D. Thermodynamic and Kinetic Considerations

The operating E_m for the primary acceptor Q seems to be ~ -130 mV (Knaff, 1975), and the maximal work expected from 1 quantum of light after photochemical conversion is about 1.2–1.3 V; therefore, we may expect the maximal potential of the P/P^+ couple to be between 0.9 and 1.2 V, depending on how much energy is lost for kinetic reasons and on work terms other than redox work, which need to be satisfied (see Crofts et al., 1971; Ross et al., 1976). The midpotential of the water/oxygen couple in equilibrium with air is 0.806 at pH 7, varying by -60 mV/pH unit at 30°C. We would expect that the water/oxygen couple would be in equilibrium with the P/P^+ couple at some pH value between ~ 5 and ~ 3, but that at pH values above ~ 5 the redox gradient between the two couples would be sufficient to ensure that the concentrations of the reactants and products offered no kinetic limitation, as long as the mean potential is considered. However, experimental evidence suggests that maximal rates of power conversion (measured as rates of phosphorylation) occur at external pH values of approximately 8.5 in conjunction with a pH gradient of >3.5 units. Under these conditions, water oxidation would be occurring in equilibrium with the internal phase at pH < 5. Rumberg and Siggel (1969), Siggel (1975), Rottenberg et al. (1972), and Bamberger et al. (1973) have measured the dependence of electron transport on pH and shown that the uncoupled rate becomes inhibited at values of pH below 5. This inhibition appears to be controlled by protonation of a group within the thylakoid, of pK ~ 5.7, associated with the electron transport chain between PQ and P-700. Electron transfer to PQ appeared to be limited by a group with a pK of ~ 4.7 in equilibrium with an internal proton. Bendall (1977) interpreted the results of Bamberger et al. (1973) as showing that, in addition, electron transport may be controlled by an external site associated with photosystem II, becoming inhibited as the external pH rises above pH 8.2. The rate of oxygen evolution also becomes inhibited at low pH, and this is associated with an inhibited rate of flow to P-680$^+$. Wraight *et al.* (1972) measured the effect of pH on the yield of oxygen from each of a series of flashes and compared these to the steady-state value. Below pH 5.5, a marked deficit of yield was apparent, which Wraight *et al.* interpreted as being due to a failure of oxidation of S_3. However, oscillations of period 4 were apparent in the yield down to pH 5, although with a modified relative yield per flash. Diner and Joliot (1976) showed a marked decline in the turnover time for the photosystem as measured by the transition $S_1^* \rightarrow S_2$ below pH 6.45 but concluded that this reflected an inhibition of the reoxidation of Q$^-$ at lower values of pH.

In Eq. (1) above four protons are released per transition of state S_4^{4+} to S_0. If we regard the equation as representing a conventional chemical reaction, we may express the thermodynamic balance by an appropriate equilibrium constant:

$$K_{eq} = \frac{[S_0][H^+]^4[O_2]}{[S_4^{4+}][H_2O]^2}$$

Recognizing that the activities of O_2 and water in equilibrium with the water-splitting reactions will not vary greatly under physiological conditions, we may write an expression,

$$[S_4^{4+}]/[S_0] = K'[H^+]^4 \qquad \text{where } K' = (1/K_{eq})([O_2]/[H_2O]^2)$$

relating the relative concentrations of states S_4 and S_0 to the concentration of protons. It is known that oxygen evolution occurs over a wide range of values of pH (from 8.5 to 5 or less) for the phase in equilibrium with these reactions. The equilibrium concentration ratio of S_4^{4+}/S_0 as expressed above would change by $>10^{14}$ over this pH range. We may assume that, in practice, the reactions occur some way away from equilibrium at neutral pH since kinetic evidence suggests that the production of oxygen occurs in a spontaneous, rapid reaction. Nevertheless, the range of values of the ratio S_4^{4+}/S_0 expected from the reaction described by Eq. (1) seems unrealistic, especially in view of the kinetic evidence for equal populations of states S_0–S_3 in the steady state. This may be seen more clearly if we include the reaction leading to formation of S_4^{4+} from S_3^{3+} and represent the overall change as a redox half cell:

$$2H_2O + S_3^{3+} \rightarrow S_0 + 4H^+ + O_2 + e^-$$

$$E_h = E^0 - 4(2.3RT/F)\text{pH} + \frac{RT}{F} \ln\left\{ \frac{[S_0]}{[S_3^{3+}]} \cdot \frac{[O_2]}{[H_2O]^2} \right\}$$

where R, T, and F are the gas constant, absolute temperature, and Faraday constant, respectively.

We can see that the operating midpotential for the oxidation of $S_3{}^{3+}$ would have to vary by -240 mV/pH unit. Assuming that in the steady state $[S_0] = [S_3]$, and taking the range of pH over which oxygen evolution occurs, this change of E_m would be 960 mV on going from pH 8.5 to 4.5; all the extra work involved in oxidizing water at lower pH would be loaded onto a single electron transition and would involve the need for an oxidizing agent with a potential of at least this much more positive than the mean midpoint for the H_2O/O_2 couple at the highest operating pH. Since this would require more energy than is available, it seems unlikely, unless very special mechanisms are invoked.

Kinetic arguments may also be given to show that a reaction mechanism that is fourth order in protons is unrealistic, although these may be annulled by proposing a multistage mechanism of proton release; a further difficulty is posed by the energetics of accumulation of four positive charges in a local environment. It may, therefore, be worth enquiring into what modifications of this reaction mechanism might be thermodynamically and kinetically more appealing.

E. Direct Measurements of H^+ Changes Associated with Photosystem II

The most detailed studies of these changes have been those of Fowler and Kok (1974, 1976; Fowler, 1977). They originally observed that, in the presence of excess methylamine or gramicidin, the release of protons on excitation with successive flashes of short duration showed a periodicity of 4, similar to that observed for oxygen evolution. They suggested that the protons were released in synchrony with the oxygen, so that the reaction of the appropriate stage of the cycle of "S" states involved could be described by Eq. (1) above.

In making this suggestion, Fowler and Kok (1974) pointed out some discrepancies between their results and those anticipated on the basis of Eq. (1). In particular, the second flash after dark adaptation (leading to the transition $S_2 \rightarrow S_3$ for the greater proportion of the states) sometimes showed a greater yield of protons than of oxygen, and the yields on other transitions did not match precisely.

The estimation of protons released inside the thylakoids due to reactions on the donor side of photosystem II is complicated by simultaneous uptake and release of protons in reactions on the acceptor side. In their first report, Fowler and Kok (1974) were able to show an unambiguous four-stage oscillation, using high concentrations of methylamine as an uncoupler, since this appeared to inhibit proton uptake associated with the acceptor reactions. Fowler and Kok (1976) and Fowler (1977) have published data dealing with flash stoichiometries of protons and electrons that highlight the difficulties of measuring the protons from water oxidation independently of other protons. In the presence of uncouplers other than methylamine, the release of protons did not appear to be in synchrony with the release of oxygen. Saphon and Crofts (1977) have extended the work of Fowler and Kok (1974, 1976) and Junge and colleagues (Ausläender and Junge, 1974, 1975), by careful use of pH indicator techniques, to observe pH changes induced by a series of flashes, both outside and inside the thylakoids of dark-adapted chloroplasts.

Figures 2–4 show traces of pH changes of chloroplasts measured using cresol red (indicating pH changes in the bulk aqueous phase) or neutral red. This latter indicator may be used to follow pH changes either in the external phase or within the chloroplasts (Ausläender and Junge, 1974, 1975) depending on the presence of buffering groups that penetrate these phases selectively. Figure 2 shows pH changes indicated by cresol red and observed in the external phase in the presence of an uncoupler (which allowed the rapid equilibration of translocated protons) when dark-adapted chloroplasts were illuminated by a series of short saturating flashes. These changes are essentially the same as those previously reported by Fowler and Kok (1976) using an electrode technique. As Fowler and Kok (1976) observed, the changes varied with the electron acceptor present, and the net pH change was as expected from the overall reaction. The extent of H^+ release after each flash varied, as can be seen most clearly in the presence of ferricyanide (Fig. 2, trace B; Fig. 3). The yields are plotted as a function of flash number in Fig. 2D and show a strong periodicity of 2 superimposed on a weaker periodicity of 4. Under similar conditions, but in the absence of uncoupling agent, the uptake of H^+ occurring outside the chloroplast could be seen (not shown). Since this showed no marked periodicity of yield (see also Fowler and Kok, 1976; Fowler, 1977), we concluded that the variability of yield in the presence of uncoupler reflects a variability in the appearance of H^+ within the chloroplasts. This is shown by direct measurement in Fig. 3. Fowler (1977) has shown (and we have confirmed these results) that the oscillation of period 2 reflects the fact that plastoquinone is reduced only after the accumulation of two charges on an intermediary acceptor, B (Bouges-Bocquet, 1973; Velthuys and Amesz, 1974), and can therefore only be reoxidized (so as to release protons inside the thylakoid) on alternate flashes. The amplitude of the oscilla-

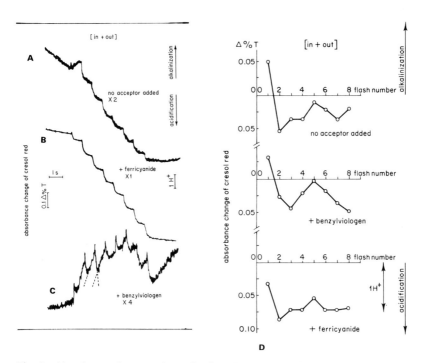

Fig. 2 Absorbance changes of cresol red in the presence of dianemycin and different electron acceptors. All traces are the average of eight separate experiments. Groups of eight flashes (15-microsecond half-band width, spaced at 0.65 second) were used to excite a fresh, dark-adapted sample for each experiment. Chloroplasts (10 μg of chlorophyll/ml) were suspended in medium containing 10 mM KCl, 50 μM cresol red, 4 μg/ml of dianemycin, and acceptor, as indicated, and adjusted to pH 7.2 with a fresh solution of NaOH. Absorbance changes were measured with a single-beam spectrophotometer linked to a minicomputer [see Saphon and Crofts (1977) for further details]. (A) No acceptor added. (B) Potassium ferricyanide added to give 0.5 mM. (C) Benzylviologen added to give 0.1 mM. (D) Change in absorbance induced by each flash plotted as a function of flash number. The changes are proportional to the net yield of protons per flash as measured from traces A–C by extrapolation to the time of the flash. From Saphon and Crofts (1977).

tions reflects the fact that in the dark state some 30–40% of reaction centers were already in the state QB^- and so reduced plastoquinone after a single flash. When account is taken of these protons, a periodicity of 4 due to release of protons in the oxygen-evolving reactions becomes more apparent. By appropriate use of acceptors and inhibitors (ferricyanide in the presence of DBMIB), the reduction (and subsequent oxidation) of plastoquinone can be prevented, and the periodicity can

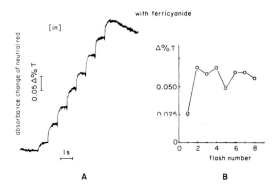

Fig. 3 Absorbance changes of neutral red due to pH changes within the chloroplasts. The absorbance changes (at 524 nm) insensitive to buffering by bovine serum albumin (2 mg/ml of BSA) but sensitive to buffering by imidazole (4 mM) plus BSA are shown. (A) The trace is the average of eight experiments, with a flash regime as in Fig. 1. Chloroplasts (20 μg of chlorophyll/ml) were suspended in 10 mM KCl, 10 μM neutral red, 3 μM valinomycin, and 0.1 mM potassium ferricyanide, and the medium was adjusted to pH 7.2 with a fresh solution of NaOH. (B) Absorbance change on each flash plotted as a function of flash number. The changes are proportional to the yield of protons inside the thylakoid per flash.

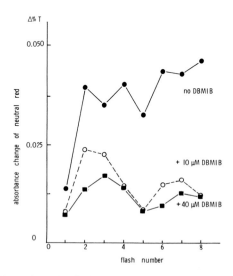

Fig. 4 Effect of DBMIB on the flash yields of protons within the thylakoids. Conditions were as for Fig. 3 except that DBMIB was added as indicated.

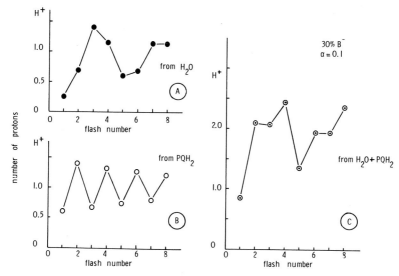

Fig. 5 Predicted flash yields of protons released inside the thylakoids. (A) Protons released on oxidation of water. (B) Protons released on oxidation of plastoquinol. (C) Sum of (A) and (B). Calculations were based on the model of Fig. 7, assuming 25% S_0, 75% S_1, 30% B^-, and $\alpha = 0.1$ using the symbols as defined by Kok *et al.* (1970) and Bouges-Bocquet (1973).

be measured directly (Fig. 4). Detailed analysis of these and similar results shows that, on illumination of dark-adapted chloroplasts, protons are released on all transitions of the S states, with the exception of the transition $S_1 \rightarrow S_2$, rather than solely or predominantly on the transition $S_3 \rightarrow S_0$ as suggested by the earlier work of Fowler and Kok (1974). Figure 5 shows a calculated set of values for the flash yields anticipated for proton release within the thylakoids due to oxidation of water and plastohydroquinone as discussed above. These values fit rather well with our observations, and indeed no other simple pattern of yield from the transitions of the S states fits adequately.

F. Effects of pH on Delayed Fluorescence in the Millisecond to Second Time Range

Wraight and Crofts (1971) observed that, in chloroplasts inhibited with DCMU, the delayed fluorescence from photosystem II was stimulated when a proton gradient was produced through the operation of photosystem I. They suggested that the stimulation of delayed fluorescence reflected the involvement of protons within the thylakoid in the equilibria of the water-splitting reactions. The report by Fowler and Kok

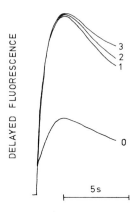

Fig. 6 Effect of flash preillumination on delayed fluorescence in the presence of DCMU. Chloroplasts (23 μg of chlorophyll/ml) were suspended in a medium containing 0.35 M sucrose, 44 mM KCl, 44 mM 4-morpholinepropanesulfonic acid, and 4.4 mM MgCl$_2$, adjusted to pH 7.2. The suspension was dark-adapted for 10 minutes before adding Na-ascorbate (to 1 mM), methylviologen (to 0.1 mM), and diaminodurene (to 0.25 mM). The chloroplasts were then exposed to from zero to three saturating flashes (15-microsecond width at half-height, 320 milliseconds between flashes), and DCMU was added (to 4 μM) approximately 1 second after the final flash. Within 2 seconds of adding the DCMU, the shutter of the phosphoroscope was opened and delayed fluorescence at 1 millisecond was recorded as described previously (Wraight and Crofts, 1971).

(1974) of a proton release from the water-splitting mechanism, which was synchronous with that of O$_2$, rendered the observation of Wraight and Crofts (1971) somewhat ambiguous. P. Joliot (personal communication) has pointed out that, since chloroplasts inhibited with DCMU are able to undergo only a single transition of S states, no stimulation of delayed fluorescence would be expected from the Fowler–Kok hypothesis on illumination of dark-adapted chloroplasts under the conditions of Wraight and Crofts (1971), since protons would not be involved in the predominant transitions (S$_1$ → S$_2$ and S$_0$ → S$_1$). J. M. Bowes and A. R. Crofts (unpublished observations) have recently made a study of the dependence on flash preillumination of these effects. They found that the production of a proton gradient through activation of photosystem I led to stimulation of delayed fluorescence in chloroplasts inhibited by DCMU added a few seconds before illumination, either after prolonged dark adaptation (20 minutes) or after illumination by one to three short saturating flashes following dark adaptation. The stimulated delayed fluorescence was considerably greater after flash preillumination, but no marked difference was observed when the number of flashes was varied (Fig. 6). These observations suggest that the internal protons are in

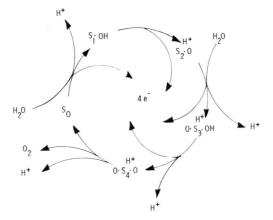

Fig. 7 Scheme showing the involvement of protons in transitions of the states of the water-oxidizing apparatus, with possible sites for entry of water into the reactions.

equilibrium with all transitions of the S states, with the possible exception of that of S_1 to S_2.

G. Protolytic Reactions on the Oxygen Side: Conclusions

It seems very likely both from theoretical considerations and from the observed effects of pH that protons are involved in the oxygen-evolving reactions in more than one state of the mechanism. Our own results suggest that protons are released on all the transitions of the S states, except possibly that of S_1 to S_2. This would indicate a mechanism in which water reacts with the apparatus in the lower S states and in which oxidized equivalents are accumulated in a complex, where bound or activated water associated with Mn at the active site undergoes transitions through the intermediate oxidation states of the water/oxygen couple. Mechanisms of this sort have been previously considered by Cheniae (1970; Cheniae and Martin, 1970) and Renger (1972b,c). Since the role of manganese is at present unknown, we show a general mechanism of this sort in Fig. 7. The well-known requirement for Cl^- in the reaction might reflect the need to compensate the positive charge of the bound proton in states S_2 to S_4.

III. THE PROTON PUMPS OF BACTERIAL PHOTOSYNTHESIS

The major review of bacterial photosynthesis edited by Clayton and Sistrom (1978) contains several articles of obvious relevance to the present topics and, in particular, a comprehensive article by Wraight *et al.* (1978). We will restrict ourselves here to a consideration of the Rhodo-spirillaceae (formerly *Athiorhodaceae*), since electron transport in this group has been most fully studied. The species studied in greatest detail are *Rhodopseudomonas sphaeroides* and *R. capsulata*, which appear to be very similar in their photosynthetic apparatus.

A. The Cyclic Electron Transport Chain

The photochemical reactions in *R. sphaeroides* and *R. capsulata* leave P-870 oxidized and the quinone–iron complex of the primary acceptor reduced. In both species the E_m for the P-870/P-870$^+$ couple is ~450–470 mV, and the E_m for the X/X$_{red}$ couple [where X is probably ubiquinone closely associated with iron (UQ · Fe)] in equilibrium with protons from the aqueous phase is approximately −60 mV at pH 7. However, no proton uptake occurs over the time range of reduction of the couple (Chance *et al.*, 1970), so it would be expected that the initial reaction would produce X$^-$, and this is supported by electron paramagnetic resonance (epr) spectroscopic and spectrophotometric evidence (Slooten, 1972; Clayton and Straley, 1970; Wraight *et al.*, 1975; Dutton *et al.*, 1973). The E_m for this couple depends on the pK for the dissociation of the protonated form:

$$XH \rightleftharpoons X^- + H^+$$

Prince and Dutton (1976b) have estimated pK's for the primary acceptors of a number of different species and shown that, although the values of E_m measured below the pK values varied widely for the primary acceptors of a number of species, the values of E_m for the couple X/X$^-$ (the values above the pK's) varied considerably less; the variation in E_m below the pK values reflected variations in pK. They pointed out that the "operating" E_m was in all cases likely to be that of the X/X$^-$ couple, the value for this being approximately −160 mV. From these results, we may conclude that the maximal redox span encompassing the components of the cyclic pathway could be from −200 to +500 mV. Of the redox components identified in *R. sphaeroides* or *R. capsulata*, the vast majority fall within this range (Dutton and Jackson, 1972; Crofts, 1974; Dutton and Wilson, 1974) so that identification of the reactants in the chain can be achieved only by combining redox measurements with kinetic measurements (Jackson and Dutton, 1973; Evans and Crofts, 1974). The components identified by epr spectroscopy are for the most part detectable only at low temperature (Dutton *et al.*, 1973; Dutton and Wilson, 1974; Prince and Dutton, 1976a), and the kinetic resolution of

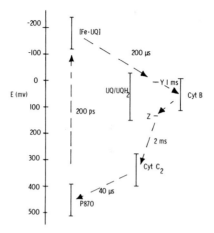

Fig. 8 Components of the cyclic electron transport chain of *R. capsulata*. The components are arranged alongside a redox scale showing the probable operating potential range. Components Y (associated with rapid H⁺ binding) and Z (associated with the reduction of cytochrome *c*) are probably both ubiquinone species, acting in the partial reactions ubiquinone/ubisemiquinone (Y) and ubisemiquinone/ubiquinol (Z). The potentials of these partial reactions are unknown. The arrows, and the times against these, show kinetically measured reactions and their half-times (the reaction center time is taken from data using *R. sphaeroides*; see the text for references).

the signals is a technically forbidding task. For this reason, most of our information about the components of the cyclic pathways has come from spectrophotometric measurements. Some of these data for *R. capsulata* are summarized in Fig. 8.

B. Topology of the Electron Transport Chain

The electron transport chain appears to be arranged anisotropically in the membrane so as to act as two proton pumps in series when electrons traverse the cyclic pathway (Crofts, 1974; Wraight *et al.*, 1978), and this arrangement is shown schematically in Fig. 9. The photochemical reaction center spans the membrane, with the reaction sites for cytochrome c_2 and for the secondary acceptor and the proton associated with its reduction on opposite sides of the membrane (Jackson and Crofts, 1971; Prince *et al.*, 1975; Dutton *et al.*, 1975). In chromatophores, cytochrome c_2 and its reaction site are inside, whereas the site of proton uptake is outside; in whole cells cytochrome c_2 is outside the cell membrane in the periplasmic space. This arrangement of the photochemical apparatus is now generally accepted, and the evidence for it has been reviewed at length elsewhere (Parsons and Cogdell, 1975; Crofts, 1974; Wraight *et*

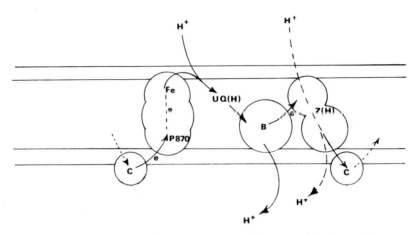

Fig. 9 Scheme showing how the components may be arranged in the membrane so as to act as two proton pumps.

al., 1978). The arrangement of the "dark" electron transport chain by which electrons return from the primary acceptor to oxidized cytochrome c_2 is much less certain, and there is some doubt as to the redox carriers contributing to the chain. However, it is clear that electron flow through this section of the chain is also linked to the pumping of protons and, through the proton gradient, to phosphorylation (Crofts *et al.*, 1975).

In the following section we will review the evidence available for the arrangement of the photochemical reaction centers and discuss our recent attempts to reconstitute the proton pumping activity of reaction centers using a liposome system.

IV. PHOTOCHEMICAL REACTION CENTERS

Several excellent reviews have covered the biophysics and biochemistry of bacterial reaction centers (Parsons and Cogdell, 1975; Sauer, 1975). We will restrict ourselves here to points pertinent to the role of reaction centers as proton pumps and avoid a discussion of those transient events and intermediates (Rockley *et al.*, 1975; Kaufman *et al.*, 1975; Tiede *et al.*, 1976; Shuvalov *et al.*, 1976; Prince *et al.*, 1976) that precede the stabilization of the energy conversion process through oxidation of P-870 and reduction of the ubiquinone–iron complex.

Most of the work reported has been done with reaction centers extracted from a carotenoid-deficient mutant of *R. sphaeroides* (R26) using LDAO (Clayton and Wang, 1971). Reaction centers from carotenoid-containing strains extracted by similar procedures are similar in all essentials, apart from the additional presence of one molecule of carotenoid per reaction center (Jolchine and Reiss-Husson, 1974, 1975). Most workers have used detergent-solubilized preparations. In these, the reactions with added c-type cytochromes (either mammalian or native) are second order and strongly dependent on ionic strength (Prince et al., 1974). In more complex preparations (Ke et al., 1970) or in those in which the reaction centers have been incorporated into liposomes (Dutton et al., 1976), the reaction may be of lower order due to binding of cytochrome c.

Reactions on the acceptor side are more complicated and depend both on the state of the complex and on the method of preparation. In preparations purified by $(NH_4)_2SO_4$ precipitation, following a single flash from the dark-adapted state, the charge is stabilized in a species absorbing at 430 nm, which is thought to be the anionic semiquinone of endogenous ubiquinone (Slooten, 1972; Wraight et al., 1975). No H^+ uptake accompanied this change. A second flash leads to the uptake of one H^+ per reaction center. In the presence of added artificial acceptor (1,4-naphthoquinone), subsequent flashes lead to the uptake of one H^+ per reaction center as the acceptor is reduced. However, in these flashes after the first, no consistent pattern of behavior for the species absorbing at 430 nm is observed. In reaction centers that have never been exposed to $(NH_4)_2SO_4$, but which have been purified by column chromatography, a more consistent pattern is observed. The first flash leads to formation of the anionic semiquinone, identified by the absorption at 430 nm and by epr spectroscopy (Wraight, 1977; Vermiglio, 1977). The second flash removes the semiquinone and leads to the uptake of a proton (Wraight, 1977) and, in the presence of added acceptor, an oscillation with a period of 2 can be observed for many flashes, with odd-numbered flashes producing the semiquinone signal and even-numbered flashes removing it. This behavior is interpreted as reflecting a mechanism for charge accumulation on the acceptor side between the one electron per molecule reaction of the primary acceptor (the tightly bound UQ · Fe), a second bound ubiquinone, and the loosely bound ubiquinone or added acceptor, the reduction of which involves two hydrogens per molecule.

Reconstitution of the Proton Pumping Activity of Reaction Centers

The suggestion that the reaction center in vivo is orientated so as to form the electrogenic arm of a proton pump carries the implication that such an activity should be shown by reaction centers appropriately incorporated into artificial lipid membranes. Skulachev and his colleagues (Skulachev, 1972; Barsky et al., 1976) have developed techniques for measuring the incorporation of electrogenic pumps into lipid membranes with high sensitivity. Using reaction centers prepared from R. rubrum, they demonstrated that, when a suspension of liposomes containing bound reaction centers was allowed to equilibrate with a planar lipid membrane separating two aqueous phases, illumination caused the transfer of charge between the two phases. The mechanism of this charge transfer was not simple, and the polarity and location of the reaction centers with respect to the membrane were not well defined. Nevertheless, the experiments provide evidence for transmembrane orientation of the reaction centers in this artificial system similar to that observed in vivo. A similar electrogenic arrangement could be inferred from the light-induced extrusion of phenyldicarbaundecarborane anion from proteoliposomes containing reaction centers. In my laboratory, Dr. H. Celis (see Crofts et al., 1977) has developed a reconstituted system that is better defined with respect to orientation and reaction mechanism.

Purified bacterial reaction centers isolated from wild-type or carotenoid-deficient strains of R. sphaeroides have been incorporated into phospholipid vesicles by a cholate dialysis procedure. The orientation of the reaction centers in the liposome membrane has been tested by observing their reactivity on illumination with added cytochrome c and with penetrating or impermeant quinone acceptors. Between 40 and 60% of the reaction centers had the reaction site for cytochrome c exposed to the external aqueous phase. Few if any reaction centers had both acceptor and donor sites available to the external aqueous phase. In the presence of added reduced cytochrome c and 1,4-naphthoquinone, illumination of liposomes containing reaction centers led to an extrusion of H^+ into the external aqueous phase (Fig. 10). The H^+ efflux had a maximal extent of 120 H^+ per reaction center and the maximal steady rate showed a turnover of 0.5 H^+ per reaction center per second. Maximal rates were observed in a K^+ (or Rb^+ or Cs^+)-containing medium in the presence of valinomycin. In the absence of valinomycin, the rate was slower, and the extent of efflux was reduced. The H^+ efflux was inhibited by o-phenanthroline, or bathophenanthroline, or by FCCP at uncoupling concentrations. No H^+ efflux was observed when 1,4-naphthoquinone sulfonate was used instead of 1,4-naphthoquinone. The light-induced H^+ efflux in the presence of cytochrome c and 1,4-naphthoquinone showed a pronounced pH dependence, with little or no efflux at pH values below 7 or above 11, and a maximal efflux at pH

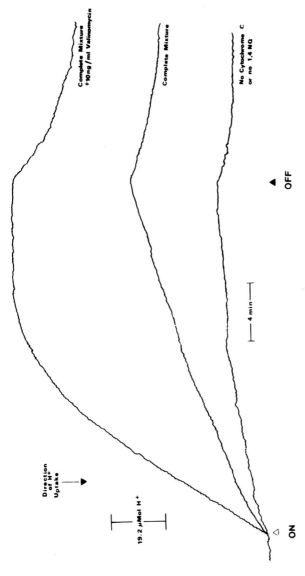

Fig. 10 Reconstituted H⁺ pump in liposomes containing reaction centers. Reaction centers were incorporated into liposomes (RC-liposomes) (Crofts *et al.*, 1977) using a protein to lipid ratio of 1 : 30 (w/w). Liposomes (0.4 μM reaction center) were suspended in a medium containing 0.25 M sucrose, 50 mM KCl, 154 μM 1,4-naphthoquinone (NQ), and 15 μM cytochrome c. The reaction vessel was flushed with N_2 to minimize pH drifts due to gaseous exchange, and the pH was adjusted to 9.5. Illumination was provided by a 50-W tungsten-filament lamp equipped with an orange filter. Where indicated, valinomycin was added or cytochrome or naphthoquinone (1,4 NQ) was omitted.

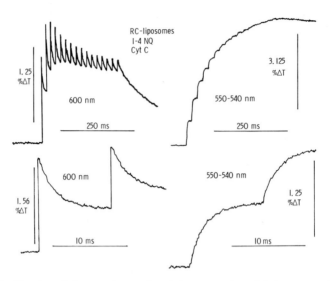

Fig. 11 Kinetics of the reaction cycle of the reconstituted H^+ pump. Absorbance changes, following illumination by xenon flashes (15 microsecond width at half-height, 85% saturating), were recorded for P-870 (measured at 600 nm) and cytochrome c (measured at 550–540 nm) for a reaction mixture like that of Fig. 10, but with 1 μM valinomycin and 200 μM 1,4-naphthoquinone present. RC, reaction center; NQ, naphthoquinone.

values between 9.5 and 10. Measurement of the kinetics of cytochrome and reaction center absorbance changes on illumination with continuous light or short flashes showed that there were at least two kinetically limiting steps in the reaction cycle. Figure 11 shows the kinetics of the redox changes of P-870 (measured at 600 nm) and of cytochrome c on illumination with a group of flashes. The decline in the level of oxidation attained by P-870 after the fourth and subsequent flashes is best explained by a failure to reoxidize the primary acceptor in the dark period between flashes. This identifies one of the limiting steps as the rate of reaction of 1,4-naphthoquinone with the endogenous acceptor; the second was the rate of reduction of cytochrome c by the reduced naphthoquinone, as seen by the very slow relaxation of cytochrome c oxidation. This latter limitation was most apparent at the lower end of the effective pH range.

A more active H^+ efflux on illumination was observed when reaction center liposomes were prepard with ubiquinone-10 as an additional constitutent of the lipid phase. In this system, no added quinone was required, but the H^+ efflux depended on the addition of N-methylphenazonium-3-sulfonate methosulfate (PMS-S) as an impermeant redox mediator between ubiquinone-10 and either externally added cytochrome c, added native cytochrome c_2, or the oxidized reaction center itself. In the absence of ubiquinone-10, the sulfonated mediator was not able to catalyze a reaction cycle of comparable rate. The pH profile and sensitivity to ionophores and inhibitors of the ubiquinone-10-containing system were similar to those of the 1,4-naphthoquinone-catalyzed cycle. The maximal extent in the former system was ~ 200 H$^+$ per reaction center, and the maximal steady rate was 2H$^+$ per reaction center per second. Kinetic studies showed that the reactivities of ubiquinone-10 with the endogenous acceptor of the reaction center and with cytochrome c (or c_2) were not rate limiting in this system (Fig. 12). An unexpected kinetic block appeared after a small number of turnovers of the reaction center; this was apparent as a failure of oxidation of reduced cytochrome c (or c_2) by the oxidized reaction center. The most rapid steady rates of H$^+$ efflux occurred when cytochrome c was omitted from the reaction mixture and replaced by higher concentrations of PMS-S.

The results indicate that, in a reconstituted system, bacterial photochemical reaction centers are able to act as the electrogenic arm of a Mitchellian proton pump in which either added 1,4-naphthoquinone or membrane-bound ubiquinone-10 acts as the H-carrying arm. Figure 13

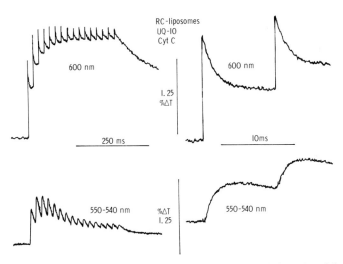

Fig. 12 Reconstituted H$^+$ pump with ubiquinone-10 (UQ-10); kinetics of the reaction cycle. Conditions were as for Fig. 11, except that liposomes were reconstituted with UQ-10 added at a ratio of 25 UQ-10 per reaction center (RC) (the ratio in the final preparation was not estimated, but it is likely to have been less than 25 : 1 since the UQ-10 was only partially taken up) and 1,4-naphthoquinone was replaced by 50 μM N-methyl-phenazonium-3-sulfonate methylsulfate.

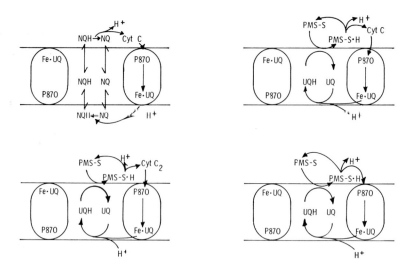

Fig. 13 Mechanisms proposed for the reconstituted H^+ pumps in reaction center-containing liposomes. PMS-S and PMS-S · H, N-methylphenazonium-3-sulfonate methylsulfonate, oxidized and reduced forms; Fe · UQ, iron–ubiquinone complex acting as the primary acceptor; NQ and NQH, 1,4-naphthoquinone, oxidized and reduced forms; UQ and UQH, ubiquinone and ubiquinol.

summarizes our idea about the mechanisms by which the reconstituted pumps might operate. The results also demonstrate that ubiquinone-10 in a membrane phase is able to catalyze a rapid H transfer across the membrane between reaction partners in the aqueous phases separated by the membrane. Several possible mechanisms may be proposed as to how this happens:

1. Both ubiquinone-10 and ubiquinol-10 may be sufficiently mobile in the lipid phase to directly transfer H across the membrane at the relatively slow rates indicated.

2. If either component is immobile, alternative mechanisms must operate. One such mechanism is H transfer through disproportionation between ubiquinone molecules within the membrane. Alternatively, some specific interaction with the reaction center protein might facilitate H transfer through an unknown mechanism.

REFERENCES

Arnason, T., and Sinclair, J. (1976). *Biochim. Biophys. Acta* **449,** 581–586.
Arntzen, C. T., and Briantais, J.-M. (1975). *In* "Bioenergetics of Photosynthesis" (Govindjee, ed.), pp. 51–113. Academic Press, New York.

Ausläender, W., and Junge, W. (1974). *Biochim. Biophys. Acta* **357,** 285–298.
Ausläender, W., and Junge, W. (1975). *FEBS Lett.* **59,** 310–315.
Babcock, G. T., and Sauer, K. (1975). *Biochim Biophys. Acta* **396,** 48–62.
Bamberger, E., Rottenberg, H., and Avron, M. (1973). *Eur. J. Biochem.* **34,** 557–563.
Barsky, E. L., Dancshazy, Z., Drachev, L. A., Il'ina, M. D., Jasaitis, A. A., Kondrashin, A. A., Samuilov, V. D., and Skulachev, V. P. (1976). *J. Biol. Chem.* **251,** 7066–7071.
Bendall, D. S. (1977). *Biochem., Ser. One* **13,** 41–78.
Bouges-Bocquet, B. (1973). *Biochim. Biophys. Acta.* **314,** 250–256.
Chance, B., Crofts, A. R., Nishimura, M., and Price, B. (1970). *Eur. J. Biochem.* **10,** 226–237.
Cheniae, G. M. (1970). *Annu. Rev. Plant Physiol.* **21,** 467–498.
Cheniae, G. M., and Martin, I. F. (1970). *Biochim. Biophys. Acta* **197,** 219–239.
Clayton, R. K., and Sistrom, W. R. (eds.) (1978). "The Photosynthetic Bacteria." Plenum, New York. In press.
Clayton, R. K., and Straley, S. C. (1970). *Biochem. Biophys. Res. Commun.* **39,** 1114–1119.
Clayton, R. K., and Wang, R. T. (1971). *In* "Methods in Enzymology (A. San Pietro, ed.), Vol. 23, pp. 696–704. Academic Press, New York.
Cramer, W. A., and Horton, P. (1975). *Photochem. Photobiol.* **22,** 304–308.
Cramer, W. A., Horton, P., and Donnell, J. J. (1974). *Biochim. Biophys. Acta* **368,** 361–370.
Crofts, A. R. (1974). *In* "Perspectives in Membrane Biology" (S. Estrada-O and C. Gitler, eds.), pp. 373–412. Academic Press, New York.
Crofts, A. R., Wraight, C. A., and Fleischmann, D. E. (1971). *FEBS Lett.* **15,** 89–100.
Crofts, A. R., Crowther, D., and Tierney, G. V. (1975). *In* "Electron Transfer Chains and Oxidative Phosphorylation" (E. Quagliarillo *et al.*, eds.), pp. 233–241. North-Holland Publ., Amsterdam.
Crofts, A. R., Crowther, D., Celis, H., Almanza de Celis, S., and Tierney, G. (1977). *Biochem. Soc. Trans.* **5,** 491–495.
Deamer, D. W., Prince, R. C., and Crofts, A. R. (1972). *Biochim. Biophys. Acta* **274,** 323–335.
Diner, B., and Joliot, P. (1976). *Biochim. Biophys. Acta* **423,** 479–498.
Dutton, P. C., and Jackson, J. B. (1972). *Eur. J. Biochem.* **30,** 495–510.
Dutton, P. L., and Wilson, D. F. (1974). *Biochim. Biophys. Acta* **346,** 165–212.
Dutton, P. L., Leigh, J. S., and Reed, D. W. (1973). *Biochim. Biophys. Acta* **292,** 654–666.
Dutton, P. L., Petty, K. M., Bonner, H. S., and Morse, S. D. (1975). *Biochim. Biophys. Acta* **387,** 536–556.
Dutton, P. L., Petty, K. M., and Prince, R. C. (1976). *Fed. Proc., Fed. Am. Soc. Exp. Biol.* **35,** 1597.
Étienne, A. L. (1974). *Biochim. Biophys. Acta* **333,** 497–508.
Evans, E. H., and Crofts, A. R. (1974). *Biochim. Biophys. Acta* **357,** 89–102.
Fowler, C. F. (1977). *Biochim. Biophys. Acta* **459,** 351–363.
Fowler, C. F., and Kok, B. (1974). *Biochim. Biophys. Acta* **357,** 299–307.
Fowler, C. F., and Kok, B. (1976). *Biochim. Biophys. Acta* **423,** 510–523.
Giaquinta, R. T., and Dilley, R. A. (1975). *Biochim. Biophys. Acta* **387,** 288–305.
Giaquinta, R. T., Dilley, R. A., Selman, B. R., and Anderson, B. J. (1974). *Arch. Biochem. Biophys.* **162,** 200–209.
Girault, G., and Galmiche, J. M. (1974). *Biochim. Biophys. Acta* **333,** 314–319.
Graber, P., and Witt, H. T. (1975). *FEBS Lett.* **59,** 184–189.
Graber, P., and Witt, H. T. (1976). *Biochim. Biophys. Acta* **423,** 141–163.
Horton, P., and Cramer, W. A. (1974). *Biochim. Biophys. Acta* **368,** 348–360.
Ikegami, I., and Katoh, S. (1973). *Plant Cell Physiol.* **14,** 829–836.
Izawa, S., and Ort, D. R. (1974). *Biochim. Biophys. Acta* **357,** 127–143.
Jackson, J. B., and Crofts, A. R. (1971). *Eur. J. Biochem.* **18,** 120–130.

Jackson, J. B., and Dutton, P. L. (1973). *Biochim. Biophys. Acta* **325,** 102–113.
Jagendorf, A. T., and Hind, G. (1963). *N. A. S.—N. R. C., Publ.* **1145,** 599.
Jolchine, G., and Reiss-Husson, F. (1974). *FEBS Lett.* **40,** 5–8.
Jolchine, G., and Reiss-Husson, F. (1975). *FEBS Lett.* **52,** 33–36.
Junge, W. (1975). *Ber. Dtsch. Bot. Ges.* **88,** 283–301.
Junge, W. and Ausläender, W. (1974). *Biochim. Biophys. Acta* **333,** 59–70.
Kaufman, K. J., Dutton, P. J., Netzel, T. L., Leigh, J. S., and Rentzepis, P. M. (1975). *Science* **188,** 1301–1304.
Ke, B., Chaney, T. H., and Reed, D. W. (1970). *Biochim. Biophys. Acta* **216,** 373–383.
Knaff, D. B. (1975). *FEBS Lett* **60,** 331–337.
Kok, B., Forbush, B., and McGloin, M. P. (1970). *Photochem. Photobiol.* **11,** 457–475.
Lien, S., and Racker, E. (1971). *J. Biol. Chem.* **246,** 4298–4307.
Neumann, J. S., and Jagendorf, A. T. (1964). *Arch. Biochem. Biophys.* **107,** 109–119.
Parsons, W. W., and Cogdell, R. J. (1975). *Biochim. Biophys. Acta* **416,** 105–149.
Prince, R. C., and Dutton, P. L. (1976a). *FEBS Lett* **65,** 117–119.
Prince, R. C., and Dutton, P. L. (1976b). *Arch. Biochem. Biophys.* **172,** 329–334.
Prince, R. C., Cogdell, R. J., and Crofts, A. R. (1974). *Biochim. Biophys. Acta* **305,** 597–609.
Prince, R. C., Baccarini-Melandri, A., Hauska, G. A., Melandri, B. A., and Crofts, A. R. (1975). *Biochim. Biophys. Acta* **387,** 212–227.
Prince, R. C., Leigh, J. S., and Dutton, P. L. (1976). *Biochim. Biophys. Acta* **440,** 622–636.
Renger, G. (1972a). *Biochim. Biophys. Acta* **256,** 428–439.
Renger, G. (1972b). *Eur. J. Biochem.* **27,** 259–269.
Renger, G. (1972c). *Physiol. Veg.* **10,** 329–345.
Rockley, M. G., Windsor, M. W., Cogdell, R. J., and Parson, W. W. (1975). *Proc. Natl. Acad. Sci. U.S.A.* **72,** 2251–2255.
Ross, R. T., Anderson, R. J., and Hsiao, T. L. (1976). *Photochem. Photobiol.* **24,** 267–278.
Rottenberg, H., Grunwalk, T., and Avron, M. (1972). *Eur. J. Biochem.* **25,** 54–63.
Rumberg, B., and Siggel, U. (1969). *Naturwissenschaften* **56,** 130–132.
Saphon, S., and Crofts, A. R. (1977). *Z. Naturforsch., Teil c* **32,** 617–626.
Sauer, K. (1975). *In* ''Bioenergetics of Photosynthesis (Govindjee, ed.), pp. 115–181. Academic Press, New York.
Schwartz, M. (1968). *Nature (London)* **219,** 915–919.
Selman, B. R., Bannister, T. T., and Dilley, R. A. (1973). *Biochim. Biophys. Acta* **292,** 566–581.
Shuvalov, U. A., Krakhmaleva, I. N., and Klimov, V. V. (1976). *Biochim. Biophys. Acta* **449,** 597–601.
Siggel, U. (1975). *Proc. Int. Congr. Photosynth. Res., 3rd, 1974* Vol. 1, pp. 645–654.
Sinclair, J., and Arnason, T. (1974). *Biochim. Biophys. Acta* **368,** 393–400.
Skulachev, V. P. (1972). ''Energy Transformations in Biomembranes.'' Nauka Press, Moscow.
Slooten, L. (1972). *Biochim. Biophys. Acta* **275,** 208–218.
Tiede, D. M., Prince, R. C., Reed, G. H., and Dutton, P. L. (1976). *FEBS Lett.* **65,** 301–304.
Trebst, A. (1974). *Annu. Rev. Plant Physiol.* **25,** 423–458.
Velthuys, B. R., and Amesz, J. (1974). *Biochim. Biophys. Acta* **333,** 85–94.
Vermiglio, A. (1977). *Biochim. Biophys. Acta.* **459,** 516–524.
von Stedingk, L.-V., and Baltscheffsky, H. (1966). *Arch. Biochem. Biophys.* **117,** 400.
Walker, D. A., and Crofts, A. R. (1970). *Annu. Rev. Biochem.* **39,** 389–428.
Witt, H. T. (1971). *Q. Rev. Biophys.* **4,** 365–477.
Witt, H. T. (1975). *In* ''Bioenergetics of Photosynthesis'' (Govindjee, ed.), pp. 493–554. Academic Press, New York.

Wraight, C. A. (1977). *Biochim. Biophys. Acta* **459,** 525–531.
Wraight, C. A., and Crofts, A. R. (1971). *Eur. J. Biochem.* **19,** 386–397.
Wraight, C. A., Kraan, G. P. B., and Gerrits, N. M. (1972). *Biochim. Biophys. Acta* **283,** 259–267.
Wraight, C. A., Cogdell, R. J., and Clayton, R. (1975). *Biochim. Biophys. Acta* **396,** 242–249.
Wraight, C. A., Cogdell, R. J., and Chance, B. (1978). *In* "The Photosynthetic Bacteria" (R. K. Clayton and W. R. Sistrom, eds.). Plenum, New York. In press.

21

The Photosynthetic
Intramembrane
Electric Field

PIERRE JOLIOT

Institut de Biologie Physico-Chimique, Fondation Edmond de Rothschild, Paris, France

The basis of the chemiosmotic hypothesis of Mitchell (1961) has been discussed in Chapter 20. Mitchell's hypothesis assumes that the formation of a membrane potential is associated with electron flow through the electron transfer chain. This chapter is concerned with the formation and the decay of the membrane potential in the case of living cells.

Figure 1 presents a model of the photosynthetic membrane based on Mitchell's chemiosmotic hypothesis. From this model, one can predict that the photochemical charge separation induces the formation of a series of dipoles at the level of each photocenter. Due to the presence of ions in the aqueous phase on both sides of the membrane, the potential is rapidly homogenized along the surface of the membrane, which then can be considered to be a charged capacitor. The formation of the transmembrane potential, which immediately follows the primary photoreaction, precedes the liberation of protons inside the thylakoid and involves several chemical steps.

Compared to other biological materials, the photosynthetic membranes have the great advantage of including several natural probes sensitive to the electric field. Using *Chlorella* cells, Duysens (1954) observed a light-induced absorption change in the blue–green region of the spectrum. Junge and Witt (1968) demonstrated that this spectral

FRONTIERS IN PHYSICOCHEMICAL BIOLOGY
Copyright © 1978 by Academic Press, Inc.
All rights of reproduction in any form reserved.
ISBN 0-12-566960-7

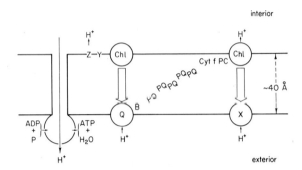

Fig. 1 Model for the photosynthetic membrane. Chl, photoactive chlorophyll; Z and Y, secondary electron donors of photosystem II; Q, primary electron acceptor of photosystem II; PQ, plastoquinone; Cyt f, cytochrome *f*; PC, plastocyanin; X, primary electron acceptor of photosystem I.

change, an absorption increase at 515 nm and an absorption decrease at 480 nm, gives a linear measurement of the membrane potential. They interpreted this absorption change as an electrochromic effect. Considering the density of photocenters on the thylakoid membrane, the membrane potential that develops after illumination by a single short saturating flash can be estimated to be about 100 mV. This potential, applied to a membrane 40 Å thick, gives rise to an electric field of about 2×10^{-5} V/cm. When the pigments included in the membrane are submitted to such a high electric field, a change in the spectroscopic properties and, more specifically, a shift in their absorption bands would be expected. The 515-nm absorption increase is generally ascribed to an electrochromic shift of the carotenoid absorption band and the 480-nm absorption decrease appears to be related to a shift in the absorption band of chlorophyll *b*.

I. FORMATION OF THE MEMBRANE POTENTIAL IN ALGAE

Most studies of the formation and the decay of the membrane potential have been performed on isolated chloroplasts. We have observed that a certain number of membrane properties are modified during the process of extracting the chloroplasts from the cells. For this reason, we have used *Chlorella* and *Chlamydomonas* to study the relaxation of the membrane potential in more detail.

In collaboration with R. Delosme (Joliot and Delosme, 1974), we studied the formation and the relaxation of the 515-nm absorption increase on dark-adapted *Chlorella* cells illuminated by a single saturating flash of 4-microsecond duration (Fig. 2). We observed first a fast increase in the absorption in less than 100 microseconds. As pointed out by Schliephake *et al.* (1968), this fast increase is induced by the charge separation occurring at the level of both photosystems. In the case of *Chlorella* cells, we observed that 60 to 80% of the fast absorption increase is linked to the photosystem I photoreaction and 20 to 40% to the photosystem II photoreaction. The fast absorption increase is followed by a slow increase ($t_{1/2} \sim 10$ to 20 milliseconds), which shows the same difference spectrum. One can conclude from the experiments in Fig. 2 that, in the case of dark-adapted algae, the formation of the membrane potential is a biphasic process.

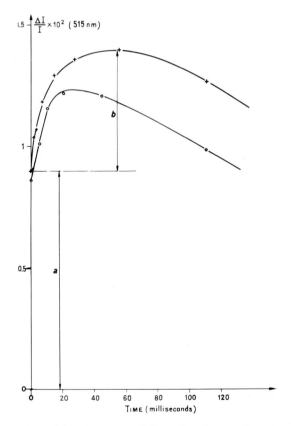

Fig. 2   Time course of the photosystem I absorption change after one saturating flash. +, *Chlorella* dark-adapted for 1 hour; o, *Chlorella* dark-adapted for 5 minutes. (a) Fast increasing phase of the absorption change; (b) slow increasing phase of the absorption change.

A. Fast Increasing Phase of the Chloroplast
Membrane Potential

Wolff *et al*. (1969) showed that the absorption increase at 515 nm is completed in less than 100 nanoseconds, but the slow increasing phase is not observed. This result is in agreement with the hypothesis of Mitchell (1961), who admits that the primary donors of photosystems I and II are located on the internal face of the membrane and that the primary acceptors are on the external face. If this hypothesis is correct, one would expect that the formation of the membrane potential is directly associated with the primary charge separation and would then occur in less than 1 nanosecond. Joliot and Joliot (1976) studied in more detail the problem of the localization of the primary acceptor and donor of photosystem II with respect to the thylakoid membrane. Measurements of the electrochromic effect were perturbed by a fast spectral absorption change that lasted about 10 microseconds due to the formation of a carotenoid triplet (Mathis and Galmiche, 1967). This spectral change showed a maximum at 515 nm, that is, at the same wavelength as the electrochromic effect, but was negligible at 480 nm. For this reason, Cox and Delosme (1976) measured the electrochromic effect at 480 nm to eliminate the contribution of the carotenoid triplet. These authors observed a biphasic decrease in the absorption, the slow phase ($t_{1/2} \sim 20$ microseconds) being exclusively linked to photosystem I. This decreasing phase is currently under study in our laboratory and will not be discussed here. Figure 3a, curve 1, shows the time course of the 480-nm absorption decrease measured after a single short saturating flash when both photosystems are active. Curve 3 was obtained when photosystem II had been irreversibly blocked by the addition of specific inhibitors. The difference between curves 1 and 3 (Fig. 3b, curve 1') gives the time course of the absorption change induced by the photosystem II reaction. As Wolff *et al*. (1969) reported, this absorption decrease appears in less than 5 microseconds, the time resolution of our method. Nevertheless, this result does not permit us to draw a conclusion about the localization of the photoactive chlorophyll in the membrane. According to Den Haan *et al*. (1974), the reduction of the photoactive chlorophyll by a secondary donor (Y in our terminology) occurs in less than 1 microsecond. However, it cannot be known whether the charge crosses the membrane during the primary photochemical step or during the transfer of the charge between the primary and secondary donors. The experiments of Den Haan *et al*. (1974) and Joliot (1977) show that, in the presence of from 1 to 10 mM hydroxylamine, the secondary donor Y is destroyed. The oxidized chlorophyll is then reduced much more slowly ($t_{1/2} \sim 20$ to 50 microseconds) by an auxiliary donor D. On the other

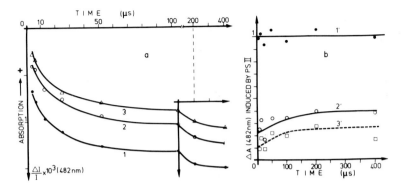

Fig. 3   Absorption change at 480 nm in *Chlorella* after a single saturating flash. (a) Curve 1, control dark-adapted algae; curve 2, 2 mM hydroxylamine; curve 3, photosystem II had been irreversibly blocked prior to the flash by adding 2 mM hydroxylamine and 20 μM dichlorophenyl dimethylurea. (b) Absorption change at 480 nm induced by photoreaction II. Curve 1', control (curve 1 minus curve 3 of a); Curve 2', 2 mM hydroxylamine (curve 2 minus curve 3 of a); curve 3', 100 mM hydroxylamine. The corresponding data are not shown in a. The absorption change for the control (curve 1') is normalized to unity.

hand, Joliot (1977) has shown that 100 mM hydroxylamine acts as very efficient donor and is able to reduce directly the photoactive chlorophyll with a half-time of about 10 microseconds. Our present knowledge of photosystem II centers is summarized in Scheme 1.

$$D$$
$$\Big| \sim\!30 \ \mu sec$$
$$Z \longleftarrow\!\text{-------}\, Y \xleftarrow{\leq 1\ \mu sec} Chl \ a_{11} - Q$$
$$\nearrow$$
$$100 \ mM \ \text{hydroxylamine}$$

Scheme 1

Y and D are two secondary donors connected in parallel on the photoactive chlorophyll. Z is the donor involved in the water-splitting process and Q is the primary acceptor. We have studied the system II photoinduced membrane potential in the presence of hydroxylamine, which makes the lifetime of the oxidized chlorophyll longer than the time resolution of our method. Figures 3a (curve 2) and 3b (curves 2' and 3') show that the membrane potential generated in the presence of 2 or 100 mM hydroxylamine is about three to five times smaller than the control. The same conclusion can be drawn from measurements performed at 515 nm. Fluorescence studies indicate that this decrease in the amplitude of the membrane potential generated by photosystem II is

not due to a decrease in the efficiency of this photoreaction. We can then
conclude that the membrane potential generated by the primary charge
separation is only 20 to 30% of that measured when the positive charge
is stabilized on the secondary donor Y. In the presence of hydroxylamine, this result implies that both chlorophyll and Q are located
close to the outer face of the membrane. Babcock and Sauer (1975) have
been led to the same conclusion by a different method. These authors
showed that, after destruction of the secondary donors, Y and Z, Mn^{2+}
acts as a very efficient donor to chlorophyll, despite the fact that the
thylakoid membrane is totally impermeable to these ions. They thus
concluded that the photoactive chlorophyll is accessible from the outside
of the thylakoid, which implies that it is located close to the outer face of
the membrane.

The following possible models for photosystem II centers take into
account the different experimental data (Fig. 4).

1. In the first model, the photoactive chlorophyll is assumed to be
always located close to the outer surface of the thylakoid. In this model,
the formation of the membrane potential should follow the non-
photochemical step $Chl^+Y \rightarrow Y^+Chl$, which is known to be very fast
[probably less than 30 nanoseconds from Mauzerall's (1972) exper-
iments] and difficult to observe by conventional spectrophotometric

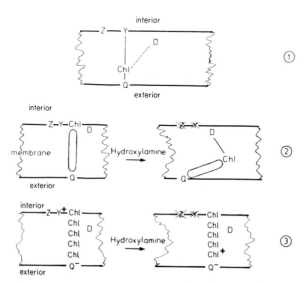

Fig. 4 Possible models for photosystem II centers. 1, 2, and 3 refer to the models
proposed in the text.

methods. The fast increase in the 515-nm absorption change observed by Wolff *et al*. (1969) is not inconsistent with this model.

2. In the second model, chlorophyll is normally located on the inner face of the membrane and, after inactivation of the secondary donors, migrates toward the outer face.

3. In the third model, we assume that the charge transfer through the membrane occurs all along a strand of pigments, for instance, chlorophylls, which establish a bridge between the two sides of the membrane. The secondary donor, Y, acts on the control algae as a very efficient well for the positive charge, which is rapidly stabilized in less than 100 nanoseconds. After the destruction of the secondary donor Y, the positive charge remains in the strand of pigments, which crosses the membrane. The delocalized membrane potential induced by both photosystems would increase the probability of the positive charge being close to the outside of the membrane.

On the basis of theoretical considerations, Hopfield (1977) showed that it is unlikely that an efficient photoreaction can occur if the primary donor and acceptor are separated by a distance as large as the membrane thickness. For this reason, several intermediary steps are probably involved, a hypothesis that favors model 2 or 3.

A more detailed analysis of the time course of the formation of the membrane potential in the nanosecond to second range would be necessary in order to choose among the different models proposed.

B. Slow Increasing Phase of the Chloroplast Membrane Potential*

As pointed out, the slow increasing phase of the membrane potential ($t_{1/2} \sim 10$ milliseconds; Fig. 2) is not observed in the case of isolated chloroplasts with broken external envelopes. In addition, as shown in Fig. 5, this phase disappears after six or seven successive flashes. It is clear, then, that this phenomenon is not linked to the flow of electrons through the main electron transfer chain. We showed that the amplitude of the slow phase remains unchanged when photosystem II is totally inhibited, which proves that this process depends only on photosystem I activity. In our laboratory, Bouges-Bocquet (1977) showed that the slow increasing phase of the membrane potential is due to electron flow through a side electron transfer chain that forms an "electrogenic loop," in Mitchell's terminology. This additional loop permits a second electron to cross the membrane in addition to the electron that crosses the

---

* Written in collaboration with R. Delosme.

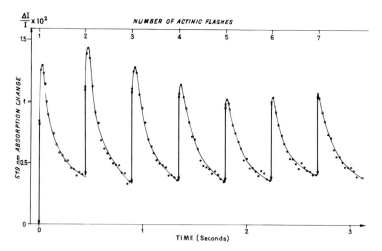

Fig. 5 Time course of the 515-nm absorption change induced by a series of saturating flashes. The arrow indicates the amplitude of the fast phase. The slow increasing phase totally disappears after seven flashes.

membrane during the primary separation. On dark-adapted algae, this side electron transfer chain is connected to the main electron transfer chain at the level of the secondary donors of photosystem I centers (cytochrome f and plastocyanin). These secondary donors can be reduced by two independent pathways: the main electron transfer chain via the plastoquinone pool and the side electron transfer chain, which is functional only at the beginning of the illumination period. The physiological role of the side electron transfer chain is not known. It is interesting to note that a slow increase in the membrane potential is observed in the case of photosynthetic bacteria and is associated with an additional electrogenic loop (Jackson and Dutton, 1973).

II. RELAXATION OF THE MEMBRANE POTENTIAL

According to the theory of Mitchell (1961), the decrease in the membrane potential is due to several processes. First, the membrane ATPase serves as a specific channel for protons and permits a proton to return from the inside to the outside of the thylakoid. This specific leak is assumed to induce ATP synthesis. Second, the membrane is slightly permeable to ions. The output of positive ions and the input of negative ions could contribute to the discharge of the membrane capacitor. Rumberg and Siggel (1968), who observed an acceleration of the decay of the membrane potential when ADP was added to the chloroplast suspension, have provided evidence that supports the first hypothesis. On the other hand, an acceleration of the decay of the membrane potential is observed when specific ionophores such as gramicidin and valinomycin are added to the chloroplasts (Junge and Witt, 1968). In the presence of KCl, the thylakoids *in vitro* are not accessible to externally added phosphorylating reagent. To study whether the decay of the membrane potential depends upon phosphorylation, we have been led to compare the relaxation of the 515-nm absorption increase in the presence of active ATPase activity to that in the absence of ATPase activity. Bennoun and Levine (1967) isolated a *Chlamydomonas* mutant (F54) that is unable to synthetize ATP. This mutant has been shown to be defective at the level of the ATPase molecule (Sato *et al.*, 1971).

Figure 6 shows the relaxation of the 515-nm absorption change after a single short saturating flash given to the mutant F54. The half-time of the decay of the absorption change is 1.4 seconds for the mutant, whereas it is 50 milliseconds for the dark-adapted wild-type algae. When wild-type algae are preilluminated under continuous light prior to the flash, the lifetime of the membrane potential is shortened to about 20

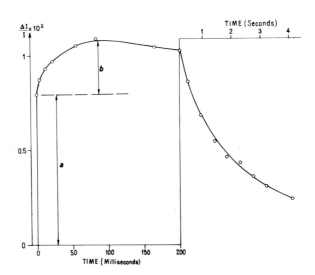

Fig. 6 Time course of the absorption after one saturating flash in dark-adapted *Chlamydomonas* mutant F54. The lifetime of the 515-nm absorption change observed with *Chlamydomonas* wild type is of the order of 50 milliseconds, that is, about 40 times faster. (a) Fast increasing phase of the absorption change; (b) slow increasing phase of the absorption change.

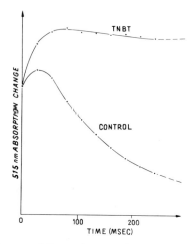

Fig. 7   Absorption change at 515 nm after a single short saturating flash in the presence or the absence of tri-*n*-butyltin (TNBT).

milliseconds, whereas the decay of the membrane potential cannot be accelerated for mutant F54 with preillumination.

Bruce Diner (personal communication) has been led to similar conclusions using a specific inhibitor of ATPase: tri-*n*-butyltin (TNBT). Figure 7 shows the relaxation of the 515-nm absorption change measured after a single flash given to dark-adapted *Chlorella* in the presence or absence of TNBT. The lifetime of the membrane potential is about 30 times slower in the presence of TNBT than it is in its absence. The lifetime of the membrane potential measured on inhibited algae is of the same order of magnitude as for the mutant F54. It is important to note that the lifetime of the membrane potential depends on the presence of active ATPase even when a single flash is given after a long dark period, that is, under conditions where no appreciable proton gradient is present. These results agree with one of Mitchell's major hypotheses: that the membrane potential, in the absence of a proton gradient, is able to induce a specific leak of protons through the ATPase channel.

A. Effect of Preillumination on Decay of the Membrane Potential*

We have studied the decay of the membrane potential induced by a single flash given after a dark period of 5, 15, or 30 minutes following a

* Written in collaboration with B. Diner.

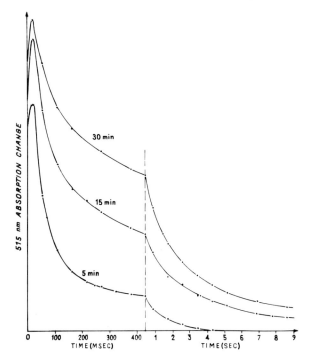

Fig. 8   Absorption change at 515 nm after a single saturating flash given to *Chlorella* dark-adapted for 5, 15, and 30 minutes. In all cases, algae had been preilluminated for 30 minutes under continuous light.

continuous preillumination (Fig. 8). In all cases, the proton gradient induced by the preillumination had been completely dissipated during the dark adaptation. One can observe that the longer the dark period, the slower the membrane potential decay, which suggests that the efficiency of the leak of protons induced by the ATPase decreases as a function of dark adaptation.

For long periods of dark adaptation, the lifetime of the membrane potential reaches a value that in certain cases can be of the same order of magnitude as that measured in the presence of inhibited ATPase. From these experiments it can be concluded that the ATPase activity is controlled by a complex mechanism that depends essentially on the conditions of preillumination. Activation of ATPase by light has already been observed for isolated chloroplasts. It is possible that this slow control of the ATPase activity depends on the fixation of a physiologic inhibitor, a process that has been described for mitochondria.

B. Kinetics of Decay of the Membrane Potential*

In the presence of active ATPase, the decay of the membrane potential is always a complex kinetic reaction. Figure 9 shows a logarithmic plot of the decay of the membrane potential after a single short saturating flash or after a weak flash given to dark-adapted algae. As pointed out, the decay is slower after a longer period of dark adaptation, but, in all cases, the decay appears to be strongly multiphasic. The comparison of the initial slopes for a saturating flash and for a weak flash suggests that the rate constant of the proton leak is an increasing function of the membrane potential. This interpretation would explain as well the nonexponential decay of the membrane potential. It then appears that the membrane potential is not only the driving force that induces the leak of protons through the specific ATPase channel, but also controls the efficiency of the leak, that is, the rate constant of the decay. A simplistic interpretation would be that the membrane potential controls the orientation of the ATPase with respect to the plane of the membrane. An ATPase channel perpendicular to the membrane would be the most efficient configuration. This hypothesis is merely speculative, and one can just as well assume that the membrane potential induces a conformational change of the ATPase that controls the activity of the enzyme.

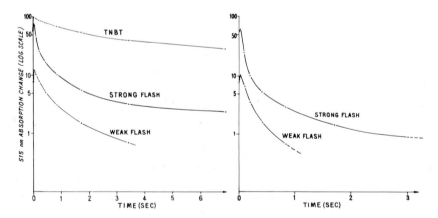

Fig. 9 Relaxation of the 515-nm absorption change after a saturating flash and after a weak flash. The absorption change is plotted on a logarithmic scale. Left, *Chlorella* dark-adapted for 15 minutes; right, *Chlorella* dark-adapted for 5 minutes. TNBT, tri-*n*-butyltin.

* Written in collaboration with B. Diner.

REFERENCES

Babcock, G. T., and Sauer, K. (1975). *Biochim. Biophys. Acta* **396,** 48–62.

Bennoun, P., and Levine, R. P. (1967). *Plant Physiol.* **42,** 1284–1287.

Bouges-Bocquet, B. (1977). *Biochim. Biophys. Acta* **462,** 371–379.

Cox, R., and Delosme, R. (1976). *C.R. Hebd. Seances Acad. Sci., Ser. D* **282,** 775–778.

Den Haan, G. A., Duysens, L. N. M., and Egberts, D. J. N. (1974). *Biochim. Biophys. Acta* **368,** 409–421.

Duysens, L. N. M. (1954). *Science* **120,** 353–354.

Hopfield, J. J. (1977). *In* "Electrical Phenomena at the Biological Membrane Level" (E. Roux, ed.), pp. 471–492. Elsevier, Amsterdam.

Jackson, J. B., and Dutton, P. L. (1973). *Biochim. Biophys. Acta* **325,** 102–113.

Joliot, A. (1977). *Biochim. Biophys. Acta* **460,** 142–151.

Joliot, P., and Delosme, R. (1974). *Biochim. Biophys. Acta* **357,** 267–284.

Joliot, P., and Joliot, A. (1976). *C.R. Hebd. Seances Acad. Sci., Ser. D* **283,** 393–396.

Junge, W., and Witt, H. T. (1968). *Z. Naturforsch. Teil B* **23,** 244–254.

Mathis, P., and Galmiche, J. M. (1967). *C.R. Hebd. Seances Acad. Sci., Ser. D* **264,** 1903–1906.

Mauzerall, D. (1972). *Proc. Natl. Acad. Sci. U.S.A.* **69,** 1358–1362.

Mitchell, P. (1961). *Nature (London)* **191,** 144–148.

Rumberg, B., and Siggel, U. (1968). *Z. Naturforsch., Teil B* **23,** 239–244.

Sato, V. L., Levine, R. P., and Neumann, J. (1971). *Biochim. Biophys. Acta* **253,** 437–448.

Schliephake, W., Junge, W., and Witt, H. T. (1968). *Z. Naturforsch., Teil B* **23,** 1571–1578.

Wolff, C., Buchwald, H. E., Ruppel, H., Witt, K., and Witt, H. T. (1969). *Z. Naturforsch., Teil B* **24,** 1038–1041.

SUBJECT INDEX

A
B
C 8
D 9
E 0
F 1
G 2
H 3
I 4
J 5